This textbook describes physical conditions in the upper atmosphere and magnetosphere of the Earth. This geospace environment begins 70 kilometres above the surface of the Earth and extends in near space to many times the Earth's radius. It is the region of near-Earth environment where the Space Shuttle flies, the aurora is generated, and the outer atmosphere meets particles streaming out of the sun. The account is introductory, at a level suitable for readers with a basic background in engineering or physics. The intent is to present basic concepts, and for that reason the mathematical treatment is not complex. SI units are given throughout, with helpful notes on units where these are likely to be encountered in the research literature. Each chapter has a reading list also.

There are three introductory chapters that give basic physics and explain the principles of physical investigation. The principal material contained in the main part of the book covers the neutral and ionized upper atmosphere, the magnetosphere, and structures, dynamics, disturbances and irregularities. The concluding chapter deals with technological applications.

This textbook is suitable for advanced undergraduate and new postgraduate students who are taking introductory courses in upper atmospheric, ionospheric, or magnetospheric physics. It is a successor to *The Upper Atmosphere and Solar–Terrestrial Relations* by J. K. Hargreaves, first published in 1979.

Cambridge atmospheric and space science series

The solar–terrestrial environment

Cambridge atmospheric and space science series

Titles in print in this series

M. H. Rees, *Physics and chemistry of the upper atmosphere*
Roger Daley, *Atmospheric data analysis*
Ya. L. Al'pert, *Space plasma, Volumes 1 and 2*
J. R. Garratt, *The atmospheric boundary layer*
J. K. Hargreaves, *The solar–terrestrial environment*
Sergei Sazhin, *Whistler-mode waves in a hot plasma*
S. Peter Gary, *Theory of space plasma microinstabilities*
Ian N. James, *Introduction to circulating atmospheres*
Tamas I. Gombosi, *Gaskinetic theory*
Martin Walt, *Introduction to geomagnetically trapped radiation*
B. A. Kagan, *Ocean–atmosphere interaction and climate modelling*

The solar–terrestrial environment

An introduction to geospace – the science of the terrestrial upper atmosphere, ionosphere and magnetosphere.

J. K. Hargreaves

University of Lancaster

Published by the Press Syndicate of the University of Cambridge
The Pitt Building, Trumpington Street, Cambridge CB2 1RP
40 West 20th Street, New York, NY 10011-4211, USA
10 Stamford Road, Oakleigh, Melbourne 3166, Australia

First published 1992
First paperback edition 1995

A catalogue record of this book is available from the British Library

Library of Congress cataloguing in publication data

Hargreaves, J. K. (John Keith), 1930 –
The solar–terrestrial environment: an introduction to geospace – the science of the terrestrial upper atmosphere, ionosphere and magnetosphere / J. K. Hargreaves.
p. cm. (Cambridge Atmospheric and Space Science Series, 7)
Includes index.
ISBN 0-521-32748-2 – ISBN 0-521-42737-1 (pbk)
1. Atmosphere, Upper. 2. Atmospheric physics. 3. Magnetosphere. 4. Ionosphere. I. Title.
QC879.H278 1992
551.5′14 – dc20 91-27182 CIP

ISBN 0 521 32748 2 hardback
ISBN 0 521 42737 1 paperback

Transferred to digital printing 2003

TAG

Contents

Preface

Almost everyone has heard about astronomy though they might not understand it, and almost everyone knows about meteorology even if they cannot spell it. This book is all about the bit in between. Primarily an introductory textbook for students with a background of basic classical physics, it endeavours to describe and explain the phenomena of the terrestrial outer atmosphere and the regions of 'space' nearest to the Earth.

As practitioners will know, this is not a part of the environment that is well known to the general public. The performance of the communications media when attempting to discuss an aurora, or describe the ionosphere, or report the effects of a magnetic storm, is ample testimony to that. Yet, while our subject is a branch of physics and also a branch of geophysics, it may properly be included amongst the environmental sciences as well. Though in the main an academic subject, it is also one which impinges on practical effects of the environment – for instance, communications technology and space activities.

The present book is a sequel to *The Upper Atmosphere and Solar–Terrestrial Relations*, which Van Nostrand Reinhold Co. Ltd. published in 1979. I would have liked to get away with merely inserting necessary corrections to the original text, but, unfortunately for me, the science of the upper atmosphere and near space has moved on apace. So I have had to add a good deal of new material, and the whole book has, in fact, been recast – though some of the original matter has been retained (with Van Nostrand Reinhold's kind permission) where it seemed appropriate.

Since the book is introductory (though intended for readers who already have a background in basic physics or engineering), the picture is painted with a broad brush. Explanations are placed in a physical context as far as possible – which means that there have to be equations – but the mathematical treatment is kept to an elementary level. Some of the material is descriptive. The intention is to inform the reader about the basic concepts and methods and to leave him or her with a good idea of what 'geospace' is all about and why it is important, and of the general state of knowledge. The book should be suitable for undergraduates after the first couple of years and for fresh graduate students, and should enable them to move on to the advanced books and the scientific literature. Professionals qualified in other fields who need information about the ionosphere, or about the effects of solar activity, for instance, should also find it useful.

The increased sophistication and greater depth of knowledge in the subject, compared with 10 years ago, have made this book more difficult to write than was my first effort. Bearing in mind the readers for whom it is mainly intended, I have constantly had to compromise between keeping the text at a suitably introductory level and being sufficiently up to date. Critics should also remember, please, that the task has to be completed within a reasonable length – or the product would come out too expensive for them to buy. It will be for the reader(s?) to judge whether the result has the right balance.

One deliberate change is that SI units are now taken as the primary system. We must still remember, nevertheless, that the enormous literature already published in c.g.s. units is not about to self-destruct, and therefore the older system has been included in a secondary role. An introductory book should lead the reader on to the advanced books and the relevent scientific papers, and this includes help with the units.

With the same thought in mind, suggestions for further reading are given after each chapter. The reading lists are in two parts: the first of books or sections of books where the presentation will be tutorial and from which the reader may verify and amplify what I have said; the second comprises mainly review papers which treat the topics in greater detail and which present the state of knowledge at the time they were written. I expect someone's favourite review will have been omitted; if so I can only plead that the lists have to be limited in length and that the selection is necessarily a personal one and in no sense definitive.

The principal material is contained in Chapters 4 to 9, which between them discuss the neutral and the ionized upper atmosphere, the magnetosphere, structures, dynamics, disturbances and irregularities. Chapter 2 summarizes points of basic physics which may or may not have been encountered before, but which are particularly important for the comprehension of the succeeding material. Chapter 3 describes the methods of geospace investigation, dwelling on the physical principles rather than the hardware. Practical applications are discussed in Chapter 10. Some paragraphs have been set in smaller type, and these can be omitted at a first reading without loss of continuity.

I have often been surprised by the degree of cooperation that goes on between scientists, who so often seem actually pleased to assist one another, expecting nothing other than reciprocation in return. I have benefited from that attitude in preparing this text. In particular I wish to thank Sa. Basu, Su. Basu, K. Bullough, M. J. Buonsanto, J. D. Craven, M. A. Clilverd, R. F. Donnelly, G. Enno, H. Gough, C. Haldoupis, M. A. Hapgood, G. Heckman, R. A. Heelis, R. B. Horne, R. D. Hunsucker, U. S. Inan, J. D. Mathews, M. H. Rees, P. H. Reiff, A. S. Rodger, H. H. Sauer, A. J. Smith, H. C. Stenbaek-Nielsen, R. D. Stewart, D. M. Willis and K. C. Yeh, each of whom has helped in some specific way, for example by providing an unpublished diagram or by helping me with some scientific point.

Last, but certainly not least, I thank the members of my family for their relative patience on the many occasions when I disappeared to the 'office' to talk to the computer.

1

The Earth in space

Far out in the uncharted backwaters of the unfashionable end of the Western Spiral arm of the Galaxy lies a small unregarded yellow sun. Orbiting this at a distance of roughly ninety-two million miles is an utterly insignificant little blue green planet whose ape-descended life forms are so amazingly primitive that they still think digital watches are a pretty neat idea.

Douglas Adams, *The Hitch-Hikers Guide to the Galaxy* (1979)

1.1 Introduction

The solar–terrestrial environment, nowadays sometimes called *geospace*, includes the upper part of the terrestrial atmosphere, the outer part of the geomagnetic field, and the solar emissions which affect them. It could be defined as that region of space closest to the planet Earth, a region close enough to affect human activities and to be studied from the Earth, but remote enough to be beyond everyday experience. Clearly, it is not the familiar atmosphere of meteorology; nor is it the inter-planetary space of astronomy, though it interacts with both. The material found there is mainly terrestrial in origin and strictly a part of the atmosphere of the Earth, though it is greatly affected by energy arriving from the Sun. Starting some 50–70 km above the Earth's surface and extending to distances measured in tens of Earth radii, geospace is a region of interactions and of boundaries: interactions between terrestrial matter and solar radiation, between solar and terrestrial magnetic fields, between magnetic fields and charged particles; and boundaries between solar and terrestrial matter, and between regions dominated by different patterns of flow.

Having an origin in the geomagnetism of the 19th century, our subject first began to develop rapidly about 60 years ago with the increasing use of radio waves in communications and the first scientific studies of the ionosphere during the 1930s. The development of radar just before and during the war of 1939–45 was technically significant, and the technology of war also brought rocketry as a tool for high altitude sounding. Then, in the late 1950s, there took place the first satellite launches which within only a few years brought a great expansion of space activity. Measuring instruments could now be placed in the media of the upper atmosphere, the magnetosphere and interplanetary space, and left there for long periods. Communications and other technological satellites began to be developed for commercial use, and it became possible for human beings to live and work in a space environment for extended periods. There have been problems, but at the present time (1990) it is safe to say that reusable vehicles (the Space Shuttle) are well established, and space station technology (Mir) is already highly developed. We may expect to see further developments in shuttle/space station technology over the next few years in each of the major space centres, and one day we shall perhaps see these competing efforts growing together into a single global enterprise.

All of this depends on a knowledge and understanding of geospace. But in addition to its importance in applications, the science is important in its own right for fundamental studies such as of the properties of tenuous atmospheres and their photochemistry, of wave propagation and of plasma physics. The medium of near space and its physics are not readily reproduced in earth-bound laboratory conditions, and to a large extent geospace provides its own laboratory.

We shall be concerned with three broad regions:

The space between Sun and Earth, across which solar–terrestrial influences propagate;

The terrestrial atmosphere, neutral and ionized, with which the solar emissions react;

The geomagnetic field external to the solid Earth, which influences the ionized atmosphere and controls the Earth's outermost regions.

1.2 The Sun and the solar wind

The rather ordinary star at the centre of the solar system establishes for each planet a radiation environment which controls its temperature and determines the rate of evolution of that planet, the composition of its atmosphere, and its suitability for life. It is our good fortune – though if it were not we should not be here to complain about it – that planet Earth is intermediate between the extreme heat of the planets closer to the Sun and the extreme cold of the outer planets. The Earth's surface temperature permits water to exist in all three phases. Life emerged in the liquid phase and proceeded to alter the composition of the atmosphere, adding oxygen to the nitrogen and carbon dioxide already present. The presence of water as vapour also provided, and continues to provide, a source of hydrogen, which, as we shall see, is important at the atmosphere's higher levels.

Thus the general level of solar radiation, combined with the distance between Sun and Earth, has largely determined the nature of the Earth's atmosphere. While long term change in this energy output may be responsible for slow climatic changes such as produced the Ice Ages, short term changes over days, weeks or a few years appear to have little climatic effect – despite strenuous efforts to discover some. At the higher levels of the atmosphere, though, the changes that accompany variations of solar activity may be large and rapid. The upper atmosphere, where most of the more energetic solar radiations are stopped, and which is heated by them, is very responsive to solar activity variations in general, as well as to the short-lived, intense and localized outbursts known as *solar flares*.

In addition to radiation the Sun also emits a stream of matter. We think of planets like the Earth as stable, self-contained bodies that do not evaporate into space to any significant extent. Not so the Sun, which is not in equilibrium and continuously loses matter as well as radiation into space. This stream of matter is the *solar wind*, which forms the second vital connection between Sun and Earth. Also important is the weak magnetic field, the *interplanetary magnetic field*, which is embedded in the solar wind and is carried with it past the Earth, where it largely determines how strongly the solar wind couples with the matter of the remote terrestrial atmosphere. Although the solar wind does not penetrate down to the ground it is highly significant in geospace; indeed, some of the most remarkable behaviour is directly attributable to the variations of the solar wind and its magnetic field. The interactions are subtle ones and we shall spend some time dealing with them.

1.3 The atmosphere and the ionosphere

Less is known about the Earth's atmosphere than many people imagine. Near the ground the atmosphere is a relatively dense gas, mainly composed of molecular nitrogen and oxygen with smaller amounts of carbon dioxide, water and various trace gases. With increasing altitude the pressure and density decline. At 50 km 99.9% of the mass of the atmosphere is below, and at 100 km all but 1 part per million. Into these rarified upper levels penetrate the ultra-violet and X-ray emissions emanating from the Sun, photons which are sufficiently energetic to dissociate and to ionize the atmospheric species, thereby altering the atmosphere's composition and heating it. The heating creates a hot upper region called the *thermosphere* which is less turbulent than the lower regions, and in which gases of different density may separate. Thus the composition of the atmosphere changes with altitude, the lighter gases, particularly hydrogen, becoming progressively more dominant.

Because of the low pressure above about 100 km, ionized species do not necessarily recombine quickly, and there is a permanent population of ions and free electrons. The net concentration of ions and free electrons (generally in equal numbers) is greatest at heights of a few hundred kilometres, and although the electron concentration may amount to only 1% of the neutral concentration the presence of these electrons has a profound effect on the properties and behaviour of the medium. This *ionosphere* is electrically conducting and can support strong electric currents. The ionized medium also affects radio waves, and as a plasma it can support and generate a variety of waves, interactions and instabilities that are not found in a neutral gas.

The upper atmosphere and ionosphere sit on the lower atmosphere, the domain of the meteorologists. We shall see that some of the behaviour of the higher regions is similar to that taught in meterology, but that there is much more besides.

1.4 Geomagnetic field and magnetosphere

As William Gilbert, physician to Queen Elizabeth I, realized 400 years ago, the Earth is itself a magnet. The geomagnetic field is generated by electric currents flowing deep within the solid Earth and to a first approximation may be represented as though due to a short bar magnet at the centre of the Earth. As a dipole field it extends beyond the planetary surface, through the troposphere on which it has no effect, and into the ionized atmosphere where its effects are considerable. The geomagnetic field affects the motions of ionized particles, and thus modifies ionospheric electric currents and the bulk movement of the plasma. The importance of the magnetic field increases with altitude as the atmosphere becomes more sparse and its degree of ionization increases. At the highest levels, more than a few thousand kilometres above the surface, all behaviour is so dominated by the geomagnetic field that this region is called the *magnetosphere*. There is no sharp boundary between the ionosphere and the magnetosphere, but between the magnetosphere and the solar wind is a boundary, the *magnetopause*, which is very significant. At this boundary energy is coupled into the magnetosphere from the solar wind, and here is determined much of the behaviour of the magnetosphere and of the ionosphere at high latitudes. In the sunward direction the magnetopause is encountered at about 10 Earth-radii, but in the anti-solar direction the magnetosphere is extended downwind in a long tail, the *magnetotail*, within which occur plasma processes of great significance for the geospace regions.

1.5 Nomenclature

The solar–terrestrial environment has many parts, which may be one reason why it has been difficult to find an apposite all-embracing appellation. Relevant material may be found in the literature under various titles. Internationally, the topic is considered a branch of geophysics and is sometimes called *external geophysics*, though in some countries there persists an outdated practice of confining the term 'geophysics' to the solid Earth (properly *internal geophysics*). The *upper atmosphere* is a term of some generality for the higher reaches of the atmosphere, though some use it to mean the neutral gas only. It should not be confused with the meteorologists' 'upper air', which is actually within the troposphere and stratosphere and therefore largely beneath the level of our considerations. The addition of 'physics' to 'upper atmosphere' obviously means that the physical processes of the region are being addressed. *Aeronomy* – literally, 'measurement of the air' – is a good modern term meaning the processes, physical and chemical, of the upper atmosphere. *Ionosphere* refers to the ionized component of the upper atmosphere, and *magnetosphere* to the outermost regions dominated by the geomagnetic field. These regions will be treated in some detail, but there is no clearly defined boundary between them. Much relevant material also appears under the heading of *space physics*, which is not unreasonable because most space data are taken not too far from the Earth in practice. We shall also use the term *geospace*, a recently coined word meaning the region of space relevant to the Earth. As an inclusive term it appears to be as good as any yet suggested.

1.6 Summary

It should be clear from the foregoing sketch that the contents of geospace are rather different from the more familiar atmospheric gas of the troposphere. In this book we shall be dealing with the physics of tenuous gases, with ionization and ionic recombination processes, with electrical conduction in a gas, with particle as well as electromagnetic radiations of various energies, and with the behaviour of a plasma permeated by a magnetic field. None of these has anything like as much significance in the atmosphere near the ground. We shall also be concerned with dynamics and transport, which are important throughout the atmosphere and magnetosphere.

It follows that the science of the solar– terrestrial environment is based principally on

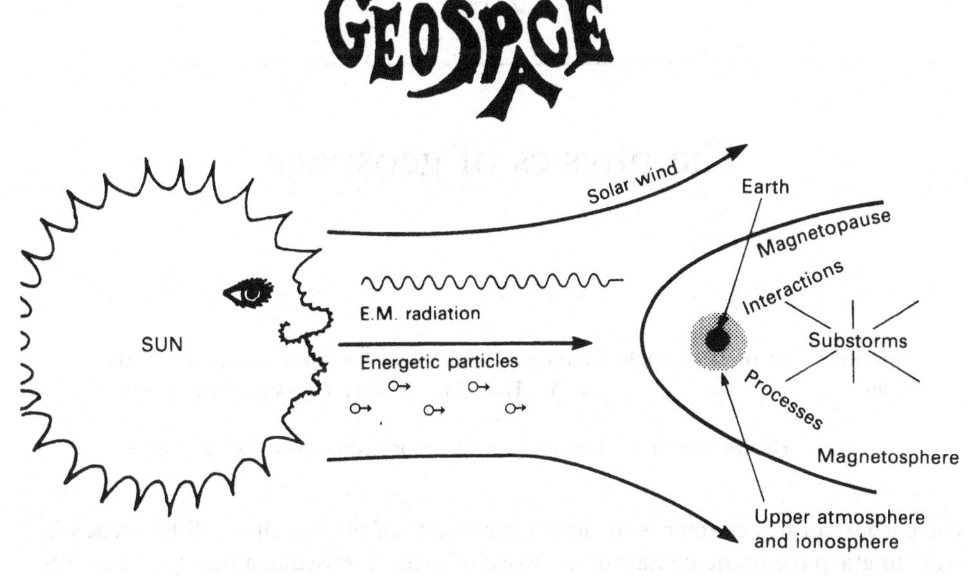

Fig. 1.1 Summary of the solar–terrestrial environment. (After a sketch by J. C. Hargreaves).

classical physics, though some knowledge of chemistry and, of course, mathematics is also needed. In Chapter 2 we shall summarize some aspects of basic physics that are particularly important for an appreciation of the more specialized material to follow.

2

The physics of geospace

Someone told me that each equation I included in the book would halve the sales. S. W. Hawking, *A Brief History of Time* (1988)

(Health warning – This chapter has more equations than any other.)

The purpose of this chapter is to summarize points of physics that will be needed in order to grasp the fundamentals of geospace science. It is assumed that the student is already familiar with basic physical concepts such as energy, temperature, quanta, waves, molecules, heat, and electric and magnetic fields – topics, it will be noted, which come mainly within the domain of classical physics. Most students of physics will have covered these areas in the first year or two of their university courses. But, like most specialities, upper atmosphere and space science have their own peculiar slant. We have to deal with a gas, and in particular with an electrified gas. We will be concerned with the propagation of waves – mainly electromagnetic waves, but some others too – in that gas. We shall need to know how a steady magnetic field affects the behaviour of gas and of waves. Energetic particles and photons will enter the gas, and their interactions have to be included. So the present chapter outlines the relevent background. Much of the material should be revision but some may be new.

It is up to the student whether to study this chapter thoroughly before tackling the subsequent ones, or merely to scan it through now in order to return for clarification later, if and when questions arise.

2.1 Useful units and fundamental constants

SI (Système International) units were internationally adopted in 1960, though in the guise of MKS (metre–kilogramme–second) they had already been creeping into use for 10 or 15 years previously. The school teachers, never slow (dare one say?) to embrace an innovation, have done their duty, taught in SI, and probably no longer even mention the older systems. Therefore – and not just because the Royal Society says so – this book is in SI. It is what the readers will expect, and properly so.

So one might think that by 1990 the changeover would have been complete and that centimetres, grammes and seconds (c.g.s.) would have followed feet, pounds and seconds (f.p.s.) into the archives of science. But not so! These changes take a long time to work through. University teachers are more conservative, and much of the delay is also due to the professional scientist no less, the very person who has created all this

Table 2.1 *Comparison of units*

Quantity	SI (MKS)	c.g.s. e.s.u.	c.g.s. e.m.u.	Relations
Time	second (s)	second (s)		
Length	metre (m)	centimetre (cm)		1 m = 100 cm
Mass	kilogram (kg)	gram (g)		1 kg = 1000 g
Force	newton (N)	dyne		1 N = 10^5 dyne
Pressure	pascal (Pa) = newton/m^2	dyne/cm^2		1 Pa = 10 dyne/cm^2
Energy	joule (J)	erg		1 J = 10^7 erg *1 eV = 1.602×10^{-19} J = 1.602×10^{-12} erg
Charge	coulomb (C)	e.s.u	e.m.u.	1 C = 0.1 e.m.u. = 3×10^9 e.s.u.
Electric potential	volt (V)	e.s.u.	e.m.u.	1 V = 10^8 e.m.u. = $\frac{1}{300}$e.s.u.
Electric field strength	volt/m	e.s.u.		1 V/m = $1/(3\times10^4)$ e.s.u.
Electric flux density (displacement)	coulomb/m^2	e.s.u.		1 C/m^2 = $12\pi\times10^5$ e.s.u.
Electric current	amp (A)	e.s.u.	e.m.u	1 A = 0.1 e.m.u. = 3×10^9 e.s.u.
Magnetic field strength	amp/m		oersted (Oe)	1 A/m = $4\pi/1000$ Oe
Magnetic flux density (induction)	tesla (T) = weber/m^2		gauss (G)	1 Wb/m^2 = 10^4 G *1γ = 10^{-9} Wb/m^2 = 10^{-5} G
Magnetic flux	weber (Wb)		maxwell (Mx)	1 Wb = 10^8 Mx
Capacitance	farad (F)	e.s.u.		1 F = 9×10^{11} e.s.u.
Inductance	henry (H)		e.m.u.	1 H = 10^9 e.m.u.

* See text

marvellous stuff, but who (naturally enough) likes to think and work in the units which he or she learned when a student. Hence many scientific papers are still being composed in c.g.s., and it goes without saying that the classic papers and books of the past stand as written, units and all. We must therefore make ourselves familiar with the c.g.s. system and be able to convert between c.g.s. and SI. In some cases the older units have a special attraction too, for example by giving numbers of a convenient magnitude – who wants to think about a million particles per cubic metre when one per cubic centimetre is the same? – and such special cases may as well be recognized and made use of. The policy of this book is therefore to emphasize SI, but also to quote, or even move into, other units where appropriate. Formulae and equations will be almost exclusively in the SI form.

Within c.g.s. there is an extra complication because there are several subsystems for

Table 2.2 *Some useful constants*

Constant	Symbol	SI (MKS)	c.g.s.
Planck	h	6.626×10^{-34} J s	6.626×10^{-27} erg s
Boltzmann	k	1.381×10^{-23} J K^{-1}	1.381×10^{-16} erg K^{-1}
Gas	R	8.315 J K^{-1} mol^{-1}	8.315×10^{7} erg K^{-1} mol^{-1}
Gravitational	G	6.673×10^{-11} N m^{-2} kg^{-2}	6.673×10^{-8} cm^{-3} g^{-1} s^{-1}
Speed of light in free space	c	2.998×10^{8} m s^{-1}	2.998×10^{10} cm s^{-1}
Electronic charge	e	1.602×10^{-19} C	4.803×10^{-10} e.s.u.
Electron mass	m_e	9.109×10^{-31} kg	9.109×10^{-28} g
Proton mass	m_p	1.673×10^{-27} kg (= 1836 m_e)	1.673×10^{-24} g
Gravitation acceleration at Earth's surface (mean value)	g	9.807 m s^{-2}	980.7 cm s^{-2}
Radius of Earth (mean)	R_E	6370 km	
Permittivity of free space	ε_0	8.854×10^{-12} F m^{-1}	1.0 (e.s.u.)
Permiability of free space	μ_0	1.257×10^{-6} H m^{-1}) (= $4\pi\times10^{-7}$ H m^{-1})	1.0 (e.m.u.)
Avagadro's number	N_A	6.022×10^{23} mol^{-1}	

(E. R. Cohen & B. N. Taylor. The fundamental physical constants. *Physics Today*, p. 8, August 1989, pt 2.)

electrical units, the most common being the electrostatic (e.s.u.), the electromagnetic (e.m.u.) units and the mixture of e.s.u. and e.m.u. usually called 'Gaussian' units. The essence of the distinction is that in e.s.u. the unit of charge is defined from the force between two charges a known distance apart, whereas in e.m.u. the unit of current is defined from the force between two currents. Basic units in these two systems differ by a factor of 3×10^{10}, the speed of light in c.g.s. In the Gaussian system (which we shall not use at all) some quantities are written in e.s.u. and others in e.m.u., and the formulae may include the speed of light 'c', explicitly. For those knowledgeable about units it should be stated that MKS formulae will be given in the 'rationalized' form; rationalized and unrationalized formulae may differ by a factor of 4π. When hunting in different books the reader would be wise to keep a weather eye open for such snares.

Table 2.1 states units and compares their values in different systems. The *electron-volt*, the *gamma* and the *ångström* are units of particle energy, of magnetic flux density and of wavelength respectively, that are widely used although not in the list of primary units. The electron-volt (eV) is the energy gained by a particle carrying one electronic charge in falling through a potential difference of one volt:

$$1\text{ eV} = 1.602\times10^{-19}\text{ joule} = 1.602\times10^{-12}\text{ erg}. \tag{2.1}$$

As a unit of flux density the gamma (γ) is of a convenient size for measuring the weak magnetic fields of the magnetosphere and inter-planetary space, as well as perturbations of the terrestrial magnetic field:

$$1\ \gamma = 10^{-5}\text{ gauss} = 10^{-9}\text{ tesla} = 1\text{ nT}. \tag{2.2}$$

However, the unit nanotesla (nT) has the same magnitude as the γ and takes but a little longer to say, so we shall prefer it.

The ångström ($\equiv 10^{-8}$ cm) is well established as a unit of wavelength, but the nanometre (nm) is nearly as convenient:

$$1 \text{ nm} = 10 \text{ Å}. \quad (2.3)$$

Table 2.2 lists some useful constants. A minor nuisance of the MKS system is that the constants ε_0 and μ_0, the permittivity and permiability of free space, have to be included in the formulae. In c.g.s. the magnetic flux density (B) in free space is made numerically equal to the magnetic field strength (H), whereas in MKS we write $B = \mu_0 H$. Similarly for the electric displacement and the electric field we put $D = \varepsilon_0 E$. The relation between ε_0 and μ_0 is

$$\varepsilon_0 \mu_0 = 1/c^2, \quad (2.4)$$

where c is the speed of light (in SI). Between c.g.s. and SI the formulae of electromagnetism differ mainly because of μ_0 and ε_0, and because of a '4π' due to rationalization. For example, the magnetic energy density in free space is $B^2/8\pi$ erg/cm^3 in c.g.s. but $B^2/2\mu_0$ J/m^3 in SI. Such pointers can be helpful when deciding which system of units an unfamiliar author may be using.

2.2 Properties of gases

2.2.1 Gas laws

The kinetic theory of gases describes the bulk properties of a gas in terms of the microscopic behaviour of its constituent molecules. Most basic are the gas laws of Boyle and Charles relating pressure (P), volume (V) and absolute temperature (T), which are generally combined into the *universal gas law*:

$$PV = RT, \quad (2.5a)$$

R being the *gas constant* with the dimensions of energy per degree per mole of gas. Equation 2.5a is for a single mole of gas, and if there are N moles we write

$$PV = NRT. \quad (2.5b)$$

Note that the mole, formerly 'gram-molecule', is a c.g.s. unit: that quantity of gas whose mass equals the gas's molecular weight in grams. In SI units we put for one mole,

$$PV = 10^3 RT, \quad (2.6a)$$

and for N moles,

$$PV = 10^3 NRT. \quad (2.6b)$$

The most useful form for our purposes, however, is in terms of *Boltzmann's constant* (k):

$$P = nkT, \quad (2.7)$$

where n is the number of molecules per unit volume, often called the *number density*. This equation is the same in both systems of units. The equivalence of Equations 2.6 and 2.7 can be demonstrated by substituting the expressions

$$R = N_A k, \quad (2.8)$$

where N_A is *Avogadro's number*, the number of molecules in one mole of gas; and

$$n = N(10^3 N_A)/V. \quad (2.9)$$

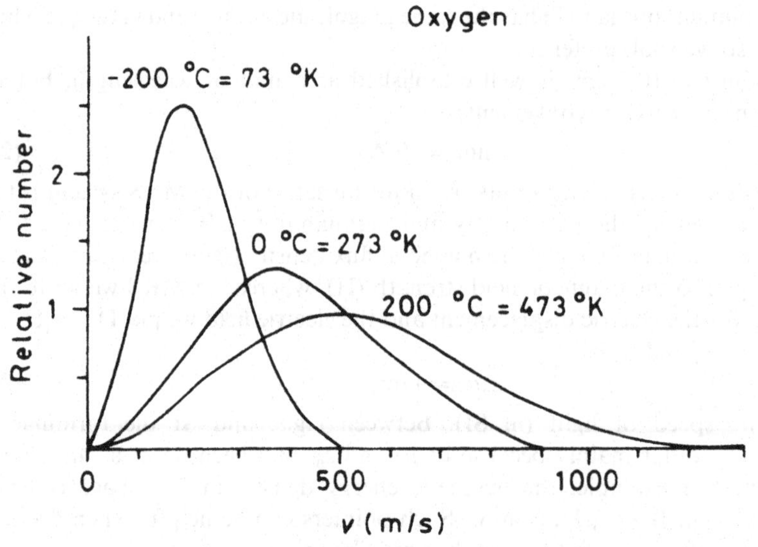

Fig. 2.1 Maxwell–Boltzmann distribution of molecular speeds in oxygen at three temperatures. (After S. Borowitz and L. A. Bornstein, *Contemporary View of Elementary Physics*. McGraw-Hill, 1968.)

The constant 10^3 arises because of the definition of the mole, and does not appear in c.g.s.

2.2.2 Thermal equilibrium

The molecules within a body of gas exchange energy by collisions and come to a state of thermal equilibrium, when the velocities are distributed according to the Maxwell–Boltzmann law:

$$N(v)\,dv = 4\pi N_T\left(\frac{m}{2\pi kT}\right)^{\frac{3}{2}} v^2 \exp\left(-\frac{mv^2}{2kT}\right) dv, \tag{2.10}$$

where $N(v)\,dv$ is the number of molecules with speeds between v and $v+dv$, N_T is the total number of molecules, m the molecular mass, T the absolute temperature and k Boltzmann's constant. The distribution is illustrated in Fig 2.1.

In a Maxwell–Boltzmann distribution the mean speed ($\bar{v}$), the root mean square speed ($v_{r.m.s.}$) and the most probable speed ($v_{m.p.}$) are given by

$$v_{m.p.} = (2kT/m)^{\frac{1}{2}},$$

$$\bar{v} = \frac{2}{\sqrt{\pi}}\left(\frac{2kT}{m}\right)^{\frac{1}{2}} = 1.128 v_{m.p.},$$

$$v_{r.m.s.} = \left(\frac{3kT}{m}\right)^{\frac{1}{2}} = 1.225 v_{m.p.}. \tag{2.11}$$

Hence the mean kinetic energy of the gas particles and the temperature are related by

$$\tfrac{1}{2}m\overline{v^2_{r.m.s.}} = \tfrac{3}{2}kT. \tag{2.12}$$

In thermal equilibrium the energy is shared also between the various components of

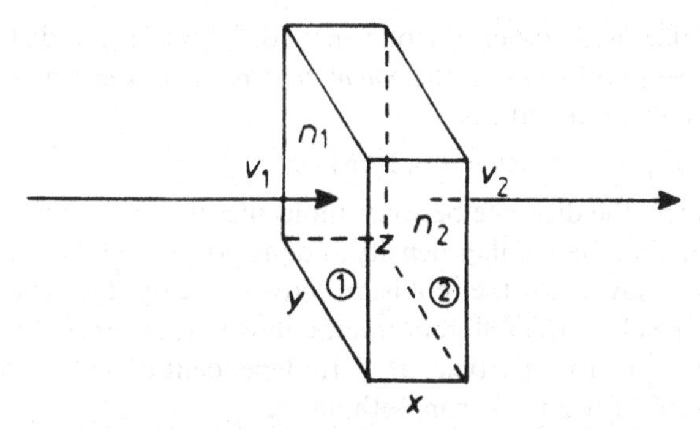

Fig. 2.2 Continuity requires that the number density, n, is related to the drift speed, v, by $\partial n/\partial t = -\partial(nv)/\partial x$.

a mixture. If two species have molecular masses m_1 and m_2, the mean square speeds $\overline{v_1^2}$ and $\overline{v_2^2}$ are related by

$$m_1\overline{v_1^2} = m_2\overline{v_2^2}.$$

This is equivalent to saying that both components come to the same temperature (Equation 2.12). The lighter gas has the higher molecular speed:

$$\overline{v_2^2}/\overline{v_1^2} = m_1/m_2. \tag{2.13}$$

2.2.3 Continuity

Superimposed on the random thermal motions there might well be a bulk motion of the gas as a whole. Continuity requires that all the molecules are accounted for, and if, for some reason, the drift velocity varies from place to place molecules will accumulate in some places and be depleted from others.

Consider an open box with sides x, y, and z, as in Figure 2.2, through which gas particles drift in the x direction only. If the particle speed and velocity are n_1 and v_1 at face 1, and n_2 and v_2 at face 2, the number entering the box in unit time is n_1v_1yz and the number leaving is n_2v_2yz. The rate of accumulation is therefore $(n_1v_1 - n_2v_2)yz$, and the rate of change of particle density within the box is $(n_1v_1 - n_2v_2)/x \rightarrow \partial(nv)/\partial x$ as the box is made infinitely small. Hence,

$$\frac{\partial n}{\partial t} = -\frac{\partial(nv)}{\partial x}. \tag{2.14}$$

In the general case with movement in three dimensions,

$$\frac{\partial n}{\partial t} = -\operatorname{div}(n\mathbf{v}). \tag{2.15}$$

2.2.4 Collisions

In a neutral gas the particles influence one another through collisions. A gas without collisions would never come to thermal equilibrium, and its physical behaviour could not be described in terms of a compressible fluid. In a gas with many collisions, both random and bulk motions are effectively communicated between the particles; the gas as a whole quickly comes to equilibrium and for many purposes may be treated as a single entity.

The simplest definition of the *collision frequency* (ν) is just the number of collisions that one particle makes with others in one second. The average distance that the particle travels between collisions is the *mean free path*. If the r.m.s. velocity is v [= $\sqrt{(3kT/m)}$], the mean free path is

$$l_f = (3kT/m)^{\frac{1}{2}}/\nu. \quad (2.16)$$

Since l_f is approximately the distance between molecules in a gas we might expect ν to be directly proportional to the number density and proportional to $T^{\frac{1}{2}}$. However, when charged particles are involved the collision cross section may be temperature dependent, and, as a result of the different interactions that occur, ν varies as T for an electron colliding with a neutral particle, as T^0 (independent of T) for an ion colliding with neutrals, and as $T^{-\frac{3}{2}}$ for an electron with an ion.

In the more distant parts of the atmosphere the mean free path becomes very long. If it exceeds the dimensions of the containing 'box' – e.g. the size of the region itself – that is a good indication that this is a region where we cannot assume thermal equilibrium or use the gas laws.

Although it is convenient to think of collision frequency as the rate of collision, there is a more rigorous definition in terms of the transfer of momentum. Consider a gas containing two species, masses m_1 and m_2, with velocities v_1 and v_2 which for simplicity we will assume to be in the same direction. If a particle of the first species collides head on with a particle of the second species, and the collision is elastic, conservation of momentum requires that the first particle gains momentum

$$2m_1m_2(v_2-v_1)/(m_1+m_2)$$

and the second particle loses the same amount. If there are ν_{12} such collisions per unit time, and n_1 particles of the first species per unit volume, the total force (total rate of change of momentum) on the first species due to collisions with the second is

$$F_{12} = n_1\nu_{12}\frac{2m_1m_2(v_2-v_1)}{(m_1+m_2)}. \quad (2.17)$$

This defines the collision frequency in terms of a force and has the advantage that it is not necessary to know in detail what happens at each collision. Defined thus, ν is the *momentum-transfer collision frequency*.

Since the forces on the two species (1 and 2) must be equal and opposite, $F_{12} = -F_{21}$, and

$$\nu_{21}/\nu_{12} = n_1/n_2. \quad (2.18)$$

We can apply the same principles to find the drag on a solid body moving through the air (Section 3.4.1).

2.2.5 Diffusion

If there is a pressure gradient within a gas, molecules will move down the pressure gradient until the pressure is equalized. At any moment the net velocity (superimposed, of course on the random thermal movements) is proportional to the gradient. Then the drift speed may be written,

$$v = -\frac{D}{n}\frac{\partial n}{\partial x}, \quad (2.19)$$

where n is the particle density and D is the *diffusion coefficient*. Diffusion in one direction only, as along a narrow column, has been assumed for simplicity. (In three dimensions, v = –(D/n)**grad** n.)

The rate of change of particle density at a given point is obtained from Equation 2.14:

$$\frac{\partial n}{\partial t} = \frac{\partial (vn)}{\partial x} = -\frac{\partial}{\partial x}\left(-D\frac{\partial n}{\partial x}\right) = +D\frac{\partial^2 n}{\partial x^2}. \tag{2.20}$$

In the three-dimensional case,

$$\frac{\partial n}{\partial t} = D \cdot \nabla^2 n.$$

Note that the diffusion coefficient, D, has the dimensions $(\text{length})^2/\text{time}$.

We can derive an expression for the diffusion coefficient of a gas at constant temperature by equating the pressure gradient,

$$\frac{\partial p}{\partial x}\left(=kT\frac{\partial n}{\partial x}\right)$$

to the drag force due to collisions, $n\nu mv$, to obtain

$$kT\frac{\partial n}{\partial x} = -n\nu mv.$$

Then, comparing with Equation 2.19,

$$D = kT/m\nu, \tag{2.21}$$

a simple but very useful formula.

If $\nu \propto nT^{\frac{1}{2}}$ (Section 2.2.4), $D \propto T^{\frac{1}{2}}n^{-1}$. Molecular diffusion in a gas proceeds more rapidly at higher temperature and at lower pressure.

The mathematical theory of diffusion has many applications beside molecular diffusion in gases, heat conduction being perhaps the most familiar. Others relevant to geospace are the diffusion of the pitch angles of geomagnetically trapped particles (Section 9.4.5), and the motion of plasma through a magnetic field (Section 2.3.6). Atmospheric gases can be mixed by turbulence as well as by molecular diffusion, the difference being that one has to treat the motions of bubbles of gas rather than individual molecules. Mixing by turbulence can also be treated by diffusion theory and an *eddy diffusion coefficient* may be defined. Unlike the molecular diffusion coefficient, it cannot be derived simply from first principles, but values can be determined from measurements. Atmospheric turbulence depends on the vertical temperature gradient, and it is clear that the relative importance of these two processes will vary with altitude. (See Sections 4.1.1 and 4.1.5 for some consequences.)

2.3 Magnetoplasma

Much of the upper atmosphere is ionized, and the ionized gas is permeated by the geomagnetic field. The combination of magnetic field and ionized gas is a *magnetoplasma*, whose unique properties underly much of the observed behaviour of the geospace region. In this section we outline some basic properties of a magnetoplasma.

2.3.1 Electric and magnetic energy

The energy of a magnetoplasma comes principally from the energy of the particles and the energy of the magnetic field – plus contributions from any electric field and wave motions that may be present – and the behaviour of the medium may well depend on which of these components is the greatest. From Equation 2.12, if the particles are at temperature T, and there are n per unit volume, their kinetic energy density is just 3nkT/2. Formulae for the magnetic and electric energy densities are readily proved.

Consider a parallel-plate capacitor C, with charge Q and potential V. The work done in charging the capacitor is

$$W = \int V\,dq = \int_0^Q (q/C)\,dq = Q^2/2C. \qquad (2.22)$$

This work may be attributed to the energy in the space between the plates. If the area is A and the separation d, the electric energy density is

$$w = \frac{W}{Ad} = \frac{Q^2}{2CAd} = \frac{\varepsilon V^2}{2d^2} = \frac{\varepsilon E^2}{2}, \qquad (2.23)$$

since $C = \varepsilon A/d$ and $V = Ed$. In vacuum (or air), $\varepsilon = \varepsilon_0$, and $w = \varepsilon_0 E^2/2$. (In c.g.s. (e.s.u) units, $w = \varepsilon E^2/8\pi$, where $\varepsilon = 1$ in air.)

The formula for magnetic energy density is derived similarly, but now in terms of a solenoid. If a solenoid of length l and area A carries a current i, the total magnetic energy is $Li^2/2$, where the inductance, $L = \mu N^2 A/l$. Then the energy density within the solenoid is

$$w = \frac{W}{Al} = \frac{Li^2}{2Al} = \frac{\mu N^2 i^2}{2l^2} = \frac{\mu H^2}{2} = \frac{B^2}{2\mu}, \qquad (2.24)$$

since the magnetic field strength in the solenoid is $H = Ni/l$, and the magnetic flux density $B = \mu H$. In a non-magnetic material $\mu = \mu_0$. (In c.g.s. (e.m.u.), $w = \mu H^2/8\pi$ and $\mu = 1$ in a non-magnetic material.)

2.3.2 Gyrofrequency

In a magnetic field charged particles tend to go round in circles, and the gyrofrequency is just the rate of this gyration. If the magnetic flux density is **B**, the charge on the particle is e and its velocity is **v**, then the Lorentz force on the particle is

$$\mathbf{F} = e\cdot\mathbf{v}\times\mathbf{B}, \qquad (2.25)$$

acting normal to both **v** and **B** (Figure 2.3). The particle is constrained to move in a curved path, and if there is no component of velocity along **B** the path will be a circle. The Lorentz force provides the centripetal acceleration v^2/r_B where r_B is the radius of the circle. If m is the mass of the particle, the gyroradius is

$$r_B = mv/Be. \qquad (2.26)$$

The period of revolution is $P = 2\pi r_B/v = 2\pi m/Be$, the angular frequency is

$$\omega_B = 2\pi/P = Be/m \text{ rad/s} \qquad (2.27)$$

and

$$f_B = Be/2\pi m \text{ Hz}. \qquad (2.28)$$

Note that r_B and ω_B depend on both the magnetic field and the mass of the particle, but the gyrofrequency is independent of the velocity. For kinetic energy, E,

$$v = (2E/m)^{\frac{1}{2}} \quad \text{and} \quad r_B = (2mE)^{\frac{1}{2}}/Be.$$

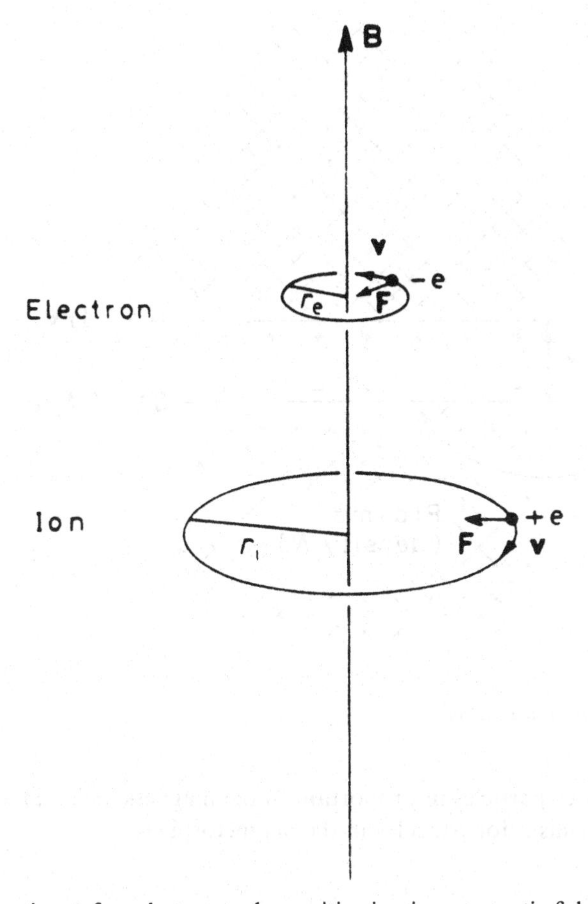

Fig. 2.3 Gyrations of an electron and a positive ion in a magnetic field.

Using subscripts i and e for ions and electrons, the energy being the same, we see that

$$\omega_i/\omega_e = m_e/m_i \tag{2.29}$$

and

$$r_i/r_e = (m_i/m_e)^{\frac{1}{2}}. \tag{2.30}$$

By way of example to illustrate typical magnitudes, take the magnetic field as 0.5 G (= 0.5×10^{-4} Wb/m²), e = 1.6×10^{-19} C and $m_e = 9.1 \times 10^{-31}$ kg. Then $\omega_e = 8.8 \times 10^6$ rad/s = 1.4 MHz. For an oxygen ion (O^+), ω_i is smaller by a factor of 29 380, giving a gyrofrequency of only 48 Hz. If E = 10 keV = 1.6×10^{-15} J, r_e = 6.7 m and $r_i(O^+)$ = 1.2 km.

2.3.3 Betatron acceleration

If the magnetic flux density is gradually increased, the gyrofrequency is increased in proportion (Equation 2.27). However, the angular momentum, $mvr_B = mv^2/\omega_B$, is conserved, and therefore the kinetic energy increases:

$$E = mv^2/2 \propto \omega_B \propto B. \tag{2.31}$$

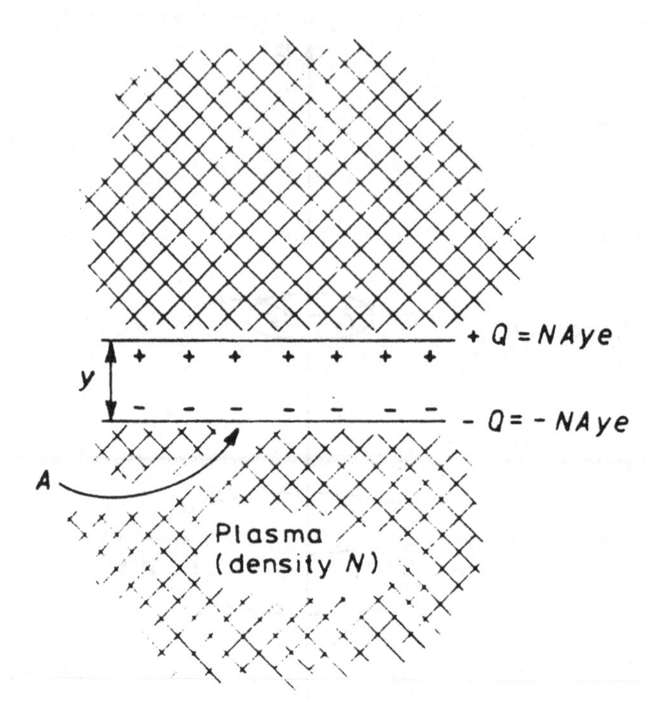

Fig. 2.4 Charge separation in a plasma.

This mechanism energizes particles in proportion to the magnetic field variation and is an acceleration mechanism for particles in the magnetosphere.

2.3.4 Plasma frequency

The *plasma frequency* is one of the basic parameters of an ionized medium, and one which is frequently encountered in the theory of radio propagation. Its significance may be appreciated by means of the following model.

Consider that over an area A the positive and negative charges are separated by a distance y (Figure 2.4). If there are N electrons per unit volume of the medium, and the same density of ions, the parallel-plate capacitor thus formed carries a total charge $Q = NAye$ and its capacitance is $\varepsilon_0 A/y$ (in SI units). The potential difference across the plates is Ney^2/ε_0, the electric field in the gap is Ney/ε_0, and therefore the force on a single electron is Ne^2y/ε_0. Therefore the restoring acceleration is $Ne^2y/\varepsilon_0 m$. Since the restoring acceleration is proportional to the separation, the charged particles when free to move will undergo simple harmonic motion with angular frequency:

$$\omega_N = \left(\frac{Ne^2}{\varepsilon_0 m}\right)^{\frac{1}{2}} \quad \text{in SI,} \tag{2.32}$$

or

$$\omega_N = \left(\frac{4\pi Ne^2}{m}\right)^{\frac{1}{2}} \quad \text{in c.g.s.} \tag{2.33}$$

Thus, ω_N is the natural oscillation frequency for electrostatic perturbations within the plasma. For numerical purposes it is often expressed as the frequency in hertz: $f_N = \omega_N/2\pi$.

Since the particle mass appears in the demominator of Equation 2.32 there is a different plasma frequency for each ion species, the frequency being lower for heavier particles. The quantities ω_N and f_N are usually reserved for electron plasma frequencies and an alternative such as Ω_N and F_N chosen for ions. The plasma frequency is proportional to the square root of the plasma density. A convenient approximate formula for the electron plasma frequency is

$$f_N = (80.5\ N)^{\frac{1}{2}}, \tag{2.34}$$

where N is in m^{-3} and f_N in Hz.

2.3.5 Debye length

If the plasma frequency represents the characteristic frequency of a plasma, the *Debye length* is the characteristic distance. Due to the thermal speeds of the particles (those of the electrons being much larger than those of the ions) the electrons and the ions separate within the plasma to an extent which is limited by electrostatic attraction. If an electron oscillates at frequency ω_N with amplitude l, its maximum velocity is $v = \omega_N l$. Now, let this speed be the r.m.s. thermal speed, $v = (kT/m)^{\frac{1}{2}}$, and therefore $l = (1/\omega_N)(kT/m)^{\frac{1}{2}}$. This amplitude of oscillation is equal to the Debye length, usually written λ_D. Substituting for ω_N from Equation 2.32,

$$\lambda_D = \left(\frac{\varepsilon_0 kT}{Ne^2}\right)^{\frac{1}{2}}. \tag{2.35}$$

In c.g.s.,

$$\lambda_D = \left(\frac{kT}{4\pi Ne^2}\right)^{\frac{1}{2}}. \tag{2.36}$$

Numerically,

$$\lambda_D = 69(T/N)^{\frac{1}{2}}\,m, \tag{2.37}$$

where T is in K and N in m^{-3}.

In c.g.s.,

$$\lambda_D = 6.9(T/N)^{\frac{1}{2}}\ cm.$$

The Debye length is best visualized as a 'sphere of influence' of individual particles. Beyond λ_D, the effect of one particle is cancelled by the collective effect of particles of opposite sign; hence it is also known as the *Debye shielding* (*or screening*) *distance*. [In the absence of other charges the potential at distance r from a charge e is $e/4\pi\varepsilon_0 r$. In a plasma it is reduced to

$$\frac{e}{4\pi\varepsilon_0 r}\cdot\exp\left(-\frac{\sqrt{2}r}{\lambda_D}\right).]$$

Within a distance λ_D the particles have to be considered individually, but over greater distances the individual particles are not seen and the plasma may be treated as a fluid. In the kinetics of the medium the Debye shielding distance marks the natural division between particle collisions on the small scale, and the collective effects on the large scale. The concept of a plasma requires that there is at least one particle in a sphere of radius λ_D. This is the *Debye sphere*.

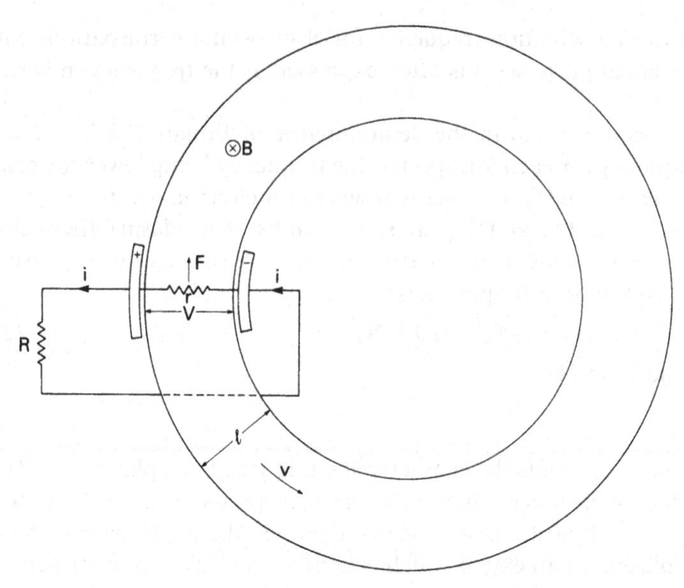

Fig. 2.5 Simple dynamo, to illustrate 'freezing' between a conductor and a magnetic field.

2.3.6 Frozen-in field

In the simple dynamo of Figure 2.5, a thin conducting annulus of width l is rotated between frictionless contacts of unit length in a transverse magnetic field of flux density B. By Faraday's law of electromagnetic induction, the potential difference generated between the contacts is the rate of change of flux, $V = Bvl$. If the circuit is now completed by an external resistance R, and the section of annulus between the contacts has resistance r, a current

$$i = V/(R+r) = Bvl/(R+r)$$

flows in the circuit. The power dissipated is

$$P = Vi = B^2v^2l^2/(R+r). \tag{2.38}$$

Since the current is flowing across the magnetic field, it experiences a force $F = Bil$, the Lorentz force, acting to oppose the rotation of the disc. Thus, mechanical work is done at a rate $Fv = Bilv = B^2v^2l^2/(R+r)$, which is equal to the electrical dissipation rate, as it should be.

Suppose now that the external circuit has zero resistance ($R = 0$) and the conductivity of the disc material is σ. Then, from Equation 2.38,

$$v^2 = rP/B^2l^2 = P/B\sigma l. \tag{2.39}$$

If $\sigma \to \infty$, $r \to 0$. In that case, if v remained finite $i \to \infty$ and the retarding force $\to \infty$, which is clearly impossible. The consequence of making the conductivity very large while the power dissipation (P) remains finite must be to make the velocity (v) very small. In the limit of infinite conductivity the only conclusion must be that the velocity is zero.

This simple model illustrates an important property of magnetoplasma which applies throughout most of geospace: when the electrical conductivity of the plasma is very large, relative motion between the plasma and the magnetic field becomes

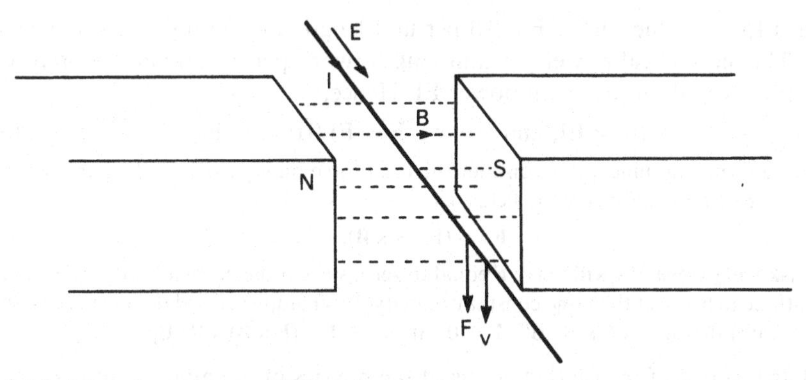

Fig. 2.6 Model to show that v = E/B.

virtually impossible. The field is then said to be *frozen in*. Similarly, a magnetic field cannot enter a region already occupied by a highly conducting plasma, and it is then *frozen out*. The frozen field concept was introduced by H.Alfvén in 1942.

The freezing of magnetic field may be proved rigorously from Maxwell's equations and is treated in texts on electromagnetic theory. We give a simplified version here.

For current density **J**, electric field **E** and magnetic induction **B**, two of Maxwell's equations, neglecting displacement current, give (SI form)

$$\nabla \times \mathbf{E} = -\frac{\partial \mathbf{B}}{\partial t},$$

$$\nabla \times \mathbf{B} = \mu_0 \mathbf{J}.$$

Ohm's law states

$$\mathbf{J} = \sigma(\mathbf{E} + \mathbf{v} \times \mathbf{B}).$$

Putting $\mathbf{v} = 0$, and eliminating **J** and **E**, we obtain

$$\frac{\partial \mathbf{B}}{\partial t} = \frac{\nabla^2 \mathbf{B}}{\mu_0 \sigma} = \varepsilon_0 c^2 \frac{\nabla^2 \mathbf{B}}{\sigma},$$

c being the speed of light.

This is a diffusion equation (Section 2.2.5) with diffusion coefficient $1/\mu_0\sigma$ having the dimensions of $(\text{length})^2/(\text{time})$. If the conductor occupies a space characterized by length L, the time taken for a magnetic field to enter or leave it is approximately $\tau = (\mu_0\sigma)L^2$. For times smaller than τ the field and the plasma can be considered to move together. For times greater than τ they can be considered to move independently. Representative values of τ are 1 second for a 1 cm copper sphere, 10^4 years for the core of the Earth, and 10^{10} years for the Sun.

2.3.7 E × B drift

If an electric field (E) is applied at right angles to the magnetic field (flux density **B**) the plasma particles drift at velocity **v** given by

$$\mathbf{v} = \frac{\mathbf{E} \times \mathbf{B}}{|\mathbf{B}|^2}. \qquad (2.40)$$

The vector **v** is normal to both **E** and **B**, and its magnitude is E/B.

This important result can be appreciated by means of a simple motor analogy. In Figure 2.6, a wire carries a current I across the magnetic flux between the poles of a magnet. We suppose that the current is associated with an electric field E. The resulting downward force on the wire is $F = IB$ per unit length and we suppose it moves with speed v. The mechanical power per unit length (work per unit time) is therefore Fv, and this must equal the electrical power EI. Hence,

$$Fv = EI, \text{ and } v = EI/F = EI/IB = E/B. \tag{2.41}$$

As a more formal argument, we consider the total force on a particle with charge e moving at velocity $\mathbf{v}$ in electric and magnetic fields, as

$$\mathbf{F} = e(\mathbf{E} + \mathbf{v} \times \mathbf{B}). \tag{2.42}$$

We then ask what value of $\mathbf{v}$ will make $\mathbf{F}$ equal to zero, since if the particle experiences no force it will continue to move at the same constant velocity. It is readily verified (by means of a simple vector diagram) that if $\mathbf{v} = \mathbf{E} \times \mathbf{B}/|\mathbf{B}|^2$, $\mathbf{F} = 0$, since $\mathbf{v} \times \mathbf{B} = (\mathbf{E} \times \mathbf{B}) \times \mathbf{B}/|\mathbf{B}|^2$.

$\mathbf{E} \times \mathbf{B}$ drift is one of the fundamentals of the physics of the magnetosphere and the ionosphere, and we will meet it again in Section 6.5 as a special case of currents and drifts in plasmas.

2.3.8 Fermi acceleration

The betatron process (Section 2.3.3) can accelerate charged particles moving at right angles to a magnetic field. For charged particles moving along a magnetic field the main acceleration process is *Fermi acceleration*. Consider a particle travelling back and forth between two 'mirrors' which are moving closer together. If the speed of the particle is v and the path has length l, there is one reflection every l/v. If the path shortens at dl/dt the particle speed increases by dl/dt at each reflection, therefore at a rate

$$dv/dt = (dl/dt)(v/l).$$

Hence,

$$dv/v = dl/l,$$

and if the particle mass is m the rate of increase of its energy is

$$\frac{dE}{dt} = \frac{d}{dt}\left(\frac{mv^2}{2}\right) = mv \cdot \frac{dv}{dt} = \frac{mv^2}{l} \cdot \frac{dl}{dt}.$$

Also, for an increment of l,

$$\frac{dE}{dl} = \frac{d}{dl}\left(\frac{mv^2}{2}\right) = mv \cdot \frac{dv}{dl} = \frac{mv^2}{l} = \frac{2E}{l}.$$

Hence, the relative increment of energy ($\delta E/E$) due to a relative increment of path shortening ($\delta l/l$) is simply

$$\frac{\delta E}{E} = \frac{2\delta l}{l}. \tag{2.43}$$

That is, if the path shortens by 5 % the energy increases by 10 %. This mechanism can energize geomagnetically trapped particles (Section 5.7.2) if the field-lines contract.

2.4 Waves

Waves are very important in geospace: not only electromagnetic waves but also waves propagating in the matter of the upper atmosphere, some being characteristic of the neutral air and some generated in the plasma. This section summarizes some properties of waves in general.

2.4.1 Phase velocity

If a source generates a displacement, $A = A_0 \cos \omega t$, which then propagates at speed v in the x direction, the displacement a distance x from the source is

$$A = A_0 \cos \omega \left(t - \frac{x}{v}\right)$$
$$= A_0 \cos \left(\omega t - \frac{2\pi x}{\lambda}\right). \qquad (2.44)$$

Here, λ is the wavelength, the distance between adjacent wave crests as measured from an instantaneous snapshot of the wave. This equation expresses the essence of a propagating wave in its simplest form.

It is sometimes more convenient to use the j notation, writing

$$A = A_0 \exp j\left(\omega t - \frac{2\pi x}{\lambda}\right), \qquad (2.45)$$

where $j = \sqrt{-1}$, and it is understood that the real part is taken (since $e^{j\theta} \equiv \cos\theta + j\sin\theta$). The angular frequency, ω, is related to the frequency in hertz (cycles per second), f, by $\omega = 2\pi f$, these frequencies being measured by a stationary observer.

In the three-dimensional case – the normal situation in space – the wavelength depends on the direction in which it is measured in relation to the direction of propagation. For example, if a plane wave travels in the y direction, the wavelength in the x direction is infinite (see Figure 4.21). λ is often expressed as a wavenumber, $k = 2\pi/\lambda$, which can be regarded as a vector along the propagation direction, whose components k_x, k_y, k_z give λ_x, λ_y, λ_z respectively.

In Equation 2.44, v is the phase velocity, $v = \lambda f$. (In three dimensions, the phase speeds in the x, y and z directions are $v_x = \lambda_x f$, $v_y = \lambda_y f$, and $v_z = \lambda_z f$. In an isotropic medium, v_x, v_y and v_z are all greater than v measured along the direction of propagation.)

2.4.2 Refractive index

The velocity depends on the nature of the medium. Thus, for an electromagnetic wave in vacuum $v = c$, the speed of light. For any other medium we write

$$v = c/n, \qquad (2.46)$$

where n is the *refractive index*. If the refractive index depends on the wave frequency the medium is said to be *dispersive*, and waves of different frequency travel at different speeds.

In some media the refractive index is complex:

$$n = \mu - j\chi. \qquad (2.47)$$

Then, using the j notation, the displacement is

$$A = A_0 \exp j\left(\omega t - \frac{n\omega x}{c}\right)$$
$$= A_0 \exp j\left(\omega t - \frac{\mu\omega x}{c}\right) e^{-\chi\omega x/c}. \qquad (2.48)$$

This wave travels at speed c/μ, and the last term represents an exponential attenuation; i.e. absorption in the medium, with the amplitude reducing a factor of e in a distance $c/\chi\omega$. Writing the attenuation term as $e^{-\kappa x}$, the absorption coefficient is

$$\kappa = \omega\chi/c = 2\pi\chi/\lambda_0, \tag{2.49}$$

where λ_0 is the free-space wavelength (c/f).

In some cases the refractive index is purely imaginary, so that $n = -j\chi$, and then

$$A = A_0 e^{j\omega t} e^{-\chi\omega x/c}. \tag{2.50}$$

This is an *evanescent wave*, which extends a distance of about $c/\chi\omega$ into the medium but does not propagate because its phase does not vary with distance. When a propagating wave is totally reflected at the interface between two media an evanescent wave exists just inside the second medium.

2.4.3 Group velocity

Fourier analysis tells us that a wave can be truly monochromatic only if it is of infinite duration. If the duration is finite (ΔT) the waves must occupy a finite bandwidth ($\Delta f \sim 1/\Delta T$) so that there may be cancellation before and after the interval ΔT. Similarly, any change of wave amplitude implies a finite bandwidth.

Several types of modulation are used in radio communications, but the simplest to explain is amplitude modulation;

$$A = A_0(1 + m\cos\omega_m t)\cos\omega_c t, \tag{2.51}$$

where ω_c and ω_m are respectively the angular frequencies of the carrier and the modulation, and m is the depth of modulation (between 0 and 1). Equation 2.51 can be written

$$A = A_0\cos\omega_c t + \tfrac{1}{2}mA_0\cos(\omega_c - \omega_m)t + \tfrac{1}{2}mA_0\cos(\omega_c + \omega_m)t, \tag{2.52}$$

showing that a modulated sinusoid is identical to a monochromatic carrier plus two monochromatic sidebands at frequencies $(\omega_c \pm \omega_m)$. If the phase or the frequency is modulated instead of the amplitude, the modulated wave can still be represented as a carrier plus a spectrum of sidebands. In the case when the wave is of finite duration, the equivalent spectrum is a continuum rather than a number of discrete lines.

The phase velocity and the refractive index introduced in Sections 2.4.1 and 2.4.2 refer to an unmodulated, and therefore monochromatic, wave of infinite duration. If a propagating wave is modulated we also need to know the velocity of the modulation. This is called the *group velocity*, and it is not necessarily equal to the phase velocity. The question is handled on the principle that each carrier and sideband travels at the velocity appropriate to its frequency. Should all these be the same, then the wave as a whole travels at the phase velocity. But this is a special case. In general the medium will be dispersive, all three of the waves in Equation 2.52 will travel at different speeds, and the modulation will travel at yet another speed, which is the group velocity.

To find the relation between group and phase velocities in a dispersive medium, consider three sinusoidal waves travelling in the x direction:

$$\left.\begin{aligned} a_0 &= \cos(\omega t - kx), \\ a_1 &= \cos[(\omega - \delta\omega)t - (k - \delta k)x], \\ a_2 &= \cos[(\omega + \delta\omega)t - (k + \delta k)x], \end{aligned}\right\} \tag{2.53}$$

where k is the wavenumber ($= 2\pi/\lambda$). k is a function of ω. The sum of these three monochromatic waves is

$$A = \cos(\omega t - kx)\,[1 - 2\cos(\delta\omega t - \delta kx)].$$

Thus, as the carrier of frequency $f(= \omega/2\pi)$ travels at speed $f\lambda(= \omega/k)$, the modulation of frequency $\delta f(= \delta\omega/2\pi)$ travels at speed $\delta\omega/\delta k$. Taking a small increment of frequency, it follows that the group velocity is

$$u = \partial\omega/\partial k. \qquad (2.54)$$

By definition the *group refractive index* is

$$n_g = \frac{c}{u} = c\left(\frac{\partial k}{\partial\omega}\right) = \frac{\partial}{\partial\omega}(n\omega) = \frac{\partial}{\partial f}(nf), \qquad (2.55)$$

where $n\ (= c/f\lambda)$ is the phase refractive index. Although derived here for a simple amplitude modulation, Equations 2.54 and 2.55 apply to any plane waves.

Some kinds of measurement give the group speed while others give the phase speed. For example, a pulse of waves travels at the group speed, but the waves within the pulse travel at the phase speed. Energy and information always travel at the group velocity. (A strictly monochromatic wave is of no use for signalling, since all the waves are identical.) Although the phase velocity of a radio wave can exceed the speed of light in the ionosphere, students of relativity can rest assured that the group velocity is always less than c.

2.4.4 Polarization

In a wave something is displaced. It might be an air molecule displaced from its rest position, the deflection of a water surface, the electric or magnetic field in an electromagnetic wave. As a rule, since space has three dimensions, the direction of the displacement has also to be specified. In a transverse wave – where the displacement is normal to the propagation direction – plane polarization, which is the simplest case, is easily described by specifying the orientation of the plane containing the displacement (e.g. vertical).

Circular polarization is a little more difficult to envisage. The displacement direction rotates by 360° for every wavelength travelled. Figure 2.7 illustrates how a circularly-polarized wave can be considered as the sum of two plane-polarized waves with a 90° phase shift. If the plane waves are not of equal amplitude or the phase difference is not 90° the resultant is elliptically polarized. Conversely, a plane-polarized wave can be considered to be composed of two circularly-polarized waves rotating in opposite directions. (See Figure 3.9.)

The polarization of a transverse wave is quantified by the ratio of the displacements in two directions at right angles, $R = A_x/A_y$. If R is real the polarization is linear, if imaginary the polarization is circular, and in the general case of a complex ratio the polarization is elliptical. In a right-handed coordinate system, with the wave travelling in the z direction, $R = +j$ means a circular wave with the displacement rotating clockwise at a fixed value of z.

Linear, circular, and elliptical transverse waves are all encountered in the theory of ionospheric radio propagation. In some types of wave – e.g. acoustic–gravity waves (Section 4.3.1) – displacements can be both transverse and longitudinal, and such a case requires a second polarization parameter, $S = A_z/A_x$.

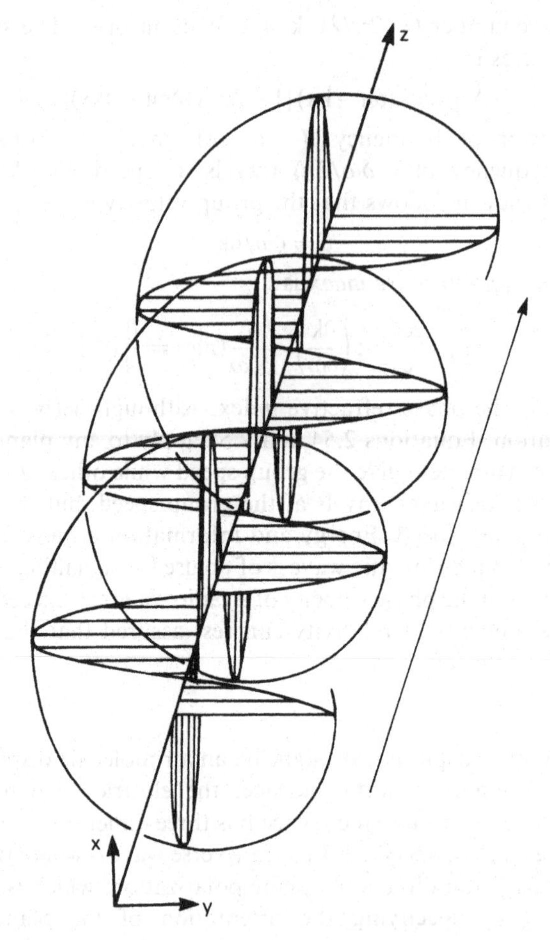

Fig. 2.7 Wave with left-handed circular polarization as the sum of two plane-polarized waves.

2.4.5 Energy density

In a sinusoidal wave the restoring force (F) is proportional to the displacement (A), F = CA, where C is a constant. The work done by increasing the displacement by dA is dW = F dA = CA dA. At maximum displacement, A_0, the potential energy is

$$W = \int_0^{A_0} dW\,dA = C\int_0^{A_0} A\,dA = CA_0^2/2. \qquad (2.56)$$

Thus the energy density of the wave is proportional to the square of the amplitude. At other phases of the cycle some of the energy is kinetic. (At zero displacement the velocity is maximum and all the energy is in the kinetic form.) If the constant, C, is defined in terms of unit volume of the medium, Equation 2.56 gives the energy density of the medium due to the wave. The energy flux is $CA_0{}^2u/2$, where u is the group velocity.

According to Equations 2.23 and 2.24, the energy density due to electric and magnetic fields is $(\varepsilon E^2 + \mu H^2)/2$, and this applies equally to the varying fields of an electromagnetic wave. It can be shown that if the electric and magnetic amplitudes are E_0 and H_0, then $E_0/H_0 = (\mu\varepsilon)^{\frac{1}{2}}$. Thus the energy density (averaged over a cycle) is

$$\frac{\varepsilon E_0^2}{2} = \frac{\mu H_0^2}{2} \tag{2.57}$$

the electric and magnetic energy densities being equal. In a propagating wave the power (W/m^2) is proportional to the square of the amplitude of the electric field and of the magnetic field.

2.5 Radio waves in an ionized medium

2.5.1 Magneto-ionic theory and the Appleton equation

The propagation of radio waves in a uniform magnetoplasma is treated by *magneto-ionic theory*. The development of this important theory came in the first part of this century, following Marconi's experiments in long distance radio communication, which, it is generally believed, achieved the first trans-Atlantic signals in 1901, and the suggestions by Kennelly and by Heaviside (independently) in 1902 that the waves were reflected from an electrified layer in the upper atmosphere. The theory did not emerge suddenly and spontaneously in the mind of one individual – such is seldom the case in science – but it evolved over a period of some 30 years from the contributions of several people. A theory by Eccles treated the layer as a metallic reflector in 1912, and Larmor treated it as a dielectric in 1924. In fact the dielectric approach is the appropriate one for most radio frequencies; absorption inside the medium is relatively small, and in most cases the wave is returned to ground not by a sharp reflection but by a process of gradual refraction. The form now generally used is mainly the work of E. V. Appleton and collaborators between 1927 and 1932.

The formula for the refractive index of an ionized medium is generally called the *Appleton equation* or the *Appleton–Hartree equation*:

$$n^2 = 1 - \frac{X}{1 - jZ - \left[\frac{Y_T^2}{2(1-X-jZ)}\right] \pm \left[\frac{Y_T^4}{4(1-X-jZ)^2} + Y_L^2\right]^{\frac{1}{2}}}. \tag{2.58}$$

n is complex in the general case: $n = \mu - j\chi$ (Equation 2.47). X, Y and Z are dimensionless quantities defined as follows:

$$\left.\begin{aligned} X &= \omega_N^2/\omega^2, \\ Y &= \omega_B/\omega, \\ Y_L &= \omega_L/\omega, \\ Y_T &= \omega_T/\omega, \\ Z &= \nu/\omega, \end{aligned}\right\} \tag{2.59}$$

Here, ω_N is the angular plasma frequency (Equation 2.32), ω_B is the electron gyrofrequency (Equation 2.27), and ω_L and ω_T are respectively the longitudinal and transverse components of ω_B with respect to the direction of propagation. That is, if θ is the angle between the propagation direction and the geomagnetic field,

$$\left.\begin{aligned}\omega_L &= \omega_B \cos\theta. \\ \text{and} \qquad \omega_T &= \omega_B \sin\theta.\end{aligned}\right\} \tag{2.60}$$

ν is the electron collision frequency and ω is the angular wave frequency. It would be quite beyond our present scope to derive the Appleton–Hartree equation; the reader is refered to the book by Ratcliffe (1959) for derivations and a detailed exposition of the applications.

Equation 2.58 looks alarmingly complicated, and indeed it does encompass the whole of radio propagation in a uniform ionized medium. The practical way of dealing with it is to consider special cases. For instance, if absorption is small or of no consequence, we put $Z = 0$. If, further, we are not concerned with the effects of the geomagnetic field, we put $Y_L = Y_T = 0$, and the expression simplifies to

$$n^2 = 1 - X = 1 - \frac{\omega_N^2}{\omega^2}, \tag{2.61}$$

which is certainly not a difficult equation and is already quite useful for understanding the basics of radio propagation in the ionosphere.

Since the ionospheric medium is non-magnetic, Equation 2.61 also expresses the dielectric constant (κ):

$$n^2 = \frac{c^2}{v^2} = \frac{\mu_0 \varepsilon}{\mu_0 \varepsilon_0} = \frac{\varepsilon}{\varepsilon_0} = \kappa. \tag{2.62}$$

If the magnetic field is included, Equation 2.58 becomes double valued, meaning that there are two values for the refractive index, as in an optically active crystal. The ionosphere is then birefringent. The two waves represented by the two refractive indices are called the *characteristic waves*, the upper sign giving the so-called *ordinary wave* and the lower sign the *extraordinary wave*. When collisions are significant ($Z \neq 0$) the refractive index is complex. Otherwise it is real or purely imaginary.

In some circumstances we may write $Y_L \gg Y_T$ (i.e. $\theta \sim 0°$) or $Y_T \gg Y_L$ (i.e. $\theta \sim 90°$), to represent waves propagating almost along or almost across the geomagnetic field. These are called the *quasi-longitudinal* (*QL*) and *quasi-transverse* (*QT*) approximations. While the use of the Appleton equation may be greatly simplified through these approximations, their ranges of application need to be carefully considered.

The polarization of a radio wave is given by another magneto-ionic equation:

$$R = \frac{j}{Y_L}\left\{\frac{Y_T^2}{2(1-X-jZ)} \mp \left[\frac{Y_T^4}{4(1-X-jZ)^2} + Y_L^2\right]^{\frac{1}{2}}\right\}. \tag{2.63}$$

Here, R is defined as the ratio $-H_y/H_x$, which for our purposes we may consider to be the same as E_x/E_y though this is not rigorously true in all cases. (Moreover, the formula depends on the definition of Y_L; in Equation 2.63 Y_L is positive if $\theta = 0$.) This equation is also double valued in the general case, one value for each of the characteristic waves. A wave of arbitrary polarization entering the ionosphere is resolved into two characteristic waves which are considered to propagate independently within the ionosphere, and are recombined on leaving it. We shall not spend time going into special cases of Equation 2.63, but the reader is invited to do so.

Further discussion of the Appleton equation will be found in Sections 3.5.1 and 3.5.2, where we consider radio methods of sounding the ionosphere.

2.5.2 Reflection of an HF radio wave from the ionosphere

The ionosphere is most important for communications in the high-frequency (HF) band, covering frequencies between 3 and 30 MHz and wavelengths between 100 and 10 m. These wavelengths are short enough that the ionospheric medium does not change much in the distance of a few wavelengths and therefore can be regarded as a number of slabs, each of which is uniform. Such a medium is often called *slowly varying*.

By this approach we can easily see why a radio signal is reflected back from the ionosphere. Figure 2.8, which assumes a flat Earth for simplicity, shows a radio signal transmitted at angle i_0 from the zenith. The ionosphere is considered as a stack of thin slabs (but each several wavelengths thick) with refractive indices n_1, n_2, n_3 etc. Applying Snell's law to each boundary,

$$\begin{aligned}
\sin i_0 &= n_1 \sin i_1 \\
n_1 \sin i_1 &= n_2 \sin i_2 \\
n_2 \sin i_2 &= n_3 \sin i_3 \\
\vdots \quad & \quad \vdots \\
n_{(r-1)} \sin i_{(r-1)} &= n_r \\
\hline
\sin i_0 &= n_r.
\end{aligned} \tag{2.64}$$

Since the plasma frequency increases with height, n becomes smaller and the ray gradually bends round until it becomes horizontal at the level where $n = \sin i_0$, and this is the reflection level. Thereafter, providing the ionosphere is horizontally uniform, the ray returns by a similar path. Taking the simple form of Equation 2.61,

$$\omega_N^2/\omega^2 = 1 - n^2 = 1 - \sin^2 i_0 = \cos^2 i_0.$$

Hence the wave is reflected where

$$\omega_N = \omega \cos i_0. \tag{2.65}$$

At vertical incidence $i_0 = 0$, and $\omega_N = \omega$ at the reflection level; i.e. the plasma frequency of the medium equals the radio frequency of the wave. This was already clear from Equation 2.61, since if $\omega_N > \omega$, n is imaginary and the wave is evanescent. Reflection occurs where $n = 0$.

Figure 2.8 illustrates another simple property of ionospheric reflection. *Breit and Tuve's theorem* states that the total group propagation time (i.e. the time taken for a pulse to travel from T to R) over the path TIR is the same as that for a pulse travelling at the speed of light over the equivalent path TER. The equivalent reflection height is always greater than the true reflection height (see section 3.5.1).

The above method of treating radio propagation in a slowly varying ionosphere is a *ray theory*, and in principle it is readily extended to ionospheres of more complicated profile where the geomagnetic field and collisions are also taken into account. In practice, because of the large number of calculations required, ray tracing is usually done with a computer program designed for the purpose. A number of these have been written, and they may be coupled to an ionospheric model for prediction purposes (see Sections 7.1.8 and 10.2.1), so as to indicate the best frequencies to use over a given circuit and the likely signal strength, polarization, angle-of-arrival, time delay, and so on.

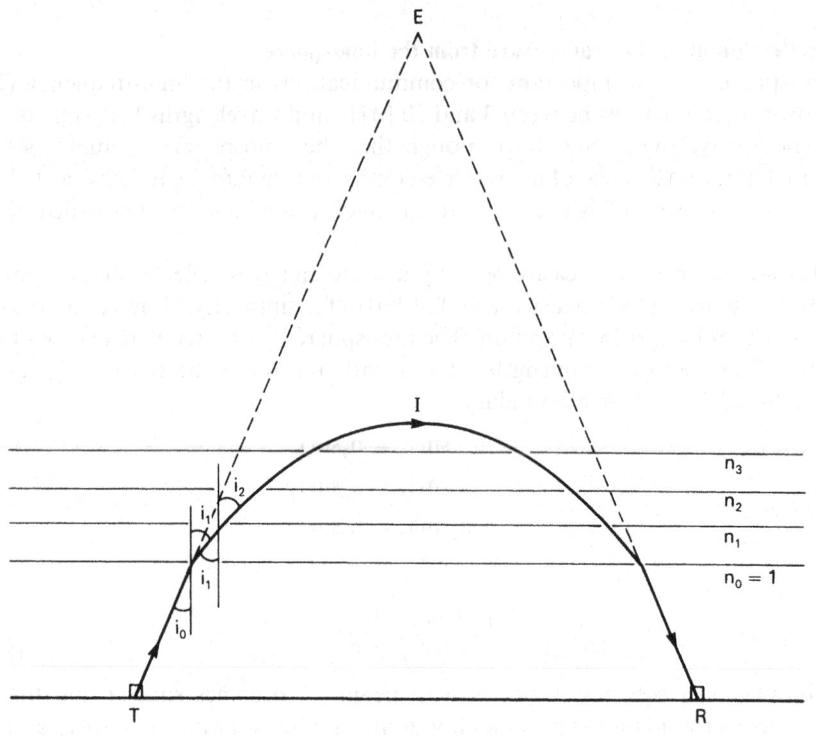

Fig. 2.8 Refraction of a radio wave in the ionosphere.

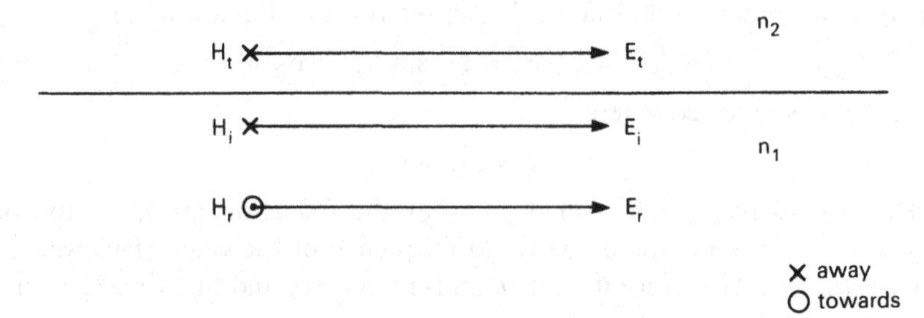

Fig. 2.9 Continuity of electric and magnetic vectors at a sharp boundary.

2.5.3 Partial reflection at a sharp boundary

Whether or not a propagation medium can be regarded as slowly varying depends on the distance over which changes occur relative to the wavelength. The other extreme is when there are discrete boundaries which are infinitely sharp relative to the wavelength; obviously, these require a different treatment.

The reflection and transmission coefficients at a sharp boundary are determined by the condition that the tangential components of the E and the H vectors must be continuous across the boundary. Consider that a linearly polarized wave in a medium of refractive index n_1 is normally incident on a boundary with a medium of refractive index n_2 (Figure 2.9). The electric vectors of the incident, transmitted and reflected waves are respectively E_i, E_t and E_r. Then, for continuity,

$$E_t = E_i + E_r. \tag{2.66}$$

For the magnetic components, however, due to the relation between the directions of E and H, H_r is opposite to H_i because the reflected wave propagates downward. Thus,

$$H_t = H_i - H_r. \qquad (2.67)$$

In a non-magnetic medium,

$$H/E = n/(\varepsilon_0/\mu_0)^{\frac{1}{2}}.$$

Therefore, by substitution,

$$E_t n_2 = E_i n_1 - E_r n_1,$$

and replacing E_t from Equation 2.66,

$$E_r/E_i = (n_2 - n_1)/(n_2 + n_1). \qquad (2.68)$$

The fraction of power reflected is

$$(E_r/E_i)^2 = (n_2 - n_1)^2/(n_2 + n_1)^2. \qquad (2.69)$$

The fraction of power transmitted is $4n_1n_2/(n_2 + n_1)^2$.

Provided the correct refractive indices are taken this formula is of wide application, to radio waves reflected at a discontinuity in the ionosphere as well as to light reflected from a block of glass. Applications are described in Sections 3.5.3 and 3.5.5.

2.5.4 Full wave solutions

Between the cases outlined in Sections 2.5.2 and 2.5.3 fall those instances where the medium changes significantly over the distance of a radio wavelength but the change is not sharp enough to count as a sharp boundary. This happens particularly at the lower end of the radio spectrum, say up to about 500 kHz, though it may also happen at higher frequencies in parts of the ionosphere where the wavelength has been lengthened because the refractive index has become small. The only correct treatment then is a *full wave solution*, in which the differential equations governing the distribution of electric and magnetic field in the spatially varying medium are solved by a numerical method. Conditions above and below the region are imposed to correspond to incident waves, and the transmitted and reflected waves may be deduced. We shall not be concerned with this level of complexity, though an example of its use is referred to in Section 3.5.3. The reader who would like to know more is referred to the book by K. G. Budden (1985).

2.6 Radio propagation through an irregular plasma

2.6.1 Introduction

We shall see that the ionosphere, like the atmosphere in general, is not strictly a uniform, horizontally stratified medium but contains irregularities of many sizes. The effects of these irregularities on propagating waves may be treated by diffraction theory.

As a wave travels through an irregular medium it will accumulate small changes of amplitude and of phase. These will not be the same at all points of the wavefront, which will therefore manifest phase and amplitude variations. According to Huygens' principle, each part of a wavefront may be regarded as a source of secondary wavelets, whose superposition builds up the wavefront at a point further along. In diffraction theory one applies this principle to determine how the amplitude and phase of a received signal are affected by passage through a region of irregularity.

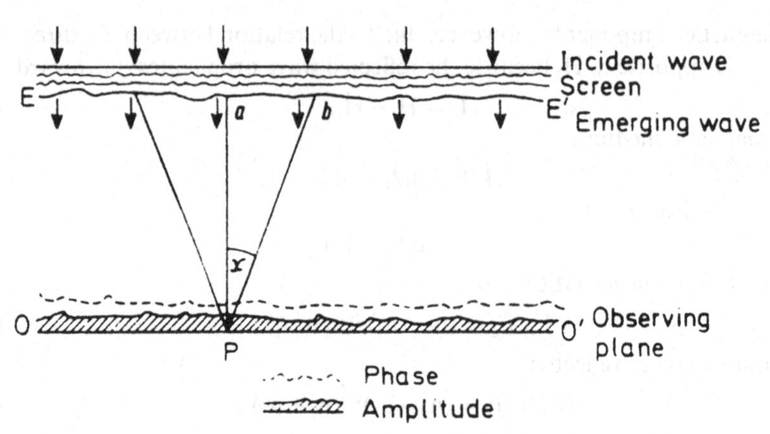

Fig. 2.10 Diffraction of a plane wave incident on a thin phase-changing screen.

It is often convenient to distinguish between 'large' and 'small' irregularities according to their size in relation to the size of the Fresnel zones (see later). Initially we are concerned with small irregularities, meaning that there are several within the distance of the first Fresnel zone.

2.6.2 Diffraction by a thin screen of weak irregularities

Phase screen

The simplest case to treat is that of a thin, shallow, phase-changing screen. In this model the irregularities are assumed to lie in an infinitely thin layer, and to introduce small (<1 radian) phase perturbations along the wavefront of a wave passing through it, as in Figure 2.10. The incident wave is plane (the source being at infinity), but the emerging wavefront is irregular. To obtain the field at a point P in the observing plane, OO′, it is necessary to sum the contributions from each point of the emerging wavefront, EE′. Since EE′ is irregular in phase, the field at OO′ will also be irregular, and in general both the phase and the amplitude are affected.

When studying ionospheric irregularities by radio methods, the general problem is to use observations at the ground to deduce as much as possible about the wavefront leaving the ionosphere, and then to use this information to learn something about the irregularities themselves. It will be apparent that there is not necessarily a one-to-one relationship between the irregularities in the ionosphere and in the wavefield at the ground, but there are, fortunately, some relationships of a statistical nature. The diffraction approach, based on Huygens' principle, is generally valid of course; but if the irregularities are large enough it is equally correct, and usually easier, to consider them as lenses and to treat the situation by geometrical optics.

Angular spectrum

To proceed we must now introduce the idea of the *angular spectrum*. Just as a wave modulated in time may be expressed as a frequency spectrum that may be derived by a Fourier transformation, so a wave modulated with respect to distance may be expressed as a spectrum in angle. A simple amplitude modulation,

$$A(t) = 1 = m\cos(2\pi t/T),$$

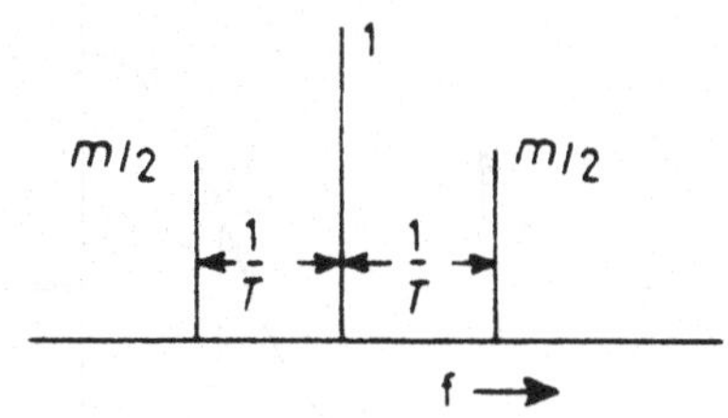

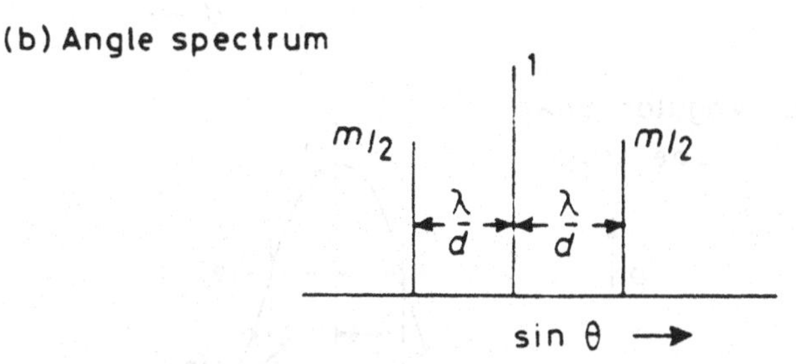

Fig. 2.11 (a) Frequency spectrum of an amplitude modulated wave. (b) Angle spectrum of a wave modulated with respect to distance.

where T is the period and m the modulation depth, is equivalent to the frequency spectrum in Figure 2.11(a), consisting of a carrier of unit amplitude and two sidebands of amplitude m/2 separated from the carrier by frequency 1/T. Similarly, if the amplitude is modulated with respect to distance as

$$A(x) = 1 + m\cos(2\pi x/d),$$

where d is the spatial period, the equivalent angular spectrum is as in Figure 2.11(b), which shows an undeviated wave of unit amplitude plus two sidewaves, amplitude m/2, at angles $\pm\sin^{-1}(\lambda/d)$, λ being the wavelength. This will be recognized as the first-order grating formula. The angle between the sidewaves and the undeviated wave is the same whether the initial modulation is of amplitude or of phase (though additional sidewaves appear if the initial modulation exceeds 1 radian.)

Spectrum of irregularities

If, now, the spatially modulated wavefront contains a spectrum of periodicities, F(d), the sidewaves also cover a spectrum, f(sinθ), where F(d) and f(sinθ) are a pair of Fourier transforms. Provided the screen is shallow – i.e. the initial modulation is only small – and since the angular spectrum received at the observing plane must be the same as that which left the screen, it follows that the spectrum of irregularities formed at the observing plane will be the same as that in the screen – provided that the comparison is made using the complex waveform, including both amplitude and phase. The problem now is that whereas it is easy to measure the amplitude of a received radio signal, it is much more difficult to measure the phase. However, it has been shown that if the observations are made sufficiently far from the screen the statistical properties of the amplitude and phase irregularities at the ground are the same as each other and the same as those at the screen. In that case, amplitude observations alone suffice to give the statistical properties of the ionospheric screen.

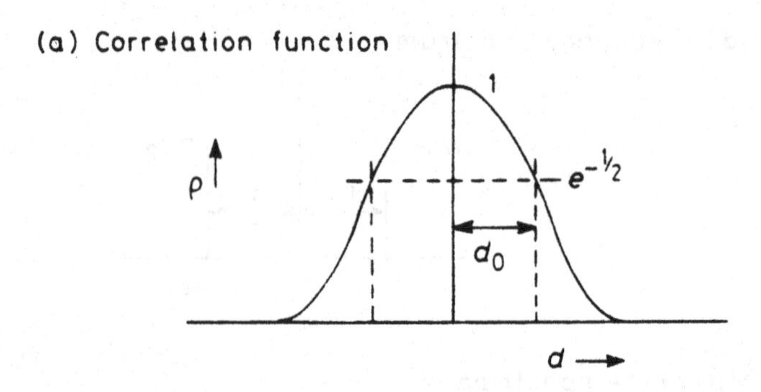

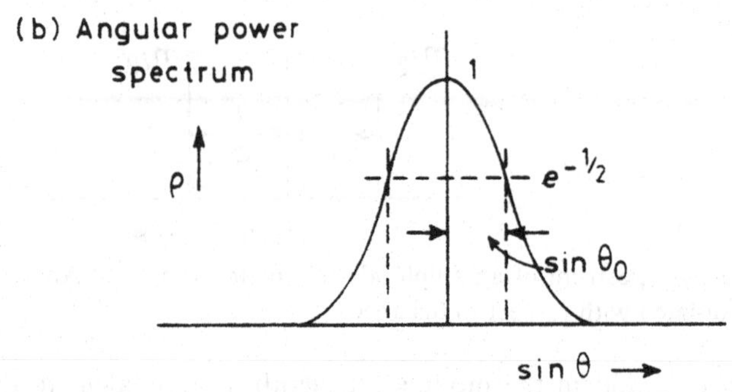

Fig. 2.12 (a) Correlation function and (b) angular power spectrum for a random diffraction screen.

Noise-like irregularities

It is unlikely that irregularities will be sinusoidal or have any other analytical form; they will more probably look like random noise. Such irregularities may be handled using the correlation function, ρ.

If a(x) are the differences of a varying quantity A(x) from its mean $\bar{A}$, and σ^2 is the variance of a

$$\text{(i.e. } \sigma^2 = \overline{(A(x)-\bar{A})^2},$$

the bars denoting averages over many values), the correlation function of A over interval y is

$$\rho = \frac{\overline{a(x)a(x+y)}}{\sigma^2}. \tag{2.70}$$

The correlation function may sometimes be assumed to have a Gaussian form,

$$\rho(d) = \exp(-d^2/2d_0^2), \tag{2.71}$$

and in this case the angular power spectrum would also be Gaussian:

$$P(\sin\theta) = \exp\left(-\frac{\sin^2\theta}{2\sin^2\theta_0}\right), \tag{2.72}$$

where

$$\sin\theta_0 = \frac{\lambda}{2\pi d_0}. \tag{2.73}$$

Thus, there is the same kind of relationship as before between angle (θ_0) and irregularity size (d_0), but in quantities specifying the widths of statistical distributions rather than individual features (Figure 2.12).

Finite source

It is well known that visible stars twinkle but planets do not, and this is because scintillations from different parts of the planetary disc are unrelated and tend to cancel. The same is true of radio scintillation. Simple considerations give the condition that the angular diameter of the source, $\alpha < d_0/D$ for scintillation to occur, where D is the distance between screen and observer, the source being at infinity.

2.6.3 Fresnel zone effects

The distance between the screen and the observer is significant because the size of the Fresnel zones depends on that distance as well as on the wavelength. Recall that, by definition, the first Fresnel zone extends to the point where the distance to the observer exceeds the minimum distance by $\lambda/2$, the resulting phase difference being 180°. Referring back to Figure 2.10, if the overhead point is 'a', we can pick a point b such that $Pb - Pa = \lambda/4$. If the screen alters the phase only, the signal at EE′ may be sketched as in Figure 2.13(a), where A is the unaffected signal and α_E the perturbation due to the screen. At point P on the observing plane, if the perturbation due to 'a' alters the phase of the signal, that due to 'b' will affect its amplitude because of the extra $\lambda/4$ travelled. The resulting signal might now look like Figure 2.13(b), with both phase and amplitude fluctations involved. The first contribution to an amplitude perturbation therefore arrives at angle

$$\chi = \tan^{-1}(\lambda/2D) \tag{2.74}$$

to the vertical. Assuming small angles, the condition for amplitude irregularities to appear in the signal from a pure phase screen is therefore

$$\theta_0 > \chi$$

$$\frac{\lambda}{2\pi d_0} > \left(\frac{\lambda}{2D}\right)^{\frac{1}{2}}$$

$$\left(\frac{\lambda D}{2}\right)^{\frac{1}{2}} > \pi d_0. \tag{2.75}$$

This says that the signal received from a phase screen will contain both amplitude and phase perturbations if the observer is sufficiently far from the screen for the first Fresnel zone to contain several irregularities of typical size. At infinity, the fading power becomes equally divided:

$$\frac{\sigma(A)}{\bar{A}} = \sigma(\phi)$$

$$= \frac{1}{\surd(2\sigma_s(\phi))} \tag{2.76}$$

where $\sigma(A)$ and $\sigma(\phi)$ are the standard deviations of amplitude and phase. Figure 2.14 illustrates how the in-phase and quadrature fluctuations depend on the distance between the screen and the observer. Here, if the screen modulates the phase initially, E_1 refers to the phase fluctuations in the received signal and E_2 to the amplitude fluctuations. The reverse is the case for an amplitude screen. When the distance D is large enough, $\overline{E_1^2} = \overline{E_2^2}$. At shorter distances the second component is not fully developed. This is often the situation in practice. If the radio wavelength, λ, is 6 m and

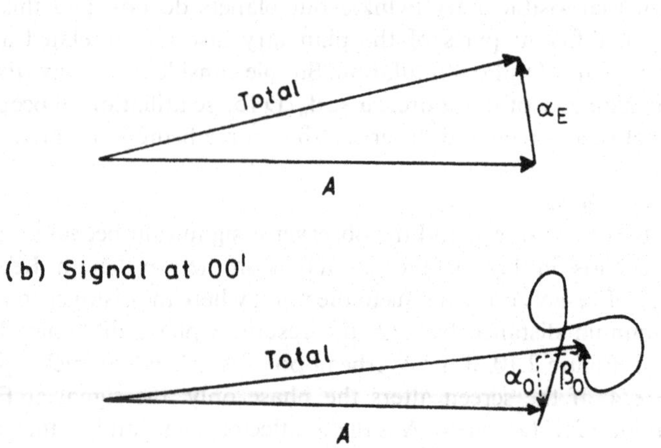

Fig. 2.13 Development of amplitude and phase perturbations from initial phase perturbations.

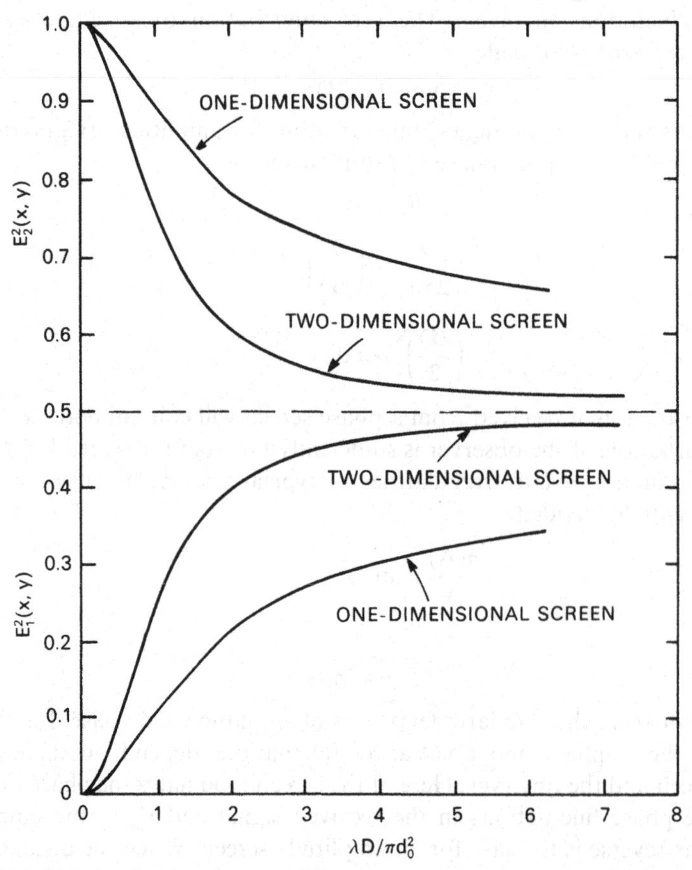

Fig. 2.14 Variation of mean square diffracted signal with distance from one- and two-dimensional random screens. (After S. A. Bowhill, *J. Res. NBS* **65D**, 275, 1961)

the irregular screen is 400 km away, the radius of the first Fresnel zone, $\sqrt{(\lambda D)} = 1.5$ km. Many of the irregularities will be larger than that and the fluctuations will not be fully developed.

If the source is not at infinity the wavefront reaching the screen is curved and the Fresnel zones are smaller. Instead of $\sqrt{(\lambda D)}$, the radius of the first Fresnel zone becomes

$$\sqrt{\left(\frac{DD'\lambda}{D+D'}\right)}$$

which is smaller, D′ being the distance from the source to the screen. In this case the development of amplitude fluctuations from a phase screen is further reduced.

The properties of a phase screen are important because the ionosphere behaves as a phase screen in most cases, and the bulk motion of the irregularities causes the signal received at a fixed place to scintillate. If, by means of a specially devised experiment (see Section 3.5.2 for example) it is possible to observe phase as well as amplitude scintillation, the Fresnel zone effect can be investigated directly by comparing the spectra of phase and amplitude fluctuations. An example is shown in Section 7.5.1.

2.6.4 Diffraction by strong irregularities

The situation becomes much more complex if the phase screen introduces deep modulation: $\sigma_s(\phi) > 1$ radian. The effect is to broaden the angular spectrum leaving the screen to

$$\theta_0 = \frac{\sigma_s(\phi)\lambda}{2\pi d_0}. \tag{2.77}$$

One consequence is that the correlation distance at the observing plane is reduced by a factor $\sigma_s(\phi)$.

2.7 More waves in plasmas

Electromagnetic waves are a phenomenon of free space but they are modified in a plasma because of the changed refractive index. In addition a plasma may support waves that do not occur in free space. Some are introduced in this section.

2.7.1 Hydromagnetic and magnetosonic waves

In a highly conducting plasma the ionized gas cannot cross the magnetic field lines (Section 2.3.6), and therefore in transverse motions the field and the plasma have to move together. Sound waves are allowed, and the magnetic field has no effect on a sound wave travelling in the magnetic field direction because the gas displacement is longitudinal. In general, therefore, if an element of magnetoplasma is displaced both the gas pressure and the magnetic-field pressure may contribute to the restoring force. The magnetic pressure makes possible a range of *hydromagnetic waves* in the ionosphere and the magnetosphere.

The basic hydromagnetic wave is the *Alfvén wave*, which propagates along the direction of the magnetic field and whose displacement is transverse to it. The nature of the Alfvén wave is simply understood by analogy with waves on a taut string. If T is the tension in the string and its mass per unit length is ρ, the speed of a transverse wave is $v = (T/\rho)^{\frac{1}{2}}$. In a magnetoplasma ρ is the mass density of the plasma ($= N_i m_i + N_e m_e$, where N_i, N_e are the number densities of the ions and the electrons, and

m_i, m_e are their masses.) The tension in the magnetic field (T) is B^2/μ_0 in SI or $B^2/4\pi$ in c.g.s. Thus, the *Alfvén speed*, is given by

$$v_A = B/(\mu_0\rho)^{\frac{1}{2}} \quad \text{in SI,} \tag{2.78a}$$

$$v_A = B/(4\pi\rho)^{\frac{1}{2}} \quad \text{in c.g.s.} \tag{2.78b}$$

The Alfvén wave can also be seen as the low frequency limit of electromagnetic waves when the ions as well as the electrons are included.

The refractive index of a plasma for the QL approximation (Section 2.5.1) when both species are included is

$$n^2 = 1 - \frac{f_N^2}{f(f \pm f_L)} - \frac{F_N^2}{f(f \mp F_L)}, \tag{2.79}$$

where F_N and F_L are respectively the plasma frequency and the gyrofrequency of the ions, both being much lower than the corresponding frequencies for electrons.

Putting $f \ll f_L$, F_L, and using the relation $f_N^2 F_L = F_N^2 f_L$ (from Equations 2.28 and 2.32),

$$n^2 = 1 + (f_N^2 + F_N^2)/f_L F_L.$$

But also, $f_N^2 \gg F_N^2 \gg F_L^2$. Hence

$$n^2 \approx 1 + \frac{f_N^2}{f_L F_L} = 1 + \frac{F_N^2}{F_L^2},$$

$$n \approx \frac{F_N}{F_L} = \left(\frac{Nm_i}{\varepsilon_0 B^2}\right)^{\frac{1}{2}}.$$

However, $Nm_i = \rho$, the density of the plasma, and $c = 1/(\varepsilon_0\mu_0)^{\frac{1}{2}}$. Therefore

$$v_A = \frac{c}{n} = \frac{B}{(\mu_0\rho)^{\frac{1}{2}}}$$

as before.

Along the magnetic field the Alfvén wave and the sound wave travel independently, the first being transverse and the second longitudinal. The speed of sound is given by the usual formula:

$$s = \left(\frac{\gamma P}{\rho}\right)^{\frac{1}{2}}, \tag{2.80}$$

where P is the gas pressure and γ the ratio of specific heats. At other angles the waves interact with each other and the gas and the magnetic pressures are involved together. These mixed waves are *magnetosonic waves*. There are two in general, and their velocities are related as in Figure 2.15. Along the field the speeds are respectively s and v_A as above, but perpendicular to the field there is only one wave whose speed is $(v_A^2 + s^2)^{\frac{1}{2}}$.

2.7.2 Whistler and ion-cyclotron waves

Whistler mode

The *whistler wave* is a circularly-polarized electromagnetic wave propagating in a plasma at a frequency below the plasma frequency. It is possible because the refractive index can be real if $f < f_L$. Putting $Z = 0$ (no collisions) and $Y_T \ll Y_L$ (QL approximation), and taking the negative sign (E-wave) in Equations 2.58:

$$n^2 = 1 + \frac{X}{(Y_L - 1)}. \tag{2.81}$$

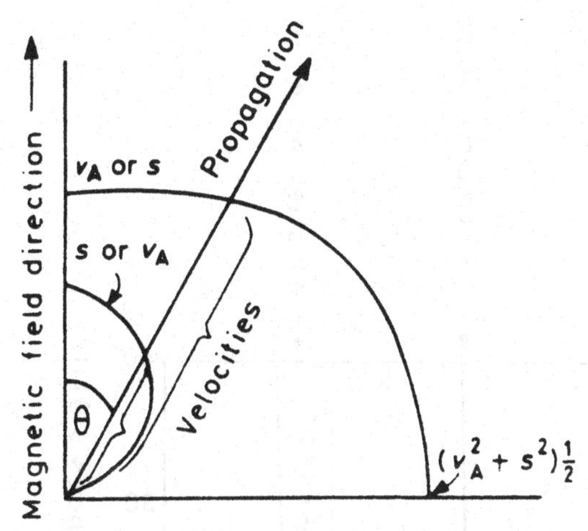

Fig. 2.15 Velocities of magnetosonic waves at various angles to the magnetic field. At angle θ to the field the speed is represented by the distance from the origin. The nature of the waves depends on whether v_A or s is the greater. (After S. J. Bauer, *Physics of Planetary Ionospheres*. Springer-Verlag, 1973)

If also $X \gg (Y_L - 1)$,

$$n^2 = \frac{X}{(Y_L - 1)} = \frac{f_N^2}{f(f_L - f)} \tag{2.82}$$

since $Y_L = f_L/f$ and $X = f_N^2/f^2$, f_N being the electron plasma frequency (Equation 2.32). The whistler mode applies to the extraordinary wave only, which is circularly polarized and rotates clockwise as seen by an observer looking along the magnetic field. (Figure 2.7 illustrates an anticlockwise wave.) There is strong dispersion, and the refractive index goes to infinity (and the phase velocity to zero) when $f = f_L$. 'Whistlers' and their application to measurements in the magnetosphere are discussed in Section 3.5.4.

Ion-cyclotron waves

In Equation 2.79, F_N and F_L, the plasma frequency and the gyrofrequency for ions, are both much lower than the corresponding frequencies for electrons because of the much larger particle mass. Ion motions have to be included for electromagnetic waves in the ULF ($f < 3$ Hz) range. At frequencies $f < F_L$, taking the O-wave (upper sign),

$$n^2 = 1 - \frac{f_N^2}{f f_L} + \frac{F_N^2}{f(F_L - f)} \tag{2.83}$$

since $f \ll f_L$. This wave rotates anticlockwise as seen by an observer looking along the geomagnetic field, and is usually called an *ion-cyclotron wave*. At frequencies below f_L the refractive index for the extraordinary wave becomes

$$n^2 = 1 + \frac{f_N^2}{f(f_L - f)} - \frac{F_N^2}{f(f + F_L)} \tag{2.84}$$

when the ion term is included. These modes have a variety of names. The O-wave or

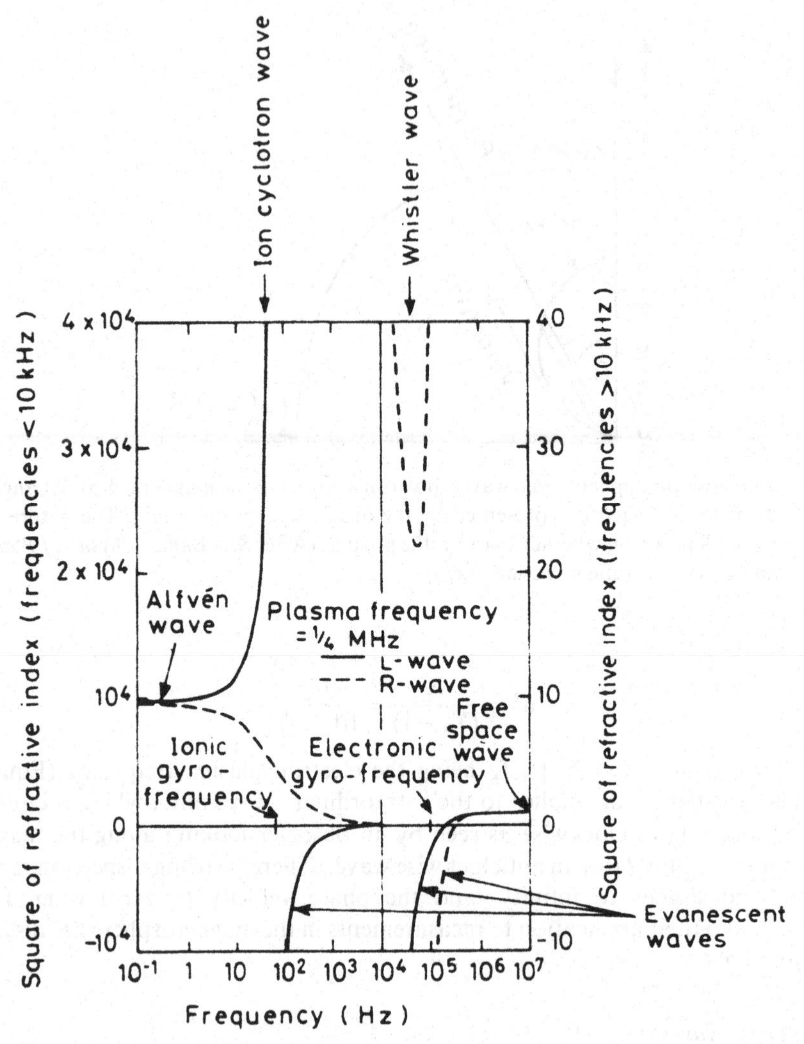

Fig. 2.16 Square of refractive index (n^2) at low radio frequencies when both electron and ion motions are included, showing the whistler and ion–cyclotron waves. (After H. G. Booker, *J. Geophys. Res.* **67**, 4135, 1962, copyright by the American Geophysical Union)

ion-cyclotron wave is also called the Pure Alfvén mode, the Slow mode or the L-wave; the E-wave may be called the Modified Alfvén mode, the Fast mode or the R-wave. These waves are illustrated in Figure 2.16, which plots the square of the refractive index as a function of wave frequency for $f_N = 250$ kHz. The Alfvén wave, the ion cyclotron wave and the whistler wave are marked, and the two polarizations are identified. The high-frequency band, for which $n^2 \sim 1$, is labelled 'free space wave'. If n^2 is negative the wave is evanescent (Section 2.4.2).

2.7.3 Electron-acoustic and ion-acoustic waves

Part of the restoring force in a magnetosonic wave is magnetic pressure. There are also important waves due to the electrostatic forces in a plasma.

In an acoustic wave the displacement is in the direction of propagation (x), and the restoring force on unit volume of gas is

$$-\frac{dP}{dx} = -\frac{d\rho}{dx}\left(\frac{dP}{d\rho}\right),$$

where P and ρ are the pressure and the density. $(dP/d\rho)$ is a property of the gas and is equal to the square of the speed of sound (s^2). (Moreover, by kinetic theory s is simply related to the speed of the gas molecules – see Equation 2.11.) Also, $d\rho = nm\,d\eta/dx$, where η is the particle displacement and n and m are the particle concentration and mass respectively. Hence the restoring force is

$$-nm\left(\frac{dP}{d\rho}\right)\frac{d^2\eta}{dx^2} = -nms^2\frac{d^2\eta}{dx^2}.$$

If the displacement η also separates charges (e), there is an additional restoring force

$$\frac{ne^2}{\varepsilon_0}\eta.$$

See Section 2.3.4. Thus, the acceleration is

$$\frac{d^2\eta}{dt^2} = s^2\frac{d^2\eta}{dx^2} - \frac{ne^2}{m\varepsilon_0}\cdot\eta.$$

Assuming a propagating wave,

$$\eta = \eta_0 \exp j\omega(t - x/v),$$

differentiating and substituting, we obtain

$$\omega^2\eta = \frac{\omega^2 s^2}{v^2} + \frac{e^2}{m\varepsilon_0}\cdot\eta,$$

from which, by rearrangement and using Equation 2.32 for the plasma frequency,

$$\frac{v}{s} = \left(1 - \frac{\omega_N^2}{\omega^2}\right)^{-\frac{1}{2}}. \tag{2.85}$$

Hence it is possible to have longitudinal waves in which gas pressure and electrostatic attraction both contribute to the restoring force. These are known as *electron-acoustic* or *electrostatic waves*.

The foregoing treatment attributes all the motion to the electrons and neglects the motion of the ions. Including both kinds of charged particle leads to a somewhat more complicated treatment and two solutions. One is

$$\frac{v}{s} \approx \frac{X}{(1-X)}, \tag{2.86}$$

where $X = \omega_N^2/\omega^2$ (Equation 2.59). This wave has infinite velocity when $\omega = \omega_N$, and this represents the plasma oscillating bodily at the plasma frequency. The electrons are displaced more than the ions and in the opposite direction.

In the other solution the electrons and ions are displaced in the same direction and the wave velocity lies between C_s and $C_s(1+T_e/T_i)^{\frac{1}{2}}$, where

$$C_s = \left(\frac{k(T_i+T_e)}{m_i}\right)^{\frac{1}{2}} \tag{2.87}$$

and is the *ion-acoustic speed*, which is close to the thermal speed of the ions – see Equation 2.11. (T_i and T_e are ion and electron temperatures and m_i the ion mass.) This wave is the *ion-acoustic* wave.

The natural turbulence of the ionized upper atmosphere due to its temperature may be regarded as a spectrum of electrostatic waves. Their role in the scattering of radio waves from the ionosphere is considered in Section 3.6.4.

Further waves are possible if the plasma contains a magnetic field. In general, the waves that exist in the medium of a plasma are called *plasma waves*.

2.8 Instabilities

2.8.1 Introduction

Instabilities may arise when – to borrow a term from radio engineering – the medium includes an element of 'positive feedback'. Suppose that the medium becomes perturbed, either by chance or by some external influence. This perturbation will in turn disturb other aspects of the medium, and if one of these consequences has the effect of enhancing the original perturbation, then that will be reinforced and will tend to grow larger. At the same time there may be other consequences tending to reduce the perturbation, but if these are less effective than those causing positive feedback then we have an *instability*.

An instability is more likely to develop if it grows rapidly, since there is then less opportunity for the energy to be diverted elsewhere. The growth rate is therefore an important consideration. Due to this competition between positive and negative factors, which may themselves be non-linear, some *threshold* may have to be exceeded before the instability can grow.

The theory of instabilities tends to be mathematically complex, and to avoid getting bogged down we shall confine ourselves to some of the simpler aspects. We will briefly discuss three instabilities which have been invoked to explain some of the natural phenomena of geospace.

2.8.2 Two-stream instability

The *two-stream instability*, also known as the *Farley–Buneman* after its discoverers, produces electrostatic waves when streams of electrons and ions differ in velocity by more than the ion-acoustic speed (Section 2.7.3). These waves propagate nearly perpendicular to the magnetic field.

The detailed theory shows that the instability applies to waves propagating within a cone of angle θ, given by

$$V_d \cos\theta = C_s(1+\psi), \tag{2.88}$$

where V_d is the relative drift speed between electrons and ions, C_s is the ion-acoustic speed (Equation 2.87), and

$$\psi = \frac{\nu_e \nu_i}{\omega_e \omega_i} \cdot \left(\sin^2 \alpha + \frac{\omega_e^2}{\nu_e^2} \cdot \cos^2 \alpha \right), \tag{2.89}$$

where α is the angle between the wave and the magnetic field, and the subscripted ν and ω are the collision and gyrofrequencies for the electrons and the ions. The cone is centered on the relative drift direction, and Equation 2.88 says that, for the instability to operate, the component of the relative drift velocity in the direction of the wave must exceed the ion-acoustic speed. By how much the latter has to be exceeded depends on the value of ψ, and hence on the propagation direction with respect to the magnetic field (α). For propagation normal to the field ($\alpha = 90°$), ψ is about 0.3 in the E region, but its value increases rapidly as α moves away from 90° because ω_e is about 100 times as large as ν_e in the E region. This is why electrostatic waves generated by the two-stream mechanism like to travel normal to the magnetic field, and why the threshold velocity difference is close to the ion-acoustic speed. If the growth of wave amplitude is written $e^{\gamma t}$, the growth rate γ includes the wavelength, and this instability grows more rapidly at shorter wavelengths.

The two-stream instability is important in the electrojets which flow in the ionospheric E region at both low and high latitudes (Sections 7.5.4 and 8.2.3).

2.8.3 Gradient drift instability

The *gradient drift* mechanism is an example of a *Rayleigh–Taylor* instability, having the general property that a reduction of total energy is achieved by interchanging two elements of a fluid. (The best known example of a Rayleigh–Taylor instability occurs when a heavier liquid overlies a lighter one.)

The gradient drift mechanism may be appreciated by means of Figure 2.17. An enhancement of plasma density extends in the x-z plane, with the density falling off in the y direction both above and below the peak. A force acts from one side and it is assumed that a small wave or kink appears spontaneously in the contours of constant plasma density. Then the plasma density in the edges of the enhancement varies (slightly) in the x direction. We must now borrow a result from Section 6.5.2 (Figure 6.19), to the effect that, provided the collision frequency is smaller than the gyrofrequency, positive ions and negative electrons move in opposite directions when acted on by a force acting at right angles to a magnetic field (B). This motion, combined with the variation of plasma density, leaves net positive and negative charge as shown, and this charge separation produces a small 'polarization electric field', E_p. Due to the presence of the magnetic field, there is now an additional force $E_p \times B$ acting in the y or the $-y$ direction, and this stabilizes the kink on the top side but increases it on the bottom side. The instability may therefore occur in the edge away from the force.

In nature, the effective force may be a wind in the neutral air or gravity. The term 'gradient drift' is sometimes reserved for the first case, which may also be called the '$E \times B$ drift' instability. This is one mechanism by which irregular structures in the F region (Section 8.2.2) break down into smaller ones, and it also produces structuring in injected ion clouds (Section 3.7.1). The case when gravity supplies the effective force may be called the 'collisional Rayleigh–Taylor' instability. This case is responsible for

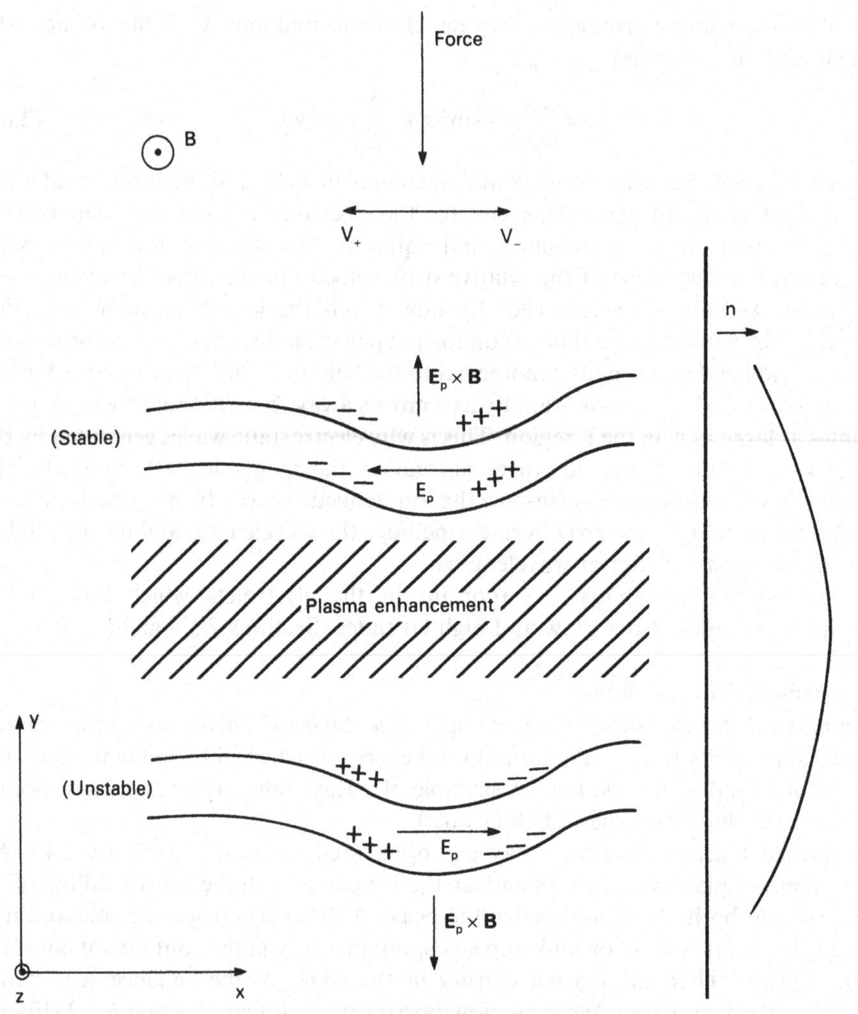

Fig. 2.17 Illustration of the gradient drift instability. In the configuration shown, the upper edge of the enhancement is stable and the lower edge is unstable. The force can be a neutral-air wind or gravity.

the production of 'bubbles' and other irregularities in the equatorial ionosphere (Section 7.5.3).

If the perturbation grows in amplitude as $e^{\gamma t}$, the growth rate, γ, is given in its simplest form by

$$\gamma = -\frac{U}{n} \cdot \frac{dn}{dy}, \tag{2.90}$$

and

$$\gamma = \frac{g}{\nu n} \cdot \frac{dn}{dy}, \tag{2.91}$$

for the two driving forces respectively, where U is the neutral-wind speed, g is the acceleration due to gravity, ν is the collision frequency, and n is the plasma density.

2.8.4 Kelvin–Helmholtz instability

The *Kelvin–Helmholtz instability* also has a commonplace manifestation. It is what causes waves on the surface of water on a windy day. Our purpose would almost be served by making that our analogy and saying no more, since when Kelvin–Helmholtz matters in geospace it is often at the interface between two media in relative motion and we can easily envisage that waves will be set up there. The instability depends on the presence of a velocity shear in the fluid flow, and although this may be due to the presence of a boundary the boundary itself is not essential. Consider that one layer of fluid is stationary while another flows over it. If a small perturbation appears in the flow, so that it becomes wavy, a centrifugal force will act outward at each bend. The bend will thereby be reinforced (positive feedback) and the instability will grow.

Real fluids have stabilizing forces as well – surface tension for a liquid, gravity in the lower atmosphere, and magnetic tension in magnetoplasma – and in these cases the velocity shear has to exceed a threshold before the instability can grow. The magnitude of the threshold depends on the detailed circumstances. In the special case of a magnetoplasma where the velocity shear is sharp (i.e. discontinuous) and the magnetic field runs parallel to the discontinuity, the critical shear is given by

$$(\Delta \mathrm{v})^2 = \left(\frac{1}{\rho_1}+\frac{1}{\rho_2}\right)\frac{(\mathrm{B}_1^2+\mathrm{B}_2^2)}{\mu_0}. \tag{2.92}$$

Here, the ρs are the densities on either side of the shear and the Bs are the magnetic flux densities. It is assumed that the perturbation has the same wavelength on each side. The threshold is independent of the wavelength but the growth rate is faster if the waves are shorter. A characteristic of the Kelvin–Helmholtz instability is that it tends to produce vortices.

This instability is believed to be important in the production of small ionospheric irregularities from larger ones (Section 8.2.2), and in the generation of certain magnetic pulsations at the magnetopause (Section 9.2.2).

Further reading

K. Davies. *Ionospheric Radio*. Peregrinus, London (1990). Chapter 1 on characteristics of waves and plasmas, Chapter 3 on magneto-ionic theory, and Chapter 7 on the amplitude and phase of radio waves.

J. A. Ratcliffe. *The Magneto-ionic Theory and its Applications to the Ionosphere*. Cambridge University Press (1959).

K. G. Budden. *The Propagation of Radio Waves*. Cambridge University Press (1985).

J. A. Ratcliffe. Some aspects of diffraction theory and their application to the ionosphere. *Rep. Prog. Phys.* **19**, 188 (1956).

J. A. Ratcliffe: *An Introduction to the Ionosphere and Magnetosphere*. Cambridge University Press (1972); specifically Chapter 6 on basic principles and Chapter 8 on electromagnetic, hydromagnetic and electron-acoustic waves.

3

Techniques for observing geospace

When you can measure what you are speaking about and express it in numbers, you know something about it; but when you cannot measure it, when you cannot express it in numbers, your knowledge is of a meager, unsatisfactory kind.

Lord Kelvin

3.1 The importance of observations

It would be difficult to overstate the importance of observational techniques in the development of the science of geospace.

Science is about the real world – the world which, most of us believe, exists outside ourselves and would continue to exist even if we were not here to see it. Our perception of that real world, however, is subjective and depends entirely on how we sense it, on the data we collect by eye or ear. This is equally true whether we observe directly, have the assistance of an instrument (telescope, ear trumpet), or take our information from the readings of a sensor (oven thermometer, particle detector on a spacecraft). The scientist's job is to make sense of such data, to fit them into existing knowledge, and to try to build up a coherent picture of the nature and working of the real, external world. Without observations we know nothing and can understand nothing. This is true of life in general and of science in particular.

The history of science plainly shows that new methods and novel techniques nearly always bring in their wake fresh advances of knowledge. Geospace is no exception. Of such an inaccessible region of the Earth we can learn very little without instruments, and the great expansion of knowledge during the last 50 years is a direct result of technological advance. Nor is the process ended. There are still desirable measurements that cannot be made for lack of a good technique, or because existing techniques are too expensive and therefore insufficiently applied.

To a large extent the development of technique parallels the advance of science. A new measurement will generally have been prompted by a scientific question, but, once in use, the new technique may well give some puzzling or unexpected results and thereby prompt further experiments. This interaction stresses the importance of the collaboration between the scientist and the engineer – a long tradition with great significance for the advance of geospace science.

Since this book is a scientific rather than an engineering text we shall not dwell on details of hardware. It is important, though, to appreciate the physical principles of devices and the philosophy of the various approaches. Generally speaking, we can go about observing geospace in three ways, which (though these terms may not be absolutely standard) we shall call 'direct', 'indirect' and 'remote' sensing.

(a) A *direct sensor* is an instrument placed in the medium to measure some property (e.g. temperature) of its immediate surroundings. Many kinds of direct sensor have been developed for satellite, rocket, or balloon vehicles, and most sensors of this kind have been derived from laboratory instruments. Direct methods are usually specific in that they respond only to that which is to be measured, but they may disturb the medium by their presence and in that event may return false readings. The need for a vehicle tends to make direct sensing relatively expensive.

(b) An *indirect sensor* is also placed within the medium. It is not instrumented with a detector (though it may carry telemetry), but properties of the medium are deduced by observing the motion of the sensor from afar. As we shall see, density, temperature and wind may be measured by indirect means. This approach also requires a vehicle, though the indirect sensor may itself be rather simple and relatively inexpensive.

(c) The advantage of *remote sensing* is that no instrument has to be placed in the medium, but properties are deduced from observations of waves that have traversed the medium, or of emissions from it. The sensing wave might be electromagnetic or sonic. If the wave is transmitted into the medium from a sender this would be *active remote sensing*, but if it arises within the medium from natural processes (an auroral emission, for example) the term *passive remote sensing* is used. A great many techniques can be described as remote sensing, so it can be risky to generalize about their advantages and disadvantages. They tend to be strong on time continuity and on economy; the equipment is not expended nor in need of recovery. In some remote techniques there can be difficulties of interpretation because the wave travels some distance through the medium and may be affected all along the path rather than at a specific point. The term 'remote sensing' is also commonly used for satellite-based observations of land terrain, cloud cover, etc. These are indeed examples of passive remote sensing, but they obviously do not come within our present scope.

3.2 Direct sensing of a gaseous medium

Direct sensing means measuring with instruments actually within the medium, as with a thermometer or a pressure gauge. In our case we have to be concerned with more than temperature, density, pressure, wind and other properties of the neutral air because the ionized part is specially interesting to us. We also have to measure over a great range of distances from the Earth's surface, in which the concentrations of various species range over a factor of 10^{16}. The changing composition is itself a matter of great interest; we are concerned not only with 'air' but also with a number of populations having individual temperatures and spatial distributions.

Major satellite missions are instrumented with a variety of devices, on the principle that if they are wisely chosen the totality may be worth more than the sum of the parts. To illustrate typical selections, the instrumentation of four important geospace missions is listed in Tables 3.1 to 3.4. We shall refer to these in forthcoming sections. The Atmospheric Explorers (Table 3.1), of which several were launched with different orbital inclinations from 1973 to 1975, were intended to study the processes of the neutral and ionized upper atmosphere related to the absorption of solar radiation; i.e.

Table 3.1 *Instruments carried on the Atmospheric Explorer (AE) satellites*

Instrument	Principal measurements
Open-source, neutral-mass spectrometer	Neutral composition, masses 1–46 a.m.u.
Closed-source, neutral-mass spectrometer	Neutral composition, masses 1–46 a.m.u.
Velocity distribution analyser	Temperature of neutral gas
Accelerometers	Total air density (heights < 400 km)
Cold-cathode gauge and capacitance manometer	Total air density
Magnetic, ion-mass spectrometer	Ion concentrations, masses 1–64 a.m.u.
Bennett type, ion-mass spectrometer	Ion concentrations, masses 0.5–72 a.m.u.
Retarding potential analyser	Ion temperature, concentration, mass and drift velocity
Cylindrical Langmuir probe	Electron temperature and concentration, and ion concentration
Spectrophotometer for solar extreme ultra-violet (EUV)	Solar radiation in wavelengths 14–185 nm
Filter photometer for solar extreme ultra-violet	Solar radiation in bands from 4 to 130 nm
Photometer for visible airglow	Intensities of airglow at 630, 557.7, 427.8, 337.1, 520, and 731.9–733 nm
Ultra-violet spectrometer	Intensities of 215 and 219 nm emission, for height profiles of NO over 80–250 km
Photoelectron spectrometer	Hot electrons in energy range 2–500 eV
Low-energy ion and electron detector	Fluxes of ions and electrons with energies 0.2–25 keV
Three-component fluxgate magnetometer	Ionospheric currents associated with small scale variations of magnetic field

(After A. Dalgarno *et al.* The Atmospheric Explorer mission. *Radio Science* **8**, 263, 1973.)

they were concerned particularly with aeronomy. The Dynamics Explorers (Table 3.2) were concerned with the electrical processes of the upper atmosphere. Two were launched together, one in an elliptical high-altitude orbit, the other in a circular low-altitude orbit. GEOS-2 (Table 3.3) was a geosynchronous vehicle launched for the International Magnetospheric Study (IMS) and it was designed to measure magnetospheric phenomena related to the behaviour of the ionosphere. HILAT (Table 3.4) was launched in 1983 into a nearly polar orbit at 800 km altitude, and its purpose was to investigate spatial irregularities in the high-latitude ionosphere. Its instrumentation was therefore designed to measure, with the maximum possible spatial resolution, the properties of the thermal plasma including its drift motions, and with observing related phenomena that may have caused the structuring.

3.2.1 Direct measurements of the neutral atmosphere

At high altitude the density and pressure of the air are similar to those in a laboratory vacuum. Direct upper atmosphere measurements are therefore based on well established techniques from the vacuum laboratory. When the instrument is carried on

Table 3.2 *Instruments carried on the Dynamics Explorer satellites*

Instrument	Principal measurements	Satellite
Three-component, fluxgate magnetometer	Magnetic field	H, L
Vector electric field instrument	Steady electric field, ± 1 V/m with 0.1 mV/m resolution Electric field variations, 4 Hz–512 kHz	L
Plasma wave instrument	Steady electric field, 0.5 mV/m–2 V/m Electromagnetic waves, 1 Hz–400 kHz Electrostatic waves, 1 Hz–2 MHz	H
Spin-scale auroral imager	3 photomers, 2 invisible, 1 in UV; 391.4, 557.7, 630, 121.6, 125–140, 145–170 nm	H
Neutral atmosphere quadrupole mass spectrometer	Composition of neutral air, 1–48 a.m.u., density 10^{10}–10^{17} m^{-3} Neutral wind normal to spacecraft motion, 10–1500 m/s Neutral air temperature, 400–2000 K	L
Wind and temperature spectrometer, using retarding potential quadrupole spectrometer	3 components of neutral wind, 10–1500 m/s Temperature, 400–2000 K Ion composition. Neutral composition 1–34 a.m.u., density 10^{10}–10^{17} m^{-3}	L
Fabry–Perot interferometer	Drift and temperature of neutral and ionic atomic oxygen	L
Retarding ion mass spectrometer	Densities of H^+, He^{++}, He^+, O^{++}, O^+ Temperature, 0–50 eV Bulk flow. Spacecraft potential, Ion composition 1–30 a.m.u.	H
Ion drift meter	Ion drift normal to spacecraft motion, 3 m/s–4 km/s Ion concentration	L
Retarding potential analyser	Ion concentration, 5×10^7–5×10^{12} m^{-3} Ion temperature, 200–10000 K Ion drift along spacecraft motion, 0–4 km/s	L
Plasma instrument	Array of 5 (H) or 15 (L) electrostatic analysers for electrons and ions, 5 eV–30 keV Pitch angle distribution (H) Bulk flow velocity (H)	H, L
Ion mass spectrometer	Ion composition, 1–138 a.m.u. Pitch angle distributions Energy range: 0–17 keV/unit charge	H
Cylindrical Langmuir probe	Electron temperature, 500–20000 K Electron density, 2×10^7–10^{12} m^{-3} Total ion density, 10^9–10^{13} m^{-3} Spacecraft potential ± 5 V	L

(After R. A. Hoffman, Dynamics Explorer Program. *EOS* **61**, 689, 1980.)

Table 3.3 *Instruments carried on the GEOS-2 synchronous satellite*

Instrument	Principal measurements
Three-component, fluxgate magnetometer	Magnetic field and its variations up to 5 Hz
Electron beam experiment	Electric field perpendicular to magnetic field. Magnetic field and its gradient. Derived from observed deflection of electron beams.
Wave-field experiment, using search-coil magnetometer, electric aerials, and VLF transmission and reception	Electromagnetic waves up to 30 kHz (magnetic) and 80 kHz (electric) Resonances of local plasma
Electrostatic analysers	Particle fluxes, 1–500 eV, and their spectra, and angular distributions Temperature of thermal plasma
Mass spectrometer using magnetic and electric deflection	Ion composition, energy spectra, and angular distribution of low energy particles, of thermal energy to 17.2 keV per unit charge
Electrostatic analysers	Energy spectra of 0.2–20 keV electrons and protons as functions of pitch angle
Electron/proton spectrometer, using magnetic deflection and solid-state detectors	Energy spectra and pitch angle distributions of electrons 20–300 keV and protons 40 keV–2.0 MeV

Table 3.4 *Instruments carried on the HILAT Satellite*

Instrument	Principal measurements
Radio beacon	Electron content Scintillation
Langmuir probe	Electron density and temperature
Retarding potential analyser	Ion density, temperature and mass
Ion drift meter	Ion speed across satellite track
Electron spectrometer	Electron fluxes in 16 channels covering 20 eV–20 keV
Three-component fluxgate magnetometer	Magnetic field
Imaging spectrophotometer	Auroral images in the ultra-violet
Downward pointing photometers	Auroral emissions at 391.4 nm and 630 nm

(After E. J. Fremouw *et al.* The HILAT program. *EOS* **64**, 163, 1983.)

a rapidly moving vehicle like a satellite, its positioning is all important. A low-altitude satellite moves at about 8 km/s, which is well above the r.m.s. speed of most gas molecules (400 m/s for oxygen). Thus if a pressure gauge points along the trajectory it sweeps up molecules in a ram effect, and if it points backward it can detect only molecules from the high energy tail of the velocity distribution. Rocket speeds are also high over most of the trajectory (escape velocity being 11 km/s), and rockets are also subject to spinning and rolling.

The AE satellites (Table 3.1) carried a pair of instruments for measuring the air pressure, covering heights from 120 to 370 km between them. The capacitance manometer, which only worked below 200 km, used the vibration of a thin membrane when the air is allowed to impinge upon it periodically. The membrane forms part of a capacitor, and the regular variation of capacitance is measured electronically. The cold-cathode gauge is a device used in laboratories which detects the ion current flowing between high voltage electrodes within the gauge. Temperature was determined from the velocity distribution of N_2 using a 'velocity distribution analyzer'. The atmospheric gas is allowed to enter a sampling chamber through an orifice, and the change of gas pressure is measured as the sampling direction moves with respect to the direction of travel as the satellite spins. Essentially, this device compares the thermal speed of the air molecules, $(2kT/m)^{\frac{1}{2}}$, with the orbital speed of the satellite, which is of course well known from the tracking data.

3.2.2 Langmuir probe and derivatives

In studying the ionized upper atmosphere we are interested, first and foremost, in the concentrations (number densities) of electrons and ions (positive and negative), in their temperatures, and in the nature of the ions. To determine the composition requires a mass spectrometer (Section 3.2.4), but much information about concentrations and temperatures can be obtained from simpler devices variously known as 'probes', 'traps' and 'analyzers', and many such instruments have been flown on rockets and on satellites over the years.

A probe projects into the medium and draws from it an electric current of electrons or ions depending on the sign and magnitude of the potential applied to it. A trap collects ions from the medium because the vehicle in orbit moves faster than the ions and so sweeps them up from its path. Additional electrodes are often incorporated to enable a more detailed analysis to be made in real time, the results being transmitted to a ground station by telemetry.

Langmuir probe

The basis of this class of device is the *Langmuir probe*, from which they are derived. What they all have in common is that the particles measured are of relatively low energy, and they are detected by the current they carry. Figure 3.1 shows the ideal current–voltage characteristic of a Langmuir probe. V_s is the *space potential*, which will serve as the reference potential for the probe. We can envisage its significance by supposing an open-wire grid to be placed in the ionized medium, and its potential (by reference to a large, distant body, such as the Earth) being varied. When the flow of electrons and ions through the holes in the grid is not altered by the grid's presence, we have found the space potential.

If the grid is now replaced by a plate at the same potential, electrons and positive ions will both be collected, the electron current proportional to $-Nev_e$, and ion current to $+Nev_i$; N is the electron and ion density (assumed equal) in the medium, e is the magnitude of the electronic charge, and v_e and v_i are the electron and ion speeds respectively. Thus the total current to the plate depends on $Ne(v_i - v_e)$ when its potential is V_s. This current is not zero, but is negative because the electron velocity is larger than the ion velocity (since the particles are lighter – Equation 2.13).

The negative current represents a preferential flow of electrons to the plate, and if

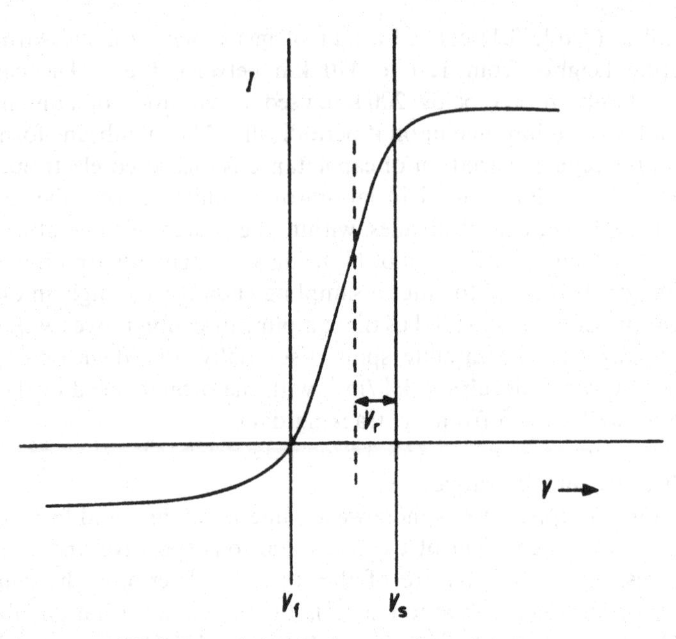

Fig. 3.1 Current–potential characteristic of a Langmuir probe.

the plate is now disconnected from the external potential it will become more negative until its potential is V_f, the *floating potential*. By the same mechanism a satellite in the ionosphere typically acquires a negative potential of about 1 V. At the floating potential the net current is zero, i.e. $e(N_i v_i - N_e v_e) = 0$, where we now allow for different ion (N_i) and electron (N_e) densities. Since it is still true that $v_e > v_i$, it follows that at the floating potential $N_i > N_e$ near the plate. The negatively charged plate is thus surrounded by a positive sheath, which effectively controls the current to the plate much like the space charge in a simple electronic diode. The thickness of the sheath is the Debye length (Section 2.3.5).

The physical basis of the Langmuir probe, then, is the different masses of the electrons and ions, and the resulting dependence of the net current on the probe potential. The characteristic shown in Figure 3.1 has three distinct regions.

(a) If the probe potential, V, exceeds the space potential, V_s, ions are repelled but electrons are attracted to the probe. The current depends on the electron density in the medium, N_e, which may therefore be determined.

(b) Between potentials V_s and V_f most of the current is due to electrons, but they are now subject to a retarding potential V_r ($= V_s - V$) which keeps away the slower electrons, and only those with $v > (2V_r e/m_e)^{\frac{1}{2}}$ reach the probe. (The condition is obtained by comparing the potential energy, eV_r, and the kinetic energy, $mv^2/2$.) In a Maxwellian distribution (Equation 2.10), the number of electrons with velocity component between v_x and $v_x + dv_x$ in the x direction is proportional to $\exp(-mv_x^2/2kT)\,dv_x$. Thus the number received by a plate in unit time, to which the current is proportional, is

$$\int_{(2V_r e/m)^{\frac{1}{2}}}^{\infty} v_x \cdot \exp(-mv_x^2/2kT)\,dv_x.$$

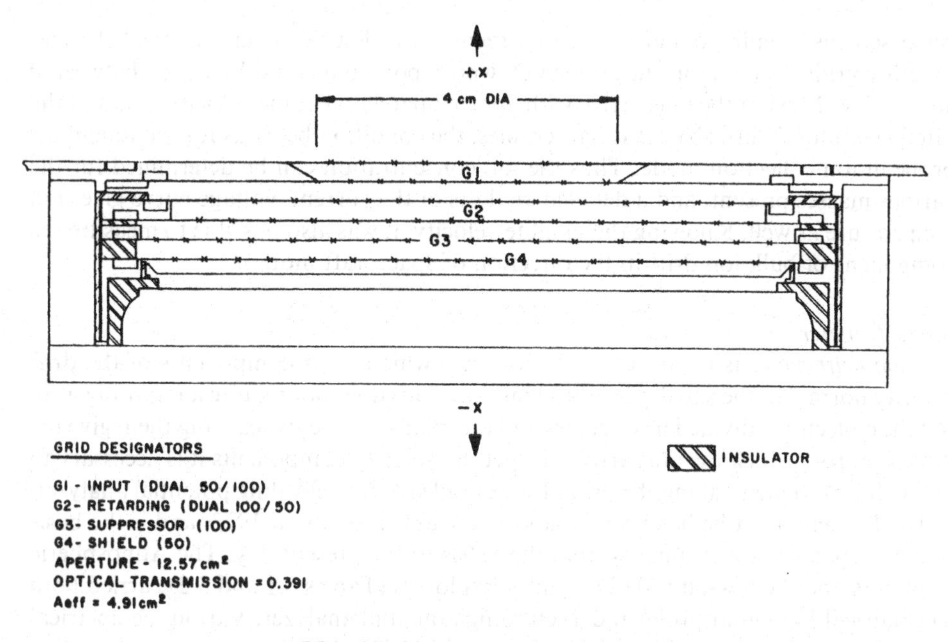

Fig. 3.2 Retarding potential analyser on Dynamics Explorer. (W. B. Hanson *et al.*, *Space Science Instrumentation* **5**, 503, 1981)

Thus, by integration,

$$I = I_0 \exp(-eV_r/kT). \tag{3.1}$$

The slope of the curve between V_f and V_s is steeper for lower temperature, and a plot of log(I) against V_r has a slope proportional to 1/T.

(c) If $V < V_f$ the electrons are repelled but ions are collected and in principle N_i (generally $= N_e$) can be found from this saturation current. However a somewhat different treatment is required because the ion speed is generally less than the speed of the spacecraft, and the collection condition is therefore $eV_r \leqslant m_i v_{sat}^2/2$. The ion masses and abundances can be determined in this mode.

An individual instrument may well be designed to measure only a part of the characteristic, and ingenious short cuts have been devised to telemeter out the essential information with maximum economy of data rate. In the Dynamics Explorer instrument, for example, noting that the slope of the characteristic depends on the temperature, the rate of sweeping the probe potential was altered so that the electrometer output (i.e. the current) always followed the same curve. The electron temperature could then be determined from the sweep rate and there was no need to telemeter the whole curve.

Retarding potential analyzer

For the study of ions, a development of the Langmuir probe involves the addition of extra grids, as in Figure 3.2, which illustrates the *retarding potential analyzer* (otherwise known as an *ion trap*) used on the low-altitude Dynamics Explorer. Some of these grids act as screens, keeping out electrons or screening the detector from electrical changes on other grids. The important grid is G2, whose potential could be varied between 0 and +32 V. Most of the ions arrive with approximately the same velocity – that of the satellite – but G2 turns back the lighter ones, the condition being as for the Langmuir probe in ion collection mode. Thus the ion concentration can be determined for the various masses present, and a detailed analysis of the current–voltage curve gives ion temperature as well. Knowing the satellite velocity, it was also possible to measure the component of bulk ion drift in the direction of spacecraft motion.

Ion drift meter

The *ion drift meter* is a variant of the ion trap which gives components of the drift velocity normal to the satellite's orbital motion. This does not require a retarding grid, but the collector is divided into sectors and the relative currents reaching them give the direction from which the ions arrive. To get the velocity components it is necessary to know the ion velocity along the orbit, for example from a retarding potential analyzer.

The Langmuir probe has been in laboratory use since about 1923, and is the basis of many spacecraft instruments from the 1960s to the present day. The Atmospheric Explorers and the low-altitude Dynamics Explorers (Tables 3.1 and 3.2) carried both a cylindrical Langmuir probe and a retarding potential analyzer. Various geometrical forms are possible. A probe may be planar, cylindrical or spherical, and a retarding potential analyzer may also assume various forms, but the principles of operation are much the same.

3.2.3 Impedance and resonance probes

Impedance probe

An alternative approach is to measure some electrical property of the medium in bulk and then obtain its plasma density from theory. At radio frequency ω rad/s, a plasma with N electrons/m^3 has a dielectric constant (Section 2.5.1) given by

$$\kappa = 1 - \frac{Ne^2}{\varepsilon_0 m\omega^2}, \qquad (3.2)$$

where e and m are the electronic charge and mass, and ε_0 is the permittivity of free space. A standard laboratory method for measuring the dielectric constant of a material is to measure the capacitance of a parallel-plate (or other form of) capacitor with a slab of material between the plates. Then (in the parallel-plate case),

$$C = \varepsilon_0 \kappa A/d,$$

A being the plate area and d the plate separation.

The same method can be applied to the ionospheric plasma, the capacitor being part of a tuned circuit whose resonant frequency therefore depends on the dielectric constant and hence on the electron density of the plasma. One technique is to control the frequency of an oscillator with the tuned circuit and to telemeter its frequency to a ground station. If the device is carried on a rocket a profile of electron density against height can be measured. Figure 3.3 shows such a flight into the auroral ionosphere. The electron density determined from the probe is compared with that from a propagation experiment (as Section 3.5.2). The *impedance probe* resolves finer structure, but is probably less accurate in absolute terms.

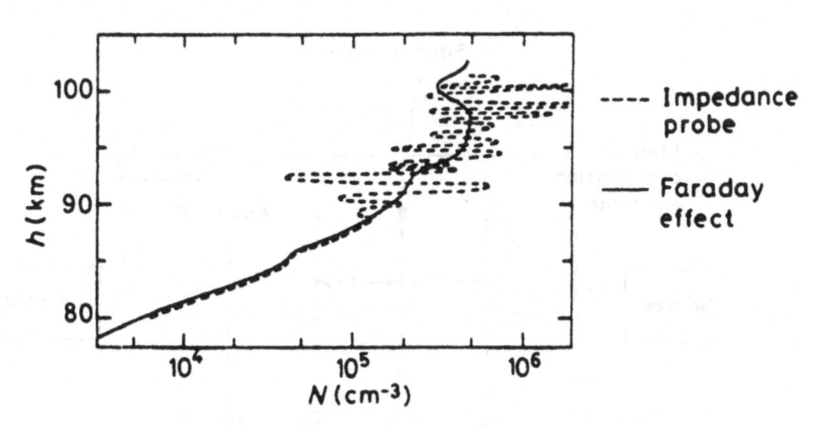

Fig. 3.3 Electron density profiles determined by two methods during a rocket flight into the auroral ionosphere. (After O. Holt and G. M. Lerfald, *Radio Science* **2**, 1283, 1967, copyright by the American Geophysical Union)

Resonance probe

A *resonance probe* uses the observation that if a radio transmitter and receiver are placed in the ionosphere there are characteristic frequencies at which the medium appears to resonate. The resonances appear as spikes on topside ionograms (Figure 3.7), and they represent a local storage of energy at frequencies where the ionosphere has a natural resonance. The fundamental frequencies are the plasma frequency,

$$\omega_N = \left(\frac{Ne^2}{\varepsilon_0 m}\right)^{\frac{1}{2}} \text{(Equation 2.32),}$$

and the gyrofrequency,

$$\omega_B = eB/m \text{ (Equation 2.27),}$$

but resonances also occur at combination frequencies such as $(\omega_N^2 + \omega_B^2)^{\frac{1}{2}}$.

The wave-field experiment on the geosynchronous satellite GEOS-2 (Table 3.3) included a resonance probe, which at that altitude had to operate in the VLF band (3–30 kHz).

3.2.4 Mass spectrometers

A mass spectrometer for the upper atmosphere starts with some advantages. If used above 100 km the spectrometer does not need a vacuum system because, with a mean free path exceeding one metre, the necessary vacuum is already present. Also, if ions are to be measured there is no need for an ion source because the sample is already ionized. But there are also some problems. One is that a magnetic deflection system, as commonly used in laboratory instruments, may not be acceptable because of the weight of the magnet and/or because the strong magnetic field interferes with other experiments. Also, some of the species to be measured are chemically reactive and likely to take part in reactions on the walls of the spectrometer should they strike it.

The *Bennett*, or *time-of-flight spectrometer* avoids the need for a magnet. To reach the detector, the ions have to pass through a number of grids, some biased by steady potentials and the central one having an oscillating potential (Figure 3.4). Between grids 1 and 2 an ion of mass m_i and charge e is accelerated to velocity $(2eV_1/m_i)^{\frac{1}{2}}$, which

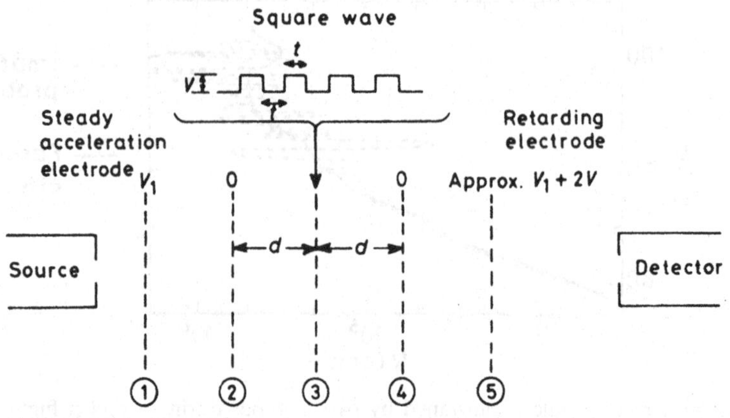

Fig. 3.4 Principle of the time-of-flight mass spectrometer.

depends on the ratio m_i/e of the ion. The oscillating potential on grid 3 ensures that ions that pass grid 2 with a velocity near to d/t (where d is the distance between grids 2 and 3, and t is the time between potential reversals) receive further acceleration. The final electrode (5) is reverse biased to reject all ions having less than the maximum possible speed. In this way, the ion beam reaching grid 2 is velocity filtered and only ions with the selected mass/charge ratio (and all mass spectrometers measure mass/charge ratio) may reach the detector.

A *quadrupole mass spectrometer*, first introduced by Narcisi and Bailey in 1965, has been used in the lower thermosphere on rockets and also higher up on satellites. The theory of this instrument is too complicated to describe in detail here, but the principle is that the ions drift in combined steady and varying electric fields in the space between four parallel cylinders. The steady and alternating fields are devised so that only ions with a certain mass/charge ratio can survive to reach the detector at the end of instrument, the others being lost to the cylinders along the way.

Inspection of our tables of specimen payloads (Tables 3.1–3.4) will show that it is quite usual to include one or more mass spectrometers, for neutral or ionized species. Each of the types we have discussed is represented there, as well as some other variants on the same principles. The problem of instrument calibration is well illustrated by the AE instrumentation (Table 3.1), which included two mass spectrometers of different design so that their performances in orbit could be compared, particularly for the measurement of reactive species like atomic oxygen. The closed-source instrument could be more certainly calibrated before launch, but since the gas enters through an orifice it has greater contact with the walls of the instrument and thus is more likely to be lost before reaching the detector. Note, further, that two kinds of ion mass spectrometer were also carried. Much of the information on neutral gas concentrations in the MSIS model (Section 4.4) came from the AE mass spectrometers.

3.3 Radiation sensors

Radiation is a major component of the Earth's environment, and radiation sensing covers many topics. We will use the term broadly, to include both electromagnetic and particle radiation, and the full range of the electromagnetic spectrum. We have to measure radiation that reaches the Earth from outside because it generates significant processes within the geospace region, and we must also cover emissions generated within the medium which may be important for diagnostic purposes.

3.3.1 Energetic particle detectors

Particles with thermal energies are generally observed by derivatives of the Langmuir probe or other devices discussed in Section 3.2. For more energetic particles several kinds of detector are available. They differ fundamentally from the techniques used for particles of lower energy in that they register the arrival of each individual particle rather than the total current collected from a large number. In the range 1 keV to 1 GeV and above the most commonly used detectors are the Geiger counter, the plastic scintillator and the channel multiplier. The *Geiger counter* relies on the electrical breakdown of a gas under a high potential gradient when an energetic particle creates an ionized track through it. In a *scintillator* the particle produces a flash of light when it enters a transparent solid, and the flash is detected by a photomultiplier tube. The *channel multiplier* is a device in the form of a long narrow tube along which a large potential gradient is applied. When an energetic particle strikes the surface secondary electrons are emitted. These are accelerated by the potential gradient and produce further secondaries in their turn on striking the surface. Thus an avalanche is generated, which is readily detected at the end of the tube. The size of the pulse is proportional to the energy of the original particle. It is usual to place some kind of energy selector before a particle detector, and collimators may be used so that the direction of arrival may be measured. On a spinning satellite, the detectors may be placed so that the spin of the vehicle carries them through a range of angles with each rotation.

The geosynchronous satellite GEOS-2 (Table 3.3) carried three energetic particle instruments to measure energy spectra and angular distributions. Two were *electrostatic analyzers*, which employed electrostatic deflection to select the energy and then counted the particles with channel multipliers. The satellite spin was used to cover a range of angles with respect to the geomagnetic field. The Dynamics Explorers and HILAT (Tables 3.2, 3.4) also carried analyzers covering energies up to 30 keV and 20 keV respectively.

3.3.2 Optical and other electromagnetic receivers

Solar spectrum

The solar spectrum in the X-ray and ultra-violet regions, at wavelengths shorter than 120 nm, is important for the formation of the ionosphere. The same band is also significant for the neutral components present above 120 or 130 km because of its heating effect and control of the temperature of the exosphere (Section 4.1.3 and Figure 4.5). Obviously, the incoming spectrum can only be measured by observing above the level where it is absorbed, and this requires a rocket or a satellite. Since the intensity of solar radiation in the extreme ultra-violet (EUV) and X-ray regions varies with solar activity (the variations being larger at the shorter wavelengths), the measurements have to be prolonged.

The instrument will be some kind of spectrometer, or a photometer consisting essentially of a photocell covered by a narrow-band optical filter. Such instruments were included on the Atmospheric Explorer satellites (Table 3.1). A solar spectrum is given as Figure 6.9.

Upper atmosphere emissions

There are many emissions from the upper atmosphere which, properly interpreted, can provide useful information about atmospheric processes. For example there are the luminous emissions (which extend into the infra-red and the ultra-violet) from the airglow and the aurora. These are dealt with in some detail elsewhere (Sections 6.4, 8.3.3). Particularly noteworthy are the auroral observations now being made from orbiting satellites (Section 8.3.2), and the use of airglow for measuring upper-atmosphere winds (Section 4.2.2).

Temperature measurement

It is also possible to determine the temperature of the air remotely from a satellite by observing the intensity of a strong emission line as a function of wavelength. The intensity of emission from a gas depends on the temperature as well as on the wavelength. Within an emission band some wavelengths are more strongly emitted than others, and by Kirchhoff's law these same wavelengths are also the most strongly absorbed. If a detector is pointed down into the atmosphere from above, the radiation it receives at some wavelength comes preferentially from one altitude. This comes about from the competing effects of emission and absorption. There will be relatively less emission higher up because of the reduced gas density, but radiation from low down suffers more absorption because it passes through a greater depth of the atmosphere. The altitude of the peak depends on the absorption coefficient, so it is possible to derive a temperature profile by observing at a number of suitable wavelengths.

We can prove these points by borrowing some results from Chapter 6. Consider an exponential atmosphere with scale height H, in which the molecular concentration is expressed by $n = n_0 \exp(-h/H)$ where n_0 is the concentration at the ground (Equation 4.4). Suppose that radiation of intensity I_h is emitted from the gas at height h. This radiation is attenuated by absorption as $dI/dh = -I\sigma n$ (Equation 6.5), where σ is the absorption cross-section per molecule; the intensity at the top of the atmosphere is therefore $I' = I_h \exp(-\sigma n H)$ from Equations 6.6, 6.7.

By Kirchhoff's law the emission from a slab of gas of thickness Δh, containing n emitting molecules per unit volume is

$$I_h = B\sigma n\,\Delta h,$$

and hence

$$I' = B\sigma n \exp(-\sigma n H)\,\Delta h.$$

Substituting for n and differentiating with respect to h shows that I' is a maximum where

$$h = H \ln(\sigma n_0 H).$$

The maximum occurs at a greater height if σ is larger. B is the Planck radiation formula which depends on the temperature. It is therefore possible to derive the temperature from the observed intensity and to relate this to a specific height, knowing the concentration and scale height of the emitting gas and the absorption cross-section as a function of wavelength.

These principles have been successfully applied in satellite experiments using the carbon dioxide band at 15 μm in the infra-red, and they enable temperatures in the stratosphere and the mesosphere to be routinely monitored. The temperature distributions in Figure 7.32 were obtained by this technique.

X-rays

X-rays are emitted from the atmosphere at high latitudes when energetic electrons arrive from the magnetosphere and are stopped by collisions with the neutral gas. These 'Bremmstrahlung' X-rays (see Section 6.2.3) are emitted over a wide range of angles. They may be detected from below by a sensor on a high-altitude balloon, or 'photographed' from above with a pinhole camera carried on a satellite. In the latter case the detector is essentially an array of proportional counters formed from two grids of parallel wires at right angles. It is filled with an inert gas such as argon, which conducts when ionized, and the numbers of photons reaching each cell of the detector can be determined from the distribution of currents in the individual grid wires. The angular resolution of such a camera is a few degrees, and the spatial resolution may vary between 200 km and 10 km depending on the altitude of the satellite. Whereas absorption in the atmosphere limits balloon-based measurements of X-rays to energies above 15–20 keV, from a satellite the measurement can be extended down to 1–2 keV. However, the cameras generally become less sensitive at energies above 30 keV.

VLF receivers

The electromagnetic environment extends also to radio frequencies. At frequencies of tens of kilohertz and below, signals are generated within the medium which are important because they give information about plasma processes in the ionosphere and, particularly, the magnetosphere. One important aspect of VLF observations from the ground is discussed in Section 3.5.4. In satellite work, which has the advantages of global coverage and of avoiding ionospheric absorption, the detector is essentially a radio receiver with some signal processing. To take an example, the low-orbit (550 km) satellite Aerial 3, launched in 1967, carried a VLF receiver to monitor and analyze the signals at 3.2, 9.6, and 16 kHz. The objective was to survey the VLF environment in the ionosphere, and, as well as magnetospheric emissions, natural emissions from thunderstorms and man-made transmissions from VLF transmitters on the ground were measured. Some key characteristics of the signals were recorded to help to indentify their nature. An example of the measurements is shown in Figure 9.11.

3.3.3 Magnetometers and electric field sensors

At the low-frequency limit of electromagnetism are steady magnetic and electric fields. The geomagnetic field has been observed from satellites for many years in order to map the field originating within the Earth, and to investigate the more distorted and variable field of the magnetosphere. The distant fields are very weak (a few nanoteslas) and require magnetically clean spacecraft. This may be difficult to achieve if the spacecraft carries other experiments, as would normally be the case.

The magnetometers carried on satellites are based on the same principles as the instruments used in ground-based magnetic observatories, the most common being the fluxgate and the alkali vapour (caesium or rubidium) magnetometers. In the *fluxgate magnetometer*, a magnetic core is driven to the point of saturation by an alternating current flowing through a winding. If a steady magnetic field is added, the core becomes biased and the periodic variation of flux through it becomes asymmetrical. The degree of asymmetry is detected by means of a second winding, and from it the magnitude of the external field can be derived. A fluxgate with three cores mutually at right angles therefore measures all three components of the steady magnetic field in which it is placed. However, it is not an absolute instrument and has to be calibrated. The *alkali-vapour magnetometer*, which exploits the Zeeman splitting of an emission line in the presence of an external magnetic field, is absolute and therefore very accurate, but it is naturally suited to measuring the total magnetic field rather than the components. All the satellites that we have taken as examples of spacecraft instrumentation (Tables 3.1–3.4) carried fluxgate magnetometers.

Electric field measurement is also an important part of some missions, particularly those concerned with the polar ionosphere where convection, and therefore the electric field, is one of the basic phenomena (Section 8.1.1). One might think that if two probes were extended from a vehicle then the potential difference between them would be easily measured and that this would immediately give the electric field in the line of the probes. But there are in fact several obstacles. One is the electric field induced by the vehicle's motion through the geomagnetic field, but this may be corrected for to the extent that the magnetic field is known. The second is the plasma sheath that forms around the vehicle since this alters the potential distribution. The low-altitude Dynamics Explorer (Table 3.2) carried an instrument to measure the electric field by means of three pairs of probes extending 11 m from the body of the vehicle so as to get outside the sheath. The inner 9 m of each was insulated and just the end 2 m was conducting. The average potential of the pair of probes gives the spacecraft potential, and the difference gives a component of the electric field in the medium. This had to be corrected for the induced electric field, which could be as large as 450 mV/m, and therefore the attitude of the spacecraft had to be accurately known. The field at geosynchronous orbit is considerably weaker, and on GEOS-2 (Table 3.3) the approach was via an electron beam technique. Much information on electric fields in the upper atmosphere has also come from the ion cloud method described in Section 4.2.2.

3.4 Indirect sensing of the neutral atmosphere

3.4.1 Falling spheres and dragging satellites

The density of the air can be determined from its effect in retarding a moving object. The air drag force on a sphere is

$$D = 0.5C_D A\rho v^2, \tag{3.3}$$

where ρ is the air density, v the speed of the sphere relative to the air, A the cross-sectional area, and C_D the drag coefficient. The form of this equation may be proved as follows. If there are n molecules per unit volume, the sphere moving at speed v encounters Anv molecules per unit time. If the change of momentum at each collision is mv, the retarding force (= rate of change of momentum) is $Anmv^2 = A\rho v^2$, where ρ is the air density. A 'drag coefficient' is needed because the sphere's effective cross-section is not the geometrical cross-section but is altered by the pattern of air flow around it.

In the *falling sphere* method, a sphere instrumented with an accelerometer and telemetry is ejected from a rocket. The sphere is retarded by an air drag force (Equation 3.3) opposing the gravitational force $m_s g$, m_s being the mass of the sphere, and hence it accelerates towards the ground more slowly than it would in a vacuum. The readings of the accelerometer are telemetered to a ground station. This method works well in the height range 90 to 130 km.

The same principle underlies the well known *satellite drag* method. Satellite tracking by optical and radio means is quite accurate enough to detect the effects of the atmosphere on the orbit of a satellite, and tracking data have been applied to the determination of air density ever since the first satellites were launched in the late 1950s. A great deal of information about the density of the thermosphere and its variations has been accumulated from satellite drag measurements over the years. The height range 200–1100 km can be covered, and as well as average density, seasonal and short-term changes, and diurnal and latitudinal variations can be investigated. The simple version of the theory, which follows, applies to a satellite in circular orbit, and assumes a spatially uniform and unchanging atmosphere.

If a satellite is subject to a drag force D, given by Equation 3.3, the rate of loss of energy is (force × distance/time)

$$Dv = C_D A \rho v^3 = -\frac{d(TE)}{dt} = -\frac{d(TE)}{dr} \cdot \frac{dr}{dt}, \tag{3.4}$$

where TE is the total energy (potential plus kinetic), r is the distance of the satellite from the centre of the Earth, and v is its velocity. The kinetic energy,

$$KE = m_s v^2/2 = GM_E m_s/2r,$$

where m_s is the satellite mass, M_E the mass of the Earth, and G the gravitational constant. (We have used the relation $v^2 = GM_E/r$, obtained by equating the centrifugal and gravitational forces on a satellite.) However, the potential energy,

$$PE = -GM_E m_s/r.$$

Hence,

$$TE = KE + PE = -GM_E m_s/2r. \tag{3.5}$$

(Note in passing that

$$|PE| = 2|KE| \text{ and } TE = -KE.$$

Also, if r decreases, the kinetic energy, and therefore the orbital velocity, increases. But the potential energy decreases twice as much, and thus the total energy decreases – as it should. The observation that a satellite speeds up as it enters the atmosphere is not, therefore, inconsistent with a loss of energy.)

From Equation 3.5,

$$\frac{d(TE)}{dr} = \frac{GM_E m_s}{2r^3} = \frac{v^2 m_s}{2r},$$

and from Equation 3.4,

$$\frac{dr}{dt} = -\frac{C_D A \rho v}{m_s}. \tag{3.6}$$

This gives the rate of decrease of satellite altitude due to air density ρ. However, the orbital period (P) is more accurately measured than the height. Using $P = 2\pi r/v$, Equation 3.6 may be converted to

$$\frac{dP}{dt} = -\frac{3\pi C_D A r \rho}{m_s}. \tag{3.7}$$

The rate of decrease in the orbital period of a satellite in circular orbit is therefore

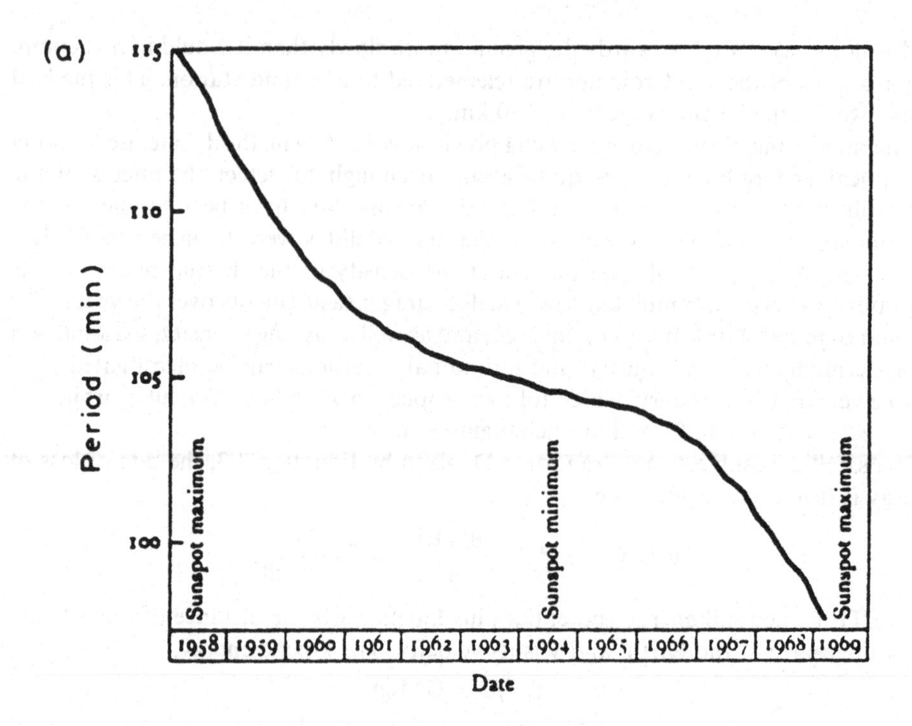

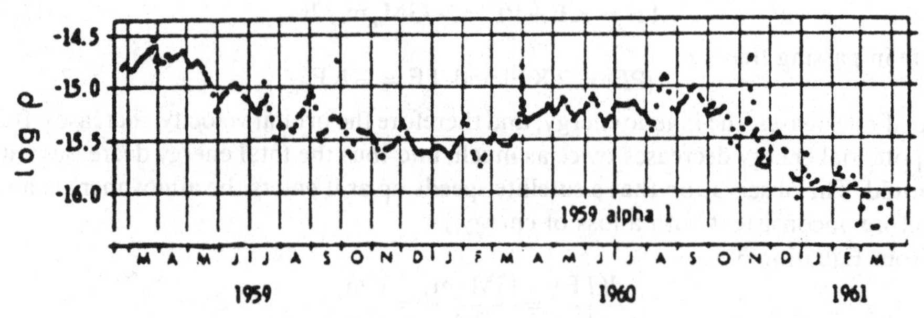

Fig. 3.5 Results of satellite drag measurements. (a) Period of satellite in its orbit, 1958–1969. The rate of decay was greater at sunspot maximum, which indicates greater air density. (b) Deduced variations of air density, showing a 27-day periodicity. (After J. A. Ratcliffe, *An Introduction to the Ionosphere and the Magnetosphere*. Cambridge University Press, 1972; and I. Harris and N. W. Spencer, in *Introduction to Space Science* (eds. Hess and Mead). Gordon and Breach, 1968)

proportional to the air density at the height of the satellite. The success of the method derives from the precision of orbital period measurement.

A more elaborate analysis is needed for the more general case of an elliptical orbit. Such an orbit covers a range of altitudes, and it is possible to determine the scale height. Moreover, since most of the drag occurs near perigee, where the air density is greatest, it is possible to determine the variation of density with time of day and with latitude. Much of our knowledge of the heating and cooling of the thermosphere over different time scales (Sections 4.2.4, 7.1.6) has come from satellite drag data. Figure 3.5 shows examples of variations due to changing solar activity. Such information is important in the management of operational satellite systems (Section 10.4.1).

It is also possible to determine the drag, and hence the air density, at satellite altitudes by means of an accelerometer carried on the satellite. The accelerometer on AE (Table 3.1) enabled air density to be measured up to 400 km; the accuracy was 1% below 200 km, falling off to 10% at 350 km.

3.4.2 The measurement of upper atmosphere winds

Winds in the neutral air are difficult to measure at high altitude, which is why our knowledge of them is incomplete. Various indirect methods have been developed, all of which make use of tracers:

radar tracking of meteor trails
Doppler shift of natural airglow emissions
visual or optical tracking of injected smoke, or luminous chemicals
radar tracking of a descending parachute or metallic foil

None of these is generally applicable, but each works in suitable conditions and over a limited height range. They are discussed in more detail in Section 4.2.2.

3.5 Remote sensing by radio waves

Some of the major techniques for studying the upper atmosphere are based on the propagation of radio waves, and a vast amount of information has been obtained by such means. A certain amount can also be learned from the performance of operational radio links, but most information of scientific value has come from techniques developed for a purpose.

The techniques generally fall into three groups.

(a) The radio wave may be totally reflected within the medium.
(b) It may pass through the medium but emerge altered.
(c) Most of the energy may travel through the medium, a small fraction being scattered or partially reflected by irregular structures.

Groups (a) and (c) involve a transmitter and receiver both below or both above the ionosphere. Greater sensitivity is needed for (c) because the returns are weaker. Techniques in group (b) normally require a source or a receiver above the ionosphere.

3.5.1 Ionospheric sounding

Ionosonde

The principle of ionospheric sounding, one of the oldest and still one of the most important techniques of ionospheric study, is to transmit a radio pulse vertically and to measure the time which elapses before the echo is received. Some readers may recognize this as radar, and they would not be wrong – except that they should remember that radar for aircraft detection was developed from the techniques of ionospheric sounding, not the other way round! Neglecting collisions ($Z = 0$) and the geomagnetic field ($Y = 0$) in the Appleton–Hartree equation (Equation 2.58), the refractive index (n) is given by

$$n^2 = 1 - X = 1 - \frac{\omega_N^2}{\omega^2}. \tag{3.8}$$

ω_N^2 is proportional to the electron density, N_e, since $\omega_N = (N_e e^2/\varepsilon_0 m)^{\frac{1}{2}}$ (Equation 2.32). As the wave penetrates further into the ionosphere the electron density increases and the refractive index therefore becomes smaller. If $\omega_N > \omega$, the refractive index is imaginary and the wave cannot propagate. Therefore the energy is reflected at the level where $\omega_N = \omega$, i.e. where the plasma frequency equals the wave frequency. Putting in values, this condition comes to

$$(f_N(\text{kHz}))^2 = 80.5 N_e(\text{cm}^{-3}) \text{ or}$$
$$(f_N(\text{Hz}))^2 = 80.5 N_e(\text{m}^{-3}). \tag{3.9}$$

The height is found from the time delay of the echo, it being assumed in the first instance that the pulse travels at the speed of light.

The equipment using these principles is an *ionospheric sounder* or *ionosonde*. A pulsed transmitter and a receiver are swept synchronously in frequency, and the echo delay time is recorded as a function of the radio frequency. Traditionally, the height was displayed against the radio frequency on a cathode ray tube, and the resulting *ionogram* (Figure 3.6) was recorded photographically. The more modern instruments, generally those dating from the late 1970s, are under the control of a microprocessor and the data are recorded digitally.

The simple interpretation of an ionogram indicated above is not quite adequate for two reasons. First, a radio pulse travels not at the speed of light, c, but at the group velocity, u, (Section 2.4.3) which is related to the speed of light by the group refractive index, n_g, i.e. $u = c/n_g$. In the ionosphere the group refractive index is given by Equation 2.55, from which it may be shown that $n_g = 1/n$. Therefore a radio pulse travels more slowly than light, and the *group retardation* in the ionosphere below the reflection level means that in general the *virtual height*, derived under the assumption of a constant velocity (c) throughout, is always too large. The cusps of Figure 3.6 are due to strong group retardation in regions where the electron density changes most gradually with height (e.g. over a local maximum). We note, also, that at the reflection level, n is zero but n_g is infinity, meaning that the forward velocity of the pulse is zero, as would be expected at a reflection. It is possible to correct for the group retardation and produce a *real-height profile*.

The ionosphere often contains several layers. The maximum plasma frequency of a given layer is the *critical frequency* of that layer. The maximum critical frequency of the whole ionosphere is the *penetration frequency*, since signals of higher frequency (if vertically incident) penetrate through the whole ionosphere to be lost to space. Conversely, only waves whose frequency exceeds the penetration frequency can be received from space by a ground-based receiver.

Ordinary and extraordinary waves

The second complication of ionogram interpretation arises from the existence of the geomagnetic field. If $Y \neq 0$, Equation 2.58 gives two values for the refractive index, implying that waves may propagate at two speeds. These are called the *ordinary wave*

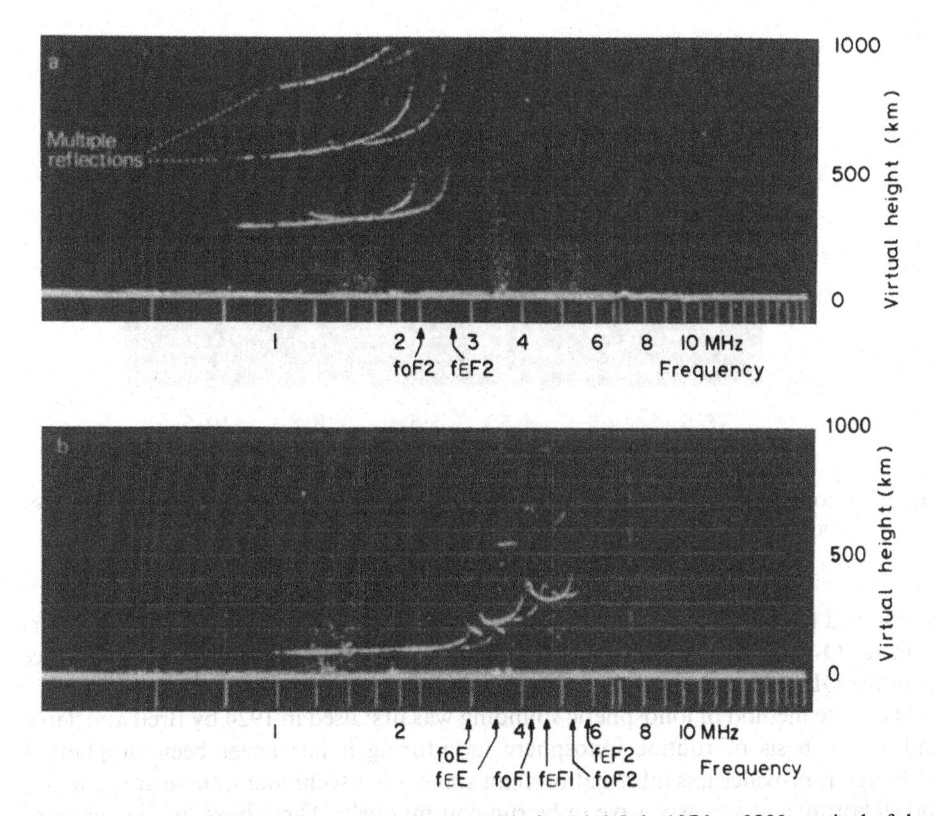

Fig. 3.6 Ionograms from Port Stanley, Falkland Islands: (a) 8 July 1976 at 0200, typical of the winter night; (b) 20 January 1977 at 1545, typical of the summer day. The symbols f_0E, f_EE, f_0F1, f_EF1, f_0F2 and f_EF2 denote the ordinary and extraordinary critical frequencies of the E, F1, and F2 layers. (Some people use X instead of E for the extraordinary ray.) (Rutherford Appleton Laboratory, UK Science and Engineering Research Council)

and the *extraordinary wave*, and their refractive indices are obtained by taking the positive and negative signs respectively. Traces due to the O- and E-waves may be seen in Figure 3.6. Further consideration of Equation 2.58 shows that the reflection conditions are

E-wave: $$\omega_N^2 = \omega(\omega - \omega_B) \tag{3.10a}$$

O-wave: $$\omega_N^2 = \omega^2 \tag{3.10b}$$

The E-wave reflection occurs according to the QL approximation (Section 2.5.1) while the O-wave condition relates to the QT approximation. If $\omega_B \ll \omega_N$, the difference between the E- and O-wave critical frequencies is

$$\omega_E - \omega_O \approx \omega_B/2.$$

Sometimes a second reflection condition,

$$\omega_N^2 = \omega(\omega + \omega_B), \tag{3.10c}$$

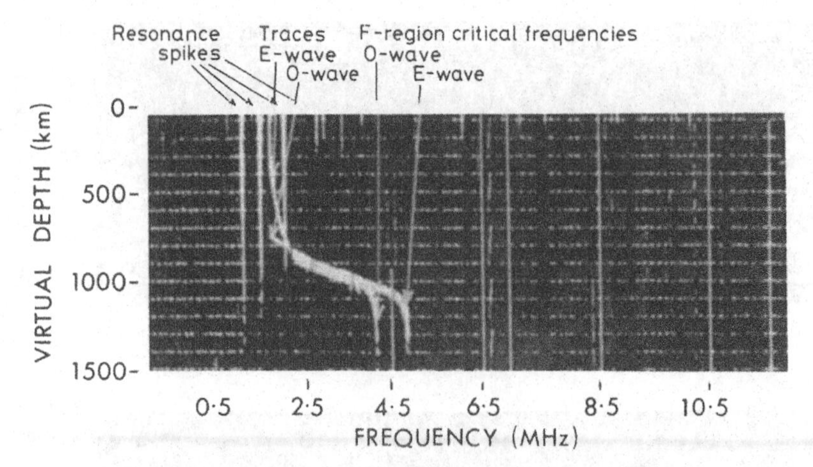

Fig. 3.7 A topside ionogram typical of middle latitudes near noon. (Rutherford Appleton Laboratory, Science and Engineering Research Council)

is observed for the O-wave, and the ionogram then shows three traces. The third trace is due to O-wave energy which has got above the QT reflection level and is returned from the QL reflection condition. The third trace is called a *Z-trace*.

The pulse method of ionospheric sounding was first used in 1924 by Breit and Tuve, and as the basis of routine ionosphere monitoring it has never been supplanted. Although it provides less information than some other techniques, these are generally too elaborate and too expensive to be run continuously. There have, moreover, been some significant improvements to ionospheric pulse sounding over the years. Digitization was mentioned above, and the microprocessor has made possible various 'advanced sounders', whose mode of operation and data display may be programmed to suit the investigation in hand. An advanced sounder will be able to measure the strength of the echo as well as its time delay, and it may have the sensitivity to detect partial reflections from the lower ionosphere (Section 3.5.5). By the use of spaced antennas, the direction of arrival of the echo can be determined – a useful facility when the ionosphere is not uniform, as in the trough regions (Section 8.1.6). With the availability of precise timing, the transmitter and the receiver can be separated and thus soundings may be taken over oblique paths.

Topside sounding

One notable development has been the *topside sounder*, which sounds the ionosphere from above. Since the electron density passes through a maximum (usually between 200 and 400 km altitude) a ground-based sounder cannot observe the 'topside' of the ionosphere – i.e. the region above the maximum. The answer is to carry a sounder on a satellite in an orbit comfortably above the peak, typically at 1000 to 1500 km. The first topside sounder was the Canadian satellite Alouette 1, launched in 1962, and there have been several others since that time though the coverage has not been continuous. A topside ionogram is illustrated in Figure 3.7. The interpretation is much as for a traditional bottomside sounding except that additional signals, the resonance spikes referred to in Section 3.2.3, appear. Although these complicate the ionogram, they are also a bonus because they give information about the local plasma in which the sounder is immersed, specifically the plasma and gyromagnetic frequencies.

Topside sounding obviously complements bottomside measurements regarding the heights measured, but it is also complementary in the matter of coverage. A ground-based sounder measures continuously but at one place only. A satellite covers a range of latitudes on each orbit, and all longitudes about twice a day – though how well local time is sampled depends on the orbit. Thus, one approach gives the best time coverage at one place, whereas the other gives the best geographical coverage though possibly at a few local times only.

Absorption

If collisions are taken into account ($Z \neq 0$), the formula for the refractive index (neglecting the magnetic field) becomes

$$n^2 = 1 - \frac{X}{1 - jZ} = 1 - \frac{\omega_N^2}{\omega(\omega - j\nu)}. \qquad (3.11)$$

Then the refractive index is complex ($n = \mu - j\chi$) and a wave propagating through the medium varies with time (t) and distance (x) as $\exp(-x\chi\omega/c)\cos\omega(t - x\mu/c)$. The first term represents the decay of the wave amplitude with distance; the absorption coefficient is $\kappa = \omega\chi/c$ (in units of nepers per metre), and the amplitude falls to $1/e$ of its original magnitude in a distance $1/\kappa = c/\omega\chi$. Substituting the expression for ω_N^2 into Equation 3.11, it can be shown that

$$\kappa = \frac{\omega}{c} \cdot \frac{1}{2\mu} \cdot \frac{XZ}{1 + Z^2} = \frac{e^2}{2\varepsilon_0 mc} \cdot \frac{1}{\mu} \cdot \frac{N\nu}{\omega^2 + \nu^2}. \qquad (3.12)$$

There are special cases which apply respectively to the lower ionosphere and to the F region.

The first applies when $\mu = 1$, and it is called *non-deviative absorption* because the velocity of the wave is not altered and so there is no bending. In that case,

$$\kappa = \frac{e^2}{2\varepsilon_0 mc} \cdot \frac{N\nu}{\omega^2 + \nu^2}. \qquad (3.13)$$

If, also, the collision frequency (ν) is small relative to the angular wave frequency (ω), the absorption coefficient is proportional to the product of electron density and collision frequency. As a practical unit of absorption it is usual to take the decibel (dB), defined from the ratio between initial (P_1) and final (P_2) powers, the power being proportional to the square of the signal amplitude (Section 2.4.5). Thus,

$$\text{Absorption A (decibels)} = 10\log_{10}(P_1/P_2). \qquad (3.14)$$

The neper and the decibel are related by

$$1 \text{ neper} = 8.69 \text{ decibels}. \qquad (3.15)$$

Putting values into Equation 3.13,

$$A = 4.5 \times 10^{-5} \int \frac{N\nu}{\omega^2 + \nu^2} dx \text{ dB}, \qquad (3.16)$$

where all quantities are in SI units. Measurements of non-deviative radio absorption are of interest in relation to radio propagation, and they provide a useful technique for monitoring variations of electron density in the lower ionosphere where the collision frequency is relatively large. In practice the two magneto-ionic modes should be distinguished because they are not absorbed to the same extent. (We should replace ω by $(\omega \pm \omega_B)$, where ω_B is the gyrofrequency. The extraordinary mode is the more strongly absorbed.)

The second special case, *deviative absorption*, applies in the F region of the ionosphere where the collision frequency is small. Putting $Z^2 \ll 1$ in Equation 3.12, and remembering that $\mu^2 = 1 - X$ if ν is small, leads to

$$\kappa = \frac{\nu}{2c}\left(\frac{1}{\mu} - \mu\right). \tag{3.17}$$

From this or from Equation 3.12, it is clear that absorption is enhanced near the reflection level where $\mu \to 0$ and the pulse is travelling more slowly.

Ionospheric radio absorption may be measured in several ways. The original method was to transmit pulses and receive the echoes after ionospheric reflection, like an ionosonde. For absorption measurement the frequency would be held constant (usually between 2 and 6 MHz) and the receiver would be designed for accurate measurement of the echo intensity over the greatest dynamic range. According to Equation 3.16, the specific absorption (that due to 1 electron/m^3) is a maximum at the height where $\nu = \omega$. It is therefore possible to get some information about the height of the absorption by using more than one radio frequency – though, again, the gyrofrequency has to be taken into account. One of the problems of pulsed absorption work is distinguishing between the deviative and non-deviative contributions. Generally, the first will come from the F region and the second from the E and D regions.

A variant of the method is to transmit a continuous wave and to monitor the intensity of the signal received some distance away. A separation of at least 100 km is required to avoid interference from the ground wave which propagates directly between transmitter and receiver. A third technique, to be described in the next section, measures the apparent intensity of the cosmic radio noise.

Ionospheric radio absorption has been measured since about 1935, and it therefore qualifies as one of the older methods of ionospheric study. Like the ionosonde it continues to be useful because, though the information provided is limited, the techniques are generally economical and suitable for continuous application in a monitoring mode.

HF Doppler

For observing small perturbations of the ionosphere the *HF Doppler* technique has proved valuable. Consider a transmitter and a receiver which are co-located so that reflection occurs at normal incidence to the layer. If the reflection level moves upward at speed w, the frequency of the refected wave is Doppler shifted by $\Delta f = -2fw/c$, where f is the transmitted frequency and c is the speed of the wave. Thus, $w = -2\lambda.\Delta f$, where λ is the wavelength. If $\lambda = 100$ m and $\Delta f = 0.5$ Hz, then $w = 100$ m/s. A typical HF Doppler system uses continuous, not pulsed, signals. The received wave is mixed with a reference signal at the transmitter frequency, and the frequency of the beat signal is measured and recorded. The measurements give the instantaneous velocity in the first instance, but this may be integrated to give the height changes. The absolute reflection height would normally be determined from an ionosonde. One of the main applications of HF Doppler has been to studies of acoustic–gravity waves in the ionosphere (Section 7.5.5).

3.5.2 Trans-ionospheric propagation

A radio wave will pass right through the ionosphere if its frequency exceeds the penetration frequency. However, provided the frequency is not too large there will be measurable effects that may give useful information. Consequently, there is a group of techniques based on *trans-ionospheric propagation*. Assuming that reception is at the ground, the source may be a transmitter aboard a satellite or an ascending rocket, or the natural radio emission from the galaxy. The Moon has also been used as a reflector for trans-ionospheric radar, in which case the signal passes through the ionosphere twice and the effects are doubled.

Carrier phase

If $X \ll 1$ and the magnetic field and collisions are neglected, Equation 3.8 for the refractive index (n) becomes

$$n = 1 - \frac{\omega_N^2}{2\omega^2}, \tag{3.18}$$

where ω_N (the angular plasma frequency) is given by

$$\omega_N^2 = \frac{Ne^2}{\varepsilon_0 m} = 3182\,N. \tag{3.19}$$

Thus, numerically,

$$n = 1 - \frac{1591\,N}{\omega^2} = 1 - \frac{40.30\,N}{f^2}, \tag{3.20}$$

where N is in m^{-3}, ω is in radians/s and f is in cycles/s. The departure of the refractive index from unity now varies linearly with the electron density and inversely with the square of the radio frequency.

If a radio wave travels a distance l in an ionized medium, its phase changes by

$$-\frac{\omega}{c}\int n\,dl = -\frac{l\omega}{c} + \frac{e^2}{2c\omega\Sigma_0 m}\int N\,dl. \tag{3.21}$$

The first term is just the phase delay due to a wave travelling a distance l at the speed of light (c). The second is a phase advance which arises because the refractive index is less than unity and the phase speed is greater than c. The quantity $\int N\,dl$ is the number of electrons in a unit column along the line of propagation, generally known as the *electron content*, I. Inserting values, the phase advance due to the medium is

$$\phi = \frac{5.308 \times 10^{-6}}{\omega} \cdot I = \frac{8.448 \times 10^{-7}}{f} \cdot I \text{ radians.} \tag{3.22}$$

It is interesting to note (though of no obvious physical significance) that the constant 8.44×10^{-7} is just the classical radius of the electron ($r_e = e^2/4\pi\varepsilon_0 m_e c^2 = 2.818 \times 10^{-15}$ m) times the speed of light ($c = 3 \times 10^8$ m/s).

To measure the phase change it is necessary to have a reference signal which is either unaffected by the medium or affected to a lesser extent. This will usually be a second transmission on a frequency different from, but coherent with, the first. For example, if coherent waves were transmitted at ω and $n\omega$, both would be received after passing through the ionosphere, and the higher frequency divided by n. The phase difference between the two signals is then

$$\begin{aligned}\Delta\phi &= \phi_1 - \phi_2 \\ &= \left(-\frac{l\omega}{c} + \frac{\omega_N^2}{2c\omega}\cdot I\right) - \frac{1}{n}\left(-\frac{ln\omega}{c} + \frac{\omega_N^2}{2cn\omega}\cdot I\right) \\ &= \left(1 - \frac{1}{n^2}\right)\frac{\omega_N^2}{2c\omega}\cdot I \\ &= \frac{n^2-1}{n^2}\cdot\frac{8.448\times10^{-7}}{f}\cdot I \text{ radians.}\end{aligned} \tag{3.23}$$

Since n is known, a measurement of $\Delta\phi$ gives a value for I.

If the electron content is $10^{17}/m^2$ (e.g. an ionosphere 250 km thick with 4×10^{11} electrons/m^3) a 100 MHz signal will be advanced in phase by 845 radians, or 134 complete cycles with respect to the free-space phase shift. The carrier phase is therefore very sensitive to changes of electron content, which by this means can be measured to an accuracy of about 0.01 % of the total. However, since phase is only measured within the range 0–2π the absolute value is ambiguous and has to be determined by other means.

Modulation phase

If the carrier (f_c) is modulated at frequency (f_m), the phase of the modulation may be measured at the ground (for instance, by reference to a coherent modulation carried at a much higher frequency), and this is given by

$$\phi_m = 8.448\times10^{-7}\cdot\frac{f_m}{f_c^2}\cdot I. \tag{3.24}$$

Since f_m/f_c can be made as small as desired, the sensitivity of ϕ_m to I can be reduced so that all values lie between 0 and 2π, thus removing the ambiguity. This variant is the *modulation phase* technique. Whereas the carrier phase is advanced in the ionosphere, the modulation phase is retarded.

Time delay

A phase delay ϕ_m corresponds to a time delay $\phi_m/2\pi f_m$ s. Thus the ionospheric time delay is

$$\Delta t = \frac{8.448\times10^{-7}}{2\pi f_c^2}\cdot I \text{ s,} \tag{3.25}$$

which is independent of the modulation frequency. A single pulse would be delayed by the same amount. If $I = 10^{17}\ m^{-2}$ and $f_c = 100$ MHz, then $\Delta t = 1.34\ \mu s$. For comparison, the time of flight through 1000 km of free space is 3333 μs. The ionospheric effect may not seem very large, but in the context of systems for position determination (see Section 10.2.3) it is significant.

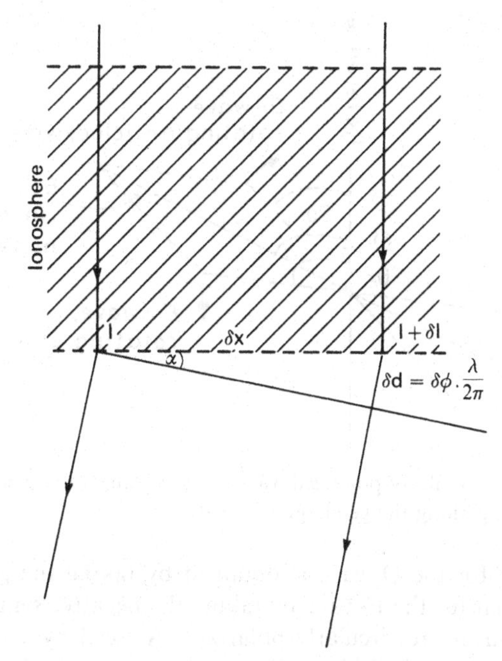

Fig. 3.8 Wedge refraction due to a spatial gradient of electron content.

Wedge refraction

If the electron content is not spatially uniform, but varies in a direction normal to the propagation direction, the result is to deviate the ray. The simplest case is *wedge refraction*, in which the electron content varies linearly with distance. Referring to Figure 3.8, if the electron content increases by δI over distance δx the phase is advanced by $\delta\phi = 8.448 \times 10^{-7} \delta I/f$ (Equation 3.22). The line of constant phase has therefore moved ahead a distance $\delta d = \delta\phi\, \lambda/2\pi$, and the phase front is rotated through angle

$$\alpha = \frac{c}{2\pi} \cdot \frac{8.448 \times 10^{-7}}{f^2} \cdot \frac{\delta I}{\delta x}. \tag{3.26}$$

One consequence is that the apparent position of sources outside the ionosphere varies regularly during the day (with sunrise and sunset), and varies irregularly at times when the ionosphere is spatially irregular. Information on these gradients can be derived from accurate measurements on radio stars.

Faraday effect

Taking account of the geomagnetic field we obtain another variation, which has been extensively used in electron content work because the measurements can be made with relatively simple equipment. Referring to Equation 2.58, we assume that $Z = 0$ and that Y is greater than zero but nevertheless much less than unity – which is realistic – and we assume, further, that propagation is directly along the direction of the geomagnetic field ($Y_T = 0$). Using Equation 2.59, we get for the refractive index,

$$n^2 = 1 + \frac{\omega_N^2}{\omega(\omega \pm \omega_B)}. \tag{3.27}$$

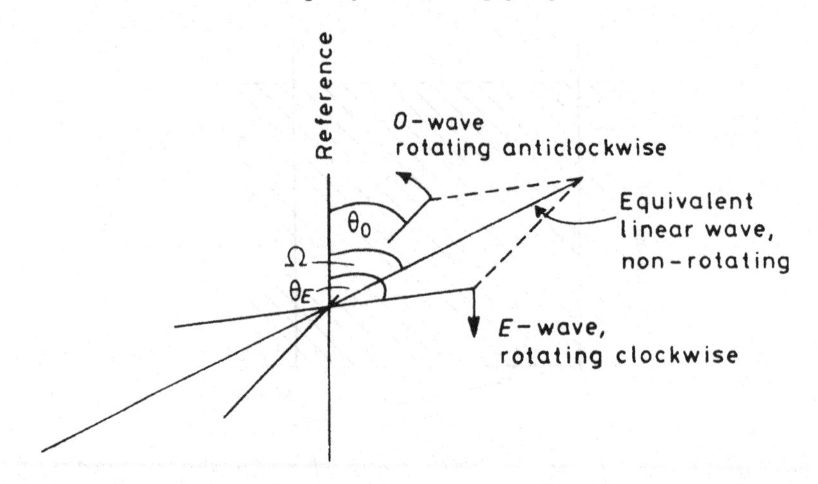

Fig. 3.9 Addition of two circularly-polarized waves to give a linear wave, as seen by a stationary observer looking along the geomagnetic field.

The refractive index for the O-wave is obtained by taking the positive sign in the denominator, and that for the E-wave by taking the negative sign. In the conditions stated, both these waves are circularly polarized. As seen by a stationary observer looking along the geomagnetic field, the O-wave rotates anticlockwise and the E-wave rotates clockwise. The sum of two circular components of equal magnitude is a wave with linear polarization, and if at some instant the electric vectors of the O- and E-waves make angles θ_O and θ_E respectively with some reference direction, the electric vector of the equivalent linear wave lies at angle

$$\Omega = (\theta_O + \theta_E)/2. \tag{3.28}$$

See Figure 3.9.

According to Equation 3.27 the two circular waves travel at slightly different speeds, depending on the ratio ω_B/ω. If we let $\theta_O = \theta_E$ at the source, then after a distance l,

$$\theta_O = \omega t + \frac{\omega n_O}{c} \cdot l,$$

and

$$\theta_E = \omega t - \frac{\omega n_E}{c} \cdot l.$$

Thus,

$$\Omega = \frac{\theta_O + \theta_E}{2} = \frac{\omega}{2c} \cdot (n_O - n_E) l. \tag{3.29}$$

When $\omega_B \ll \omega$, which is true at radio frequencies exceeding 50 MHz, Equation 3.27 gives

$$(n_O - n_E) = \frac{\omega_N^2 \omega_B}{\omega^3}. \tag{3.30}$$

Substituting in Equation 3.29,

$$\Omega = \frac{1}{2c} \cdot \frac{\omega_N^2 \omega_B}{\omega^2}. \tag{3.31}$$

This means that the polarization angle changes as the wave propagates through the ionized medium. This is the *ionospheric Faraday effect*. Substituting the expressions for ω_N and ω_B, inserting values, and allowing the electron density to vary along the path,

$$\Omega = \frac{2.97 \times 10^{-2}}{f^2} \int H_L N \, dl \text{ radians,} \tag{3.32}$$

where N is the electron density in m^{-3}, dl is the element of path in m, f is the radio frequency in Hz, and H_L is the longitudinal component of the geomagnetic field in A/m. The use of H_L means that ω_B has been replaced by ω_L (the quasi-longitudinal approximation), a modest extension of the argument to allow for propagation somewhat across the geomagnetic field. In the Faraday effect, the QL approximation actually has wide validity, to within a few degrees of the normal to the magnetic field.

In Equation 3.32 the electron content is weighted by the geomagnetic field, but to the extent that the geomagnetic field at ionospheric heights is known this may be taken into account. Measurements of the polarization angle of the signal received from a satellite transmission may thus be converted to values of electron content. The method is both convenient and inexpensive, requiring only a receiver with rotating antenna (or its electronic equivalent) – though it is also necessary that a suitable satellite beacon is within view. The Faraday technique has been widely applied to ionospheric monitoring over the years. It catches sudden perturbations and wave-like fluctuations, as well as observing ionospheric storms and the regular diurnal and seasonal variations. Many of the phenomena we shall discuss in later chapters owe something to electron content data.

The Faraday method is not exact, however. Variations in the height of the ionization alter the magnetic field weighting, and derived values of electron content can thus be in error by 10 or 20%. Also, one has to decide to what altitude the measured electron content applies, since the ionization actually extends far from the Earth's surface. If the phase and the polarization methods are used together and the transmitter is far out (at geosynchronous distance, $6.6R_E$, for instance), something may be learned about the height distribution of the ionization as well as its integrated value.

Slab thickness

Electron content values are often combined with ionosonde data. In particular, the ratio of electron content to peak electron density,

$$\tau = I/N_m, \tag{3.33}$$

is called the *slab thickness*. This is the thickness of a hypothetical ionosphere with uniform electron density N_m and content I. The slab thickness is a useful parameter of the ionosphere and it depends physically on the temperature and the ion composition.

Absorption

In the third important technique based on trans-ionospheric radio propagation a stable receiver is used to monitor the intensity of a signal passing through the ionosphere and thereby determine the total absorption. Referring to Equation 3.16, the integral is taken along a straight line through the ionosphere. It would be possible to use a satellite transmission as the source, but nature has provided a convenient one in

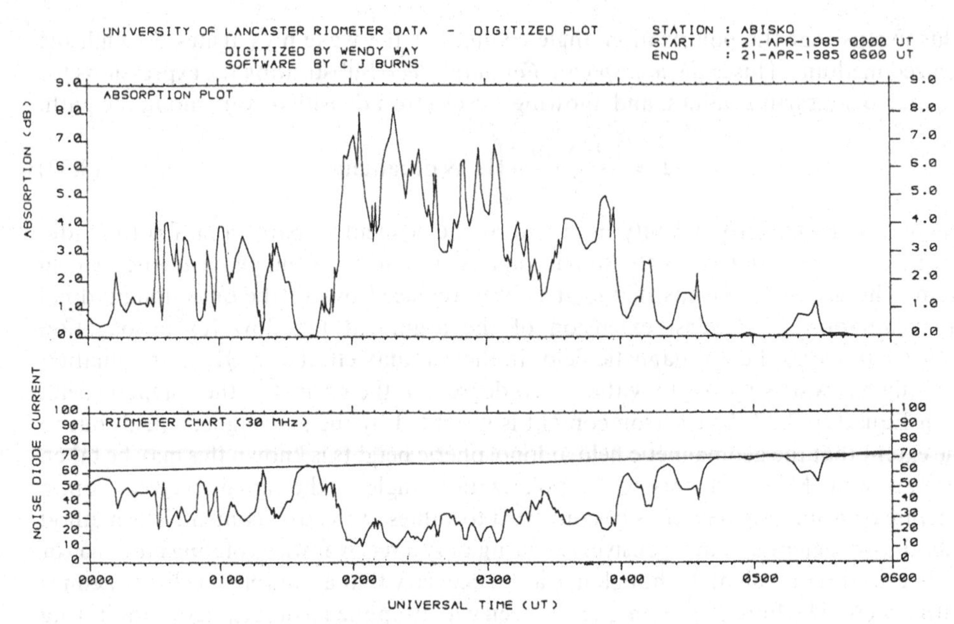

Fig. 3.10 Auroral radio absorption at 30 MHz over a 6-hour period. On the riometer chart (lower panel) the 'noise diode current' is proportional to the received cosmic noise power, and the straight line is the 'quiet day curve' representing the power that would be received in the absence of absorption.

the form of radio stars and hot gas in the galaxy, which together make up the *cosmic radio noise*. The stable receiver required to monitor the received intensity of the cosmic noise is a *riometer* (Relative Ionospheric Opacity METER). Figure 3.10 illustrates a riometer record and the absorption variations derived from it. The riometer has proved most useful at high magnetic latitudes, where absorption is greatest and most sporadic (Sections 8.3.6, 8.5). For practical reasons most riometers use a relatively small antenna with a wide beam, and therefore the measurement averages over about 100 km of horizonal distance. Systems with better spatial resolution are gradually gaining in popularity, however. Since the cosmic noise is not coherent, having the statistical properties of wide-band noise and therefore no defined phase, it is not suitable as a source for electron content measurement.

Rocket methods

The methods of trans-ionospheric propagation can be adapted to studying the lower ionosphere by sending the source up on a rocket. Because the signal need not penetrate the denser parts of the ionosphere its frequency can be reduced to make the observations more sensitive. By recording as the rocket ascends and descends, the electron density and the collision frequency can be determined as a function of height. (See Figure 3.3.) There are several versions of this approach, including observations of the field due to a VLF transmitter (Section 3.5.3).

3.5.3 VLF propagation

Sharp boundary treatment

To handle the propagation of radio waves in the very low frequency (VLF) band, 3–30 kHz, we have to take a different approach. At these frequencies the wavelength (10–100 km) is so long that the ionosphere changes considerably within the space of one wavelength, and an approaching VLF signal sees the base of the ionosphere as a sharp boundary. In the lower ionosphere the collision frequency (ν) is about $2 \times 10^6\ \mathrm{s}^{-1}$ (at 70 km), which is greater than the wave frequency (ω). Thus, from Equation 2.58, neglecting Y,

$$n^2 = 1 - j\frac{X}{Z} = 1 - \frac{\omega_N^2}{j\omega\nu}. \qquad (3.34)$$

The ionosphere now behaves as a metal rather than a dielectric, having conductivity

$$\sigma_i = \frac{Ne^2}{m\nu} = \frac{\varepsilon_0 \omega_N^2}{\nu}. \qquad \text{(See Equation 6.66).}$$

The lower ionosphere can be studied by observing the amplitude and phase of the VLF signals received from senders at various distances. From the data an effective reflection height and conductivity are readily deduced, and the variations which take place diurnally and during disturbances can be monitored. Reflection coefficients are typically 0.2 to 0.5.

VLF propagation over distances of more than a few hundred kilometres is generally treated theoretically in terms of a waveguide bounded by the ground below and the ionosphere above, both being partially conducting. This waveguide also exhibits electromagnetic resonance at 7 or 8 Hz and harmonics. The resonance can be exited by a flash of lightning, and the response, which decays after a few cycles, may be observed using magnetic detectors. The damping rate gives the 'Q' of the circuit, and this varies with perturbations of the lower ionosphere.

Boundary not sharp

Although it makes a convenient first approximation, the assumption of a sharp boundary is not strictly true. The boundary problem can be solved exactly by computer methods, and this can be done for any assumed vertical profile of electron density and collision frequency, and for any radio frequency, polarization and angle of incidence. The method is to regard the ionospheric medium as a succession of slabs, each thin enough to be considered uniform. Maxwell's equations are solved for each step, and thereby the electric and magnetic fields are both derived in detail over the whole height range, many points being taken within each wavelength. If the wave is incident from below some energy is reflected at the boundary, some is absorbed, and some is transmitted. One condition, therefore, is that at a sufficiently great height there can be only an upward propagating wave. Below the ionosphere the wave-fields are resolved into two waves, one propagating upward and the other downward, the incident and the reflected waves respectively. The relations between the incident and reflected waves give the required set of complex reflection coefficients, of which there are four in general so as to include polarization changes on reflection as well as the amplitude and the phase changes. Transmission coefficients may be derived similarly.

Solutions derived in this way are known as *full-wave solutions*. The method is different from that of the Appleton–Hartree magneto-ionic theory, which, using the concept of refractive index, requires that the medium is either uniform or only 'slowly varying' – in practice that it changes only slightly in the distance of a wavelength. The full-wave treatment is more exact and more general, but also less convenient. Its solutions become identical to the Appleton–Hartree characteristic waves (Section 2.5.1) when the medium is slowly varying.

3.5.4 Whistlers

However the boundary is treated, it is clear that some of the radio energy incident on it passes through into the ionosphere. Although its radio frequency is less than the plasma frequency of the medium, this wave can propagate in the ionosphere in the *whistler mode*, as follows.

We take the quasi-longitudinal approximation to the Appleton–Hartree equation (Equation 2.58), put $X/Y_L \gg 1$ to indicate wave frequencies much smaller than the plasma frequency but not too small in relation to the gyrofrequency, and neglect collisions ($Z = 0$). Then,

$$n^2 = -\frac{X}{(1 \pm Y_L)}. \tag{3.35}$$

By taking the negative sign in the denominator we get n^2 positive, and therefore a real refractive index and a propagating wave, provided $Y_L > 1$. Substituting the expressions for X and Y_L,

$$\mu^2 = n^2 = \frac{f_N^2}{f(f_L - f)}. \tag{3.36}$$

The velocity of a pulse is determined by the group refractive index,

$$\mu_g = \frac{d(\mu f)}{df}$$

from Equation 2.55. Whence,

$$\mu_g = \frac{f_N f_L}{2f^{\frac{1}{2}}(f_L - f)^{\frac{3}{2}}}. \tag{3.37}$$

Because μ depends on f, the medium is dispersive; the velocity depends on the radio frequency. At the lower end of the frequency range ($f \ll f_L$),

$$\mu_g \approx f_N / 2f^{\frac{1}{2}} f_L^{\frac{1}{2}}. \tag{3.38}$$

There is some dependence on the direction of propagation with respect to the geomagnetic field, as a result of which signals tend to follow the geomagnetic field direction. Another, and possibly stronger, influence is ducting by field-aligned irregularities in the ionization. Consequently, whistlers not only propagate through the ionosphere but also can follow the geomagnetic field from one hemisphere to the other and may return after reflection there.

A natural source of VLF signals is the lightning flash, which radiates a short radio pulse (or *spheric*) covering a wide spectrum of frequencies. Due to the dispersion, the higher of the frequencies within the audible range (say < 10 kHz) arrive first in the opposite hemisphere (Equation 3.38) and the signal is received as a falling tone; hence the name *whistling atmospheric*, now generally shortened to *whistler*. The association with spherics was established by H. Barkhausen and by T. L. Eckersley some 70 years

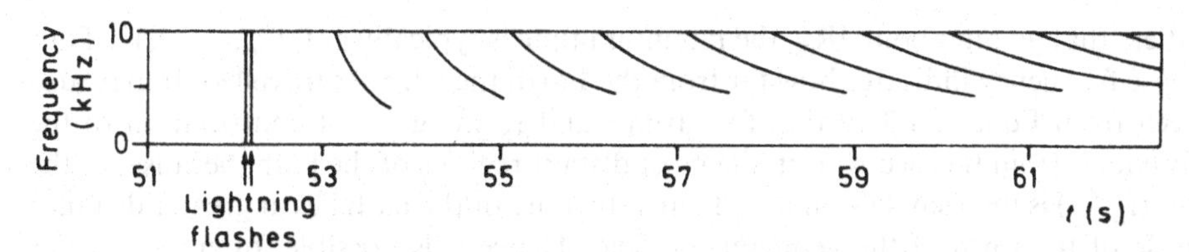

Fig. 3.11 A long train of whistler echoes, in which the dispersion increases at each reflection. (After R. A. Helliwell, *Whistlers and Related Ionospheric Phenomena*. Stanford University Press, 1965)

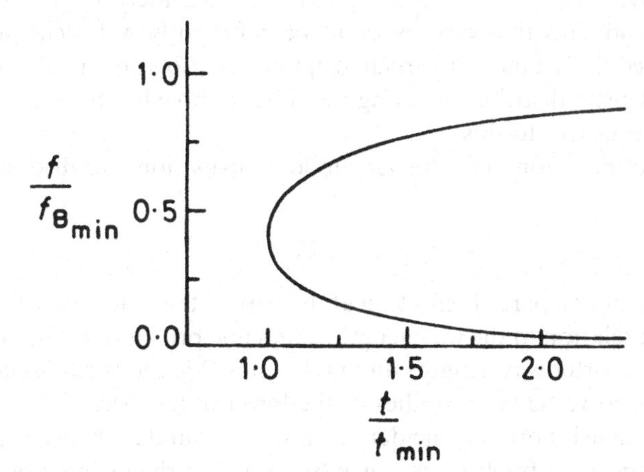

Fig. 3.12 Theoretical normalized dispersion curve for a nose whistler. (D. L. Carpenter and R. L. Smith, *Reviews of Geophysics* **2**, 415, 1964, copyright by the American Geophysical Union)

ago, but that the whistling characteristic is due to inter-hemispheric propagation was not appreciated until the work of L. R. O. Storey in the early 1950s. When several reflections back and forth occur the dispersion accumulates, as illustrated by the frequency–time curves in Figure 3.11. Whistler mode energy is also launched by VLF transmitters used for communications, and it plays a major role in wave–particle interactions in the magnetosphere where the waves are generated by natural processes within the magnetosphere (Chapter 9).

Observations of whistler dispersion provide an important ground-based method for measuring electron densities in the magnetosphere. From Equation 3.37, the total propagation time is

$$t = \frac{1}{c}\int \mu_g \, ds = \frac{1}{2c}\int \frac{f_N f_L}{f^{\frac{1}{2}}(f_L - f)^{\frac{3}{2}}} ds, \tag{3.39}$$

where c is the speed of light and the integration is along the magnetic field-line. For given f_N, which depends on the electron density, the travel time will be large if f is small or if $f \sim f_L$. Between these values of f is one for which the travel time is a minimum. This is characteristic of all whistlers, though not all show the minimum in practice. Those which do are called *nose whistlers*, and the frequency at the minimum is the *nose frequency* (Figure 3.12). The detailed form of the curve depends on the distribution of electron density along the path, and Figure 3.12 is for one particular model. For whistlers that go out about $4R_E$, the minimum time is typically at $f/f_{Bmin} \sim 0.42$. Thus the nose frequency indicates how far from the Earth the whistler travelled. It may also be seen from Equation 3.39 that for given f and f_N the greatest contribution to the delay comes from the electrons in the most distant regions of the path, because (1) the factor $(f_L - f)$ is then smallest and (2) relatively more of the path lies at greater distance because of the form of the geomagnetic field. Hence it is possible, by nose whistler analysis, to determine the electron density in the region where the whistler crossed the equatorial plane. The nose frequency gives the field-line along which the signal travelled and the time delay gives the electron density near the equatorial plane.

In general, the analysis requires that three points be accurately read from the frequency–time curve, and thus the trace needs to be sufficiently well defined. The accuracy can be improved if the time of the related spheric is also known. Although a model for the electron density distribution along the tube of force has to be assumed, the results are not very sensitive to this.

Results and further applications of whistler-mode propagation are discussed in Sections 5.6.2 and 9.2.1.

3.5.5 Partial reflections

We saw that VLF waves suffer a partial reflection at the base of the ionosphere because the boundary is sharp in relation to the wavelength. The same principle applies within the ionosphere at rather shorter wavelengths and makes possible the *partial-reflection technique* which has proved valuable in studies of the lower ionosphere.

If an electromagnetic wave is normally incident on a sharp boundary between media of refractive indices n_1 and n_2, a fraction $(n_1 - n_2)^2/(n_1 + n_2)^2$ of the incident power is reflected (Section 2.5.3). Up to about 100 km altitude the turbulence of the atmosphere produces spatial irregularities from which partial reflections may be received at medium frequencies, approximately over the band 2–6 MHz (wavelengths 0.5–1.5 km). The echoes are weak, amounting to a mere 10^{-3} to 10^{-5} of the amplitude of total reflections, and thus a transmitter of high power (up to 100 kW) and a large antenna array are needed; hence relatively few partial-reflection systems have been constructed.

The essence of the method is as follows. If partial reflections are received from a height h, the wave has twice traversed the ionosphere below h and will have been partly absorbed there. The electron density profile is determined by measuring how the absorption varies with height. Since it would not be possible to secure an adequate calibration in absolute terms, the usual approach is to transmit pulses of circularly-polarized E- and O-waves alternately and to compare the amplitudes of the received echoes. This ratio is given by

$$\frac{A_E}{A_O} = \frac{R_E}{R_O}\exp\left[-2\int_0^h (\kappa_E - \kappa_O)\,dh\right]. \tag{3.40}$$

Here, R_E/R_O is the ratio between the theoretical reflection coefficients at a sharp boundary (generally known as the *Fresnel reflection coefficients*) in a magneto-ionic medium, the refractive indices being taken from the Appleton–Hartree equation (2.58); and the κs are the absorption coefficients for the two modes which also may be expressed in terms of electron density, geomagnetic field, and collision frequency. The experimental method therefore comes down to comparing measured values of A_E/A_O with the theoretical R_E/R_O. For purposes of the computation the geomagnetic field

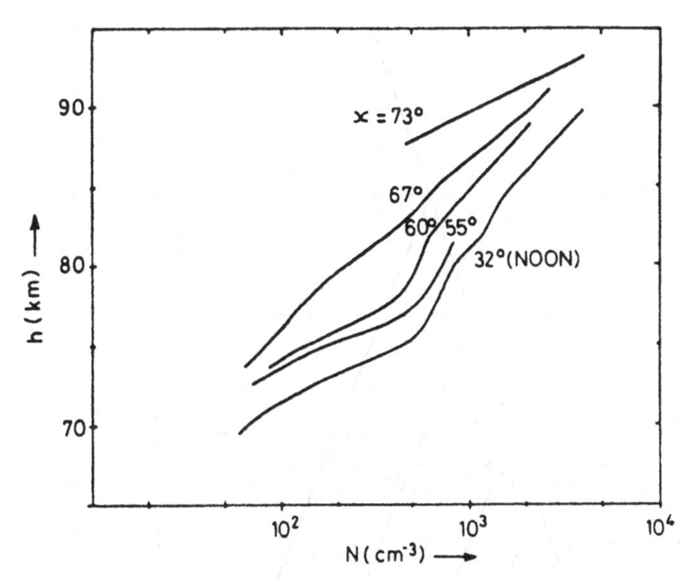

Fig. 3.13 D-region profiles of electron density, measured by the partial-reflection technique at various solar zenith angles. Measurements in Crete on the afternoon of 11 September 1965. (Reprinted with permission from E. V. Thrane *et al.*, *J. Atmos. Terr. Phys.* **30**, 135, copyright (1968) Pergamon Press PLC)

intensity is well known, and the collision frequency depends on the air pressure and can be obtained from a model. If the experiment measures phases as well as amplitudes both collision frequency and electron density can be derived.

The value of the partial-reflection method is that it enables the ionosphere to be observed in a continuous manner over the height range 60–90 km. This is difficult to achieve otherwise, particularly at middle and low latitudes, because the electron density is too small to be detected with an ionosonde and satellites cannot remain in orbit so low down. (Even with a powerful incoherent scatter radar (Section 3.6) long integration is needed.) The example in Figure 3.13 includes electron densities as small as 10^8 m^{-3} (100 cm^{-3}).

3.6 Scatter radar techniques

3.6.1 Volume scattering

A scatter radar detects energy scattered from within a medium when spatial variations or irregularities are present. This is quite different from the action of an ionosonde (which depends on the gradual bending of the ray within a medium of gradually varying refractive index) or of a radar for detecting a hard target (where the echo comes from a discrete surface). The process of volume scattering is essentially one of partial reflection from the refractive index discontinuities due to the irregularities. (Refer to Section 3.5.5 for the theory of partial reflection.)

Whether ionized or not, the atmosphere contains irregularities of many different sizes, and we may imagine that at each boundary a small fraction of the incident energy is scattered in all directions. Signals scattered from irregularities spaced by half a

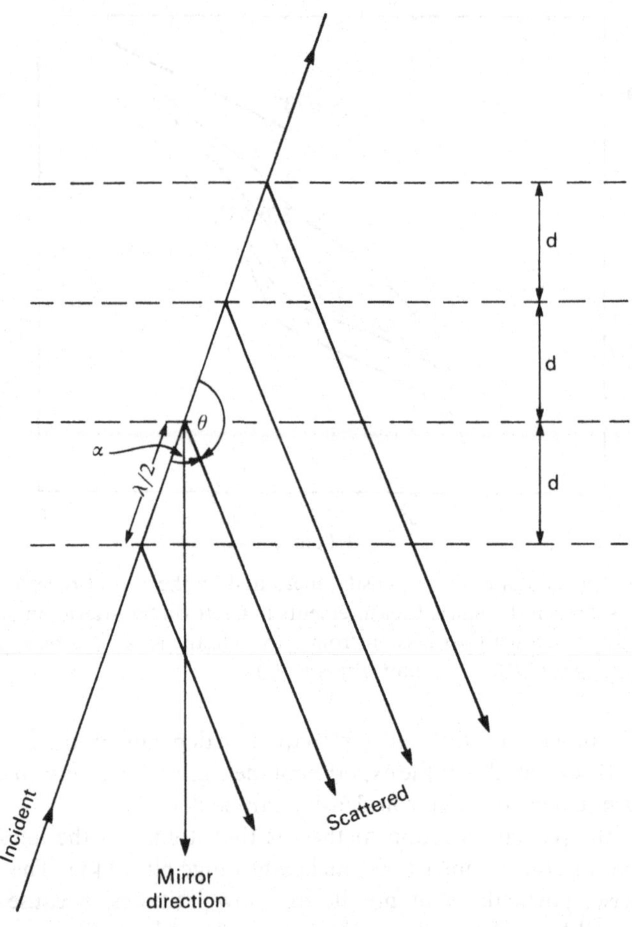

Fig. 3.14 Volume scattering. A signal of wavelength λ, scattered through angle θ, selects spatial period $d = \lambda/2(\sin\theta/2)$.

wavelength will reinforce in the direction back to the radar and thus, despite the weakness of each scattering, they can add up to a signal strong enough to be detected. It is not necessary for the scattering structures to be regularly spaced; a radar of wavelength λ will effectively select the spatial component of period $\lambda/2$, ignoring the others. Scatter in other directions, as determined by the relative locations of transmitter and receiver, will select some other spatial period. Referring to Figure 3.14, the signals from two scattering planes reinforce when their path difference is λ. Therefore the separation between the planes is

$$d = \frac{\lambda}{2\cos(\alpha/2)} = \frac{\lambda}{2\sin(\theta/2)}, \tag{3.41}$$

where θ is the angle between incident and scattered waves. The normal to the scattering plane is called the 'mirror direction'.

Instead of a discrete echo, volume scattering produces a continuous, noise-like return which has to be 'gated' to select the signal from each range. For example, gating

Table 3.5 *Scatter radars*

Type	Scatter mechanism	Region
Coherent	Turbulent irregularities of electron density	E or F regions
Incoherent	Thermal fluctuations in electron density	D, E or F regions
MST	Turbulence of the neutral air	Troposphere to Mesosphere
Lidar	Rayleigh scatter, Raman scatter	Troposphere to Mesosphere

every 10 μs gives a 'range gate' of $3 \times 10^5 \times 10 \times 10^{-6} \times 0.5 = 1.5$ km (since we can assume electromagnetic propagation at the speed of light). The gate at 1 ms thus corresponds to a range of 150 km. The intensity of the return will be related to the strength of the scattering mechanism, and its frequency will show Doppler shift according to the line-of-sight (in the case of backscatter) velocity of the scatterers. The shape of the echo spectrum may also contain valuable information about the medium, such as its temperature.

3.6.2 Coherence

Several types of scatter radar are in use for upper atmosphere research, and Table 3.5 summarizes their salient features. The first three use radio, the fourth light.

The terms 'coherent' and 'incoherent' are somewhat loosely applied here. What coherence really means is the stability of the medium – i.e. how quickly it changes – in relation to the radar's ability to resolve those changes. All fluids are in reality partly coherent, and there will be some 'coherence time' (between zero and infinity) for each situation. Thus, one system might be able to receive echoes from a scattering region so rapidly that little change occurs between successive echoes, while with another there might be a large variation between the arrival of one echo and the next. If the coherence is high, successive echoes (or echo samples, depending on the details of the technique) have the same amplitude and phase, and thus may be added coherently, so improving sensitivity. If incoherent, successive returns are unrelated in amplitude and phase, and may be added only in terms of their powers. This also improves the sensitivity, but to a lesser extent.

By convention, radars detecting the thermal fluctuations of the medium, where the coherence time is relatively short, have been called 'incoherent', while those using structures within the medium, which tend to vary more slowly, are called 'coherent'. But it is worth remembering that in each case the coherence time may be one of the most important measurements to be made, and to do this the radar must in fact bridge the gap between coherence and incoherence.

3.6.3 Coherent scatter radar

A radar designed to receive echoes from physical stuctures within an ionized medium is usually called a *coherent scatter radar*. Somewhat different systems are required for the E and F regions. In the E region they are used to observe the *radio aurora*, the

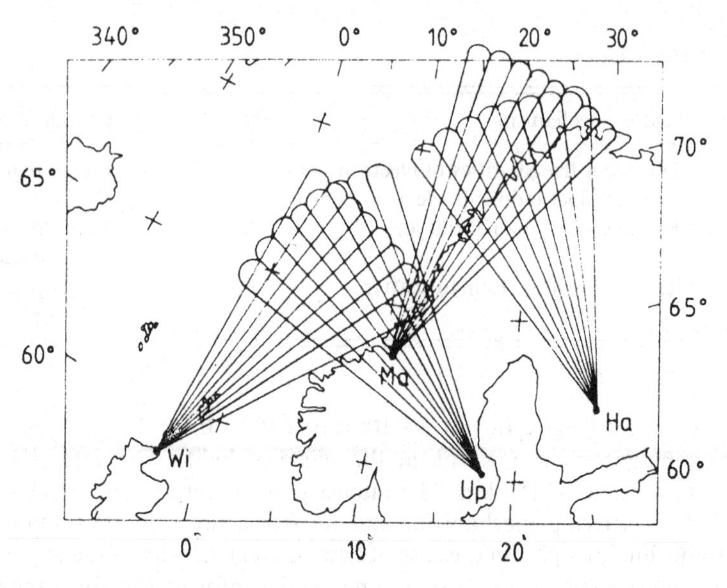

Fig. 3.15 Coherent radars in northern Europe. The German radar operates from Hankasalm (Ha) and Malvik (Ma), the British one from Uppsala (Up) and Wick (Wi). (After E. Nielsen and J. D. Whitehead, *Advances in Space Research* **2**, 131, 1983)

ionization associated with auroral activity. It is now known that these echoes come from irregularities in the auroral ionization, which form when an electrojet (Section 8.4.2) flows in the auroral E region, generating instabilities by the two-stream and the gradient drift mechanisms (Sections 2.8.2 and 2.8.3). The waves thus produced within the medium travel almost perpendicular to the geomagnetic field, an important consequence being that the irregularities are field-aligned and the echoes are strongly aspect sensitive. To obtain the strongest echoes the radar must be in the plane normal to the direction of the geomagnetic field at E-region heights (100–110 km). Since the field is more vertical than horizontal at high latitude, the radar has to be several hundred kilometres away on the equatorward side to meet the condition of normal incidence. In the equatorial regions, similar instabilities form in the equatorial electrojet (Section 7.5.4) and similar techniques can be used to observe them. The observing geometry is different of course, since the geomagnetic field is almost horizontal in the equatorial zone.

One useful configuration has two radars looking into a common viewing area from different directions, as in Figure 3.15. Each radar has an antenna that forms several beams, which in combination with the range gating divide the viewing area into a number of 'boxes'. The radars operate independently but simultaneously, and each determines the line-of-sight velocity from the Doppler shift of the echo. This kind of system provides distributions of both the echo strength and the two-dimensional velocity within the common viewing area. Auroral echoes may be received over a wide range of radio frequency, but these twin systems generally operate in the VHF band at about 70 MHz.

F-region irregularities are also field-aligned because of plasma diffusion along the direction of the geomagnetic field (Section 7.5.1), but it is not possible to satisfy the

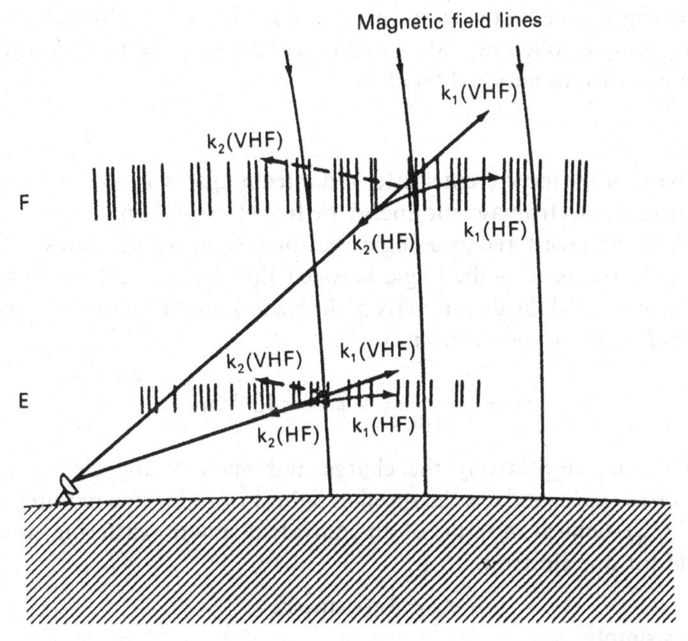

Fig. 3.16 The geometry of VHF and HF scattering by field-aligned irregularities. k_1 and k_2 are the incident and reflected waves. The ionospheric bending at HF makes it possible to observe F-region irregularities at high latitudes. (After R. A. Greenwald *et al.*, *Radio Science* **20**, 63, 1985, copyright by the American Geophysical Union)

normality condition for the high-latitude F region. The solution is to operate in the HF band (e.g. at 8–20 MHz), when the wave is refracted in the ionosphere until it meets the irregularities at right angles (Figure 3.16). The interpretation is not as straightforward as for an E-region coherent radar because of the curved path followed by the HF ray. These F-region systems are particularly valuable for their ability to measure the convection of the ionosphere in the polar regions (see Section 8.1.1).

3.6.4 Incoherent scatter radar – principles

Incoherent scatter radar is a relatively young technique for observing the ionosphere. It is a ground-based, remote sensing technique which over the last 25 years or so has been developed into a powerful and flexible tool able to measure many ionospheric quantities and some of the properties and behaviour of the neutral atmosphere as well. Compared with the ionosonde (Section 3.5.1) – which has also become more sophisticated over the years – it retains two further advantages. It is not restricted to the region below the level of peak electron density (or above it, in the case of a topside sounder) but can observe both sides of the peak simultaneously. Also, because the antenna has to be large relative to the radio wavelength, it produces a narrow beam and achieves far better spatial resolution. The principal disadvantage is that an incoherent scatter radar has to work with a very weak signal. It therefore requires a transmitter of high power, a large antenna, and the most sensitive receiver and sophisticated data processing available, all of which add up to a major facility and considerable expense. The first such facility was constructed in 1960–61. By 1989 the number of operating incoherent scatter radars had grown only to five, and these are generally national or international facilities.

Physical basis

The physical basis of the incoherent scatter radar technique is *Thomson scatter*, the re-radiation of incident electromagnetic energy by free electrons. In 1906 J. J. Thomson showed that the energy scattered by a single electron is $(r_e \sin\psi)^2$ per unit solid angle per unit incident flux, where ψ is the angle between the electric vector of the incident electromagnetic wave and the direction from the electron to the observer. The quantity r_e is the *classical radius of the electron*,

$$r_e = \frac{e^2}{4\pi\varepsilon_0 m_e c^2} = 2.18 \times 10^{-15} \text{ m}, \tag{3.42}$$

where e and m_e are respectively the charge and mass of the electron, ε_0 is the permittivity of free space, and c is the speed of light. For radar one requires the 'radar cross-section' (σ_R), defined in terms of an equivalent sphere scattering equally in all directions, and for one electron,

$$\sigma_R = 4\pi(r_e \sin\psi)^2, \tag{3.43}$$

which becomes simply

$$\sigma_R = 4\pi r_e^2 = 1.0 \times 10^{-28} \text{ m}^2 \tag{3.44}$$

for radar backscatter.

If there are 10^{12} electrons/m^3 in the scattering region, and a volume of 1.4×10^{11} m^3 is sampled at one time (for example, the medium within a cylinder 3 km diameter and 20 km long), the total scatter cross-section is 1.4×10^{-5} m^2, equivalent to a sphere slightly more than 2 mm in radius. In the lower ionosphere there may be only 10^{10} electrons/m^3 observed over a volume of 0.78×10^9 m^3 (e.g. cylinder 1 km diameter, 1 km long), giving a cross-section of 0.78×10^{-9} m^2, equivalent to a sphere of radius 0.016 mm. (See Figure 3.17.)

Spectrum

Thus a radar must be extremely sensitive to detect the incoherent scatter signal. The possibility of doing so was raised by W. E. Gordon in 1958 and achieved in practice by K. L. Bowles in the same year. It was found, however, that the spectrum was much narrower than had been expected – a discrepancy that for once benefited the experimenter. Since the electrons move at thermal velocities in the medium it was expected that the scattered signal would show a spectrum of Doppler shifts with half-width of about 0.71 Δf_e, where Δf_e is the Doppler shift $(2v/\lambda)$ due to an electron approaching the radar at the mean thermal speed: $v = (2kT_e/m_e)^{\frac{1}{2}}$. Thus, one would expect

$$\Delta f_e = \frac{(8kT_e/m_e)^{\frac{1}{2}}}{\lambda}, \tag{3.45}$$

where λ = radio wavelength, T_e = temperature of the electron gas, m_e = electron mass, and k = Boltzmann's constant. Putting in values,

$$\Delta f_e = 11 T_e^{\frac{1}{2}}/\lambda \text{ kHz}, \tag{3.46}$$

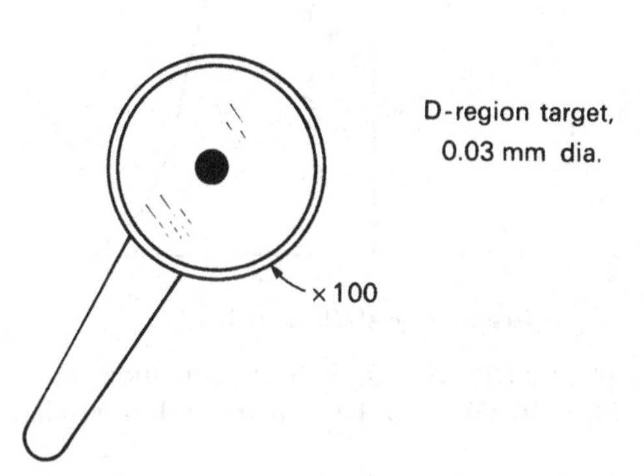

Fig. 3.17 Hard targets equivalent to typical incoherent scatter from the ionosphere.

where λ is in m. If $T_e = 1600$ K, and $\lambda = 75$ cm (f = 400 MHz), $\Delta f_e \sim 600$ kHz. The actual spectrum is some 200 times narrower than this. The narrower spectrum makes the signal easier to detect because a narrower receiver bandwidth can be used, reducing the noise level. Further, there is a major scientific benefit in that the spectrum is more complicated and contains more information about the ionosphere.

The immediate explanation for the narrow incoherent scatter spectrum is that although the scattering comes from the electrons, the motion of the electrons is controlled by the ions. This reduces the spectral width by a factor $(m_i/m_e)^{\frac{1}{2}}$, where m_i and m_e are the ion and electron masses. If the ion is O^+ the ratio is 170. The extent to which the ions control electron motion depends on the radio wavelength in relation to the Debye length (Section 2.3.5):

$$\lambda_D = \left(\frac{\varepsilon_0 k T_e}{Ne^2}\right)^{\frac{1}{2}} = 69\,(T_e/N)^{\frac{1}{2}}\ \text{m}. \qquad (3.47)$$

In the ionosphere λ_D varies from a few millimetres ($T_e \sim 1500$ K, $N \sim 10^{12}$ m^{-3}, as in the F region) to about a centimetre ($T_e \sim 200$ K, $N \sim 10^{10}$ m^{-3}, as in the D region). Each ion may be considered to influence electron motions within a distance λ_D. If the radio wavelength is smaller than the Debye length the motions seen are controlled by the electron temperature and mass. If $\lambda > \lambda_D$, the scattering is due mainly to irregularities of electron density around, and controlled by, the ions.

All practical incoherent scatter radars are designed primarily to observe the ion line, and to a first approximation it can be considered that they see an assembly of irregularities, each the size of a Debye length moving at the ionic thermal speed. Thus,

$$\Delta f_i = \frac{(8kT_i/m_i)^{\frac{1}{2}}}{\lambda} = 65\,T_i^{\frac{1}{2}}/\lambda \text{ Hz} \qquad (3.48)$$

if the ions are O^+. If $T_i = 1600$ K and $\lambda = 75$ cm, $\Delta f_i \sim 3500$ Hz.

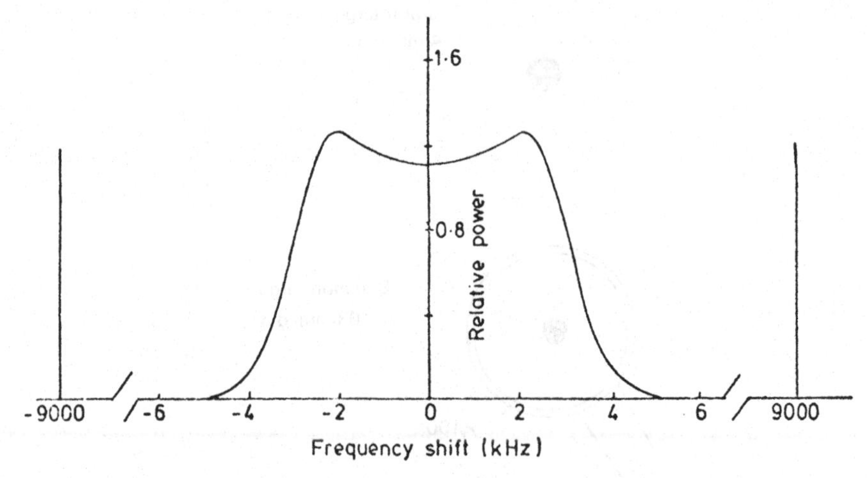

Fig. 3.18 Ion spectrum and plasma (electron) lines for the conditions $N = 10^{12}\ m^{-3}$, $T_e = T_i = 1000$ K, $\lambda = 1$ m, $M_i = 16$. (W. J. G. Beynon and P. J. S. Williams, *Rep. Prog. Phys.* **41**, 909, 1978)

If $\lambda < \lambda_D$, the scattering comes mainly from the electrons not controlled by ions, and this gives the *electron lines*. In the general case both are present, and the ratio of power in the electron and ion lines is

$$\frac{\text{Electron line power}}{\text{Ion line power}} = \alpha^2(1+\alpha^2+T_e/T_i), \tag{3.49}$$

where

$$\alpha = 4\pi\lambda_D/\lambda. \tag{3.50}$$

A typical spectrum is shown in Figure 3.18.

A more exact treatment of the spectrum (which is somewhat complicated, but we can give a simple version) begins by considering the fluctuations within the medium as a spectrum of waves. Ion-acoustic and electron-acoustic waves were introduced in Section 2.7.3. They are analogous to acoustic waves in a neutral gas, but now the electrostatic forces are also included. These waves occur over a wide range of wavelengths and propagate in all directions. Backscatter will occur from those ion-acoustic waves whose wavelength equals half a wavelength of the incident radio wave. Two such waves exist, one moving away from the transmitter and the other towards it, and they impose Doppler shifts of $-F(\Lambda)$ and $+F(\Lambda)$ respectively, where

$$F(\Lambda) = \frac{1}{\Lambda}\left[\frac{kT_i}{m_i}\cdot\left(1+\frac{T_e}{T_i}\right)\right]^{\frac{1}{2}}, \tag{3.51}$$

the expression in square brackets being the velocity, which is close to the speed of the ions. This leads to a spectrum with two lines whose separation is proportional to $(T_i/m_i)^{\frac{1}{2}}$. If $T_e = T_i$ and $\Lambda = \lambda/2$, this formula is the same as that obtained by more elementary considerations as Equation 3.48.

This spectrum, consisting of two discrete lines, bears little resemblence to that shown in Figure 3.18. The reason is that the lines are greatly broadened by the process of *Landau damping*,

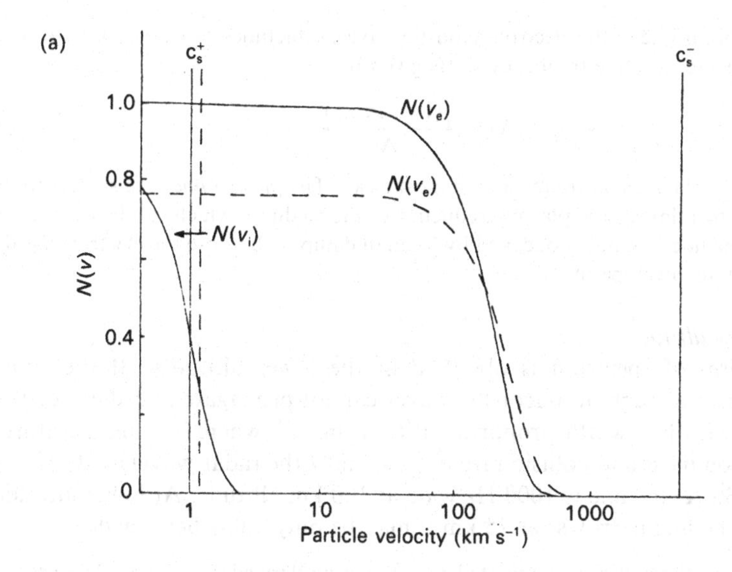

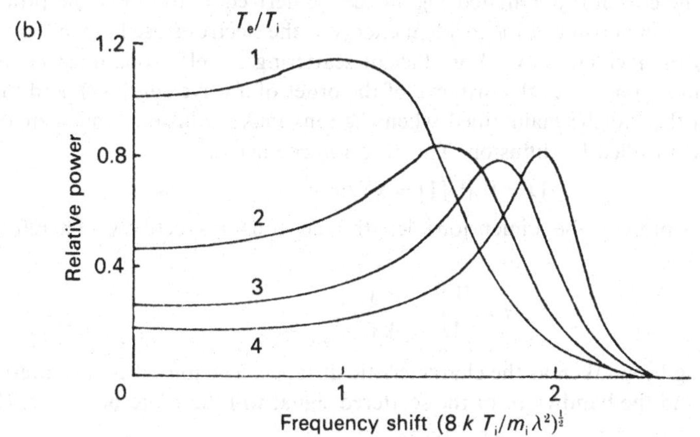

Fig. 3.19 (a) Velocity distribution of thermal electrons $N(v_e)$ and thermal ions $N(v_i)$, and the phase velocity of electron–acoustic (c_s^-) and ion–acoustic (c_s^+) waves. $N(v_e)$ and $N(v_i)$ are normalized so that $N(0) = 1$ in each case. $N = 10^{12}$ m^{-3}, $\lambda = 1$ m, $M_i = 16$, $T_i = 1000$ K, $T_e = 1500$ K (———), $T_e = 2500$ K (---------). (b) Dependence of ion spectrum on T_e/T_i. (After W. J. G. Beynon and P. J. S. Williams, *Rep. Prog. Phys.* **41**, 909, 1978)

which concerns the exchange of energy between a wave and a particle travelling at nearly the same speed. The relevant principle is that energy is transferred from the faster to the slower (Figure 3.19a). Since the ion-acoustic speed comes within the range of thermal ion speeds Landau damping may occur. The rate of energy transfer depends on the gradient of the ion speed distribution around the ion-acoustic speed – i.e. on whether there are more particles taking energy from the wave than giving to it. For example, if the ratio T_e/T_i increases, the ion-acoustic velocity (Equation 3.51) increases and moves into a part of the ion speed distribution that is less steep. The damping effect then decreases and the ion line becomes more sharply double peaked. Hence the shape of the line depends on the ratio T_e/T_i, which may thereby be determined (Figure 3.19b).

Returns also occur due to the electron-acoustic waves, sometimes called *plasma waves*. They produce positive and negative frequency shifts given by

$$F(\Lambda) = \frac{f_p(1+12\pi^2\lambda_D^2)^{\frac{1}{2}}}{\Lambda}, \tag{3.52}$$

almost equal to f_p (the plasma frequency) in most cases. The *plasma lines* are shifted from the radar frequency by (almost) the plasma frequency of the medium, which can be as much as 10 MHz. The plasma lines are not broadened by Landau damping because the electron speeds are not large enough to envelope them.

Spectrum at low altitude

A different form of spectrum is observed at the lower altitudes. If the collision frequency is great enough, ion-acoustic waves cannot propagate and the spectrum is single peaked with line width proportional to $T/m_i\nu_i\lambda^2$, where T = temperature, m_i and ν_i are the ion mass and collision frequency, and λ the radar wavelength. If $\lambda = 70$ cm and T = 230 K, the line is 1000 Hz wide at 100 km altitude. At lower altitudes ν_i increases and the line narrows; at 75 km it may be only a few hertz wide.

The linewidth in the collision dominated region can be derived as follows. The process of scattering is a re-radiation of some of the incident energy as the electrons oscillate in the electric field associated with the incident wave. For efficient scattering the electrons must be able to oscillate in a coherent manner over the distance of the order of a wavelength (λ), and thus we ask over what times this can be maintained when the ions make collisions and their overall motion is therefore controlled by diffusion. The diffusion coefficient

$$D = [L]^2/[T] = kT/m_i\nu_i, \tag{3.53}$$

where [L] and [T] represent the dimensions length and time respectively. Therefore the characteristic time

$$\tau = \frac{[L]^2}{D} \sim \frac{m_i\nu_i}{kT}\lambda^2,$$

the wavelength having been taken as the characteristic distance. The ion motion is interrupted every τ on average, and the bandwidth of the scattered signal will therefore be $\sim 1/\tau$. Hence,

$$\Delta f \sim kT/m_i\nu_i\lambda^2.$$

Such an argument is not, of course, rigorous, but it is dimensionally correct and brings out the essential dependencies. The spectrum has the Lorentz form

$$S(f) = \frac{A}{1+f^2/(\Delta f)^2}. \tag{3.54}$$

It is a much simpler spectrum than that shown in Figure 3.18, and is fully described by its half-width:

$$\Delta f = 16\pi kT/\lambda^2\nu_i m_i \tag{3.55}$$

and any Doppler shift that may be present. The spectrum is broadened if negative ions are present, and Equation 3.55 also assumes that the electron and ion temperatures are equal.

3.6.5 Incoherent scatter radar – measurements

Incoherent scatter radar can provide a wealth of information, though not all of it at the same time, and the art of using the technique is to contrive experiments that give the maximum amount of information relevant to the problem in hand. The following quantities may be measured directly.

Electron density:
– From the total power returned from the scattering region, since the small amounts of power scattered from each electron are added. The scattering cross-section per unit volume is σN, where $\sigma = \sigma_e(1 + T_e/T_i)^{-1}$, assuming all the power is returned in the ion line.
– Also from the frequency offset of the plasma lines. This is potentially a very accurate method, though the plasma lines are not always strong enough to be detected.
– Also from the Faraday Effect (Section 3.5.2) if the polarization can be measured.

The last two are absolute methods. Otherwise the radar has to be calibrated against another instrument such as an ionosonde.

T_e/T_i ratio:
– From the ratio of peaks to dip in the spectrum (in the ion-acoustic region).

T_i/m_i ratio:
– From the separation of the peaks (in the ion-acoustic region). If m_i is known (e.g. from a model) T_i can be found, and thence T_e.

T_e, T_i:
– In the collision region, from the linewidth if m_i is known.

m_i:
– A mixture of two ions also affects the detailed shape of the spectrum (ion-acoustic region), and in principle it is possible to identify the transitions O^+ to H^+, and O^+ to (O_2^+, NO^+).

Plasma velocity:
– Bulk plasma drift in the line of sight produces a Doppler shift of the whole spectrum. By observing in three directions or using a tristatic radar the drift can be determined in three dimensions. In the lower ionosphere the drift will be the same as the motion of the neutral air. In the F region the horizontal plasma drift is due to an electric field (see Section 6.5.4) and hence the electric field can be measured.

Relative motion of ions and electrons:
– This makes the spectrum asymmetrical, one peak being higher than the other.

The following quantities may be derived indirectly.

Conductivities:
– In the E region the electrical conductivity can be calculated from measurements of electron density, using model values of collision frequency.

Neutral wind:
– The same as the plasma velocity in the E region and below. In the F region the wind component along the geomagnetic field moves the plasma along the field. Diffusion is another cause of field-aligned flow, and this component has to be taken into account.

Neutral air temperature:
– Related theoretically to electron and ion temperature, and to electron, O and N concentrations in the F region. Exospheric temperature can be estimated.

Gravity waves:
– Detected from height variation of horizontal velocities and their variations with time.

Ion mass:
– In the collision region, from the height variation of spectral width.

Electron precipitation:
– In the auroral zone, from the electron density profile during a precipitation event, using a model of the neutral atmosphere.

The principal incoherent scatter facilities currently (1989) operating are listed in Table 3.6. Between them they cover a wide latitude range from the magnetic equator to the polar cap, though there is a marked concentration into longitudes 50–80° W. Jicamarca, Arecibo and Millstone all opened in the early 1960s. The installation at Sondre Stromfjord was formerly (until 1982) at Chatanika in Alaska. The EISCAT (European Incoherent SCATter) radar first operated in 1983.

A scatter radar may be monostatic or multistatic. In the monostatic mode the transmitter and receiver are installed at the same site. The transmitter is pulsed, and the length of the pulse determines the height resolution. Returns come from the whole thickness of the ionosphere, but with appropriate time delays. Thus, by gating the echo it is possible to measure over a wide range of heights virtually simultaneously. In bistatic or multistatic mode a continuous wave (or a long pulse) is transmitted, and the echoing region is selected by pointing all the antennas towards a common volume. Only one height can be studied multistatically at a time. The combination of monostatic and tristatic operation, such as is possible with the EISCAT UHF system, is a powerful arrangement since three-dimensional plasma velocity can be measured in the common volume while other measurements are taken monostatically over a range of heights.

3.6.6 MST radar

The techniques of incoherent scatter radar can also be applied to the neutral atmosphere because the turbulence within the homosphere – below about 100 km – is able to scatter radio waves, though by a different mechanism from that which gives incoherent scatter returns from the ionosphere. Scattering from the neutral atmosphere occurs from irregularities of temperature and humidity, the two quantities which determine the radio refractive index of air.

Since the refractive index (n) is almost unity, it is expressed in the form $N = (n-1) \times 10^6$, where N is called the *refractivity* and is given by the formula

$$N = \frac{77.64p}{T} + \frac{37.34 \times 10^4 e}{T^2}, \tag{3.56}$$

Table 3.6 *Incoherent scatter radars*

Facility, location	Geographic coordinates Lat.	Long.	Mode (1)	Power (MW) (2)	Frequency (MHz)	Antenna	L (3)
Jicamarca, Peru	11.6° S	76.5° W	M	6	50	Broadside array, 290 × 290 m, limited steering	1.1
Arecibo, Puerto Rico	18.3° N	66.8° W	M	2	430	Spherical dish, 300 m diameter, limited steering	1.4
Millstone Hill, USA	42.6° N	71.5° W	M	5	440	68 m parabola, fixed 46 m parabola, fully steerable	3.2
EISCAT (UHF)							
Tromsø, Norway	69.6° N	19.2° E	T	2	933	32 m parabolas, all	6.5
Kiruna, Sweden	67.9° N	20.4° E		RX only		fully steerable	5.6
Sodankylä, Finland	67.4° N	26.6° e		RX only			5.2
EISCAT (VHF)							
Tromsø, Norway	69.6° N	19.2° E	M	5	224	Parabolic cylinder, 120 m × 40 m Steerable 30° N to 60° S of zenith	6.5
Sondre Stromfjord, Greenland	67.0° N	51.0° W	M	5	1290	27 m parabola, fully steerable	14

(1) M: Monostatic, T: Tristatic
(2) Nominal peak power, not necessarily used in all experiments
(3) A measure of the geomagnetic latitude, explained in Section 5.7.2 and Equation 5.43

where p is the total air pressure in millibars, e is the partial pressure of water vapour in millibars and T is the absolute temperature in degrees Kelvin. The two terms of Equation 3.56 are often referred to as the 'dry' and 'wet' terms respectively.

Echoes from the neutral air were first observed as unwanted low-altitude returns in incoherent scatter masurements, but now radars have been constructed whose primary work is to investigate the Mesosphere, the Stratosphere and the Troposphere (hence 'MST'). Like incoherent scatter systems, MST radars are built on the grand scale, with powers measured in megawatts and antennas measured in hundreds of metres. An installation which operated in Alaska had an antenna of more than 12 000 dipoles. The best operating frequency seems to be about 50 MHz; with wavelength 6 m such waves are preferentially back-scattered from atmospheric irregularities having a spatial period of 3 m. In the upper mesosphere, above 80 km, MST radar echoes are received from irregularities in the electron density, again arising from the turbulence. (Compare with the partial reflection technique, Section 3.5.5.)

The strongest echoes are obtained from the tropopause up to about 12 km and from

60 to 75 km, depending on the sensitivity of the radar. There are daily and seasonal variations. Scientifically, these studies are aimed at wind measurements, the wind being derived from the Doppler shift of the echo, at wavelike perturbations such as gravity waves (Section 4.3), and at investigations of the scale and nature of the turbulence itself. Considering the inherent difficulty of observing this region of the atmosphere by other means, MST radar technique appears likely to make important contributions to meteorology as well as to upper-atmosphere physics.

3.6.7 Lidar

The ideas of scatter radar may be applied also in the visible region of the electromagnetic spectrum, though the techniques are quite different. Light is scattered in the atmosphere by particles much smaller than the wavelength, a process generally called *Rayleigh scattering*. In clean air the scattering is by individual molecules, and thus the atmospheric density can be measured by observing how the scattered intensity varies with height. From the hydrostatic equation (Section 4.1.2), it is possible to obtain the pressure and temperature as well (Equations 2.7, 4.1) provided the composition is known.

Searchlights were used in the first experiments and heights up to 70 km were explored. Lasers have extended the range to 100 km, and they made lidar possible, in which the laser is pulsed and the height is determined from the time delay as with a radar. With a pulse duration of a few microseconds the height resolution is less than 1 km. The energy transitted in the pulse might be 10 J. For reception, an optical telescope with the maximum light gathering power is required, and the detector will be a photmultiplier which may be cooled to reduce its internal noise.

In addition to air density from the back-scattered intensity, temperature may be deduced from the Doppler broadening of the returned spectrum, and the component of the wind in the line of sight can be found from its mean Doppler shift. Raman scattering, in which the received signal is shifted in frequency by an amount characteristic of the scattering molecule, can be applied to determine the vertical profile of a specific component. Water vapour can be measured in this way to heights of several kilometres. An enhanced return is obtained by transmitting at a specific resonance line, where both absorption and emission are unusually strong. The energy received at the line centre comes from a scattering volume relatively close to the lidar, whereas that at the line edges, where the absorption coefficient is less, comes from further away. An analysis of the shape of the line may thus be interpreted as a profile of concentration against height.

3.7 Ionospheric modification

Although most upper atmosphere studies are conducted by observing natural phenonema, a range of modification experiments has been developed. In most instances the modification achieved is limited in extent and short-lived in duration, but, relying mainly on traditional diagnostic methods, it is possible to observe the medium's response to the introduced perturbation. Thus, a start, at least, has been made to controlled experimentation in geospace. The experiments may involve the injection of cold plasma, an electron beam, heat, electromagnetic waves, or water. We will outline each of these techniques below.

3.7.1 Plasma and beam injection

Plasma clouds

Generally speaking, the natural plasma of the magnetosphere and the solar wind is invisible. Plasma injection was developed to render it visible from the ground, initially so that plasma motions in the magnetosphere or the solar wind could be observed visually. Barium is the element most widely used in twilight conditions. It ionizes quickly in sunlight, eventually (after a few minutes) showing up against the dark sky as a violet or purple cloud due to resonance emissions. The development and motion of the coloured cloud are recorded photographically. Since the ion drift is in response to an electric field (See Sections 2.3.6 and 6.5.4) this has become an important technique for measuring electric fields in the upper atmosphere.

There are two methods of creating the cloud. In one, a canister is filled with barium (Ba) grains and copper oxide (CuO) powder. When the rocket is at the correct altitude, an igniter starts a chemical reaction between Ba and CuO. The heat of the reaction bursts the seal of the canister and the excess barium is expelled. The barium includes a small amount (1–5%) of strontium as impurity, and this serves a useful purpose becomes it remains neutral and may be seen as a blue or green cloud. This serves as a tracer of the neutral air motion (Section 4.2.2). Lithium has also been used as the neutral tracer.

The second method employs high explosive to shoot a jet of barium gas in a selected direction. This is achieved by shaping the explosive charge – the *shaped charge technique*. The jet may travel at more than 10 km/s, and heights of tens of thousands of kilometres can be reached. One application is tracing the geomagnetic field.

Barium cloud measurements have proved most valuable in the ionosphere of the polar cap and the auroral zone, the equatorial region, the magnetosphere including the tail, and the solar wind.

Another aspect of plasma injection experiments is to observe how the clouds expand and change shape. Following a barium/strontium injection, the neutral strontium cloud expands uniformly in all directions, whereas the ionized barium cloud elongates along the geomagnetic field. In addition, the ionized cloud frequently distorts and nearly always develops striations and smaller blobs a few hundred metres across. These structures arise because the addition of plasma modifies the electrical conductivity, which in turn alters the electric field locally and thereby the plasma motion. If a neutral wind is present structuring may be caused by the gradient drift mechanism (Section 2.8.3). Figure 3.20 illustrates one possible sequence based on observations. Since they modify the original situation these effects are undesirable in tracing work, but, on the positive side, they provide a new technique for investigating the electrical structure of the upper atmosphere and the relations between electric fields and currents therein. Experiments in 1984–85 succeeded in creating an artificial comet when barium was released in the solar wind.

Beams

Controlled plasma experiments are also performed by injecting a beam of charged particles from a rocket or a spacecraft into the surrounding medium. A complex range of phenomena have been observed. In some cases the beam retains the character of a stream of single particles, whose motion depends on the electric and magnetic fields

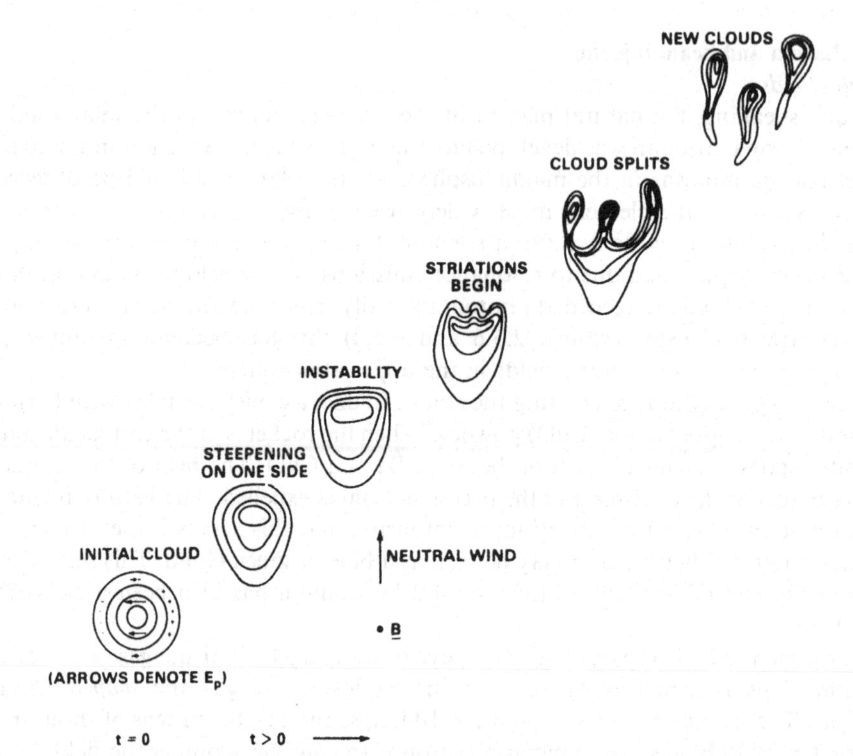

Fig. 3.20 The evolution of a barium cloud under the influence of a neutral wind. (S. T. Zalesak, in *Solar–Terrestrial Physics* (eds. Carovillano and Forbes). Reidel, 1983. Reprinted by permission of Kluwer Academic Publishers)

present, and then it may be applied to magnetic field mapping and to the measurement of electric fields. In other cases the beam experiences collective processes which alter its energy and density, and in these conditions the phenomena of plasma turbulence can be studied. Experiments have been made in various parts of the ionosphere and magnetosphere into widely varying conditions of ambient gas pressure and electron density, using beams from 1 to 10^3 mA and energies from below 100 eV to 40 keV. The principal topics addressed have been the effects of spacecraft charging, the fundamental physics of beam–plasma interactions – particularly the mechanisms that convert particle energy to wave energy – and the application of particle beams as probes or for other purposes in space. We shall not be concerned further with these techniques in this book; the interested reader is referred to the review literature for further information.

3.7.2 Heat injection

Cross modulation

Not all modification experiments are newly devised. The first instance was discovered accidentally many years ago when it was called the *Luxemberg effect*. Radio Luxemberg was a commercial radio station, which broadcast to the United Kingdom at a time when on-shore stations were not licensed to advertise. Since it was in another country, and some distance from its intended listeners, Radio Luxemberg

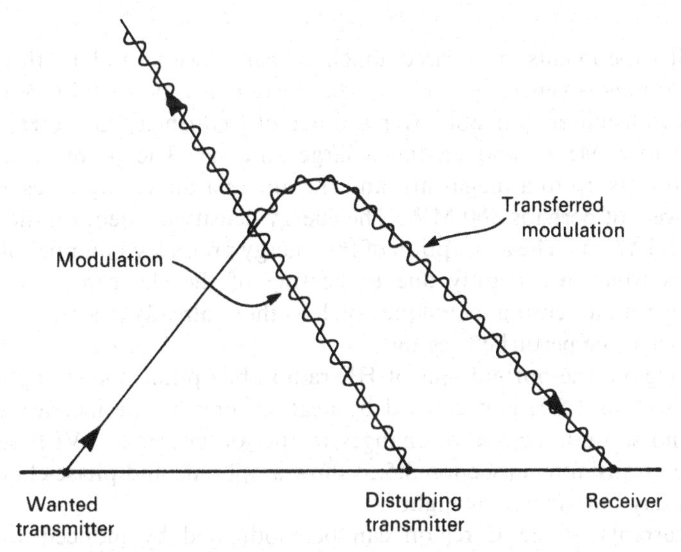

Fig. 3.21 Luxemberg Effect. (After L. G. H. Huxley and J. A. Ratcliffe, *Proc. Inst. Elec. Engrs.* **96-III**, 433, 1949)

was equipped with a high power transmitter and a directional antenna. Thus, though such was not the primary intention, it achieved a high power density in the ionosphere. The Luxemberg effect was discovered in 1933, when the programmes of Radio Luxemberg were heard as a background to other broadcasts in the medium-wave band. The mechanism was that temporary heating of the ionosphere by the high power 'disturbing signal' altered the local collision frequency, changing the absorption of 'wanted' radio signals that were passing through the same volume of the ionosphere and thus imposing on them an additional modulation. (See Figure 3.21.) In the form of the *cross-modulation* technique (also known as *wave interaction*), the Luxemberg effect has been applied in investigations of the lower ionosphere since the mid-1950s. Here, the wanted and disturbing signals are both transmitted in the HF band (probably between 2 and 3 MHz) and both are pulsed, the disturbing signal being pulsed at half the rate of the wanted signal. By adjusting the timing between these two trains of pulses it is possible to arrange that every second wanted pulse, after reflection from the E region, meets an upgoing pulse at a known height. The effect of the disturbing signal may then be seen by comparing the amplitudes of the odd and even pulses. The usual application is to the measurement of electron density as a function of height. The profile of collision frequency has to be obtained from a model, and a theoretical treatment is required for the rate of change of absorption coefficient with temperature. In principle, the collision frequency may be determined from the rate at which the temperature decays to its original value. The disturbing signal is likely to have a power of about 50 kW, and the electron temperature may be enhanced by 20% in the experiments. (Recall, from Equation 3.13, that the absorption depends on N_e multiplied by a function of ν, and that absorption increases with ν if $\nu < \omega$, and decreases if $\nu > \omega$. If $\nu = \omega$ there is no effect.)

Heating

More recent experiments have used much higher powers, and to these the term *ionospheric heating* is generally applied. The 'heater' is a powerful CW (continuous wave) radio transmitter, tuneable over a range of frequencies in the HF band – for example 2.5 to 8 MHz – and having a large antenna. The power will be several hundred kilowatts up to a megawatt, and the antenna directivity gives an 'effective radiated power' of perhaps 300 MW. The energy density incident on the ionosphere can approach 1 W/m^2. The absorption of this energy produces a number of interesting consequences which are mainly due to heating of the electrons. The effects are diagnosed by remote sensing techniques such as those already described, or by flying a rocket through the perturbed region.

In the D region, the normal seat of HF radio absorption, the strength of partial reflections (Section 3.5.5) is decreased by heating, first by the increase of collision frequency and second because of changes to the ion chemistry. VLF signals from ground-based transmitters (Section 3.5.3) show amplitude and phase changes, which are small by day but greater by night.

Electric currents in the E region can be modulated by induced variations of conductivity. (Heating alters the electron temperature, thus the recombination rate, thus the electron density, thus the conductivity, and thus the current – see Section 6.5.5.) The rate of modulation is limited by the natural time constant of the electron density; at a sufficiently low frequency the electrojet can be made to radiate at the modulation frequency.

In the E and F regions the heating wave is absorbed near the level where the wave frequency equals the plasma frequency of the medium. A number of phenomena have been observed, from the generation of small scale irregularities and striations to large 'holes' in the ionosphere, which generally are due to various kinds of plasma instability. Irregularities may develop that can be detected with a coherent radar observing normal to the geomagnetic field (Section 3.6.3), or spread-F (Section 7.5) may occur on ionograms. An exploring HF signal may experience anomalous absorption, which is due to a conversion of electromagnetic energy to other forms – e.g. electrostatic or ion-acoustic waves (Section 2.7.3) – in the density gradients of F-region striations. Incoherent scatter plasma and ion lines (Section 3.6.4) can be stimulated, and secondary electromagnetic waves may be emitted away from the frequency of the primary wave. Airglow emissions (Section 6.4) are excited, for example the oxygen line at 630 nm, as the neutral atoms are excited by collision with the hot electrons.

3.7.3 Wave injection

Another variety of active experiment involves transmitting electromagnetic waves of very low frequency from the ground. Like whistlers, these waves propagate through the ionosphere and into the magnetosphere, where they may stimulate wave–particle interactions (to be discussed in Chapter 9). There is now no question of high power (VLF antennas being inherently inefficient) and these are not heating experiments. Communications transmitters in the VLF band must also cause effects, but the point of this line of work is to learn about the mechanisms of wave–particle interaction through controlled experiments.

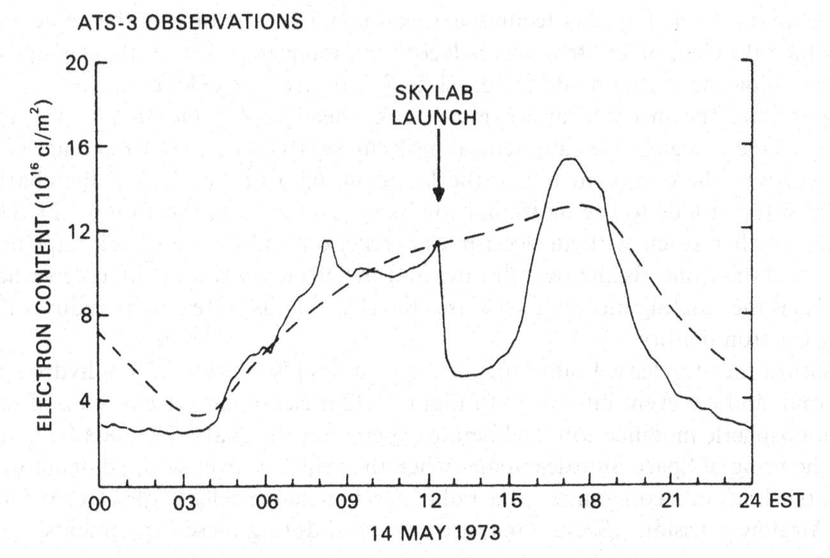

Fig. 3.22 Electron content measurements from Sagamore Hill, Massachusetts, showing the depletion due to the launch of Skylab. The dashed line shows the behaviour expected for that day. (After M. Mendillo *et al.*, *Science* **187**, 343, 1975)

The most notable facility is at Siple station in the Antarctic, at a magnetic latitude (~ 60°) where natural whistlers and emissions are plentiful. The main problem with a VLF station is the size of the antenna which has to be very large because of the long wavelength – for instance, the wavelength is 100 km if the frequency is 3 kHz. The Siple antenna is in the form of a crossed dipole by means of which any desired polarization can be transmitted, and each monopole of the structure is 21 km long. The ice improves the efficiency by raising the antenna 2 km above the true ground, that being the thickness of the ice sheet (which is non-conducting). The transmitter can be tuned between 1 and 6 kHz, and various forms of modulation are available. Although the transmitter is rated at 150 kW, much less than that is radiated because the antenna efficiency is still only 1 to 5%.

By analogy with 'heating', in which the particles of the medium become more disorganized, VLF wave injection can be regarded as a 'cooling' experiment, since it tends to produce greater organization in the particle motions, reducing the entropy. Some results will be mentioned in Section 9.4.3.

3.7.4 Water and hydrogen injection

As the ionosphere can be enhanced by injecting extra plasma, so it can also be depleted by introducing material that speeds up the rate at which the natural plasma recombines. The result may be called an *ionospheric hole*, a term intended to convey not that the ionosphere is entirely removed but that the depletion is localized in extent.

The effect of water and hydrogen injection was first seen in electron content measurements during the launch of the Skylab orbiting laboratory in 1973 (Figure 3.22). The Saturn 5 rocket which carried Skylab passed some 200 km from the line of propagation between a geosynchronous satellite and a ground station monitoring the electron content by the Faraday technique (Section 3.5.2). The time of the depletion, during which the electron content was halved, corresponded well with the passage of the rocket, allowance being made for the diffusion time of the exhaust gases.

To appreciate the mechanism we must look ahead to Chapter 6 and the ion chemistry of the F region. Certain neutral molecules (H_2O, H_2, CO_2, SF_6) react very rapidly with O^+, the dominant ion of the F region, with the result that the charge moves from the atomic to the molecular ion as in reaction 6.44 (Section 6.3.2). The molecular ion then reacts with an electron as in reaction 6.43. In the F region the first of these reactions controls the overall rate, and the efficiency of that first step when H_2O or H_2 is the reacting molecule increases the electron loss rate and so reduces the resulting electron density.

The Saturn rocket released more than a ton of gases (70% water, 30% hydrogen) each second, and the event illustrated in Figure 3.22 is clearly an extreme case of this type of ionospheric modification. Deliberate experiments have also been made. Some require the firing of Space Shuttle engines when the vehicle is over selected diagnostic facilities (such as an incoherent scatter radar), and some the release of water from a rocket. Airglow emissions (Section 6.4) are enhanced during these experiments.

Further reading

E. Burgess and D. Torr. *Into the Thermosphere – the Atmospheric Explorers*. NASA (Document NASA SP-490), NASA Scientific and Technical Information Division, Washington DC (1987).

L. L. Lazutin. *X-ray Emission of Auroral Electrons and Magnetospheric Dynamics*. Springer-Verlag, Berlin and New York (1986). Chapter 1, Balloon experiment technique.

J. A. Ratcliffe. *An Introduction to the Ionosphere and the Magnetosphere*. Cambridge University Press, Cambridge, England (1972); specifically Chapter 9 on exploring the ionosphere with radio waves and Chapter 10 on measurements using space vehicles.

R. D. Hunsucker. *Radio Techniques for Probing the Terrestrial Ionosphere*. Springer-Verlag, Heidelberg (1991).

J. A. Ratcliffe. *The Magneto-ionic Theory and its Applications to the Ionosphere*. Cambridge University Press, Cambridge, England (1959).

K. Davies. *Ionospheric Radio*. Peregrinus, London (1990). Chapter 4 on radio sounding, Chapter 8 on earth-space propagation, and Chapter 14 on ionospheric modification.

J. M. Kelso. *Radio Ray Propagation in the Ionosphere*. McGraw-Hill, New York (1964).

A. Giraud and M. Petit. *Ionospheric Techniques and Phenomena*. Reidel, Dordrecht, Holland (1978); particularly Part 2, The techniques of ionospheric measurements.

R. A. Helliwell. *Whistlers and Related Ionospheric Phenomena*. Stanford University Press, Stanford, California (1976).

A. S. Jursa (ed.). *Handbook of Geophysics and the Space Environment*. Air Force Geophysics Laboratory, US Air Force. National Technical Information Service, Springfield, Virginia (1985). Section 10.1 on measuring techniques, Section 10.2 on long-wave propagation, and Section 10.10 on modification experiments.

R. A. Greenwald. New tools for magnetospheric research. *Reviews of Geophysics* **21**, 434 (1983).

Special issue on Dynamics Explorer. *Space Science Instrumentation* **5**, No 4 (1981).

R. A. Hoffman. The magnetosphere, ionosphere, and atmosphere as a system: Dynamics Explorer 5 years later. *Reviews of Geophysics* **26**, 209 (1988).

A. P. Willmore. Exploration of the ionosphere from satellites. *J. Atmos. Terr. Phys.* **36**, 2255 (1974).

G. Pfotzer. History of the use of balloons in scientific experiments. *Space Science Reviews* **13**, 199 (1972).

N. F. Ness. Magnetometers for space research. *Space Science Reviews* **11**, 459 (1970).

A. Pedersen, C. A. Cattell, C.-G. Falthammar, V. Formisano, P.-A. Lindqvist, F. Mozer and R. Torbert. Quasistatic electric field measurements with spherical double probes on the GEOS and ISEE satellites. *Space Science Reviews*. **37**, 269 (1984).

K. Bullough, A. R. W. Hughes and T. R. Kaiser. VLF observations of Aerial III. *Proc. Roy. Soc.* **A311**, 563 (1969).

J. V. Evans. Some post-war developments in ground-based radiowave sounding of the ionosphere. *J. Atmos. Terr. Phys.* **36**, 2183 (1974).

URSI Handbook of Ionogram Interpretation and Reduction. Report UAG-23, World Data Center A (Nov 1972). Also, Report UAG-23A (July 1978).

K. Davies. Recent progress in satellite radio beacon studies with particular emphasis on the ATS-6 radio beacon experiment. *Space Science Reviews* **25**, 357 (1980).

Special issue on Topside Sounding. *Proc.IEEE* **59**, 859 (1969).

A. D. M. Walker. The theory of whistler propagation. *Reviews of Geophysics and Space Physics* **14**, 629 (1976).

S. Kato, T. Ogawa, T. Tsuda, T. Sato, I. Kimura and S. Fukao. The middle and upper atmosphere radar: first results using a partial system. *Radio Science* **19**, 1475 (1984).

Special section on Technical and Scientific Aspects of MST Radar. *Radio Science* **25**, 475- (1990).

W. J. G. Beynon and P. J. S. Williams. Incoherent scatter of radio waves from the ionosphere. *Reports on Progress in Physics* **41**, 909 (1978).

Special issue on Active Experiments in Space Plasmas. *J. Atmos. Terr. Phys.* **47**(12), Dec 1985.

G. Haerendel. Active plasma experiments. Chapter 11 in *The Solar Wind and the Earth* (eds. S-I. Akasofu and Y. Kamide), Terra Scientific Publishing Company, Tokyo, 1987.

T. N. Davies. Chemical releases in the ionosphere. *Reports on Progress in Physics* **42**, 1565 (1979).

J. R. Winckler. The application of artificial electron beams to magnetospheric research. *Reviews of Geophysics*. **18**, 659 (1980).

P.A. Bernhardt, L. M. Duncan and C. A. Tepley. Artificial airglow excited by high-power radio waves. *Science* **242**, 1023 (1988).

R. A. Helliwell. VLF wave stimulation experiments in the magnetosphere from Siple Station, Antarctica. *Reviews of Geophysics*. **26**, 551 (1988).

M. Mendillo. Ionospheric holes: a review of theory and experiments. *Advances in Space Research*. **8**, 51 (1988).

4

The neutral atmosphere

...this most excellent canopy, the air, look you, this brave o'erhanging firmament, this majestical roof fretted with golden fire, why, it appears no other thing to me than a foul and pestilent congregation of vapours.

W. Shakespeare, *Hamlet*, Act II Scene (ii)

4.1 Vertical structure

4.1.1 Nomenclature of atmospheric vertical structure

The static atmosphere is described by the four properties, pressure (P), density (ρ), temperature (T) and composition. Between them these properties determine much of the atmosphere's behaviour. They are not independent, being related by the universal gas law which may be written in various forms (Equations 2.5–2.7). For our purposes the form

$$P = nkT, \qquad \text{(Equation 2.7)}$$

where n is the number of molecules per unit volume, is particularly useful. The quantity 'n' is properly called the *concentration* or the *number density*, but *density* alone is often used when the sense is clear.

The regions of the neutral atmosphere are named according to various schemes based in particular on the variations with height of the temperature, the composition, and the state of mixing. Figure 4.1 illustrates the most commonly used terms. The primary classification is according to the temperature gradient. In this system the regions are 'spheres' and the boundaries are 'pauses'. Thus the *troposphere*, in which the temperature falls off at 10 K/km or less, is bounded by the *tropopause* at a height of 10–12 km. The *stratosphere* above was originally thought to be isothermal, but in fact is a region where the temperature increases with height. A maximum, due to heating by ultra-violet absorption in ozone, appears at about 50 km and this is the *stratopause*. The temperature again decreases with height in the *mesosphere* (or *middle atmosphere*) to a minimum at the *mesopause* at 80–85 km. With a temperature around 180 K this is the coldest part of the atmosphere. Above the mesopause, heating by solar ultra-violet radiation ensures that the temperature gradient remains positive, and this is the *thermosphere*. The thermospheric temperature eventually becomes almost constant at a value that varies with time but is generally over 1000 K. This is the hottest part of the atmosphere.

Though the classification by temperature is generally the most useful, others based on the state of mixing, the composition or the state of ionization are indispensable in

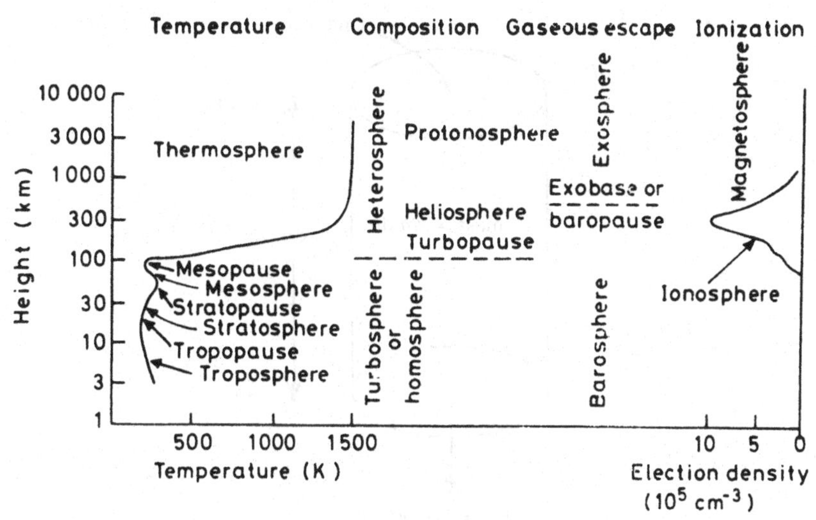

Fig. 4.1 Nomenclature of the upper atmosphere based on temperature, composition, mixing and ionization.

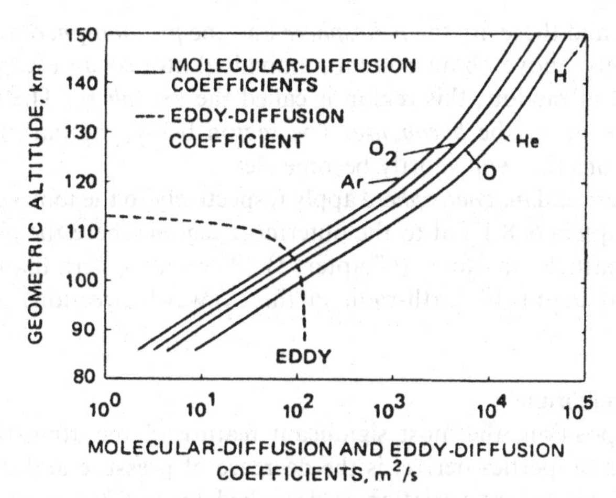

Fig. 4.2 Comparison of coefficients for molecular and eddy diffusion. (A. D. Richmond, in *Solar–Terrestrial Physics* (eds. Carovillano and Forbes). Reidel, 1983. Reprinted by permission of Kluwer Academic Publishers)

context. The lower part of the atmosphere is well mixed, with a composition much like that at sea level except for minor components. This is the *turbosphere* or *homosphere*. In the upper region, essentially the thermosphere, mixing is inhibited by the positive temperature gradient, and here, in the *heterosphere*, the various components may separate under gravity and so the composition varies with altitude. The boundary between the two regions, which occurs at about 100 km, is the *turbopause*. Above the turbopause the gases can separate by gaseous diffusion more rapidly than they are mixed by turbulence – compare the relative magnitudes of the coefficients of molecular diffusion and eddy diffusion in Figure 4.2.

Within the heterosphere provision is made for regions where the dominant gas is

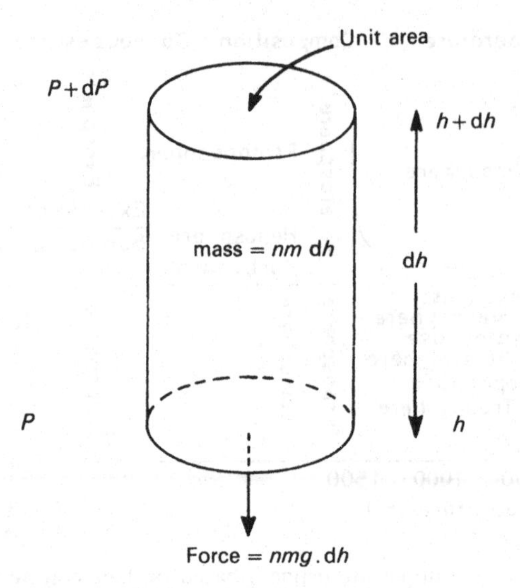

Fig. 4.3 Forces acting on an atmospheric parcel of gas.

helium or hydrogen, and these are the *heliosphere* and the *protonosphere* respectively. From the higher levels, above about 600 km, individual atoms can escape from the Earth's gravitational attraction; this region is called the *exosphere*. The base of the exosphere is the *exobase* or the *baropause*. The region below the baropause is the *barosphere* – for reasons that will shortly become clear.

The terms *ionosphere* and *magnetosphere* apply respectively to the ionized regions of the atmosphere (Chapters 6–8) and to the outermost region where the geomagnetic field controls the particle motions (Chapter 5). The outer termination of the geomagnetic field (at about 10 earth-radii in the sunward direction) is the *magnetopause*.

4.1.2 Hydrostatic equilibrium

Apart from its composition, the most significant feature of the atmosphere, from which so many of its properties derive, is the decrease of pressure and density with increasing altitude. This height variation is described by the *hydrostatic equation*, sometimes called the *barometric equation*, which is derived as follows.

If a gas contains n molecules (or atoms if the gas be atomic) per unit volume, each of mass m, a cylinder of unit cross-section and height dh contains total mass nm dh (Figure 4.3). Due to gravity (g) the cylinder experiences a downward force nmg dh, which at equilibrium is balanced by the pressure difference between the lower and upper faces of the cylinder (dP). Thus

$$(P+dP)-P = -nmg\,dh$$

and hence

$$dP/dh = -nmg.$$

But, by the gas law (Equation 2.7), P = nkT. Hence

$$\frac{1}{P}\cdot\frac{dP}{dh} = -\frac{mg}{kT} = -\frac{1}{H}, \tag{4.1}$$

where the expression kT/mg has the dimension of length and is defined as the *scale height*, H. If H is constant, Equation 4.1 is easily integrated to

$$P = P_0 \exp(-h/H), \tag{4.2}$$

where P_0 is the pressure at height $h = 0$. As the vertical distance in which the pressure changes by a factor of e (= 2.718), the scale height is a critical property of any atmosphere. H is greater if the gas is hotter or lighter.

Writing Equation 4.1 in terms of n,

$$\frac{dn}{n} + \frac{dT}{T} = -\frac{dh}{H}, \tag{4.3}$$

and if both H and T are constant with height,

$$n = n_0 \exp(-h/H), \tag{4.4}$$

where n_0 is the number density at $h = 0$.

Even if g, T, and m change with height (which in reality they do), H has a local value defined by Equation 4.1 in terms of the relative variation of pressure with height. In the terrestrial atmosphere the scale height increases with height from about 5 km at height 80 km to 70–80 km at 500 km, as the mean molecular mass decreases and the temperature increases. A useful approximate formula, accurate to about 10% over 200–900 km is

$$H\,(\text{km}) \sim T(\text{K})/M \text{ (mass units)}. \tag{4.5}$$

Instead of Equation 4.2 it is often convenient to write

$$\frac{P}{P_0} = \exp\left(-\frac{(h-h_0)}{H}\right) = e^{-z}, \tag{4.6}$$

where $P = P_0$ at the height $h = h_0$, and z is the *reduced height* defined by

$$z = (h - h_0)/H. \tag{4.7}$$

The hydrostatic equation can also be written in terms of density (ρ). If T, g, and m are constant over at least one scale height the equation is essentially the same in terms of P, ρ and n, since $n/n_0 = \rho/\rho_0 = P/P_0$. The ratio k/m can also be replaced by R/M, where R is the gas constant and M the molecular weight.

Whatever the height distribution of the atmospheric gas, its pressure P_0 at height h_0 is just the weight of gas above h_0 in a column of unit cross-section. Hence

$$P_0 = N_T mg = n_0 k T_0, \tag{4.8}$$

where N_T is the total number of molecules in the column above h_0, and n_0 and T_0 are the concentration and the temperature at h_0. Therefore,

$$N_T = n_0 k T_0/mg = n_0 H_0, \tag{4.9}$$

H_0 being the scale height at h_0. Equation 4.9 says that if all the atmosphere above h_0 were compressed to density n_0 (that already applying at h_0) then it would occupy a column extending just one scale height. Note also that the total mass of the atmosphere above unit area of the Earth's surface is equal to the surface pressure divided by g.

The acceleration due to gravity varies with altitude as $g(h) \propto 1/(R_E + h)^2$. The effect of changing gravity may be taken into account by defining a *geopotential height*

$$h^* = R_E h/(R_E + h). \tag{4.10}$$

A molecule at height h over a spherical Earth has the same potential energy as one at height h* over a flat Earth with gravitational acceleration g(0).

Within the homosphere, where the atmosphere is well mixed, the mean molecular mass is used to determine the scale height and the height variation of pressure. In the heterosphere, the partial pressure of each constituent is determined by the molecular mass of that species. Each species takes up an individual distribution and the total pressure is the sum of the partial pressures in accordance with Dalton's law.

4.1.3 The exosphere

Definition

In deriving the hydrostatic equation we treated the atmospheric gas as a compressible fluid whose temperature, pressure and density are related by the gas law. This is only valid if there are sufficient collisions between the gas molecules for a Maxwellian velocity distribution to be established. With increasing height the pressure and the collision frequency decrease, and at about 600 km the distance travelled by a typical molecule between collisions, the mean free path, becomes equal to the scale height. At this level and above we have to regard the atmosphere in a different way, not as a single fluid but as an assembly of individual molecules or atoms, each following its own trajectory in the Earth's gravitational field. This is the region called the *exosphere*.

While the hydrostatic equation is strictly valid only in the barosphere, it has been shown that the same form may still be used if the velocity distribution is Maxwellian. This is true to some degree in the exosphere, and the use of the hydrostatic equation is commonly extended to 1500–2000 km, at least as an approximation.

However this liberty may not be taken if there is significant loss of gas from the atmosphere, since more of the faster molecules will be lost and the velocity distribution will be altered thereby. Naturally, the lighter gases, helium and hydrogen, are affected most. This brings us to the question of the escape of gas from a planetary atmosphere.

Gaseous escape

Neglecting collisions, a particle of mass m and vertical velocity v_z will escape from the gravitational field if $mv_z^2/2 > mgr$, g being the gravitational acceleration and r the distance of the particle from the centre of the planet; i.e. if the particle's kinetic energy exceeds its gravitational potential energy. Thus, the escape velocity is given by

$$v_e^2 = 2gr. \tag{4.11}$$

At the Earth's surface the escape velocity is 11.2 km/s. By kinetic theory (Section 2.2.2), the r.m.s. thermal speed of gas molecules depends on their mass and temperature, and for speeds in one direction, i.e. vertical:

$$m\overline{v_z^2}/2 = 3kT/2. \tag{4.12}$$

Thus, corresponding to an escape velocity there can be defined an *escape temperature*, having the values given in Table 4.1 for common atmospheric gases. At 1000–2000 K, exospheric temperatures are much less than the escape temperature, but loss of gas, if any, will be mainly at the high speed end of the velocity distribution. In fact, loss is insignificant for O, slight for He, but significant for H. Detailed computations show

Table 4.1 *Escape temperatures*

Atom	0	He	H
Escape Temp (K)	84 000	21 000	5 200

that the resulting vertical distribution of H departs significantly from the hydrostatic at distances more than one Earth radius above the surface, but for He the departure is small.

4.1.4 Heat balance and vertical temperature profile

The atmosphere's temperature profile results from the balance between sources of heat, loss processes and transport mechanisms. The total picture is complicated, but the main points are as follows.

Heat production and loss

The troposphere is heated by convection from the hot ground, but in the upper atmosphere there are four sources of heat:

(a) Absorption of solar ultra-violet and X-ray radiation, causing photodissociation, ionization and consequent reactions that liberate heat.
(b) Energetic charged particles entering the upper atmosphere from the magnetosphere (Section 8.3).
(c) Joule heating by ionospheric electric currents (Section 8.4).
(d) Dissipation of tidal motions and gravity waves (Sections 4.2 and 4.3) by turbulence and molecular viscosity.

Generally speaking, the first source is the most important, though (b) and (c) are also important at high latitude. The contribution from dissipated motions is uncertain in magnitude but is estimated at about 0.7 mW/m^2 from gravity waves and about the same again from tides. Source (a) provides much more than this. Most radiation of wavelength less than 180 nm is absorbed by N_2, O_2 and O. Some energy is re-radiated, but about half, on average, goes into local heating.

Basically, this heating occurs because a photon absorbed to dissociate or ionize a molecule or an atom generally has more energy than that needed for the reaction, and the excess appears as kinetic energy of the reaction products. A newly created photoelectron, for example, may have between 1 eV and 100 eV of kinetic energy, which subsequently becomes distributed throughout the medium by interactions between the particles. This secondary excitation can be optical, electronic, vibrational, rotational, or by elastic collision, depending on the energy. This last process distributes energy less than 2 eV, and since it operates mainly between electrons they remain hotter than the ions. It can be assumed that in the ionosphere the rate of heating in a given region is proportional to the ionization rate.

The temperature maximum at the stratopause is due to the absorption of 200–300 nm (2000–3000 Å) radiation by ozone (O_3) over the height range 20 to 50 km. Some

18 W/m^2 is absorbed in the ozone layer. Molecular oxygen (O_2), which is relatively abundant up to 95 km, absorbs radiation between 102.7 and 175 nm, much of this energy being used to dissociate O_2 to atomic oxygen (O). This contribution amounts to some 30 mW/m^2. Radiation of wavelengths shorter than 102.7 nm, which is the ionization limit for O_2 (see Table 6.2), is absorbed to ionize the major atmospheric gases O_2, O, and N_2 over the approximate height range 95–250 km, and this is what heats the thermosphere. The amount absorbed is only about 3 mW/m^2 at solar minimum (more at solar maximum), but a small heat input may raise the temperature considerably at great height because the air density is small. Indeed, the heating rate and the specific heat are both proportional to the gas concentration at the greater altitudes, and in this case the rate of temperature increase is independent of height.

At high latitude, heating associated with the aurora – items (b) and (c) – is important during storms. Joule heating by electric currents is greatest at 115–130 km. Auroral electrons heat the atmosphere mainly between 100 and 130 km.

The principal mechanism of heat loss from the upper atmosphere is radiation, particularly in the infra-red. The oxygen emission at 63 μm is important, as are spectral bands of the radical OH and visible airglow (Section 6.4) from oxygen and nitrogen. The mesosphere is cooled by radiation from CO_2 at 15 μm and from ozone at 9.6 μm, though during the long days of the polar summer the net effect can be heating instead of cooling.

Heat transport

The third significant factor in the heat balance of the upper atmosphere is the movement of heat from one level to another by transport processes. Conduction, convection and radiation all come into play:

(a) Radiation is the most efficient process at the lowest levels. The atmosphere is in radiative equilibrium between 30 and 90 km.
(b) Molecular conduction is more efficient in the thermosphere (above 150 km); here the thermal conductivity is large because of the low pressure and the presence of free electrons. Its large thermal conductivity ensures that the thermosphere is isothermal above 300 or 400 km, though the temperature varies greatly with time. At the base of the thermosphere heat is conducted down towards the mesosphere.
(c) *Eddy diffusion*, or convection (Section 2.2.5) is more efficient than conduction below the turbopause. Thus, below about 100 km eddy diffusion carries heat from the thermosphere into the mesosphere. This is a major loss of heat from the thermosphere but a minor source for the mesosphere.
(d) Transport by large-scale winds can affect the horizonal distribution in the thermosphere.
(e) *Chemical transport* of heat occurs when an ionized or dissociated species is created in one place and recombines in another. The mesosphere is heated in part by the recombination of atomic oxygen created at a higher level.

The balance between these various processes produces an atmosphere with two hot regions, one at the stratopause and one in the thermosphere. There is a strong daily variation of thermospheric temperature (Figure 4.4) and the sunspot cycle also has a marked effect (Figure 4.5).

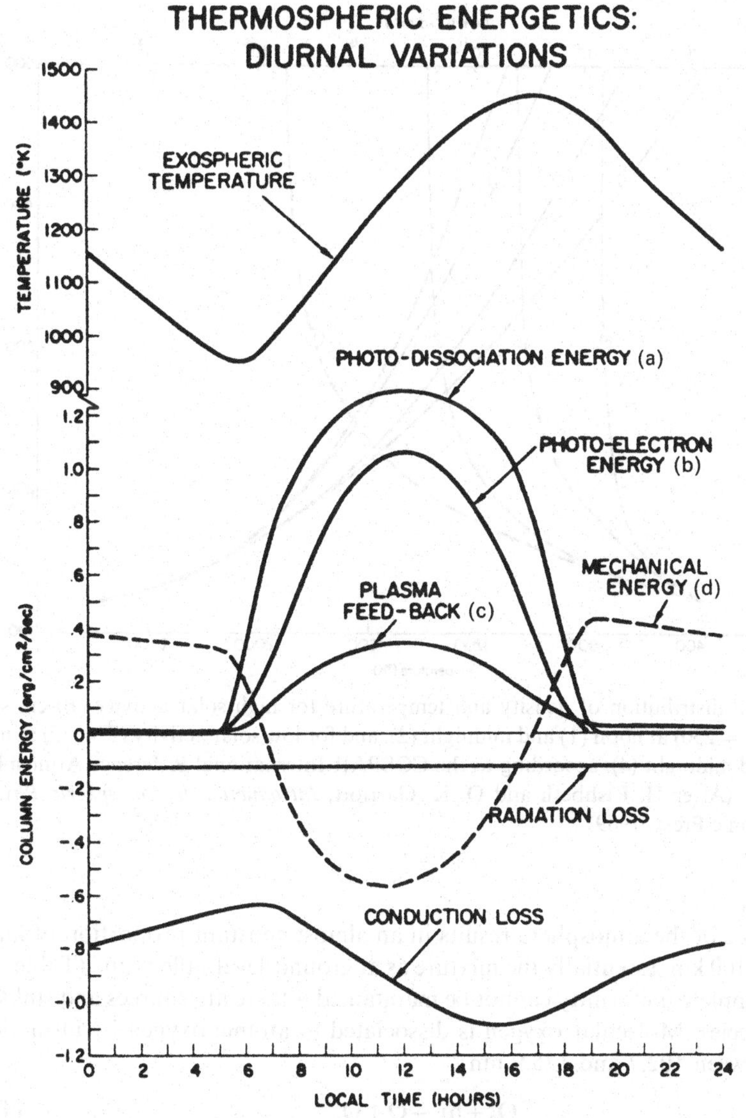

Fig. 4.4 Diurnal variation of exospheric temperature and the main causes of heating and loss (all summed above 120 km). Heating is due to molecular dissociation (a), hot electrons resulting from ionization (b), collisions and chemical reactions (c), and atmospheric expansion and contraction (d). Loss by radiation is due to infra-red emissions. (After S. J. Bauer, *Physics of Planetary Ionospheres*. Springer-Verlag, 1973)

4.1.5 Composition

The upper atmosphere is composed of various major and minor species. The former are the familiar oxygen and nitrogen in molecular or atomic forms, or helium and hydrogen at the greater heights. The minor constituents may be present as no more than mere traces, but they can exert an influence far beyond their numbers. They include ozone, oxides of nitrogen, alkali metals, carbon dioxide, and water.

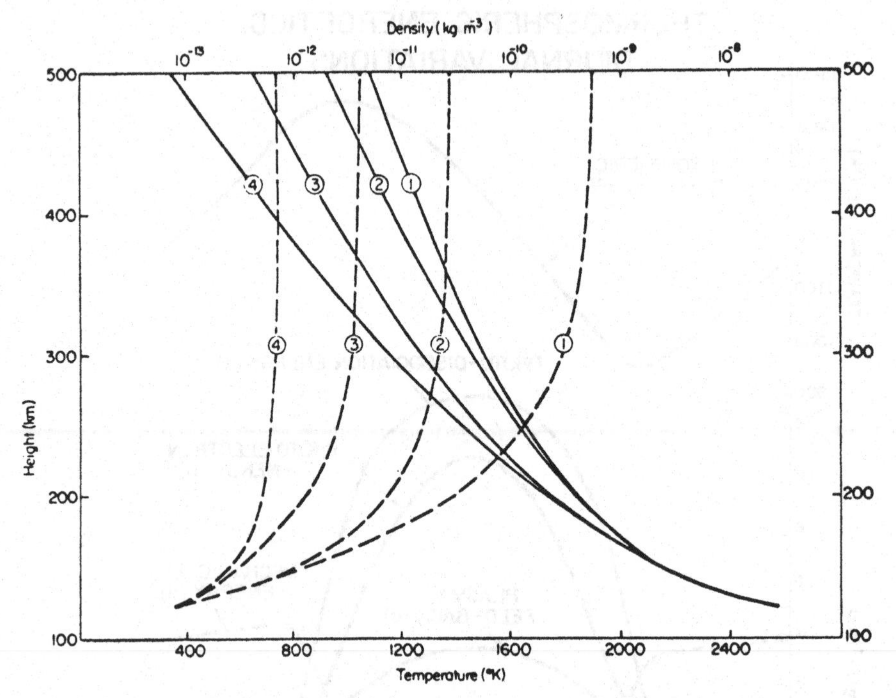

Fig. 4.5 Vertical distribution of density and temperature for high solar activity (10-cm solar flux, F = 250) at noon (1) and midnight (2), and for low solar activity (F = 75) at noon (3) and midnight (4), according to the COSPAR International Reference Atmosphere (1965). (After H. Rishbeth and O. K. Garriott, *Introduction to Ionospheric Physics*. Academic Press, 1969)

Major species

The turbulence of the atmosphere results in an almost constant proportion of major species up to 100 km, essentially the mixture as at ground-level called 'air' (Table 4.2). However, complete uniformity cannot be maintained if there are sources and sinks for particular species. Molecular oxygen is dissociated to atomic oxygen by ultra-violet radiation between 102.7 and 175.9 nm:

$$O_2 + h\nu \rightarrow O + O, \tag{4.13}$$

where $h\nu$ is a quantum of radiation. One of the atomic oxygen atoms is in an excited state. An increasing amount of O appears above 90 km and the atomic and molecular forms are present in equal concentrations at 125 km; above that the atomic form increasingly dominates. Nitrogen, however, is not directly dissociated to the atomic form in the atmosphere, though it does appear as a product of other reactions.

Above the turbopause, where mixing is less important than diffusion, each component takes an individual scale height depending on its atomic or molecular mass ($H = kT/mg$). Because the scale heights of the common gases vary over a wide range ($H = 1$, $He = 4$, $O = 16$, $N_2 = 28$, $O_2 = 32$) the relative composition of the thermosphere is a marked function of height, the lighter gases becoming progressively more abundant as illustrated in Figure 4.6. Atomic oxygen dominates at several

Table 4.2 *Composition of the atmosphere at ground level*

Molecule	Mass	Volume %	Concentration per cm^3
Nitrogen	28.02	78.1	2.1×10^{19}
Oxygen	32.00	20.9	5.6×10^{18}
Argon	39.96	0.9	2.5×10^{17}
Carbon dioxide	44.02	0.03	8.9×10^{15}
Neon	20.17	0.002	4.9×10^{14}
Helium	4.00	0.0005	1.4×10^{14}
Water	18.02	Variable	

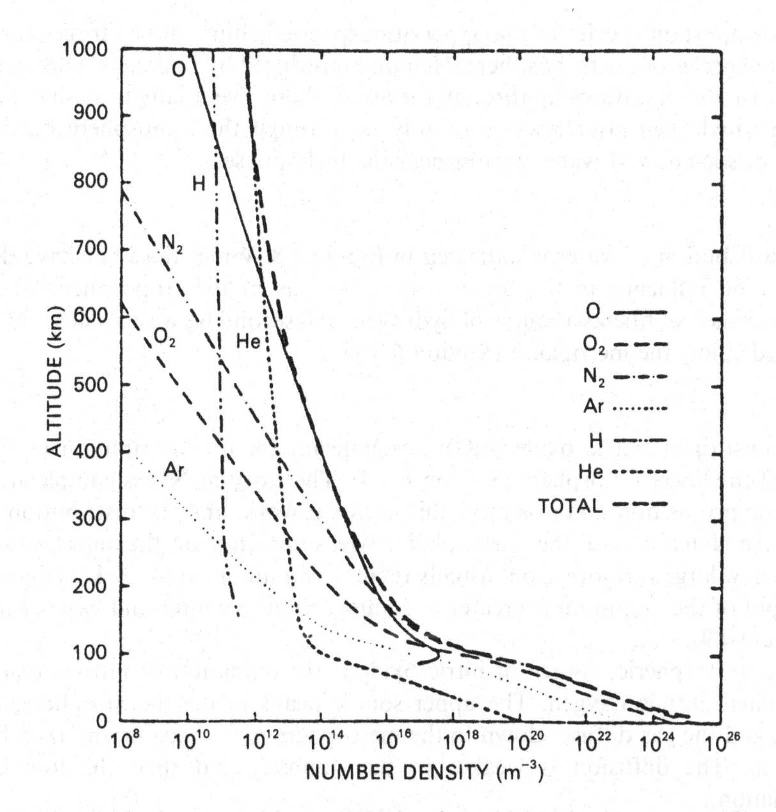

Fig. 4.6 Atmospheric composition to 1000 km for a typical temperature profile. (After A. D. Richmond, in *Solar–Terrestrial Physics* (eds. Carovillano and Forbes). Reidel, 1983. Reprinted by permission of Kluwer Academic Publishers)

hundred kilometres. Above that is the heliosphere where helium is the most abundant, and eventually hydrogen becomes the major species in the protonosphere. Because the scale height also depends on the temperature, so do the details of the composition, as may be seen from Figure 4.7. Note that the protonosphere starts much higher in a hot exosphere, and the heliosphere may be absent from a cool one.

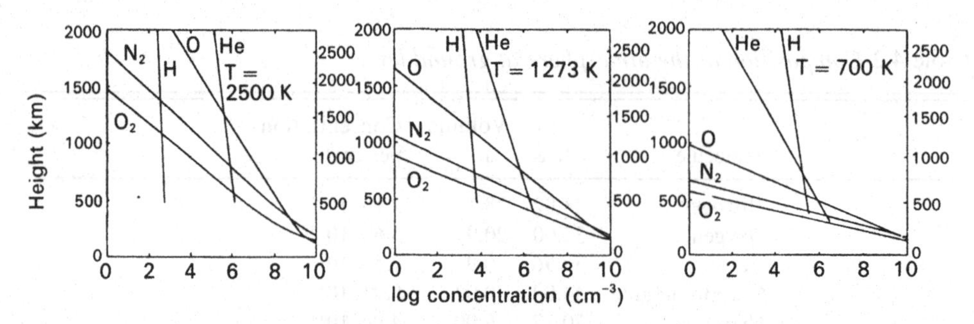

Fig. 4.7 Effect of temperature on exospheric composition. Geopotential heights are given on the left of the diagrams, geometric heights on the right. (From L. G. Jacchia, *COSPAR International Reference Atmosphere*. North-Holland, 1965. Elsevier Science Publishers)

Two of the important species of the upper atmosphere, helium and hydrogen, are, at most, minor species of the troposphere. Helium is produced by radioactive decay in the Earth's crust, and it diffuses up through the atmosphere, eventually to escape into space. Atomic hydrogen also flows constantly up through the atmosphere but its source is the dissociation of water vapour near the turbopause.

Water vapour

The height distribution of water is indicated in Figure 4.8. Water does not have the same dominating influence in the upper atmosphere as in the troposphere. It is important nevertheless, first as a source of hydrogen, and second because it causes ions to be hydrated below the mesopause (Section 6.3.4).

Nitric oxide

The minor constituent, nitric oxide (NO), is significant for its contribution to the production of the lower ionosphere (Section 6.3.4). The story of NO is complicated because several production and loss mechanisms are at work, and the distribution is affected by the dynamics of the mesosphere. Measurements of the nitric oxide distribution show large variations, but usually there is a minimum at 85–90 km (Figure 4.8). The depth of the minimum is greater in summer than in winter and varies with latitude (Figure 4.9).

The lower, stratospheric, source of nitric oxide is the oxidation of nitrous oxide (N_2O) by excited atomic oxygen. The upper source peaks in the thermosphere, at 150–160 km, and the gas diffuses down to the mesosphere by molecular and then by eddy diffusion. The diffusion is weaker in the summer, and then the loss by photodissociation,

$$NO + h\nu \rightarrow N + O, \tag{4.14}$$

and by recombination,

$$NO + N \rightarrow N_2 + O, \tag{4.15}$$

aided also by the effect of the low mesopause temperature, are sufficient to create the minimum.

The main thermospheric sources involve neutral or ionized atomic nitrogen, for example

$$N^* + O_2 \rightarrow NO + O, \tag{4.16}$$

where the * indicates an excited state. The production of these atomic nitrogen species

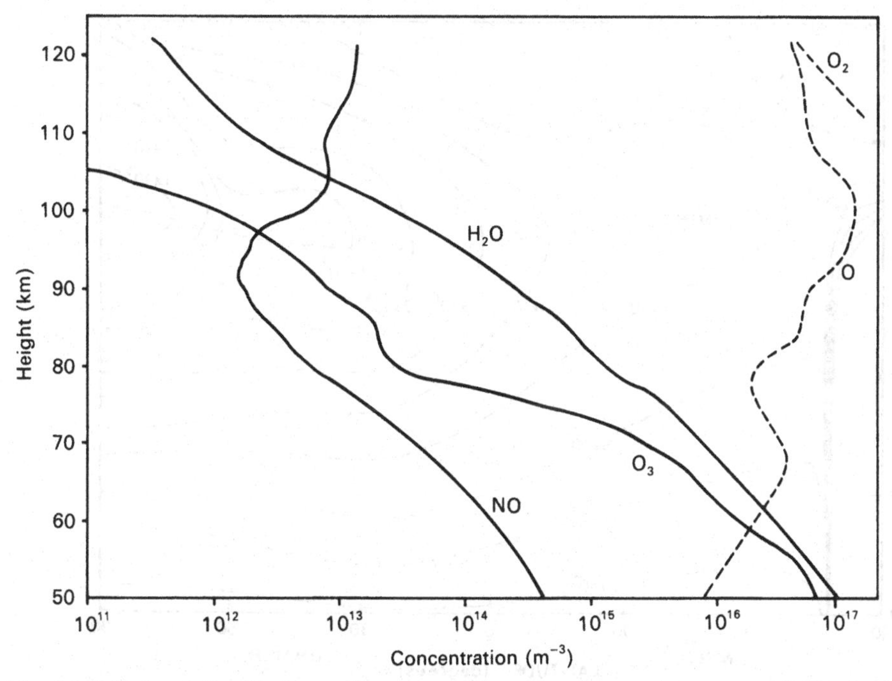

Fig. 4.8 Calculated distributions of important minor constituents for noon at 45° N latitude. O_2 and O are shown for comparison. (After *Handbook of Geophysics and the Space Environment*, AFGL, 1985; from T. J. Keneshea *et al.*, *Planetary and Space Science* **27**, 385, 1979)

is closely linked to ionization processes (Section 6.2), and it is estimated that 1.3 NO molecules are produced on average for each ion. The concentration of nitric oxide therefore varies with time of day, latitude and season. It is three to four times greater at high latitude than at middle latitude, and more variable. The production rate increases dramatically during particle precipitation events.

Ozone

Relatively low in the atmosphere, between 15 and 35 km, ozone is produced by the reaction

$$O + O_2 + M \rightarrow O_3 + M. \quad (4.17)$$

This is a *three-body reaction*, in which the third body, M, serves to carry away excess energy. Such a reaction needs a relatively high pressure because its rate depends on the probability that three particles come together at the same time. The atomic oxygen required is produced by the dissociation of O_2, as in reaction 4.13, though at this lower altitude the effective radiation is in the 200–240 nm band.

Once formed, the ozone may be dissociated again by radiation over a wide range of wavelengths, but particularly in the band 210–310 nm,

$$O_3 + h\nu \rightarrow O_2 + O, \quad (4.18)$$

or destroyed by the reaction

$$O_3 + O \rightarrow O_2 + O_2. \quad (4.19)$$

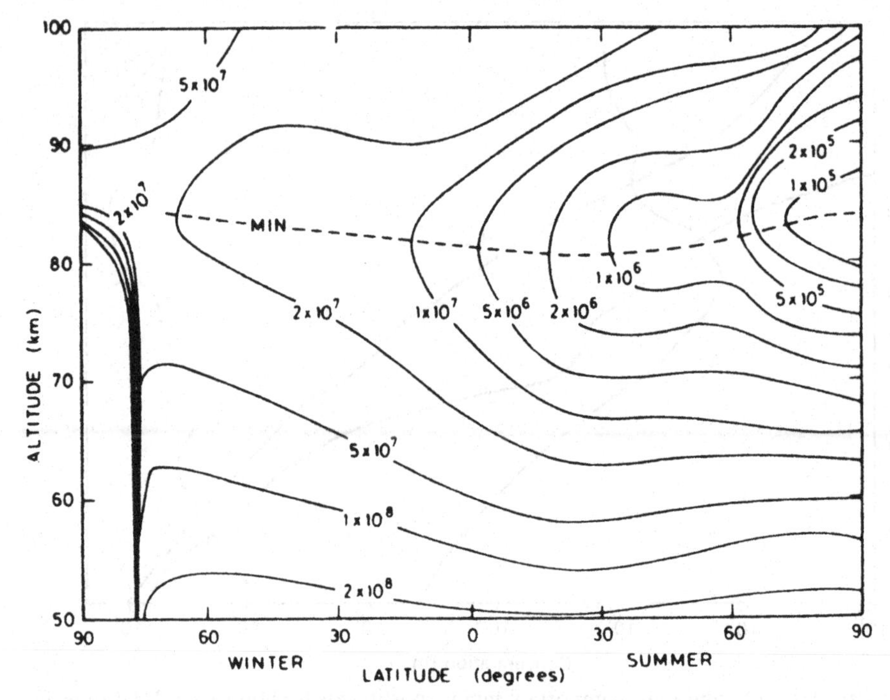

Fig. 4.9 Height and latitude distribution of nitric oxide in the middle atmosphere, in units of cm^{-3}. (From *Handbook of Geophysics and the Space Environment*, AFGL (1985); after G. Brasseur, *Middle Atmosphere Handbook* 10 (1984), published by SCOSTEP)

Only a small concentration of ozone remains on balance, amounting to less than 10 parts per million of the total gas concentration. The *ozonosphere* peaks near 20 km.

Ozone's importance to the atmosphere is that its strong absorption of UV provides a powerful heat source for the stratosphere. It also has a biological importance because the radiation that it absorbs would be harmful to life were it to reach the Earth's surface at sufficient intensity. The amount of ozone in the atmosphere varies with the latitude and the season, but since the early 1980s particularly strong seasonal variations have been observed, especially in the Antarctic, accompanied by a progressive reduction from year to year. These changes have been attributed to the action of atomic halogens, mainly chlorine (Cl):

$$\begin{array}{c} Cl + O_3 \rightarrow ClO + O_2 \\ \underline{ClO + O \rightarrow Cl + O_2} \\ O_3 + O \rightarrow 2O_2 \end{array} . \qquad (4.20)$$

The second stage can involve NO instead of O, but in either case the net result is to remove ozone. The chlorine atom is regenerated and so acts merely as a catalyst. The chlorine atoms are believed to originate in chlorofluorocarbon compounds such as $CFCl_3$ and CF_2Cl_2, of which several thousand tons are released into the atmosphere each year and which are known to have long lifetimes (e.g. 75–150 years). Figure 4.10 illustrates the so called *ozone hole* as observed from an Antarctic base.

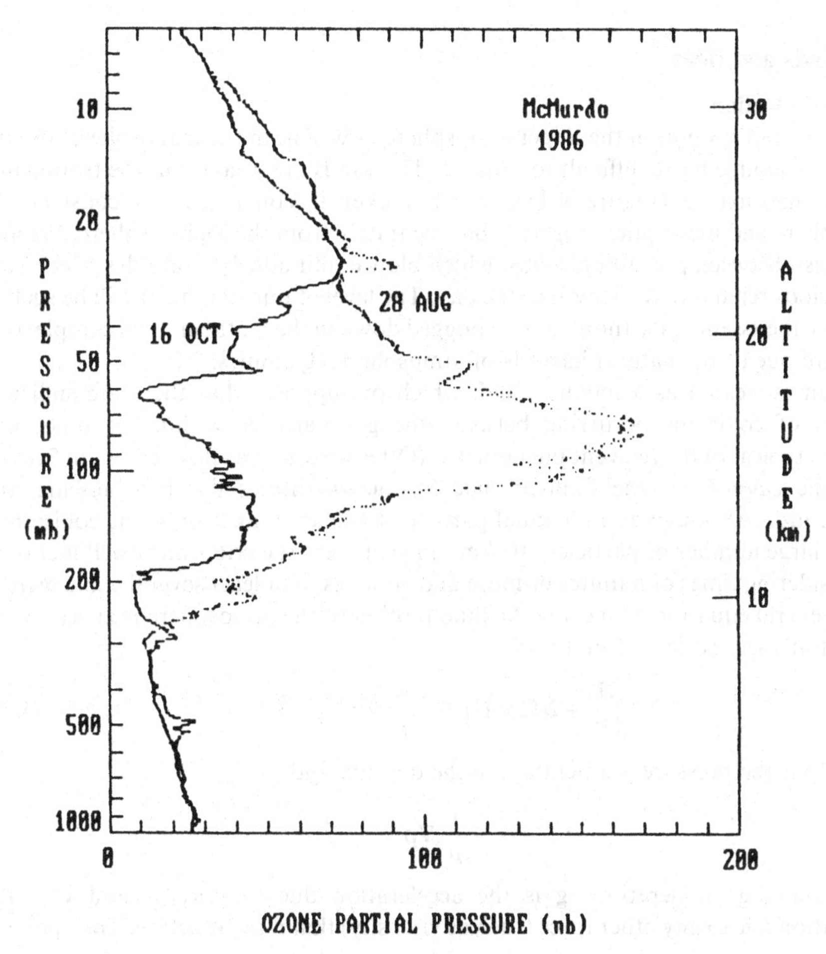

Fig. 4.10 The dreaded ozone hole: observations of the vertical profiles of ozone over McMurdo Station, Antarctica, on 28 August 1986 and 16 October 1986. (S. Soloman, *Reviews of Geophysics* **26**, 131, 1988, copyright by the American Geophysical Union)

Ozone also reacts with nitric oxide:

$$\begin{array}{c} O_3+NO \rightarrow O_2+NO_2 \\ \underline{O+NO_2 \rightarrow O_2+NO} \\ O_3+O \rightarrow 2O_2 \end{array} \qquad (4.21)$$

The net result, in the presence of atomic oxygen, is again a catalytic conversion of ozone back to molecular oxygen. Thus, the ozone concentration is also affected by the natural production of nitric oxide as discussed above.

Metals

Metallic atoms are introduced to the atmosphere in meteors, whose flux over the whole Earth amounts to 44 metric tonnes per day. In the ionized state, metals such as sodium, calcium, iron and magnesium are significant in the aeronomy of the lower ionosphere (Sections 6.3.2, 6.4, 7.1.4 and 7.3).

4.2 Winds and tides

4.2.1 Introduction

There are wind systems in the upper atmosphere as well as in the troposphere, though they are of course more difficult to observe. The winds are weakest at the tropopause and the mesopause (Figure 4.11), which makes it convenient to consider the stratosphere and mesosphere together but separately from the thermosphere. We also distinguish between prevailing winds, which blow continuously, and tides which vary with periods related to the length of the day. The level of our treatment will be such as to outline the essentials without getting bogged down in the mathematical complexities which are one of the natural hazards of atmospheric dynamics.

The air is treated as a moving fluid, which presupposes that there are sufficient numbers of collisions occurring between the gas particles within the times and distances typical of the relevent phenomena. (Otherwise, as was pointed out in Section 4.1.3, one cannot use the familiar macroscopic quantities density, pressure and temperature.) Although an individual particle at 400 km makes only one collision in 3 h, the large number of particles (10^{13} m^{-3}) means that the criterion is well met if we are considering times of minutes or more and volumes of at least several cubic metres.

The general equation governing the fluid motion of the atmosphere is an extension of Newton's second law of motion:

$$\frac{D\mathbf{U}}{Dt}+2(\mathbf{\Omega}\times\mathbf{U}) = -\frac{1}{\rho}\cdot\nabla p+\mathbf{g}+\mathbf{F}, \tag{4.22}$$

where ∇p is the pressure gradient and ρ the density, and

$$-\frac{1}{\rho}\cdot\nabla p$$

is the resulting acceleration; $\mathbf{g}$ is the acceleration due to gravity, and $\mathbf{F}$ is the acceleration due to any other force (such as viscosity) that may be acting. The operator

$$\frac{D}{Dt}=\frac{\partial}{\partial t}+(\mathbf{U}\cdot\nabla),$$

where $\mathbf{U}$ is the velocity, and it includes both changes within the fluid with time ($\partial/\partial t$), and those due to advection ($\mathbf{U}\cdot\nabla$). Advection represents the changes seen by a stationary observer that are due to the movement of a fluid containing spatial variations, whereas the first term gives the changes that would be seen by an observer moving with the fluid. The second term of Equation 4.22 is the Coriolis acceleration, Ω being the Earth's angular velocity. Coriolis acceleration arises from the rotation of the coordinate system, since we view matters as observers on the rotating Earth. It is important for the slowest variations, those having frequencies less than 2Ω.

Among the forces in Equation 4.22, gravity, acting downward, is relatively large, and vertical motions involve significant changes in potential energy. Consequently, we can generally assume that on the scales of interest here the principal motions will be confined to the horizontal plane.

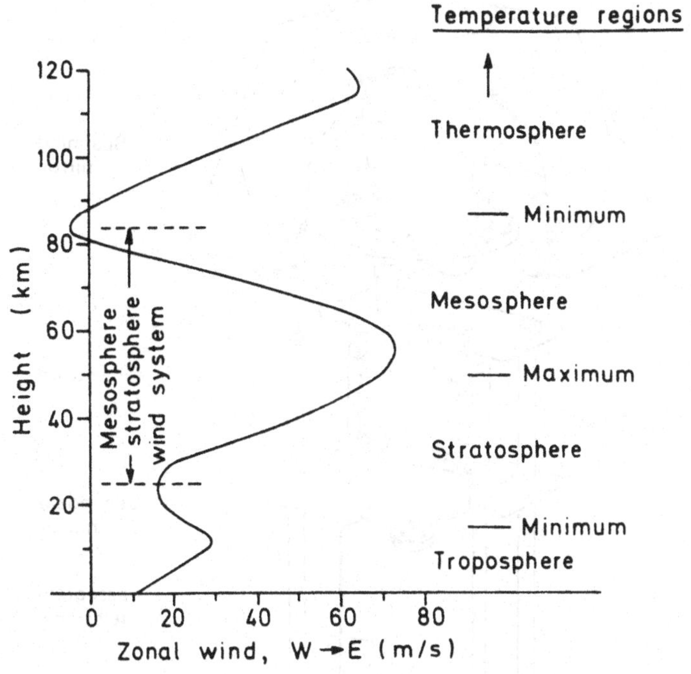

Fig. 4.11 General magnitude of zonal winds in the upper atmosphere, illustrated by a height profile for 45° N in January. (*COSPAR International Reference Atmosphere*. Akademie-Verlag, 1972)

4.2.2 The measurement of high-altitude winds

The techniques of upper atmosphere observation were discussed in Chapter 3, and some of them may be applied to measuring motions in the upper atmosphere. The most obvious approach is to use a tracer, something present naturally in the medium or injected for the purpose, whose motion may be observed from the ground. Tracers come into our category of 'indirect sensors'.

Natural tracers

Meteor trails provide a good example of a tracer present naturally. As a meteor ablates in the atmosphere, part of the dissipated energy goes to ionize the air in a column around the meteor. Most trails are formed in the height range 75–105 km, and they can be tracked by radar as they blow along in the wind.

Regions of airglow emission (Section 6.4) also move with the neutral air. By observing the apparent wavelength of a prominent airglow line it is possible, from the Doppler shift, to determine the component of velocity in the line of sight. This has become a well established technique for winds in the thermosphere, using a Fabry–Perot interferometer (Figure 4.12) to find the Doppler shift of the 630 nm oxygen line. In a typical instrument the separation of the Fabry–Perot plates is varied cyclically with a period of a few seconds by means of a pressure cell or a magnetostrictive device, so as to scan the interferometer through the airglow line. For a neutral air wind of 50 m/s the Doppler shift of the 630 nm line is only 10^{-4} nm, yet

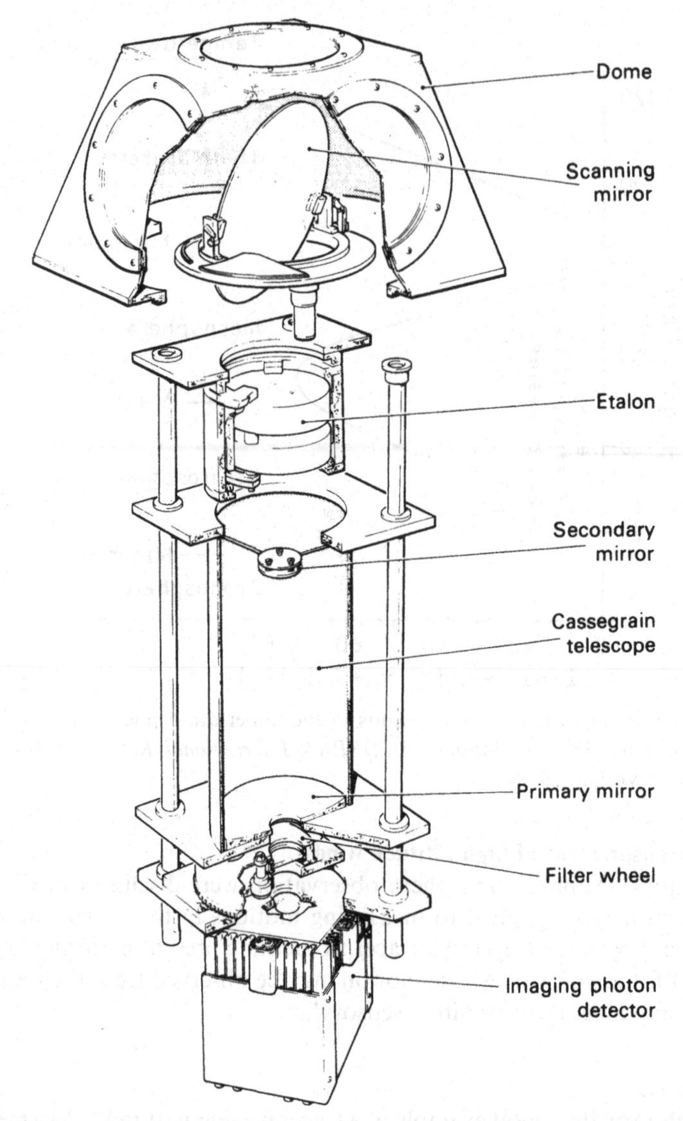

Fig. 4.12 A Fabry–Perot interferometer for determining thermospheric motions from the Doppler shift of an airglow line. The dome has several windows so that velocity components can be measured in several directions. The etalon plates are 15 cm in diameter with 15 mm separation, and the inner surfaces are polished flat to better than 3 nm. The whole instrument is about 1 m long. (A. S. Rodger, British Antarctic Survey, private communication)

the velocity can be measured to 1 m/s with a modern instrument. The horizontal wind vector is determined by observing in two directions more or less at right angles. The width of the line gives the air temperature.

Injected tracers

Moving on to injected tracers, various kinds can be inserted into the medium from rockets. The puff of smoke from a grenade can be followed visually from the ground. Metal foil, or a parachute carrying a reflector, can be tracked by radar, giving the wind as a function of height as the target falls. Each technique works only over a restricted range of heights. A parachute will not open above about 60 km, and foil, often known as *chaff*, disperses in the turbulence and wind shears at about the same height. In the earlier experiments the chaff was made from thin wires cut to resonate with the wavelength of the tracking radar. However, this tended to aggregate into clumps during the descent; more recent experiments have used metallized plastic film almost a centimetre wide, and this material appears to work well up to 95 km.

Most measurements of winds in the thermosphere have made use of chemicals ejected from an ascending rocket. A trail of sodium is luminous at twilight by resonant scattering of sunlight at 589 nm, and since the trail then shows up against a dark sky its movement can be tracked by photography with spaced cameras. In the minutes after release, irregularities – evidence of turbulence and waves – form in the trail and serve as reference points for triangulation so that heights and velocities can be determined from the photographs. Trimethyl aluminium (TMA) is used in a similar manner, but this chemical reacts with atomic oxygen and the trail is luminous at night, not just at twilight. Figure 4.13 shows photographs of a luminous aluminium trail released from a rocket between altitudes of 105 and 129 km over Alaska. The deduced wind pattern is also shown. The TMA payload can be made sufficiently robust to be fired from a gun, and this remarkable technique has been used, though not from many sites, to inject TMA as high as 140 km from a modified naval gun. One advantage claimed for the gun over the rocket is the precision of ballistics compared with the uncertainties of rocketry, which means that only a small landing area is needed.

A powerful technique, which enables both the neutral-air wind and the electric field to be measured, makes use of a mixture of barium and strontium. If a cloud containing these elements is released at twilight the barium ionizes to Ba^+ but the strontium remains neutral. The neutral strontium appears blue or green and the ionized barium is violet or purple, and thus they may be distinguished in colour photographs. After release the barium cloud starts to elongate along the magnetic field direction whereas the strontium cloud remains spherical. The two clouds separate, since the neutral one moves with the neutral air, but the ionized one drifts in any electric field that may be present due to $\mathbf{E} \times \mathbf{B}$ drift (as discussed in Section 6.5.4). Typical speeds for the violet and green clouds at middle latitude are 80 and 40 m/s respectively.

Remote methods

Sound propagation provides a method for measuring winds in the mesosphere and stratosphere. The speed of sound depends on the air temperature ($s = (\gamma P/\rho)^{\frac{1}{2}}$, from Equation 2.80, $\propto T^{\frac{1}{2}}$ for constant composition), but this is of course relative to the air. The velocity measured by observers at the ground, for example with spaced microphones, is therefore the vector sum of the wind and the sound velocities. The technique is to eject grenades from an ascending rocket, and then both wind and temperature can be determined as a function of height.

A radio propagation technique for measuring the movement of ionization

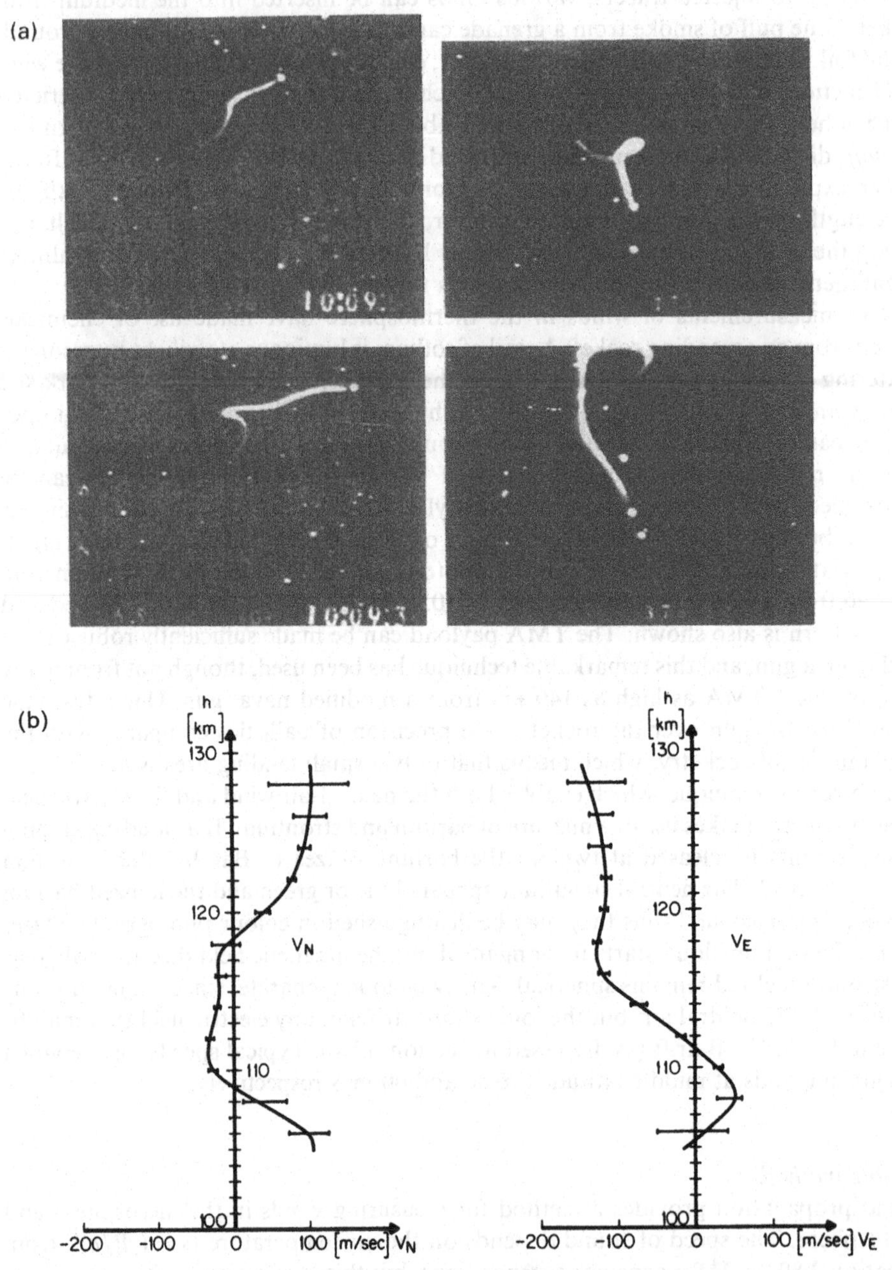

Fig. 4.13 (a) A luminous trail created 105–129 km over Alaska by burning aluminium-doped propellant from a rocket. The trail was observed for 5 minutes with auroral TV cameras at three sites, and the motion of the trail was deduced by triangulation. The pictures shown are from two sites (left and right) and for two times (above and below) 30 seconds apart. (b) The meridional and zonal neutral wind profile deduced from the observations. (H. C. Stenback-Nielsen, private communication)

irregularities will be described in Section 7.5.2. In the lower ionosphere (D and E regions) it may be assumed that the velocity so determined is the same as that of the neutral wind, but for the upper thermosphere (F region) it gives the electric field.

In the next section we shall see that in the stratosphere and mesosphere the wind profile is related to latitude and longitude variations of the temperature profile. Winds may thus be computed from temperature data, though, since it depends on the Coriolis force, the procedure cannot be expected to be accurate at low latitude.

4.2.3 Winds in the stratosphere and mesosphere

Coriolis force dominates the steady horizontal flow of air in the stratosphere and the mesosphere. The pressure gradient is north–south due to the variation of solar heating with latitude, and it results in an east–west wind: westward in summer, eastward in winter. Balancing the pressure gradient against the Coriolis force gives

$$\frac{1}{\rho R_E}\cdot\frac{\partial P}{\partial \theta} = 2\Omega U \sin\theta, \tag{4.23}$$

where θ is the latitude and R_E is the radius of the Earth. Hence the wind speed is determined by the latitude, θ, the air density, ρ, and the latitudinal pressure gradient. This is a *geostrophic wind.*

Since the pressure gradient is just a consequence of the temperature gradient, the theory of the *thermal wind,* which is covered in textbooks of meteorology, applies. Considering the north–south direction (y), it is easily shown from the hydrostatic equation (4.2) that the pressure and temperature gradients are related by

$$\frac{\partial P}{\partial y} = P\cdot\frac{h}{HT}\cdot\frac{\partial T}{\partial y} \tag{4.24}$$

at height h, where H is the scale height and T is the temperature (assumed here not to vary with height). This pressure gradient is balanced by the Coriolis force $\rho U_x f$ associated with an east–west wind U_x, where f is the standard Coriolis parameter $f = 2\Omega \sin\theta$. Thus,

$$U_x = -\frac{P}{\rho f}\cdot\frac{h}{HT}\cdot\frac{\partial T}{\partial y} = -\frac{hg}{fT}\cdot\frac{\partial T}{\partial y}. \tag{4.25}$$

In the more general case where T varies with height the relation is

$$\frac{\partial}{\partial z}\left(\frac{U_x}{T}\right) = -\frac{g}{fT^2}\cdot\frac{\partial T}{\partial y}. \tag{4.26}$$

The significance of this relation is that the prevailing winds of the stratosphere and mesosphere are determined by the thermal structure of that region, and either may be deduced if the other is known. Figure 4.14 illustrates consistent distributions of zonal wind and temperature. The wind speeds are generally between 10 and 100 m/s.

4.2.4 Thermospheric tides

Tidal principles

At greater heights, tides become increasingly important.

A tide may be either solar or lunar, according to whether the period is related to the solar or the lunar day. A tidal perturbation must travel westward with respect to the

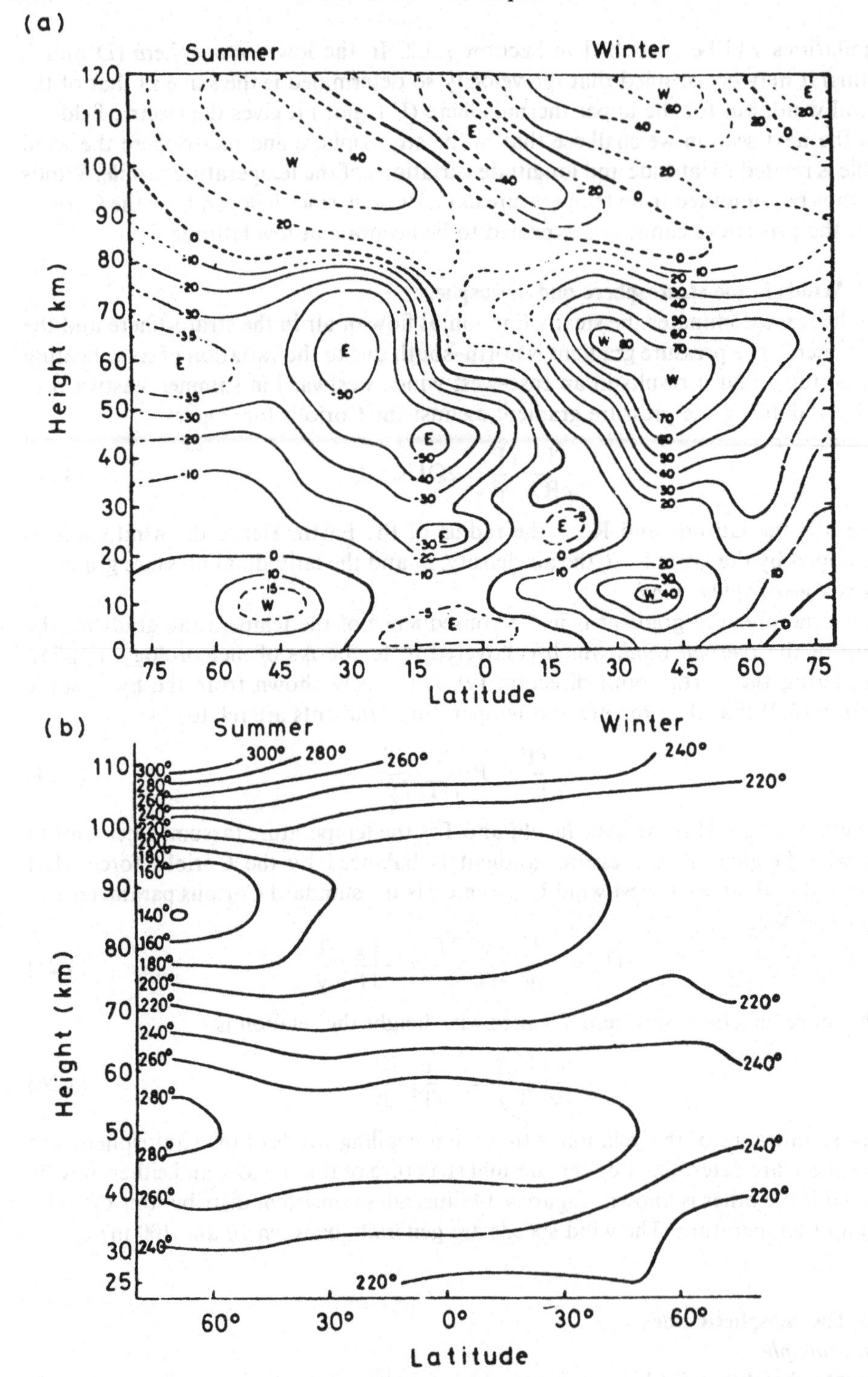

Fig. 4.14 Height–latitude cross-sections of (a) zonal wind and (b) temperature in stratosphere and mesosphere. (*COSPAR International Reference Atmosphere*. Akademie-Verlag, 1972)

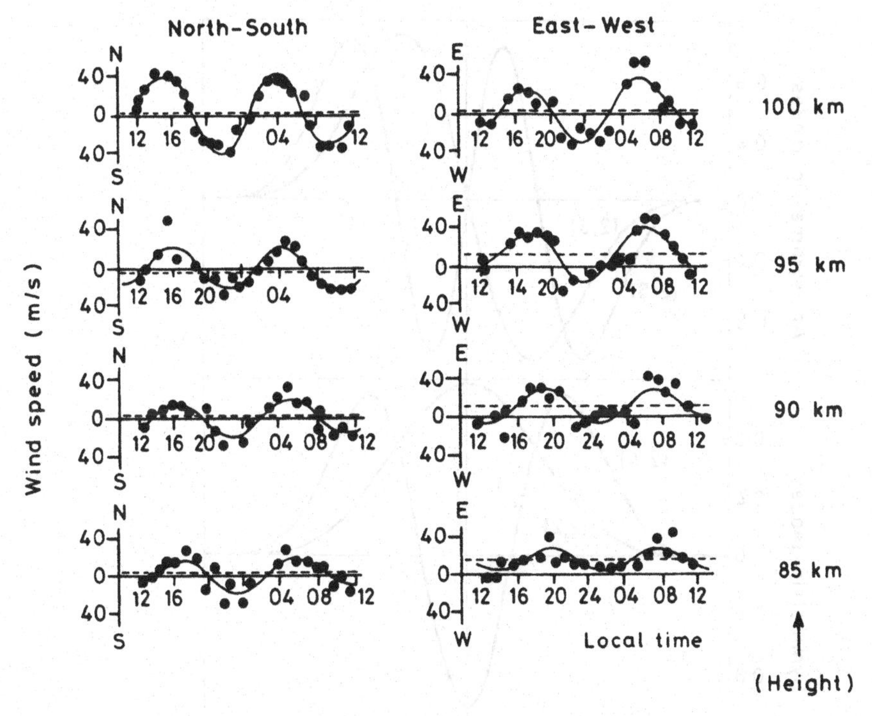

Fig. 4.15 Meridional and zonal winds in the mesosphere, according to meteor radar observations. (After J. S. Greenhow and E. L. Neufeld, *Phil.Mag.* **1**, 1157, 1956)

Earth's surface so that it remains synchronized with the position of the Sun or the Moon. For example, the solar diurnal tide has a period of 24 hours and covers the Earth's circumference at the observer's latitude in 24 hours. Such a tide has 'wave number 1'. The semi-diurnal tide has a period of 12 hours but also covers the circumference in 24 hours and has 'wave number 2'. (Wave number = Earth's circumference/wavelength.)

At upper mesosphere levels the tide is semi-diurnal and its amplitude is at least as large as the prevailing wind (tens of metres per second). See Figure 4.15. In the thermosphere a 24-hour tide dominates, and velocities can be up to 200 m/s. Both of these tides are predominantly solar, though a small lunar tide can be recognized in ionospheric data.

To understand the tidal motions of the atmosphere it is necessary to appreciate its natural modes of oscillation, and then to consider how effectively these may be stimulated by the forcing agent. The various modes of oscillation can be described by *Hough functions* (Figure 4.16), showing the latitudes of nodes and antinodes. The modes are designated (m,n) where m is the number of cycles per day and there are (n − m) nodes between the poles (not counting those at the poles). For example, mode (2,4) has a period of 12 hours and there are three antinodes, one at the equator and two at middle latitudes.

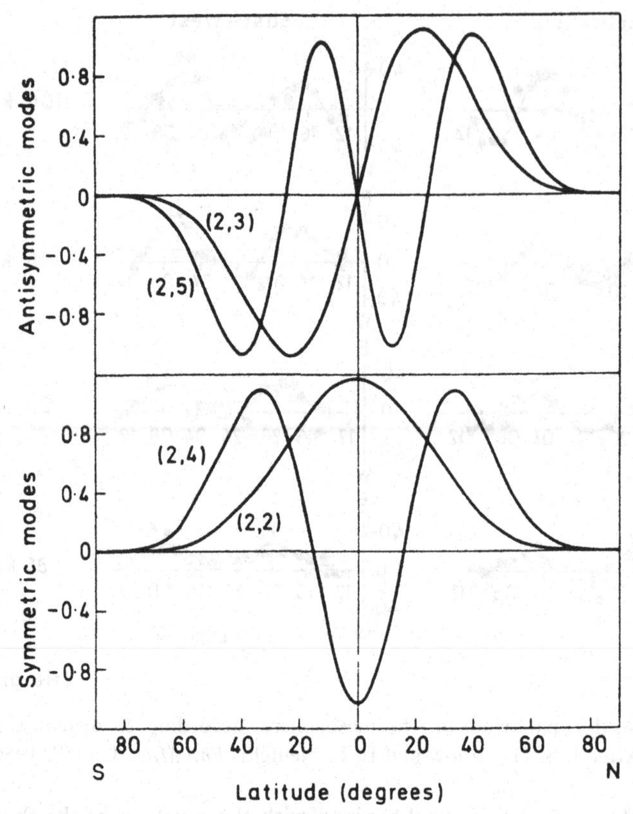

Fig. 4.16 Hough functions for the two antisymmetric and symmetric modes of lowest order for the semi-diurnal tide. (After J. V. Evans, *Proceedings of International Symposium on Solar–Terrestrial Physics*, Boulder, Colorado, 1976)

Solar tides

Solar heating would be expected to excite most strongly modes having an antinode at the equator. But another consideration is the vertical structure of the tide. Some tidal modes are propagating waves with a vertical wavelength, and others are trapped modes which do not propagate vertically. Modes whose vertical wavelength is small relative to the vertical extent of the excitation will suffer self-interference, and consequently will be weak. In the stratosphere and mesosphere the diurnal tide (1,1) has a short vertical wavelength of only 30 km, and is therefore not the main component. The semi-diurnal tide (2,2), however, has long vertical wavelength and is the dominant solar tide, even at the surface. Figure 4.17 illustrates the heating due to water vapour and ozone below 100 km. In the stratosphere/mesosphere tide the air flow is governed by the balance between the tide-raising force and the Coriolis force, as in equation 4.23.

The diurnal solar tide dominates in the thermosphere above 250 km, being excited by the absorption of EUV radiation at ionospheric levels. (Between 100 and 250 km both diurnal and semi-diurnal components can be present.) The air flow in the thermosphere is strongly influenced by the phenomenon of ion drag, the retarding

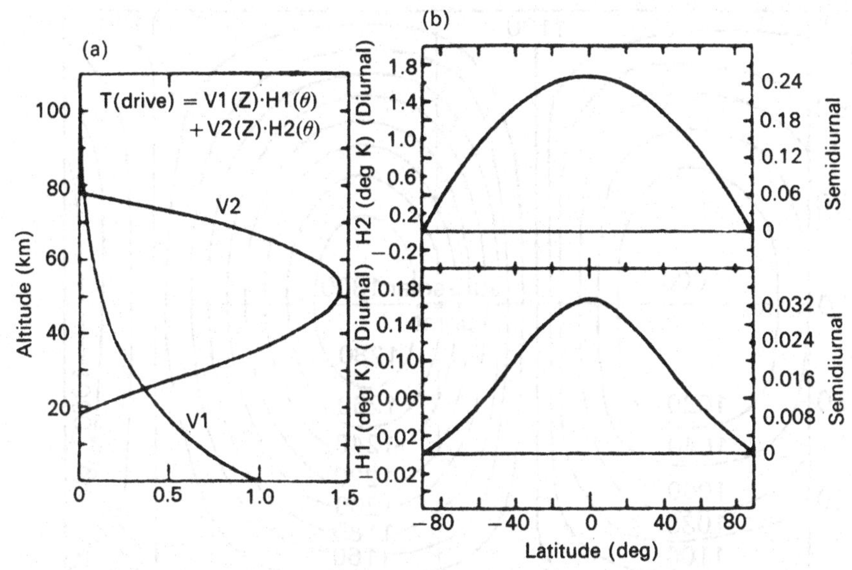

Fig. 4.17 (a) Vertical distribution of thermal excitation due to water vapour (V1) and ozone (V2) on a relative scale. (b) Latitudinal distribution of the diurnal and semi-diurnal heating components due to water vapour (H1) and ozone (H2), given as the amplitude of the temperature variation. The total energy driving the tide is given by V1·H1+V2·H2 for a selected height and latitude. (After S. Chapman and R. S. Lindzen, *Atmospheric Tides*. Reidel, 1970. Reprinted by permission of Kluwer Academic Publishers)

force due to collisions with the ions which are themselves inhibited from crossing the geomagnetic field (Sections 2.2.4 and 7.2.4). In the thermosphere this drag exceeds the Coriolis force and the tidal equation becomes

$$\mathrm{d}\mathbf{U}/\mathrm{d}t = -\nu_{ni}(\mathbf{U}-\mathbf{v}_i)-\nabla\psi, \qquad (4.27)$$

where ν_{ni} is the rate of collision of a given neutral particle with ions and $\mathbf{v}_i$ is the ion velocity. $\nabla\psi$ is the force producing the tide. Equation 4.27 does not include viscosity, which provides another retarding force. As in the atmosphere near the ground, the effect of a large drag force is to make the air flow almost directly down the pressure gradient instead of along the isobars.

The pressure gradients in the thermosphere are related to the temperature distribution (Figure 4.18), having maxima and minima over the equator, respectively near 1500 and 0400 LT (local time). The wind computed from this distribution is shown in Figure 4.19. The wind is smaller by day (e.g. 40 m/s) than by night (120 m/s) because the drag force is larger in the more intense daytime ionosphere.

Like other upward propagating waves, a propagating tide increases in amplitude with height. This is because, in the absence of dissipation, the energy density must be constant. Thus if ρ is the air density and U is the amplitude of the air motion due to the tide, $\rho U^2/2$ is constant and thus $U \propto \rho^{-\frac{1}{2}}$. At ground level the fractional pressure variation is about 10^{-3} and the tidal wind is about 0.05 m/s. At 100 km the pressure variation is 10% and the corresponding air speed 50 m/s. The tides are even stronger in the thermosphere.

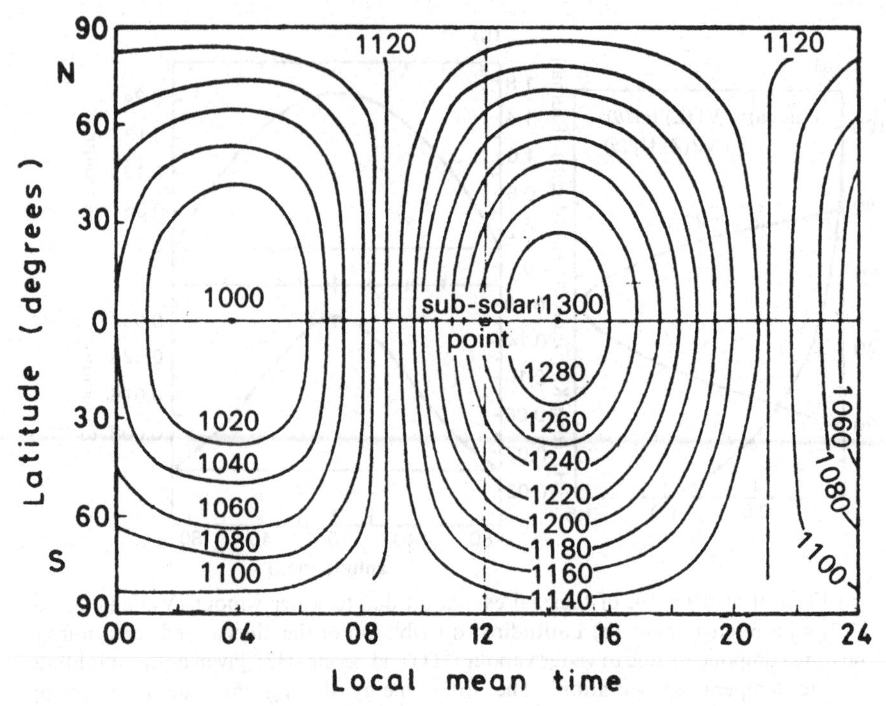

Fig. 4.18 Global temperature distribution in the thermosphere (in K) at the equinox in conditions of medium solar activity. No account is taken of any high-latitude sources of heat, such as energetic particles. (After H. Kohl and J. W. King, *J. Atmos. Terr. Phys.* **29**, 1045, copyright (1967) Pergamon Press PLC; data from J. G. Jacchia, *Smithsonian Contrib. Astrophys.* **8**, 215, 1965)

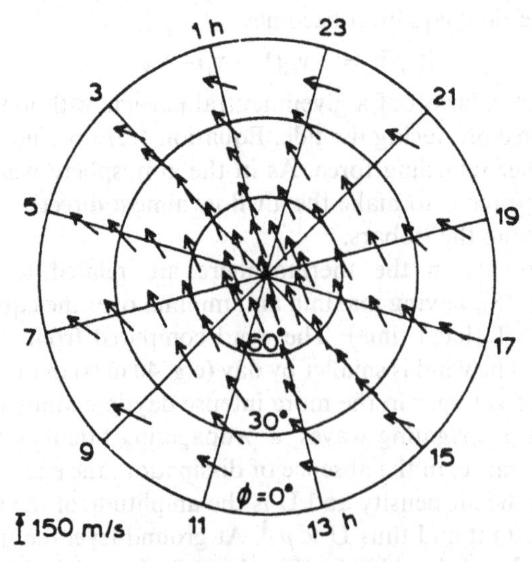

Fig. 4.19 Winds at 300 km, computed from the temperature distribution of Fig. 4.18. (Reprinted with permission from H. Kohl and J. W. King, *J. Atmos. Terr. Phys.* **29**, 1045, copyright (1967) Pergamon Press PLC)

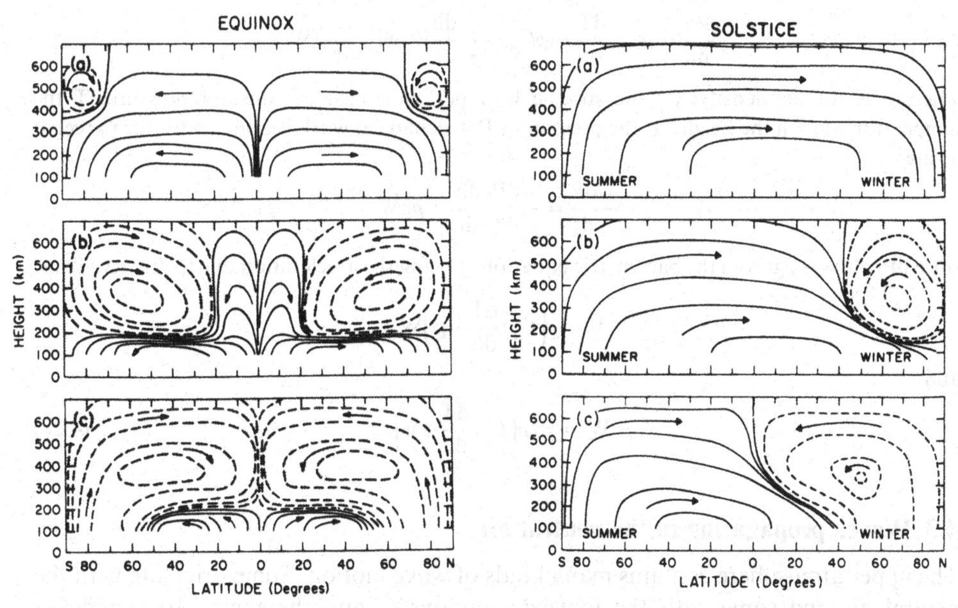

Fig. 4.20 Effect of high-latitude heating on the circulation of the thermosphere, at equinox and at solstice, in modifying the basic Hadley cells. Panels (a) show the meridional circulation under magnetically quiet conditions, (b) are for average activity levels (10^{11} J/s), and (c) are for a geomagnetic substorm (10^{12} J/s). (R. G. Roble, *The Upper Atmosphere and Magnetosphere*. National Academy of Sciences, Washington DC, 1977)

Since the ionosphere affects the thermospheric wind, it is only fair to allow the wind to have some effect on the ionosphere. The greatest effect is a raising or lowering of the ionosphere according to the wind direction; this point will be discussed in Section 6.5.3.

Effect of a high-latitude heat source

The thermospheric wind is a response to the heating of the upper atmosphere, and thus it will be altered if additional sources of heating appear. During magnetic storms (Section 8.4) the auroral zones are heated by electrojets and by energetic particles arriving from the magnetosphere. Electrojets can dissipate as much as 0.5 W/m² in the ionospheric E region during a severe disturbance, and the auroral particles that stop in the same region will contribute some smaller amount. The energy flux of solar EUV which drives the normal tide is much smaller than this, about 0.5 mW/m² above 120 km. Computations (Figure 4.20) show how the wind in the thermosphere can be drastically altered by this energy source at high latitudes.

The vertical wind, W, caused by a local energy input, E, may be calculated as follows:
Consider a parcel of air of unit volume, which rises at velocity W and receives energy E per second from an external source. The energy input (a) heats the gas and (b) provides the mechanical work done by the gas on its surroundings as it expands in moving to regions of lower pressure. The heating rate

$$\frac{dQ}{dt} = \rho C_p \cdot \frac{dT}{dt} = \rho C_p \cdot \frac{dT}{dh} \cdot \frac{dh}{dt} = \rho C_p \cdot \frac{dT}{dt} \cdot W,$$

where ρ is the air density, C_p the specific heat per unit mass at constant pressure, T the temperature, and h the height. If the pressure is P, mechanical work is done by the air parcel at a rate

$$-\frac{dP}{dt} = -\frac{dP}{dh} \cdot \frac{dh}{dt} = \rho g W$$

since $dP/dh = -\rho g$ from the barometric equation, g being the acceleration due to gravity. Thus,

$$E = \rho\left(C_p \cdot \frac{dT}{dh} + g\right)W,$$

and

$$W = E/\rho\left(C_p \cdot \frac{dT}{dh} + g\right). \tag{4.28}$$

4.3 Waves propagating in the neutral air

The upper atmosphere contains many kinds of wave motion. Some originate with the neutral air and some with the ionized component, and there may be significant interaction between the neutral and the ionized components. One important type is the *acoustic–gravity wave* in the neutral air. These waves have periods of minutes and tens of minutes, wavelengths of tens and hundreds of kilometres, and they travel at speeds related to the speed of sound at the relevent level of the atmosphere.

4.3.1 Theory of acoustic–gravity waves

In an acoustic wave the force restoring a displaced particle towards its original position is the change of pressure associated with the compression of the medium. In a gravity wave, the simplest example of which occurs on the surface of an ocean, the restoring force is the action of gravity on a displaced body of fluid. Ocean waves can exist at the surface because of the abrupt change of density there. In the atmosphere the density varies gradually with altitude, and in this situation gravity waves may exist within the medium. In general both the compressional and the gravitational forces should be taken into account, and all these waves belong to the more general acoustic–gravity class. For these waves the effects of the Earth's rotation and curvature can be neglected.

The idea of an acoustic wave is generally familiar, but most people find the gravity wave rather baffling on first acquaintance because of its unfamiliar dispersion characteristics. For the simplest case, which assumes

(a) only small variations of pressure and density,
(b) no loss of energy (i.e. zero viscosity), and
(c) a two-dimensional plane-wave solution of form $\exp j(\omega t - k_x x - k_z z)$,

the dispersion relation is

$$\omega^4 - \omega^2 s^2(k_x^2 + k_z^2) + (\gamma - 1)g^2 k_x^2 + \omega^2 \gamma^2 g^2/4s^2 = 0. \tag{4.29}$$

In this equation, ω is the angular frequency of the wave, s is the speed of sound, γ is the ratio of specific heats for the atmospheric gas, g is the acceleration due to gravity,

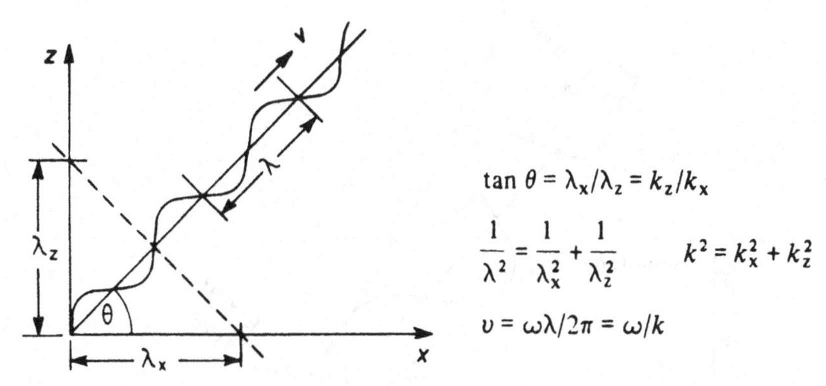

Fig. 4.21 Relationships between wavelength, velocity and propagation angle.

and k_x and k_z are the wavenumbers ($k = 2\pi/\lambda$) in the horizonal and vertical directions. There is no need to invoke the y direction because there is no asymmetery in the horizontal, and thus a two-dimensional (plus time) treatment suffices.

Equation 4.29 expresses the permitted relations between the wave's frequency and its horizontal and vertical wavelengths. As may be seen from Figure 4.21, these quantities define the phase behaviour of the wave fully. Phase propagation is at elevation angle

$$\begin{aligned}\theta &= \tan^{-1}(\lambda_x/\lambda_z) \\ &= \tan^{-1}(k_z/k_x),\end{aligned} \tag{4.30}$$

the wavelength (λ) is given by

$$\begin{aligned}1/\lambda^2 &= 1/\lambda_x^2 + 1/\lambda_z^2 \\ &= (k_x^2 + k_z^2)/(2\pi)^2\end{aligned}$$

and therefore

$$\lambda = 2\pi/(k_x^2 + k_z^2)^{\frac{1}{2}}, \tag{4.31}$$

and thus the phase velocities in the direction θ and in the horizontal and vertical respectively are

$$\begin{aligned}v &= \omega\lambda/2\pi = \omega k, \\ v_x &= \omega/k_x, \\ v_z &= \omega/k_z.\end{aligned} \tag{4.32}$$

If $g = 0$, Equation 4.29 becomes

$$s = \omega/(k_x^2 + k_z^2)^{\frac{1}{2}} = \omega\lambda/2\pi, \tag{4.33}$$

which is the dispersion relation for a sound wave, in which the phase velocity is independent of direction.

Substituting

$$\omega_a = \gamma g/2s \tag{4.34}$$

and

$$\omega_b = (\gamma - 1)^{\frac{1}{2}} g/s \tag{4.35}$$

into Equation 4.29 and rearranging, gives

$$k_z^2 = \left(1 - \frac{\omega_a^2}{\omega^2}\right)\frac{\omega^2}{s^2} - k_x^2\left(1 - \frac{\omega_b^2}{\omega^2}\right). \tag{4.36}$$

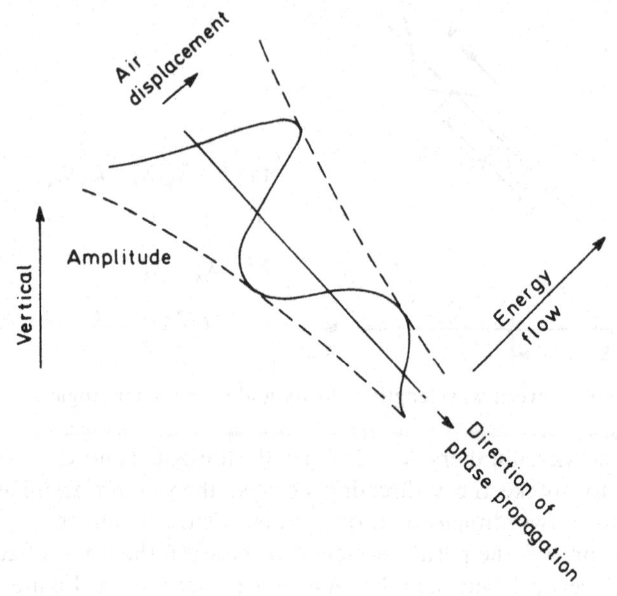

Fig. 4.22 A simple gravity wave.

Then, putting $\omega^2 \ll s^2 k_x^2$, which removes the effect of compressibility,

$$k_z^2 = k_x^2\left(\frac{\omega_b^2}{\omega^2} - 1\right), \tag{4.37}$$

which represents a pure gravity wave.

The wave can only propagate if k_x and k_z are real and positive, requiring that $\omega < \omega_b$. From Equation 4.30 it is seen that the angle of propagation is

$$\theta = \tan^{-1}\left(\frac{\omega_b^2}{\omega^2} - 1\right)^{\frac{1}{2}}. \tag{4.38}$$

In fact the negative square root should be taken, since phase propagation is downward – a point that does not emerge from our simplified discussion. In an acoustic wave the speed is determined by properties of the propagation medium, and selecting a frequency fixes the wavelength but not the direction of travel. A gravity wave is rather different. Here, the frequency fixes the propagation angle (Equation 4.38) but neither the velocity nor the wavelength. However, v, ω and λ are related by Equations 4.32.

Figure 4.22 illustrates a simple gravity wave. At the lowest frequencies the air particles move perpendicular to the direction of phase propagation, and the energy also travels at right angles to the phase velocity. The wave amplitude increases with height, varying as the square root of the air density to maintain the energy flux at a constant value (viscous effects being neglected).

The full range of acoustic–gravity waves (within our assumptions) is included in Equation 4.29. In general the air particles move in elliptical trajectories, combining the longitudinal displacement typical of an acoustic wave with the transverse displacement of a gravity wave. Since k_x and k_z must both be real in a propagating wave, either $\omega > \omega_a$ or $\omega < \omega_b$, and these are called the acoustic and gravity ranges respectively. If

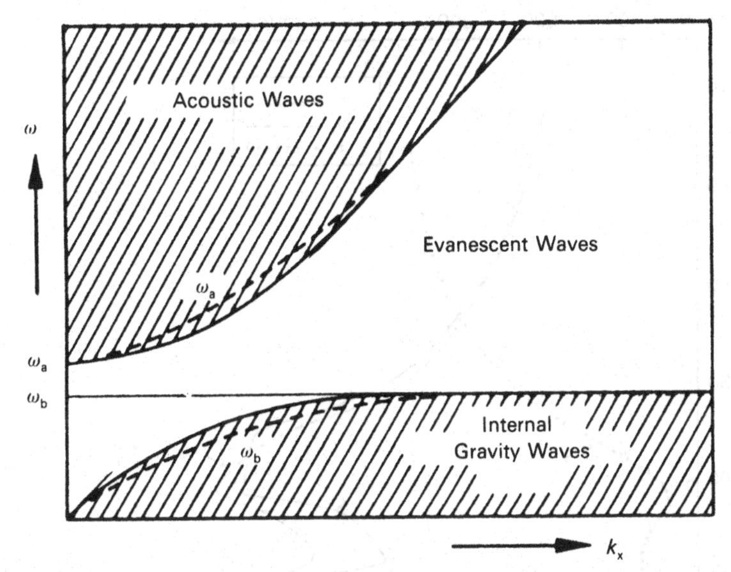

Fig. 4.23 The acoustic, evanescent, and gravity regimes of acoustic–gravity waves. The dashed lines show the effect of neglecting gravity and compressibility respectively. (After J. C. Gille, in *Winds and Turbulence in Stratosphere, Mesosphere and Ionosphere*, (ed. Rawer). North-Holland, 1968. Elsevier Science Publishers)

$\omega_b < \omega_a$, waves with frequency between these two values are evanescent and do not propagate. Figure 4.23 illustrates these three regimes on a diagram of wave frequency against horizontal wave number.

The substitutions of Equations 4.34 and 4.35 can be justified by dimensional arguments. Comparing the first and fourth terms of Equation 4.29 shows that $\gamma g/2s$ has the dimensions of frequency, and comparing the second and third terms shows that $(\gamma-1)^{\frac{1}{2}}g$ has the dimensions of ωs. The parameters ω_a and ω_b have physical significance as the *acoustic resonance* and the *buoyancy* or *Brunt–Vaisala resonance* respectively. The first is the resonant frequency of the whole atmosphere in the acoustic mode, when the restoring force is due to compression, and the second is the natural resonance of a displaced parcel of air within the atmosphere when buoyancy is the restoring force. Typical values of the acoustic and Brunt frequencies are shown in Figure 4.24. At ionospheric levels (Section 7.5.5) waves with period longer than 10–15 minutes are likely to be gravity waves, and any with periods of only a few minutes are probably acoustic.

Although acoustic–gravity waves are a phenomenon of the neutral atmosphere, the motions can be communicated to the ions and the electrons through collisions. How they respond depends on the altitude: at the lower levels, where collisions are frequent, it may be assumed that the ionization moves in the same way as the particles of the neutral air, but at higher levels ion motion is inhibited across the geomagnetic field and the ionospheric response is then a biased one.

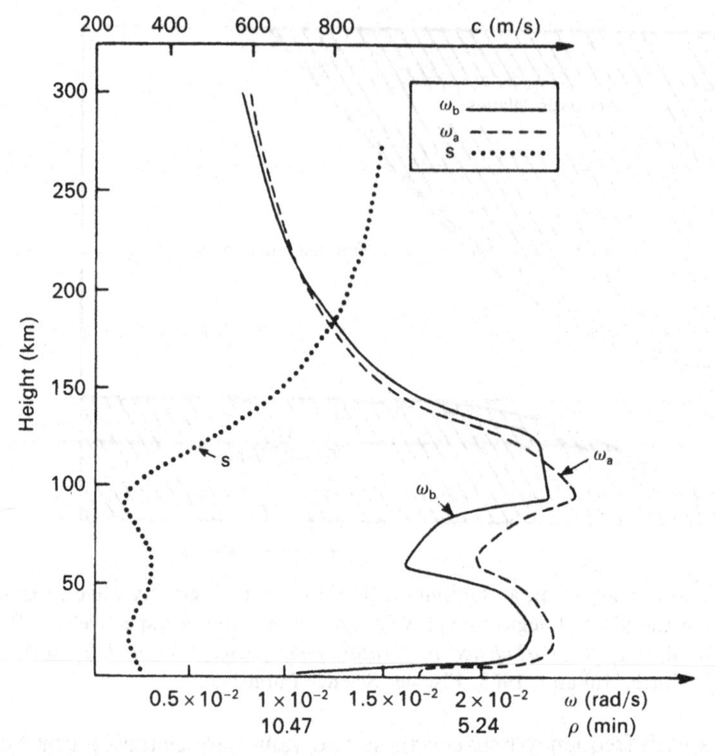

Fig. 4.24 Height profiles of the speed of sound (s), the Brunt frequency (ω_b) and the acoustic cut-off frequency (ω_a) in a realistic atmosphere. (After I. Tolstoy and P. Pan, *J. Atmos. Sci.* **27**, 31, 1970)

4.3.2 Observations

Despite the bias of the ionospheric response, most of the available information about acoustic–gravity waves in the upper atmosphere has come from ionospheric observations, in which the perturbation caused by a passing gravity wave is called a *travelling ionospheric disturbance*. This aspect is discussed in Section 7.5.5. Direct evidence is rather more restricted but is convincing none the less. Gravity waves are seen in photographs of noctilucent clouds, occurring at heights of 80–90 km, as wave-like structures resembling waves on the surface of an ocean (Figure 4.25), and in the distortions of meteor trails and of vapour trails released from rockets (Section 4.2.2). They have also been detected *in situ* by satellite-borne mass spectrometer and accelerometer instruments.

Most acoustic–gravity waves detected in the upper atmosphere were generated at some lower level. An exception is the *infrasonic waves*, having periods of a few seconds to a minute, which can be detected by pressure sensors (microbarographs) at the ground. One interesting class of infrasonic event is related to auroral activity and appears to travel from the auroral zone. These events are thought to originate in heating and electrodynamic forces in the auroral electrojet (Section 8.4.2) at heights near 110 km.

Fig. 4.25 This way up: noctilucent clouds over northern Canada. (From postcard based on photograph by B. Fogle.) Turn upside down for a view of waves approaching a beach.

4.4 Standard atmospheres and models

From time to time the known parameters of the atmosphere are collated so as to provide, in a convenient form, the latest information about key quantities such as pressure, temperature, density, and composition. The data may be presented as tables according to the major external factors of latitude, season, time of day, sunspot number, and so on. Such a compilation is generally known as a *standard atmosphere*. It may also take the form of a computer program which recalls the required values from a data base or, in some cases, computes them from first principles. In this case the term *model* is commonly used.

The modelling approach has recently (mid-1980s) come into prominence for two reasons. First, it is scientifically worthwhile because there is now sufficient information about many aspects of the solar–terrestrial environment. Second, we now have the computers to make it both feasible and available to all. Modelling is a natural development in the progress of science, whereby attempts are made to put together the available knowledge and comprehension of a topic so that the information can be recalled at will and even used for predictions.

Notable 'standard atmospheres' include the US Standard Atmosphere issued first in 1962 and again in revised forms in 1966 and 1976, and the COSPAR International Reference Atmosphere (CIRA) of which there are versions dated 1961, 1965 and 1972.

Basically there are two kinds of model. An *empirical model* is made up of accumulated facts and figures. It is essentially a data collection, but with the dependencies on independent variables expressed. In some cases these dependencies can be represented by mathematical expressions, but this does not alter the empirical nature of the model. An empirical model contains no physics, and its success depends on how reproducible are the quantities represented; for example, on whether the atmospheric temperature is always the same at some given height under the same conditions of season, time of day, etc. Obviously some component of 'noise' will always be present as well, and the model is only useful as long as the noise is small compared with the actual variations.

A *mathematical model*, on the other hand, is composed of a set of equations representing the basic physics of the situation, and these are solved (usually by numerical techniques) to obtain the quantities of interest. The computation of winds in the neutral air is an example of mathematical modelling, and this method is also important in ionospheric studies. The power of mathematical modelling is that one obtains a mathematically correct result for a given input, and computations may also be made for conditions that have not been observed – for instance, exceptionally high sunspot numbers that occur only infrequently. There are some dangers in the mathematical approach because some of the relevant processes might not be known, or some of the input parameters may be uncertain. Mathematical precision does not of itself guarantee physical accuracy, and validation against the real world is all important.

In addition to purely empirical and purely mathematical models, some are mixed, with both empirical and mathematical parts. In these, some quantities may be calculated while others are taken from a data base, or computed values may be used to fill observational gaps in a data set.

In much modelling and computational work on the upper atmosphere, the neutral atmosphere is likely to be involved to a greater or lesser extent, and thus models of the neutral atmosphere (including tabular standard atmospheres) are of basic importance. The Standard Atmospheres, mentioned above, are published as books of tables showing temperature, density, pressure, scale height, composition, wind etc. Some of this is available as a computer program, the MSIS (Mass-Spectrometer-Incoherent-Scatter) model, which was developed at the NASA Goddard Space Flight Center. It deals with the neutral atmosphere from 50 to 2500 km altitude, and is based on data from rockets, satellites, and incoherent scatter radar. MSIS is available on computer diskette and it provides the number densities of various neutral species, the total mass density and the temperature for stated conditions of time of day, season, latitude, etc.

Further reading

T. E. Van Zandt. The neutral atmosphere, Chapter III-3 in *Physics of Geomagnetic Phenomena* (eds. S. Matsushita and W. H. Campbell). Academic Press, New York and London (1967).

H. Rishbeth and O. K. Garriott. *Introduction to Ionospheric Physics*. Academic Press, New York and London (1969); specifically Chapter 1 on the neutral atmosphere.

R. C. Whitten and I. G. Poppoff. *Fundamentals of Aeronomy*. Wiley, New York, London, Sydney and Toronto (1971). Chapter 3 on the structure of the neutral atmosphere.

I. Harris and N. W. Spencer. The Earth's atmosphere, Chapter 2 in *Introduction to Space Science* (eds. W. N. Hess and G. D. Mead). Gordon and Breach, New York, London and Paris (1968).

S. J. Bauer. *Physics of Planetary Ionospheres.* Springer-Verlag, Berlin, Heidelberg and New York (1973). Chapter 1 on neutral atmospheres.

T. E. Van Zandt and R. W. Knecht. The structure and physics of the upper atmosphere, Chapter 6 in *Space Physics* (eds. D. P. Le Galley and A. Rosen). Wiley, New York, London and Sydney (1964).

M. Nicolet. The structure of the upper atmosphere, Chapter 10 in *Geophysics* (*Vol* 1): *Sun, Upper Atmosphere and Space.* M.I.T. Press, Cambridge, Massachusetts (1964).

G. Brasseur and S.Solomon. *Aeronomy of the Middle Atmosphere.* Reidel, Dordrecht, Boston, Lancaster and Tokyo (1984).

G. M. Keating and J. S. Levine. Response of the neutral atmosphere to variations in solar activity, in *Solar Activity Observations and Predictions* (eds. P. S. McIntosh and M. Dryer). MIT Press, Cambridge, Massachusetts (1972), p 313.

A. S. Jursa (ed.). *Handbook of Geophysics and the Space Environment.* Air Force Geophysics Laboratory, US Air Force. National Technical Information Service, Springfield, Virginia (1985). Chapter 14, Standard and Reference Atmospheres; Section 17.5 on thermospheric winds; Chapter 21, Atmospheric Composition.

S. Chapman and R. S. Lindzen. *Atmospheric Tides.* Reidel, Dordrecht (1969).

S.-I. Akasofu and S. Chapman. *Solar-Terrestrial Physics* Oxford University Press, Oxford, England (1972); Chapter 4 on upper atmosphere dynamics.

A. D. Richmond. Thermosphere dynamics and electrodynamics, in *Solar–Terrestrial Physics* (eds. R.L.Carovillano and J.M.Forbes). Reidel, Dordrecht, Boston and Lancaster (1983).

S. Kato. *Dynamics of the Upper Atmosphere.* Reidel, Dordrecht, Boston and London (1980).

T. Beer. *Atmospheric Waves.* Wiley, New York (1974).

K. Rawer. Modelling of neutral and ionized atmospheres, in *Encyclopaedia of Physics XLIX/7, Geophysics III part VII* (ed. K. Rawer). Springer-Verlag, Heidelberg, New York and Tokyo (1984).

H. J. Fair and B. Shizgal. Modern exospheric theories and their observational relevance. *Reviews of Geophysics* **21**, 75 (1983).

H. U. Widdel. Foil clouds as a tool for measuring wind structure and irregularities in the lower thermosphere (92–50 km). *Radio Science* **20**, 803 (1985).

T. N. Davies. Chemical releases in the ionosphere. *Reports on Progress in Physics* **42**, 1565 (1979).

M. A. Geller. Dynamics of the middle atmosphere. *Space Science Reviews* **34**, 359 (1983).

T. L. Killeen and R. G. Roble. Thermosphere dynamics: contributions from the first 5 years of the Dynamics Explorer program. *Reviews of Geophysics* **26**, 329 (1988).

R. D. Hunsucker. Atmospheric gravity waves generated in the high-latitude ionosphere: a review. *Reviews of Geophysics* **20**, 293 (1982).

D. C. Fritts. Gravity wave saturation in the middle atmosphere: a review of theory and observations. *Reviews of Geophysics* **22**, 275 (1984).

5

The solar wind and the magnetosphere

Magnus magnes ipse est globus terrestris. (The terrestrial globe is itself a great magnet.)
W. Gilbert, *De Magnete* (1600)

5.1 Introduction

The magnetosphere is the region of the terrestrial environment where the geomagnetic field exerts the dominating influence. Generally (though there can be exceptions) the energy density of the magnetic field exceeds that of the plasma:

$$B^2/2\mu_0 > nkT \tag{5.1}$$

(in SI units) where B is the magnetic flux density, k is Boltzmann's constant, and the plasma contains n particles per cubic metre at temperature T. The ratio of energy density between particles and magnetic field, usually written β, is an important parameter. On its low altitude side the magnetosphere merges imperceptibly into the ionosphere (Chapters 6 and 7). There is no sharp boundary and the traditional separation between magnetosphere and ionosphere owes more to semantics than to physics. The outer boundary of the magnetosphere is determined by its interaction with the solar wind. The magnetosphere can thus be regarded as the outermost part of the Earth.

Present knowledge of the magnetosphere is the result of more than a quarter century of space exploration. Only a limited amount can be learned from the ground (though there are some important ground-based techniques) and thus magnetospheric physics could only be a modern science. Ionospheric science is older because many of the observations can be made from ground level.

Given the existence of the geomagnetic field, the form and structure of the magnetosphere are determined largely by emissions from the Sun, and the magnetosphere responds readily and rapidly to changes in the various solar emissions. Since solar activity is continually varying, the magnetosphere also changes from day to day, and from hour to hour. We therefore begin our account at the Sun.

5.2 Solar radiations

The Sun is a typical star, emitting electromagnetic radiation over a wide spectral range, a continuous stream of plasma, and bursts of energetic particles. The visible surface, the photosphere, approximates to a black body at 6400 K. Its radius is about 700 000

Fig. 5.1 Solar eclipse showing the corona. (G. Newkirk, High Altitude Observatory, Boulder, Colorado)

km and above it are regions that are transparent to light: the chromosphere and the corona. The chromosphere may be seen during eclipses; it extends some 2000 km above the photosphere and has a temperature up to 50 000 K. The corona at some 1.5×10^6 K is observable for more than 10^6 km but in fact has no apparent termination. At the Earth's orbit the coronal temperature is about 3×10^5 K. Figure 5.1 is a photograph of the corona taken during a solar eclipse.

The rotation of the Sun on its axis is important in relation to geophysical phenomena because the Sun's active regions are not uniformly distributed. The rotation period increases with latitude from 25.4 days at the equator to 33 days at latitude 75°. The intensity of certain of the emissions received at Earth therefore varies with the solar rotation, a typical average period for geophysical purposes being 27 days.

5.2.1 Solar electromagnetic radiation

The Sun emits 4×10^{26} W of electromagnetic radiation, irradiating the Earth with almost 1.4 kW/m^2, about half of which reaches the surface. Corresponding to its black body temperature, the spectrum peaks in the visible and the intensity falls away both at shorter wavelengths (ultra-violet, X-rays and γ-rays) and at the long wavelength end (infra-red and radio). We shall be much concerned with the variations of solar

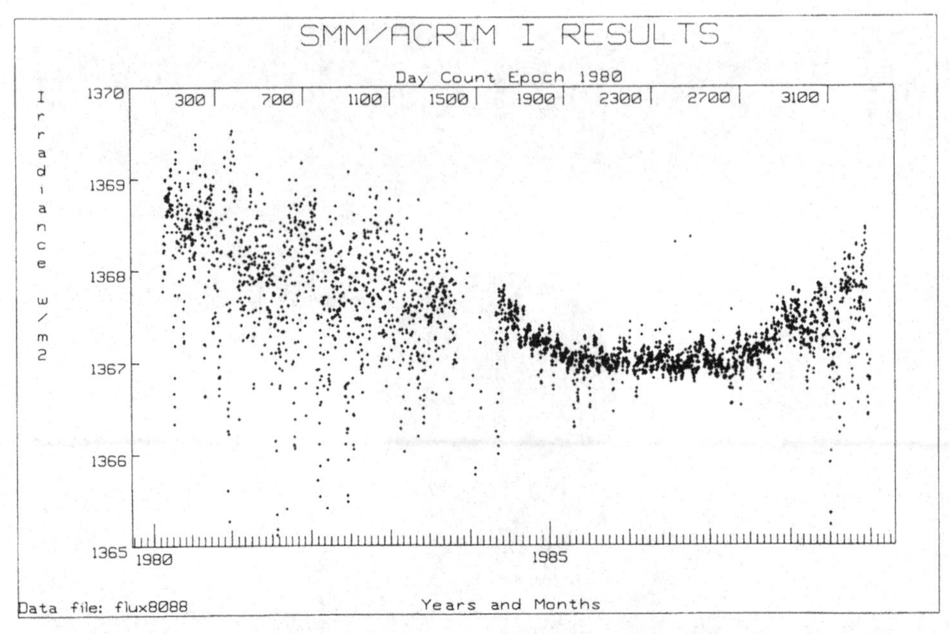

Fig. 5.2 Flux of solar radiant energy detected with the 'active cavity radiometer irradiance monitor' aboard the Solar Maximum Mission. (R. C. Willson, private communication)

emissions, but in fact they are remarkably constant in the visible, to within 1% – see Figure 5.2 – except for the local brightenings called solar flares. Of particular ionospheric interest is the short wavelength end comprising the ultra-violet and the X-rays. These emissions come from the chromosphere and the corona. They do not reach the Earth's surface but have been extensively observed from high altitude by means of rockets and satellites. In addition to a continuum of radiation there are some strong discrete lines. This part of the spectrum is more variable than the visible, particularly at the short-wave end. The region of the solar spectrum up to 3.2 μm is shown in Figure 5.3. The region of greatest concern to us will be that below 100 nm (1000 Å, 0.1 μm), which is far to the short-wave side of the peak.

5.2.2 The phenomenon of the solar flare

A solar flare is a sudden brightening of a small area of the photosphere that may last between a few minutes and several hours. Flares are classified on a scale of 1 to 4 according to the area of the brightening when the Sun is viewed in Hα light (656.2 nm). In addition, very small brightenings are designated S for *subflare*. Flares tend to develop near the long lasting dark areas called *sunspots*, and the regions of enhanced Calcium-K and Hydrogen-alpha emission known as *plages*.

Solar flares are important in solar–terrestrial relations because they are the sources of sporadic particle and electromagnetic emissions that affect the Earth's upper atmosphere, but there is still much uncertainty about their mechanism. Observations show that a flare contains both hot and cool plasma. The flare seen optically is relatively cool at about 10^4 K, the main part being in the lower 5000 km though there can be significant upward extension into the low corona. The visible brightening is due

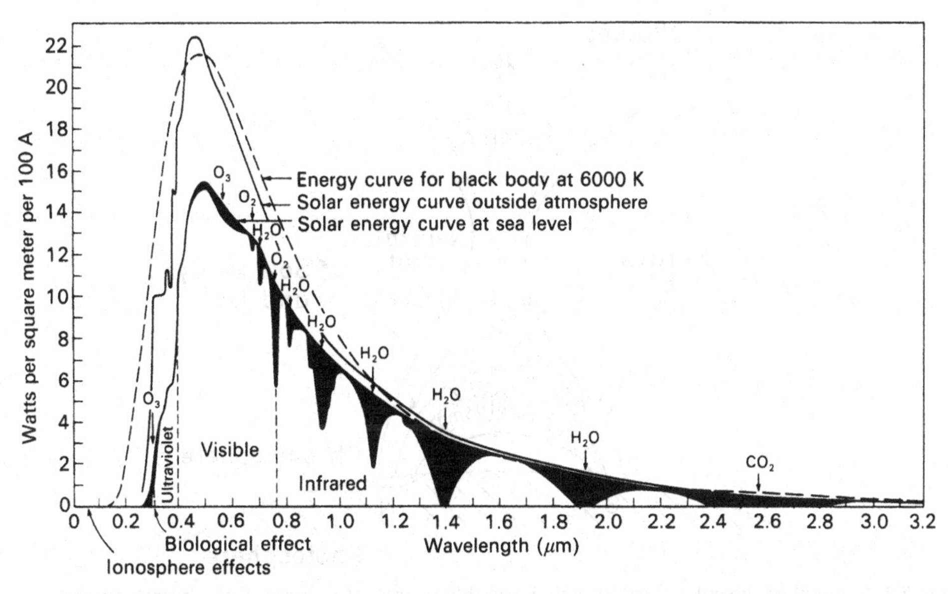

Fig. 5.3 The spectrum of solar electromagnetic radiation reaching the Earth's surface (including the effects of major absorbing species), and outside the atmosphere. The 6000 K black body curve is shown for comparison. The principal ionospheric effects are below 0.1 μm (100 nm). (After J. C. Brandt and P. W. Hodge, *Solar System Astrophysics*. McGraw-Hill, 1964)

to an increase of plasma density from 10^4 to 10^7 electrons/m^3; the region seen may cover one thousandth of the visible solar disc yet be only 20 km thick. It seems clear that the plasma has been greatly compressed, and previously occupied a volume of some 10^{14} km^3. High energy plasma produced at the same time has a temperature of 10^7–10^8 K and electron density about 10^2 m^{-3}.

A typical flare releases 10^{25} J (10^{32} erg) of electromagnetic energy and probably emits another 3×10^{25} J as particles. It is fairly certain that this enormous amount of energy is stored in magnetic fields and released when the fields are annihilated. The plausibility of this idea is readily demonstrated from the magnitude and volume of the fields concerned. Considering that the magnetic fields near sunspots amount to hundreds of gauss and that the volume before compression is some 10^{14} km^3, a simple calculation shows that the total magnetic energy comfortably exceeds the required 4×10^{25} J.

Although the build up of energy in a given region takes about a day, when the energy is released this occurs in only a few hundred seconds. Some 'triggering' process must be involved and the magnetic fields must have taken on a configuration from which annihilation may occur. The exact form of that configuration is not certain, but one possibility is shown in Figure 5.4 on which are marked possible sources of various emissions that are of interest in solar–terrestrial relations. In this model the conversion of magnetic energy begins at a neutral point above two sunspots of opposite polarity. As the field collapses, particles are energized and travel away from the neutral point along the field-lines. Some protons move outward and eventually may cause a *solar proton event* (Section 8.5) at the Earth. Some of the inward travelling electrons are

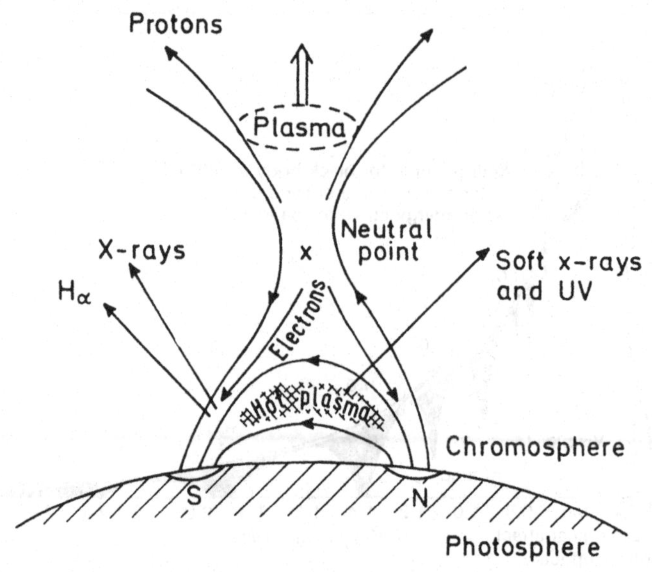

Fig. 5.4 A model of the solar flare, showing possible sources, for some of the known products. (After J. H. Piddington, *Cosmic Electrodynamics*. Wiley, 1969)

stopped in the denser chromosphere and there produce Bremsstrahlung X-rays (Sections 6.2.3 and 8.3.7), which on reaching the Earth cause sudden disturbances of the lower ionosphere (Section 7.1.5). In the same region of the chromosphere hydrogen is ionized by electron collisions and on recombining emits the Hα line, observed from Earth as the *Hα flare*. In the region of the flare the corona is heated and emits soft X-rays and ultra-violet light, both having terrestrial consequences. Plasma ejected above the flare enhances the *solar wind* (Section 5.3). Many of the solar radio bursts (Section 5.2.3) can also be explained by such a model.

Because of the uncertainties the details of Figure 5.4 should not be taken literally, but more as an indication of the kinds of mechanism likely to be involved. It is nevertheless of interest that there are striking similarities between the structures in the solar flare and in the *substorm* in the terrestrial magnetosphere (Section 5.9).

5.2.3 Radio emissions from the Sun

Thermal emissions

The Sun is the source of several kinds of radio emission, reflecting different aspects of solar activity. A thermal emission is present at all times, corresponding to the temperature of the emitting region. The level from which an emission can leave the Sun depends on the electron density, since an ionized medium is only transparent to waves whose frequency exceeds the plasma frequency (Section 2.3.4). In fact, most of the radiation at a given frequency originates near the level where that is the plasma frequency. 10-cm waves come from the upper chromosphere, and metre waves from the corona. The corresponding temperatures are 75 000 K and 10^6 K respectively, and these determine the emission intensities.

Sporadic emissions

In addition, four types of sporadic emission occur, known as Types I to IV.

Type I *noise storms* lasting from hours to days are made up of intense, sharp, spikes. They come from small areas near sunspots, and frequently start when a flare occurs. They are observed at metre wavelengths and are probably generated by electron streams in the magnetic fields above sunspots.

The most intense of the solar emissions are related to solar flares and are designated Type II or Type III. The latter consist of a number of sharp bursts at the start of a flare, and there is rapid outward movement of the disturbance as evidenced by the variation of radiofrequency and as observed by scanning interferometers. Impulsive bursts of microwaves and hard X-rays (10–100 keV) usually occur at the same time. Type II bursts last 5 to 30 minutes and may be observed when large flares occur. They are again at metre wavelengths, but are slower than Type III and probably due to a shock wave travelling out from the flare. The velocity, 1000 km/s, is typical of a fast solar wind stream (Section 5.3.5).

Type IV is a burst of long duration that follows some large flares and covers a wide band of frequencies. It is attributed to synchrotron emission from high energy particles in solar magnetic fields. Various sub-classes of Type IV have been related to proton emission, and to expanding loops of the solar magnetic field.

5.2.4 Solar activity cycles

The Sun undergoes periodic variations on several time scales, the principal one being the 11-year cycle. If sunspots are observed in 'white light' – that is, with a telescope not including a coloured filter – they are seen as dark areas, tending to occur in groups and lying mainly between solar latitudes 5° and 30°. The *Wolf sunspot number* is defined as

$$R = k(f+10g), \tag{5.2}$$

where f is the total number of spots seen, g is the number of disturbed regions (either single spots or groups of spots) and k is a constant for the observatory, related to the sensitivity of the observing equipment. Started by Wolf in 1848, R is still the principal index used for quantifying solar activity, and Figure 5.5 shows some recent solar cycles in terms of R. Often, some smoothing is applied; to emphasize the trends it is usual to plot the monthly mean values (as in Figure 5.5), and these may then be averaged over 12 months and plotted as a running mean, R_{12}. (The average is obtained by taking values for 13 months centred on the month of interest, but the end months are given half weight.) The strongest cycle on record was cycle 19 which peaked in 1958 when R reached a value of 200; a more typical maximum would be 100–120. At sunspot minimum, such as in 1954, 1965, 1976 or 1986, the sunspot number falls below 20 and sometimes below 10. The typical cycle is not symmetrical; the time from minimum to maximum is 4.3 years, and the time from maximum to minimum is 6.6 years on average. It is observed that cycles which rise more rapidly tend to reach higher maxima.

Sunspot numbers are available from direct observations going back to 1700, though they are only considered 'good' since 1818 and 'reliable' since 1848. By other means it has been possible to carry the data back for some 2000 years and Figure 5.6 shows one such estimate. Note the Sporer and Maunder minima of the 15th and 17th

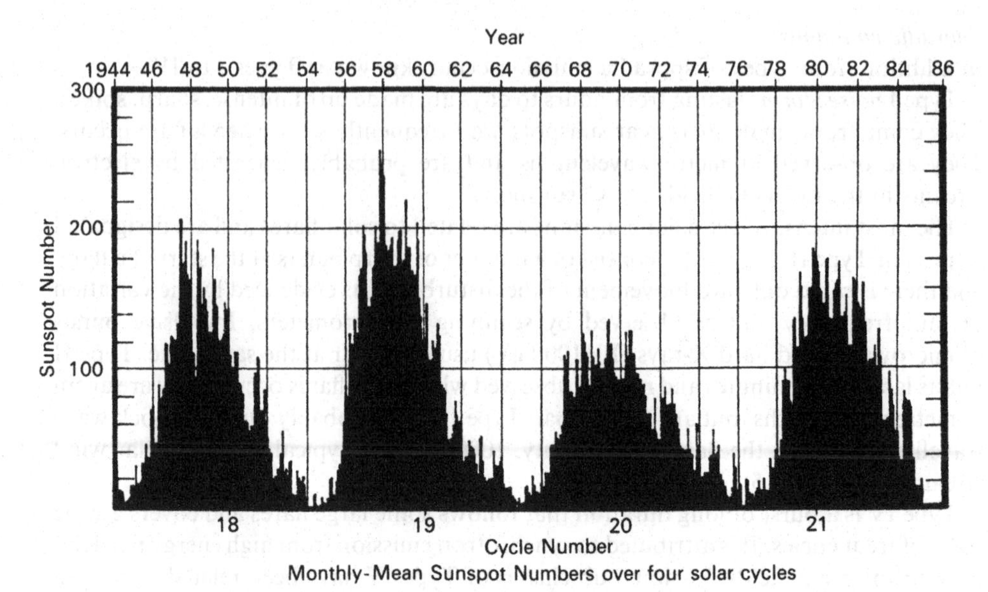

Fig. 5.5 Monthly-mean sunspot numbers over four solar cycles. (*Annual Report for FY* 1984, National Geophysical Data Center, Boulder, Colorado)

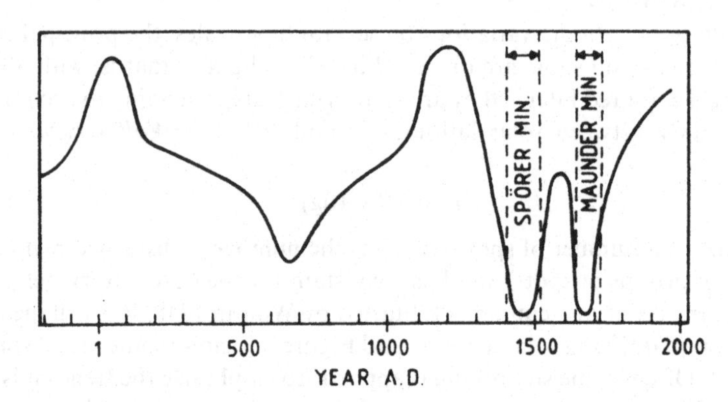

Fig. 5.6 The mean solar activity during the last 2000 years. (After D. W. Hughes, *Nature* **266**, 405, copyright © 1977 Macmillan Magazines Ltd)

centuries when solar activity was exceptionally low for some decades. These serve to remind us how short is the time for which we have had direct observations, and warn us not to assume that patterns observed during this brief period of the Sun's existence are necessarily typical.

Even within modern times the data show that all cycles are not equally strong, the largest being four times as great as the smallest. This implies a longer term modulation of the basic 11-year cycle, and spectral analysis of the time series of sunspot numbers shows that there are other significant periodicities, notably at 57 and 95 years. Further, the 11-year cycle is not particularly stable but varies between 9 and 14 years. The stronger cycles seem to arrive relatively earlier. One reason for studying sunspot data

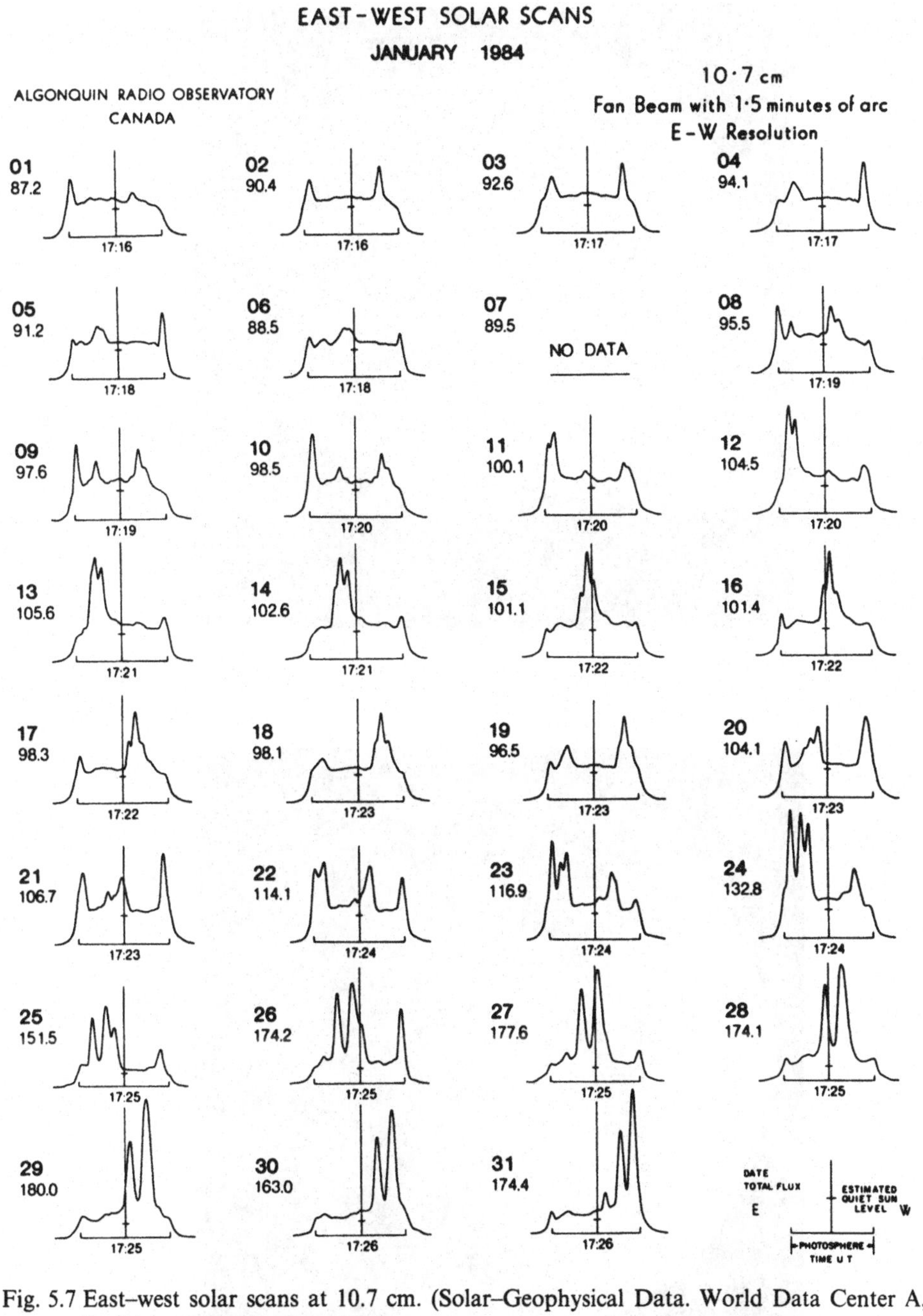

EAST-WEST SOLAR SCANS
JANUARY 1984

ALGONQUIN RADIO OBSERVATORY
CANADA

10·7 cm
Fan Beam with 1·5 minutes of arc
E-W Resolution

NO DATA

DATE
TOTAL FLUX
ESTIMATED QUIET SUN LEVEL
E
W
PHOTOSPHERE
TIME U T

Fig. 5.7 East–west solar scans at 10.7 cm. (Solar–Geophysical Data. World Data Center A, Boulder, Colorado, Feb. 1984)

in the long term is to aid the prediction of the timing and strength of future cycles (Section 10.5.3).

Some evidence (particularly from the study of Sun–weather relationships) has

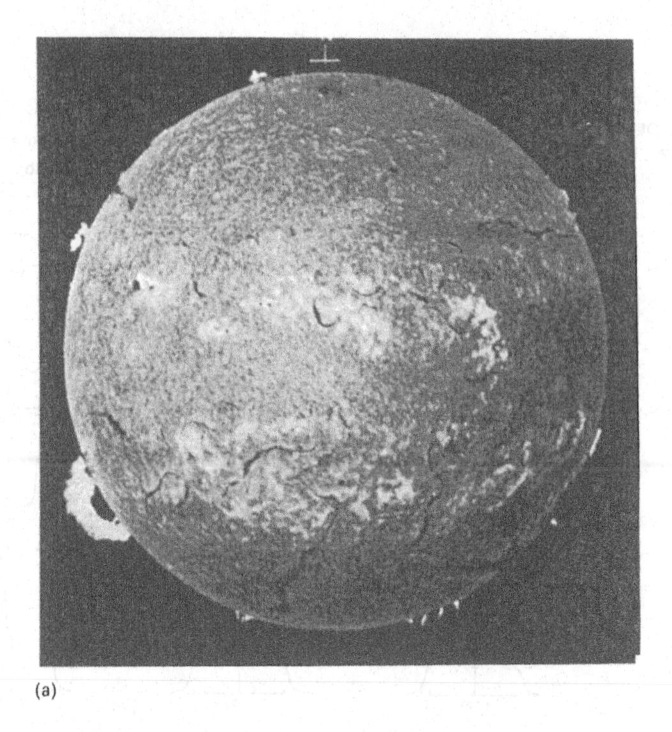

(a)

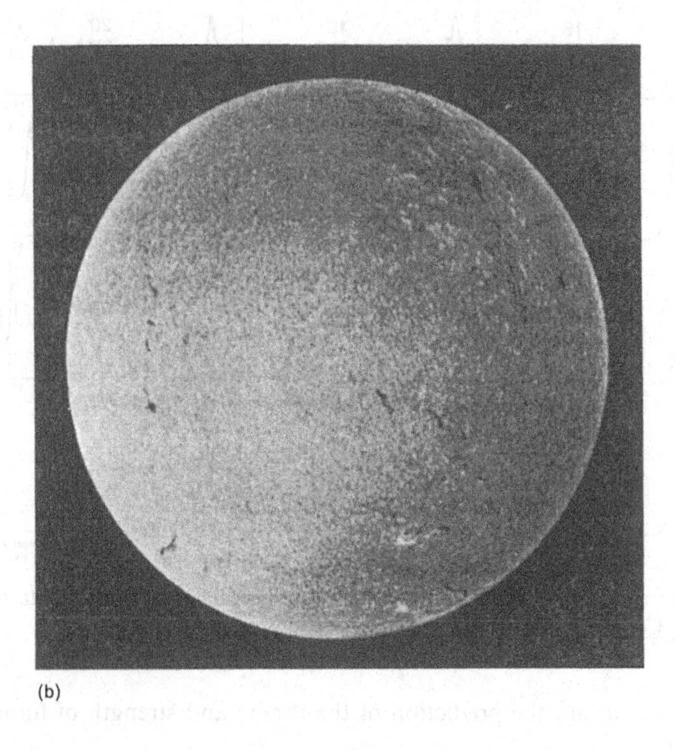

(b)

Fig. 5.8 For legend see facing page.

drawn attention to the 22-year cycle, or *double sunspot cycle*. The Sun's general magnetic field reverses direction as each new cycle develops, and some studies on the relation between solar activity and certain terrestrial phenomena have indicated periodicities of 22 rather than 11 years. In contrast to the instability of the 11-year cycle, the 22-year cycle is quite stable, which supports the view that this, not 11 years, is really the fundamental period of solar activity.

The incidence of solar flares is another measure of solar activity and is related to the sunspot number by

$$N_f = \alpha(\bar{R} - 10), \tag{5.3}$$

where N_f is the number of flares observed during one solar rotation (27 days), $\bar{R}$ is the mean sunspot number and α is an observatory constant of value between 1.5 and 2.

The flux of 10-cm radio emissions provides the third main indicator of general solar activity. This quantity, the *10-cm flux*, is routinely monitored and reported in *flux units* of 10^{-22} W/m^2-Hz. This flux is highly correlated with the sunspot number and is particularly valuable as an observation not affected by the weather. Scans of the solar disc using a narrow beam antenna are able to map the positions of active regions in some detail, though not, of course, with the resolution of an optical telescope. Figure 5.7 shows an assembly of daily east–west scans in which the motion of active regions due to solar rotation can be clearly seen.

The appearance of the Sun in Hα light under active and quiet conditions is illustrated in Figure 5.8.

Satellite measurements indicate that the Sun's total electromagnetic emission varies less than 0.1 % over a solar cycle. The measurements of Figure 5.2 were fitted by a sine wave of period 10.95 years and amplitude 0.04 % of the mean irradiance. Correlation of irradiance (S) with sunspot number (R) and 10-cm flux (F) in flux units gives the relations

$$S = 1366.82 + 7.71 \times 10^{-3} R, \tag{5.4a}$$

$$S = 1366.27 + 8.98 \times 10^{-3} F. \tag{5.4b}$$

5.2.5 Proton emissions

Occurrence

Some flares produce a flux of energetic (1–1000 MeV) protons sufficient to cause a *proton event* at the Earth (Section 8.5). The association between large flares and proton events is not complete, since many large flares are not followed by proton events and some of the latter cannot be associated with any known flare. Nevertheless, the occurrence of proton events detected at the Earth depends strongly on the phase of the sunspot cycle (Figure 5.9). Over the long term there are six events each year, but there may be ten or more events in an active year and none at all (or, more likely, one or two) in a quiet one. Although the occurrence correlates with sunspot number in a general way the pattern changes from cycle to cycle. Some cycles show a tendency for relatively

Fig. 5.8 Appearance of the Sun in Hα light when (a) active and (b) quiet. The sunspot numbers on 17 August 1980 and 19 August 1984 were 190 and 9 respectively, and the 10-cm radio fluxes were 197.7 and 74.4. (*Proceedings of the Workshop on Solar–Terrestrial Predictions*. Meudon, 1984)

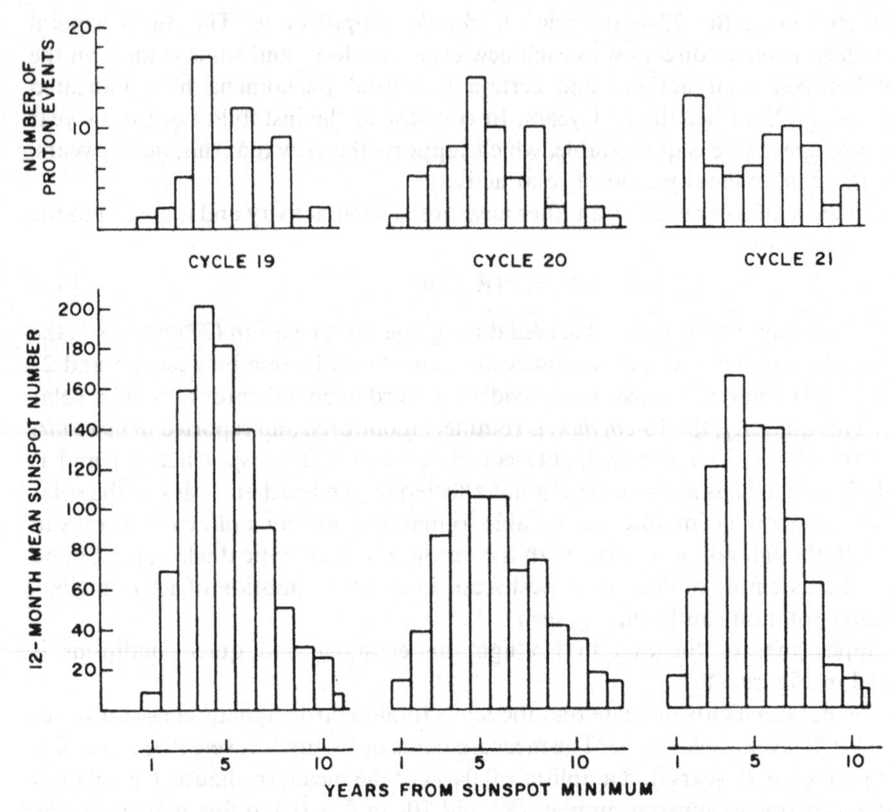

Fig. 5.9 Occurrence of proton events during solar cycles 19 (May 1954–Oct 1964) 20 (Nov 1964–June 1976) and 21 (July 1976–Sept 1986). Some preference for the declining phase can be seen in cycles 19 and 21. (D. F. Smart and M. A. Shea, *J. Spacecraft and Rockets* **26**, 403, 1989)

more events to occur in the declining phase, but it would not be safe to rely on this for prediction purposes. The occurence of proton events also tends to be episodic, and some 20 % of events come from active solar regions that produce more than one event. When viewing such data over a number of years it should be remembered, too, that detection techniques have improved with time.

Connection with radio noise

There is a correlation between the occurrence of proton events and type IV radio bursts, which makes it possible to predict the flux of protons exceeding 10 MeV in energy and also the proton spectrum from the spectral characteristics of the radio burst. The bursts which correlate with proton events are those having a U-shaped spectrum. In a typical example, the intensity of the radio burst will be about 10 000 flux units at frequencies of a few hundred megahertz and again around 10 000 MHz, but at frequencies between, say near 1000 MHz, will be a minimum down to 100–1000 flux units. Although the mechanism is not certain, the correlation indicates a close connection between the acceleration of protons and electrons (the source of synchrotron radiation) in the proton flare. Its practical importance is that some characteristics of the ensuing proton event can be predicted before it reaches the Earth.

5.3 The solar wind

5.3.1 Discovery

The story of the solar wind begins with the work of S. Chapman and V. C. A. Ferraro on magnetic storms (Section 7.4.2). Some storms commence with a sharp increase in the geomagnetic field as recorded with ground-based magnetometers. By way of explanation, Chapman and Ferraro suggested in 1931 that they are caused by streams of corpuscles ejected from the Sun, to reach the Earth about a day later. There, unable to penetrate the geomagnetic field, it would be deflected around and, at the same time, cause some compression of the field, which would be detectable at the ground.

It was generally assumed that the streams were released from flares and were therefore intermittent. In 1951 L. Biermann, studying comet tails, concluded that radiation pressure, the traditional explanation, could not account for the observations. He inferred that there must also be pressure from a stream of particles from the Sun, and that this stream must be present all the time. He deduced a velocity of about 500 km/s for the stream – which turns out to have been a surprisingly good estimate.

Theoretical studies by Chapman and by E. N. Parker subsequently proved that, unlike the Earth's atmosphere, the solar corona is not in hydrostatic equilibrium but expands continuously, with matter leaving the Sun and streaming out into space. Parker called this flow the *solar wind*.

The existence of the solar wind was verified by space probes in the early 1960s. It was detected by the Russian satellites Luniks 2 and 3 and by the American vehicle Explorer 10 which measured a speed of 300 km/s. But most notably it was Mariner 2, *en route* for Venus, which observed that the wind blew constantly during the whole four months of the flight. It was this observation which confirmed the solar wind as a permanent feature of the solar system.

5.3.2 Theory of the solar wind

It is possible, from optical measurements, to estimate the electron density in the solar corona at various distances from the Sun. If it is assumed that the material of the corona is moving outward, the velocity can then be calculated from the conservation of flux. That is, for particle density n and velocity v at distance r, then nr^2v is constant. From this, the outward acceleration can be calculated. It is found that beyond 15 or 20 solar radii the convective acceleration, acting outward, exceeds the gravitational acceleration acting inward. Matter therefore leaves the Sun continously beyond a certain distance.

In the simplest treatment the solar corona is assumed to contain one type of particle with constant temperature throughout. The flow is outward and steady; solar rotation, magnetic fields and viscosity are neglected. By continuity,

$$\rho v r^2 = \text{constant}, \tag{5.5}$$

where ρ is the gas density, v the velocity, and r the radial distance from the Sun. Considering the forces acting on a unit volume of gas,

$$\rho v \cdot \frac{dv}{dr} + \frac{dP}{dr} + \frac{GM_s \rho}{r^2} = 0, \quad (5.6)$$

where P is the gas pressure, G the gravitational constant, and M_s the solar mass. Also,

$$P = \frac{kT}{m} \cdot \rho \quad (5.7)$$

from the gas law, where k is Boltzmann's constant, T the absolute temperature, and m the particle mass. Substituting from Equation 5.7, integrating Equation 5.6, and putting the velocity and the radial distance in dimensionless form by means of $U = v/(2kT/m)^{\frac{1}{2}}$ and $\Pi = rkT/mGM_s$ ($\sim r/8R_s$), Equations 5.5 and 5.6 become

$$\ln \rho = -\ln U - 2\ln \Pi, \quad (5.8)$$

$$U^2 - \ln U = 2\ln \Pi + \frac{1}{\Pi} + \text{constant}. \quad (5.9)$$

The constant of integration is found as follows. Differentiating Equation 5.9 gives

$$\left(2U - \frac{1}{U}\right)\frac{dU}{d\Pi} = \frac{1}{\Pi}\left(2 - \frac{1}{\Pi}\right). \quad (5.10)$$

The left-hand side is zero if $U = 1/\sqrt{2}$, and since $dU/d\Pi$ cannot be infinite it follows that $\Pi = \frac{1}{2}$ at that point. These corresponding values of U and Π enable the constant in Equation 5.9 to be found, and thus the velocity to be expressed as a function of distance.

From a theoretical treatment, Parker in 1958 deduced solar wind velocities between 260 and 1160 km/s at the distance of the Earth for coronal temperatures between 5×10^5 and 4×10^6 K. The value $U = 1/\sqrt{2}$ corresponds to a velocity $v = (kT/m)^{\frac{1}{2}}$, which is the acoustic speed and occurs in this model at $r = 4\ R_S$, R_S being the solar radius. The corona streams out subsonically to this distance but supersonically beyond it. Later treatments include a varying temperature, viscosity, a magnetic field, and more than one kind of particle.

5.3.3 Properties of the solar wind

There have been many observations of the solar wind since its discovery in the early 1960s. Measurements of the positive ions using plasma probes find that the flux of particles with energy exceeding 25 eV is between 2×10^{12} and 7×10^{12} particles/m^2-s. Although most ions are protons (H^+) there is an α-particle (He^{++}) component typically amounting to 5% though exceptionally up to 20% of the total. Heavier atoms total perhaps 0.5%, and, in contrast to the light ions, these are not fully ionized. The number density of positive ions varies between 3 and 10 cm^{-3} (3×10^6 to 10^7 m^{-3}), the most typical value being 5 cm^{-3}, and there is a similar number of electrons for bulk neutrality. The mean mass of solar wind particles is therefore almost half that of the proton, about 10^{-27} kg. Fluctuations as great as a factor of ten can occur in times of minutes and hours, implying irregularities within the solar wind over distances of 10^5 km and more.

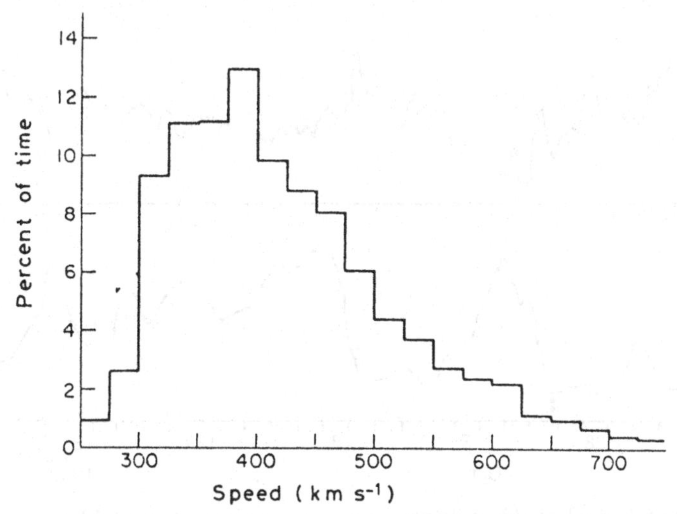

Fig. 5.10 Speed of the solar wind: a histogram of measurements between 1962 and 1970. (J. T. Gosling, in *Solar Activity Observations and Predictions* (eds. McIntosh and Dryer). MIT Press, 1972)

At the distance of the Earth's orbit the speed of the solar wind is usually between 200 and 700 or 800 km/s (Figure 5.10), on which is superimposed a random component of temperature 10^5 K. The energy is more directed than random, and the plasma is considerably cooler than its directed velocity would suggest. The energy carried is about 10^{-4} W/m^2, approximately a tenth of that in the extreme ultra-violet region of the solar spectrum.

The variability in the density and speed of the solar wind is shown over one solar rotation in Figure 5.11. There is a degree of anticorrelation between density and speed.

The solar wind is the principal medium by which the activity of the Sun is communicated to the vicinity of the Earth, and it is extremely important in solar–terrestrial relations. The interaction depends on a weak magnetic field carried along by the plasma.

5.3.4 Interplanetary magnetic field and sector structure

Form of IMF

The solar wind carries with it a weak magnetic field amounting to a few nanoteslas (= a few gamma). The field is 'frozen in' to the plasma because the electrical conductivity is very large (Section 2.3.6), and the plasma controls the motion of the total magnetoplasma because it has the greater energy density: $nmv^2/2 > B_s^2/2\mu_0$, where n is the particle density, m the particle mass, v the solar wind velocity, and B_s the magnetic flux density. In the solar wind the kinetic energy of the particles exceeds the energy density of the magnetic field by a factor of about eight.

The *interplanetary magnetic field* (*IMF*) was discovered in 1963 with a very sensitive magnetometer on IMP-1 (the Interplanetary Monitoring Platform), an eccentric orbit satellite with apogee at 32 R_E. Although the solar wind flows out almost radially from

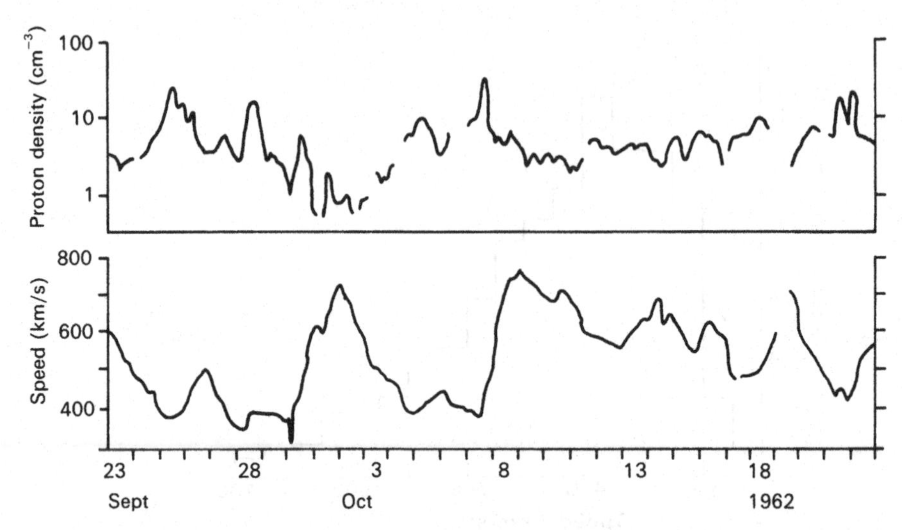

Fig 5.11 Density and speed of the solar wind over one solar rotation in 1962, measured by Mariner-2. (After H. Rosenbauer, in *Solar System Plasmas and Fields* (eds. Lemaire and Rycroft). *Advances in Space Research* **2**, 47, copyright (1982) Pergamon Press PLC)

the Sun, the solar rotation gives the magnetic field a spiral form, as in Figure 5.12a. This effect can be simulated by any amateur space scientist who waters his/her garden with a hosepipe. By turning round while watering the garden it is seen that the jet of water follows a spiral path though the trajectories of individual drops are still radial. Therefore the spiral form of the IMF is said to result from the *garden hose effect*! It happens that at the orbit of Earth the field lines run at about 45° to the radial direction: the radial and the east–west components of the IMF are therefore about equal in magnitude.

Sectors

In direction the IMF can be either inward or outward with respect to the Sun, and one of the most remarkable of the early results, and a fact of great significance, was that distinct sectors may be recognized, the field being inward and outward in alternate sectors. Figure 5.12b shows some of the original measurements, where four sectors – two inward and two outward – were present. However this is not always the case because the sector structure evolves with time. Sometimes there are only two sectors, and sometimes the sectors are not all of the same width.

Obviously, though the IMF originates in the Sun and is strictly part of the main solar magnetic field, the observations in the ecliptic plane between 0.3 and 10 AU show a form that is not dipolar. (The complex fields of the photosphere can be neglected here because they extend out only a couple of solar radii.) The resolution of this anomaly is to imagine a current sheet in or near the equatorial plane which effectively divides the outward field (above the plane) from the inward field (below it) as in Figure 5.13. If the solar magnetic dipole is tilted from the rotation axis, the current sheet will be

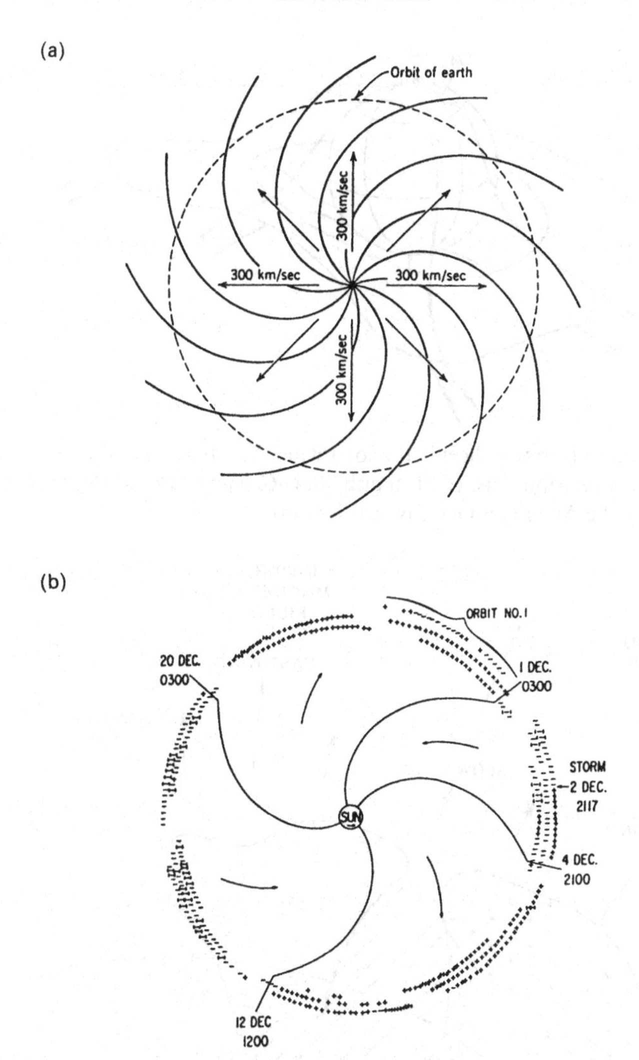

Fig. 5.12 (a) Form of the interplanetary magnetic field in the solar equatorial plane, corresponding to a solar wind speed of 300 km/s. (T. E. Holzer, *Solar System Plasma Physics, Vol I.* North-Holland, 1979, p. 103. Elsevier Science Publishers) (b) Sector structure of the solar wind in late 1963, showing inward (−) and outward (+) IMF. (J. M. Wilcox and N. F. Ness, *J.Geophys. Res.* **70**, 5793, 1965, copyright by the American Geophysical Union)

tilted from the ecliptic plane and a spacecraft near the Earth will observe a two-sector structure as the Sun rotates. When more than two sectors are seen it is thought that the current sheet has developed undulations as in the skirt of a pirouetting ballerina; hence the concept of Figure 5.13 is often known as the *ballerina model.* Most solar wind measurements have been near the ecliptic plane, but spacecraft venturing out of the ecliptic plane have observed the disappearance of the sector structure – which is consistent with the ballerina model.

It is also possible that the more complex sector structures are due to a more

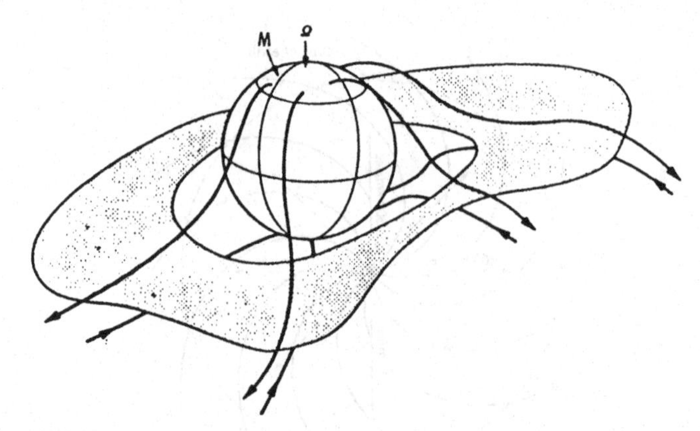

Fig. 5.13 Ballerina model of current sheet in the solar wind. M is the axis of the current sheet and Ω is the Sun's rotation axis. (E. J. Smith, *Rev. Geophys. Space. Phys.* **17**, 610, 1979, copyright by the American Geophysical Union)

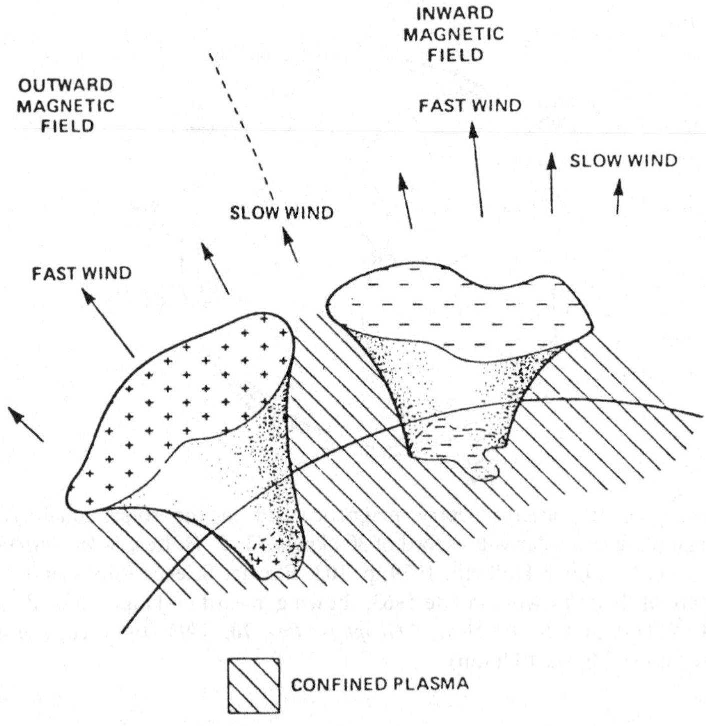

Fig. 5.14 Sketch of coronal holes as the source of high speed solar wind. The + and − indicate the polarity of the coronal magnetic field in the holes, and the shaded region outside the holes represents plasma confined by closed magnetic field. (A. Barnes, *Rev. Geophys. Space. Phys.* **17**, 596, 1979, copyright by the American Geophysical Union)

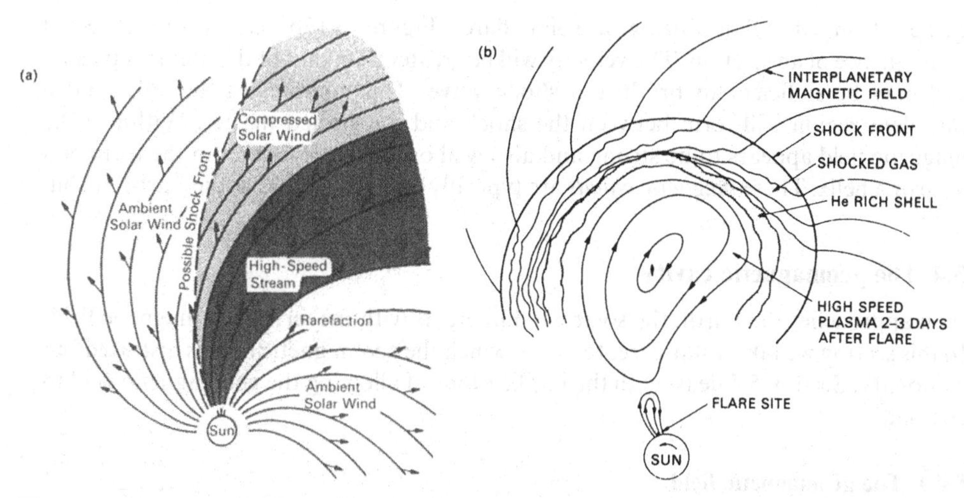

Fig. 5.15 (a) Interaction of a steady, localized stream of high speed plasma with slower ambient solar wind. (T. E. Holzer, in *Solar System Plasma Physics, Vol I.* North-Holland, 1979, p. 103. Elsevier Science Publishers) (b) High speed plasma from a solar flare driving an inter-planetary shock. The ejected plasma contains an ordered magnetic field, but the magnetic field is turbulent between the shock and the ejecta. (Reprinted with permission from L. F. Burlaga, in *Solar System Plasmas and Fields* (eds. Lemaire and Rycroft). *Advances in Space Research* **2**, 51, copyright (1982) Pergamon Press PLC)

complicated solar field, which when the Sun is active could be due to localized plasma ejections, for instance from flares.

5.3.5 The coronal hole and fast solar streams

A link between the solar wind and a particular feature of the corona was discovered by the Skylab missions of May 1973 to February 1974. A so-called *coronal hole* emits less light at all wavelengths than adjoining regions, but it is most marked in an X-ray photograph where it appears as a black area. Coronal holes are regions with abnormally low density where the magnetic field has a single polarity – all inward or all outward. This is open magnetic field, going out into interplanetary space rather than returning to the Sun, and it is rapidly diverging (Figure 5.14).

The hole is the source of fast solar-wind streams in which the speed exceeds 700 km/s. The speed is greater from a larger hole. Less than 20 % of the solar surface is composed of coronal holes, and they are more numerous during the declining phase of the sunspot cycle.

The fast streams interact with the slower solar wind as in Figure 5.15a, compressing the magnetic field and the plasma ahead and sometimes, though not always, creating a shock front. The compressed plasma is heated. A rarefaction follows. Within the stream the magnetic field maintains the same polarity (inward or outward) and is the same as in the corresponding coronal hole. The fast streams from coronal holes co-rotate with the Sun and can persist for several rotations. They are the probable cause of recurring geomagnetic storms (Section 8.4).

A more extreme perturbation of the solar wind may be caused when plasma is ejected from the Sun during a solar flare. Figure 5.15b shows the expected consequence of an ejection. The velocity will be greater than that of the ambient plasma and may be sufficient to produce a shock wave. This compresses the IMF and a turbulent region will form between the shock and the ejected matter. Within it the magnetic field appears to be strong and also well ordered, sometimes in the form of a loop or a helix. These *magnetic clouds* are typically 0.25 AU across at the Earth's orbit.

5.4 The geomagnetic cavity

As it approaches the Earth, the solar wind interacts with the terrestrial magnetic field. In this section we take a static treatment in which the geomagnetic field is distorted but stationary. Section 5.5 deals with the implications of allowing the geomagnetic field to circulate.

5.4.1 The geomagnetic field

Dipole field

To a first approximation the geomagnetic field at and close to the planet's surface can be represented as a dipole field. The poles of the dipole are at geographic latitudes and longitudes 79° N, 70° W, and 79° S, 70° E. The magnetic flux density is given by

$$B(r, \lambda) = \frac{M}{r^3}(1 + 3\sin^2\lambda)^{\frac{1}{2}}, \tag{5.11}$$

where M is the dipole moment, r the geocentric radial distance, and λ the magnetic latitude. This is accurate to about 30% at points within 2 or 3 Earth-radii of the surface. Although not very accurate, the dipole form is useful for making approximate calculations about the inner magnetosphere and we therefore summarize its main properties.

The field components in the r and λ directions are

$$B_r = -\frac{2M\sin\lambda}{r^3}, \tag{5.12}$$

$$B_\lambda = \frac{M\cos\lambda}{r^3}. \tag{5.13}$$

Thus,

$$\frac{B_\lambda}{B_r} = \frac{r\,d\lambda}{dr} = -\frac{1}{2\tan\lambda}$$

and, by integration,

$$r = r_0\cos^2\lambda. \tag{5.14}$$

For a selected value of r_0 this is the equation of a field-line, the locus of the force acting on a single north pole. If $\lambda = 0$, $r = r_0$; r_0 is thus the radial distance to the field-line over the equator, and its greatest distance from the Earth.

It is convenient in the magnetosphere to use the radius of the Earth, R_E, as the unit of distance. Putting $r/R_E = R$,

$$B(R, \lambda) = \frac{0.31}{R^3}(1 + 3\sin^2\lambda)^{\frac{1}{2}} \text{ gauss} \tag{5.15}$$

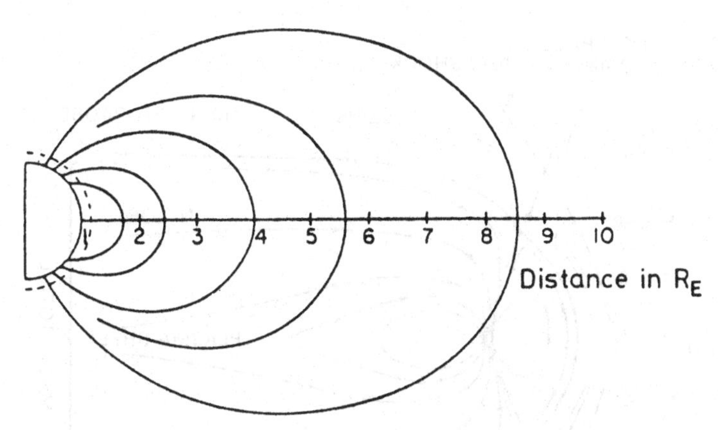

Fig. 5.16 Dipolar field-lines. (D. L. Carpenter and R. L. Smith, *Reviews of Geophysics* **2**, 415, 1964, copyright by the American Geophysical Union)

0.31 G ($= 3.1 \times 10^{-5}$ Wb/m^2) being the flux density at the Earth's surface at the magnetic equator. In these terms, the field-line equation becomes

$$R = R_0 \cos^2 \lambda, \tag{5.16}$$

R and R_0 being measured in Earth-radii. The latitude where the field-line intersects the Earth's surface is given by

$$\cos \lambda_E = R_0^{-\frac{1}{2}}. \tag{5.17}$$

Figure 5.16 shows dipolar field-lines.

Field expansion

The displaced dipole model, in which the dipole is displaced 400 km from the centre of the Earth, is a closer approximation to the real field than is the centered dipole. However, for accurate work (not too far above the surface) the field is expressed as a series of spherical harmonics in which the magnetic potential, V, is given by

$$V = \sum_{n=1}^{\infty} (R_E/R^{n+1}) \sum_{m=0}^{n} (g_n^m \cos m\phi + h_n^m \sin m\phi)\, p_n^m(\cos\theta), \tag{5.18}$$

where g_n^m and h_n^m are coefficients, $p_n^m(\cos\theta)$ are associated Legendre functions, and R, θ and ϕ are spherical polar coordinates. ($R = r/R_E$, θ = co-latitude, ϕ = longitude.) In this formula, n is the order of the magnetic multipole; $n = 1$ for dipole, $n = 2$ for quadrupole, but $n = 0$ for monopole is not included. Also, $p_n^m = 0$ if $m > n$. The potential due to a dipole is thus $V = (R_E/R^2) g_1^0 \cos\theta$.

The coefficients are derived by fitting the expression to measurements of the magnetic elements on the global scale, using magnetometers both on the ground and on satellites. The expression is applied as a computer program, the flux density being obtained as the gradient of the magnetic potential. Because the geomagnetic field changes with time – *the secular variation* – a fresh set of coefficients, relating to a specific epoch, is published from time to time. Such representations are accurate to about 0.5 % at and near the surface. The terms of higher order become less important at greater distances and the field tends to become more dipolar. However beyond 3 or 4 R_E a gross distortion due to the solar wind takes over.

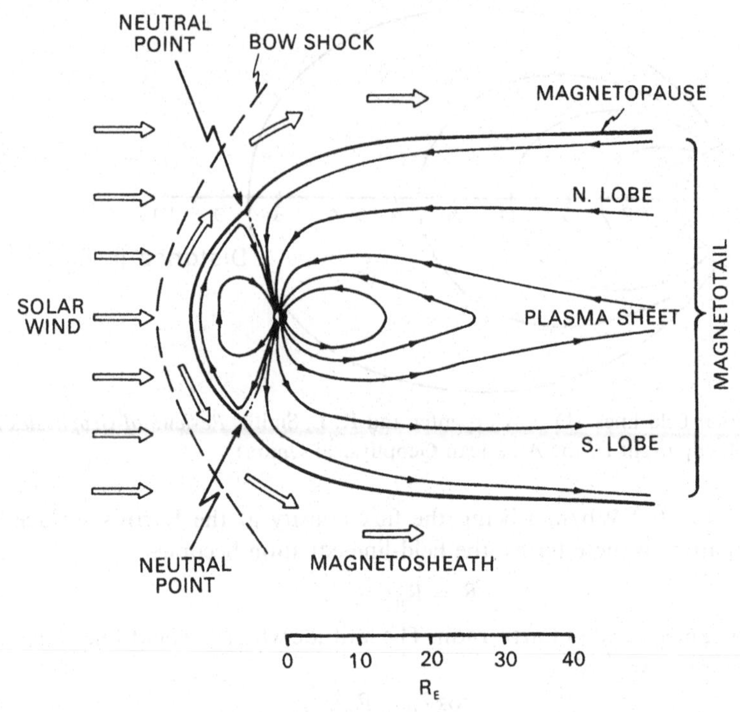

Fig. 5.17 Cross-section of the geomagnetic cavity and external plasma flow, showing the magnetopause and the shock. (Adapted from V. M. Vasyliunas, in *Solar–Terrestrial Physics*. Reidel, 1983, p. 243. Reprinted by permission of Kluwer Academic Publishers)

The cavity

The geomagnetic cavity, illustrated in north–south section in Figure 5.17, is formed because the solar wind is not able to penetrate the geomagnetic field, being swept around it and at the same time distorting the field. This kind of behaviour was foreseen by Chapman and Ferraro as long ago as 1930 in their study of magnetic storms (Section 5.3.1). In modern terms the solar wind, because of its high electrical conductivity, is 'frozen out' of the geomagnetic field (Section 2.3.6).

For a simple analogy consider a conducting plate in a strong laboratory magnetic field. As soon as the plate is moved, electric currents are induced in it and these currents create another magnetic field which, by Lenz's law, interacts with the original field to oppose the original motion. In the hypothetical limit when the plate has infinite conductivity, the induced current is also infinite, the restoring force is infinite, and therefore no relative motion is possible. The conductivity of the solar wind is very large and penetration of the geomagnetic field is virtually prevented. Section 2.3.6 gives a mathematical treatment based on Maxwell's equations, and details of the interaction are given in later sections of this chapter.

5.4.2 The magnetopause

Pressure balance

The form of the boundary between the geomagnetic field and the incident solar wind can be deduced from the pressure balance across the boundary. When the system is in equilibrium the pressure of the solar wind outside is at every point of the surface equal to that of the magnetic field inside.

If the solar wind contains N particles/m^3, each of mass m kg, travelling at velocity v m/s and striking the surface at angle ψ from the normal, then $Nv \cos \psi$ strike each m^2 of the surface per second. Assuming specular reflection, each particle changes momentum by $2mv \cos \psi$. The total rate of change of momentum is then $2Nmv^2 \cos^2 \psi$ N/m^2, which has to be equated to the magnetic pressure $B^2/2\mu_0$. All species within the solar wind contribute but the protons have greatest effect. A simple calculation along these lines readily gives a realistic distance for the position of the boundary (10 R_E) along the Earth–Sun line and enables one to estimate how it varies if the solar wind changes. A full computation is more complicated since the orientation of the boundary at each point is not known at the outset and an iteration procedure is required.

The resulting boundary, the *magnetopause*, is indicated in Figure 5.17. The geomagnetic field is severely distorted. Note in particular that

(a) Field-lines originating at low latitude still form closed loops between northern and southern hemispheres, though there can be some distortion from the dipole form.
(b) Lines emerging from the polar regions are swept back, away from the Sun; some of these would have connected on the day side in a dipole field.
(c) Intermediate between these regions are two lines, one in each hemisphere, that go out and meet the magnetopause on the day side, though in fact their flux density falls to zero as they reach it; here, neutral points are formed.

The magnetopause has a finite thickness, though it is thin (~ 1 km) in comparison with the size of the magnetosphere.

Image dipole

Useful approximate solutions for the shape of the magnetopause may also be obtained by the *image dipole* method, which replaces the dynamic pressure of the solar wind by the magnetic pressure of an image dipole of moment M_I placed parallel to and distance d from the Earth's dipole, M_E. The fields due to both dipoles are added and the distorted field-lines associated with M_E – the two fields do not interconnect – are taken to represent the geomagnetic field within the magnetopause. Those associated with the image have no physical significance. A satisfactory model is given by $M_I = 28\ M_E$; $d = 40\ R_E$. The method only works for the 'front' of the magnetosphere, the side towards the Sun.

Observations

In spacecraft observations the magnetopause is recognized by a sharp drop in magnetic field strength and the appearence of turbulence. Figure 5.18 shows measurements from one of the first crossings of the magnetopause. Here the solid line gives the theoretical variation of the field strength with distance in the absence of a boundary, and it is noted that the observed field is twice the theoretical value at the

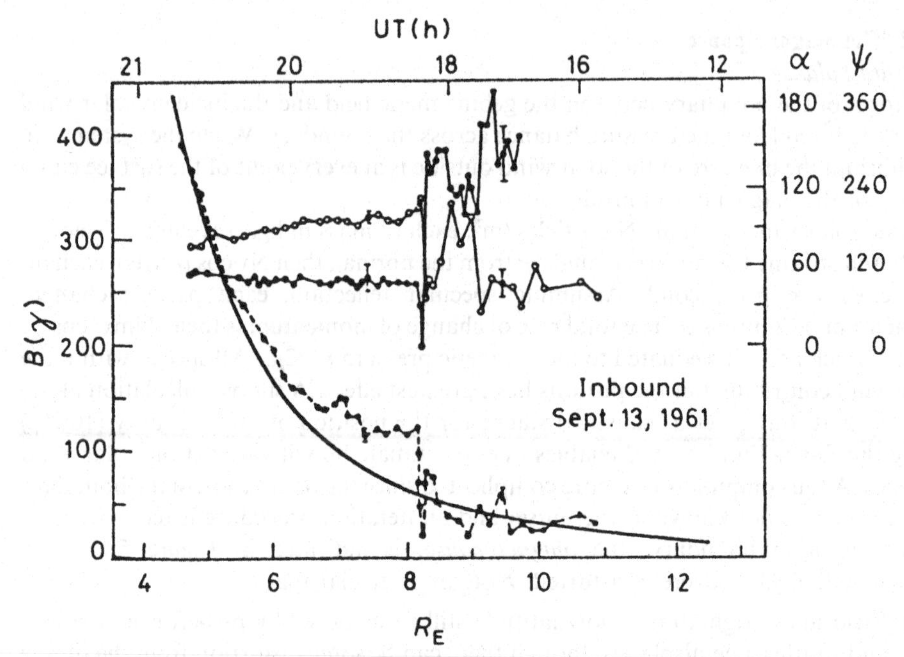

Fig. 5.18 The boundary of the magnetosphere, observed at 8.2 R_E by Explorer XII on 13 Sept 1961. Outside the boundary the field is weaker but more turbulent. (L. J. Cahill and P. G. Amazeen, *J. Geophys. Res.* **68**, 1835, 1963, copyright by the American Geophysical Union)

boundary (here just beyond 8 R_E). This is as expected, as may be easily seen from the image dipole model. (The fields due to M_E and to M_I are equal at the boundary.)

Generally good agreement is obtained between calculated and observed shapes of the magnetopause, and Figure 5.19 illustrates this in the ecliptic plane. These measurements, having come from one spacecraft, were made over a period of time and the overall result is thus a long term average. In fact the magnetosphere varies considerably in response to changes in the solar wind.

5.4.3 The magnetosheath and the shock

The region immediately outside the magnetopause is not in fact typical of the solar wind though it is composed mainly of solar material. As shown in Figures 5.17 and 5.19, a shock front is formed in the solar wind 2 or 3 R_E upstream of the magnetopause. In the region between the shock and the pause, which is known as the *magnetosheath*, the plasma is turbulent.

Tenuous though it may be by any ordinary standard, the magnetosphere is a relatively solid object in comparison with the solar plasma. Further, its velocity is 'supersonic' at the orbit of Earth, meaning that the velocity exceeds that of any waves that can propagate within it. In the solar wind the speed of hydromagnetic waves (Section 2.7.1), that is, the Alfvén speed, given by

$$v_A = \frac{B}{(\mu_0 \rho)^{\frac{1}{2}}}, \quad \text{(Equation 2.78a)}$$

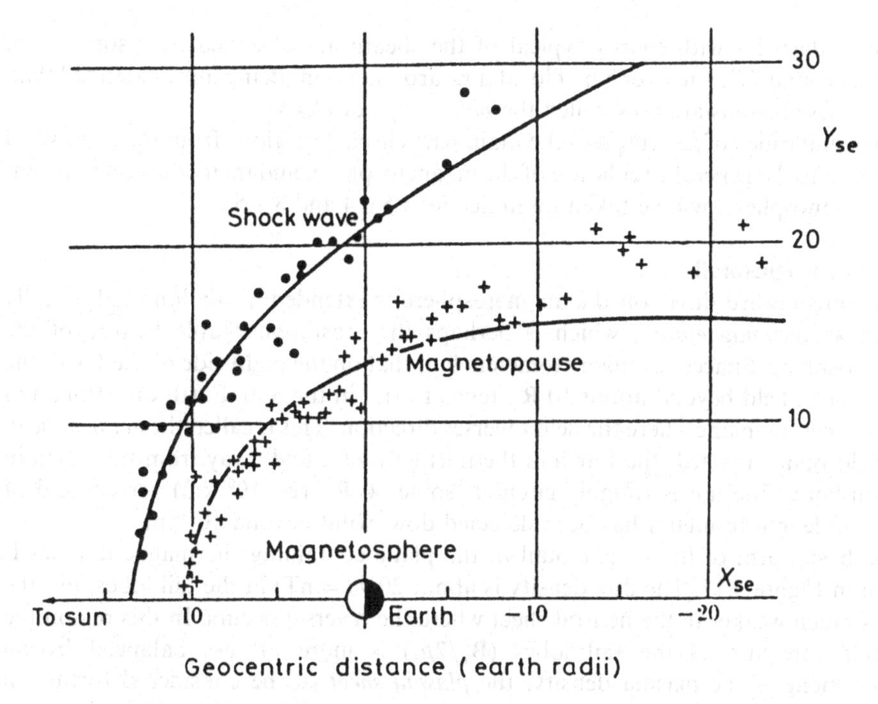

Fig. 5.19 Observed and calculated positions of the magnetopause and the shock in the solar–equatorial plane. (After N. F. Ness *et al.*, *J. Geophys. Res.* **69**, 3531, 1964, copyright by the American Geophysical Union)

where B is the magnetic flux density and ρ the particle density (in kg/m³), is about 50 km/s. For a solar wind speed of 400 km/s, therefore, the *Alfvén Mach number* is 8.

The situation is therefore similar to that of any solid object, a bullet or a fast aircraft for example, flying through the air supersonically, and a shock wave is formed. The existence and location of the shock were predicted from theory in the early 1960s and subsequently verified by observation. A shock wave represents a discontinuity in the medium of the solar wind, created when information about the approaching obstruction is not transmitted ahead into the medium. In crossing the shock, solar wind plasma is slowed down to about 250 km/s and the corresponding loss of directed kinetic energy is dissipated as thermal energy, increasing the temperature to 5×10^6 K. Magnetosheath plasma is therefore slower than the solar wind proper but 5–10 times hotter. This accounts for the irregular flow directions outside the magnetopause in Figure 5.18.

5.4.4 The polar cusps

The simple models of the magnetosphere predict two neutral points on the magnetopause where the total field is zero. These points connect along field-lines to places on the Earth's surface near $\pm 78°$ magnetic latitude. These are the only points that connect the Earth's surface to the magnetopause and all the field from the magnetopause converges to those two points. They are therefore regions of great interest where solar wind particles (from the magnetosheath) can enter the magnetosphere without having to cross field-lines. There is good direct evidence that this happens. Particles with energy typical of the sheath are observed over some 5° of latitude around 77°, and over 8 h of local time around noon. Being more extended than points, these regions are now called the *polar cusps* or *clefts*.

The significance of the cusps as the main particle entry regions from the solar wind in relation to the general circulation of the magnetosphere, and their effect on the high-latitude ionosphere, will be taken up in Sections 5.6.4 and 8.1.5.

5.4.5 The magnetotail

In the antisunward direction the magnetosphere is extended into a long tail, usually known as the *magnetotail*, which is perhaps the most remarkable feature of the magnetosphere. Spacecraft magnetometers find that on the night side of the Earth the geomagnetic field beyond about 10 R_E tends to run in the Sun–Earth direction, and there is a central plane where the field reverses direction. This is called the *neutral sheet*. The field points towards the Earth in the northern *lobe*, and away from the Earth in the southern. The tail is roughly circular, some 30 R_E (2×10^5 km) across, and of uncertain length though it has been detected downwind beyond 10^7 km.

The basic form of the magnetotail in the plane containing the magnetic poles is shown in Figure 5.17. The flux density is about 20 γ (= nT) in the tail lobes, but the field is much weaker in the neutral sheet where the reversal occurs. In this region the magnetic pressure of the tail lobes ($B_T^2/2\mu_0$) is more or less balanced by an enhancement of the plasma density, the *plasma sheet* (to be considered further in Section 5.6.3). But in fact the tail, like the whole magnetosphere, is dynamic and it forms an essential part of the magnetospheric circulation.

5.5 Circulation of the magnetosphere

5.5.1 Circulation patterns

We have seen that the general form of the magnetopause can be derived by assuming that the solar wind exerts a pressure against the surface but otherwise flows past smoothly. This must be a dubious assumption, however, because most objects flying through a fluid experience some friction, and we might expect something similar at the magnetopause as the solar wind travels past super-Alfvénically. This is important because any friction at the boundary would provide a mechanism for transferring energy from the solar wind into the magnetosphere.

Viscous interaction

In 1961, W. I. Axford and C. O. Hines suggested that momentum is transferred from the solar wind into the magnetosphere by some unspecified process equivalent to friction, and that the whole system was thereby caused to circulate. Experimental evidence was adduced from a study of the S_q^p *current system*, whose existence may be inferred from observations with magnetometers at medium and high latitudes. S_q^p is the polar part of the basic S_q system (Section 7.2.2), the upper atmosphere current related to the solar day under quiet conditions. The basic form of S_q^p, illustrated in Figure 5.20, is a current flowing over the poles from night to day, with return currents at lower latitudes. It is characteristic of the dynamo region of the ionosphere (Section 7.2.1) that positive ions move with the neutral air but electrons with the magnetic field. S_q^p can therefore be explained by a motion of magnetic field-lines opposite to the

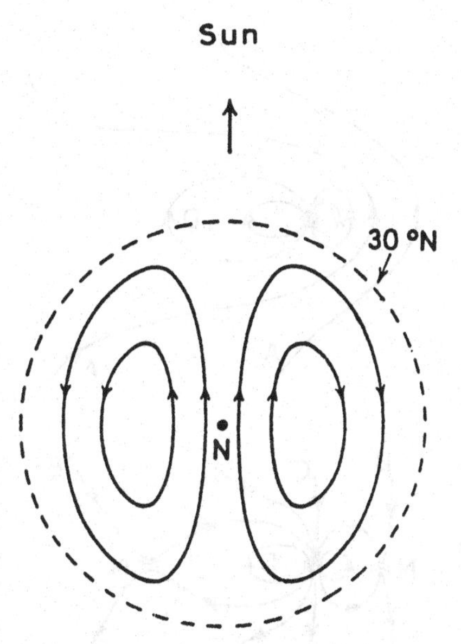

Fig. 5.20 S_q^p current system due to the motion of the feet of magnetospheric field-lines over the pole. (J. A. Ratcliffe, *An Introduction to the Ionosphere and Magnetosphere*. Cambridge University Press, 1972)

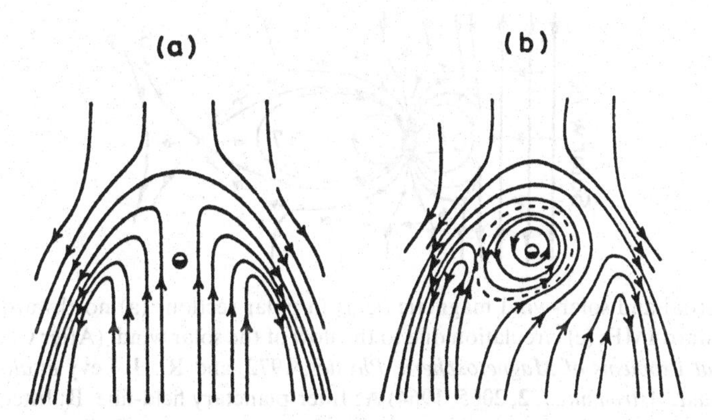

Fig. 5.21 Patterns of magnetospheric circulation in the equatorial plane: (a) due to 'friction' at the magnetopause; (b) including the effect of the Earth's rotation. (After A. Nishida, *J. Geophys. Res.* **71**, 5669, 1966, copyright by the American Geophysical Union)

current flow in Figure 5.20. Projected to the magnetosphere, this implies that the field-lines circulate over the poles from the day to the night sectors of the Earth with a return flow around the dawn and dusk sides. The resulting circulation of the magnetosphere is similar to that within a falling raindrop, in which the fluid is swept back at the surface and returns inside the drop. Figure 5.21a shows the magnetospheric circulation pattern in a section through the equatorial plane. Figure 5.21b takes account of the Earth's rotation which carries the inner part of the magnetosphere with it.

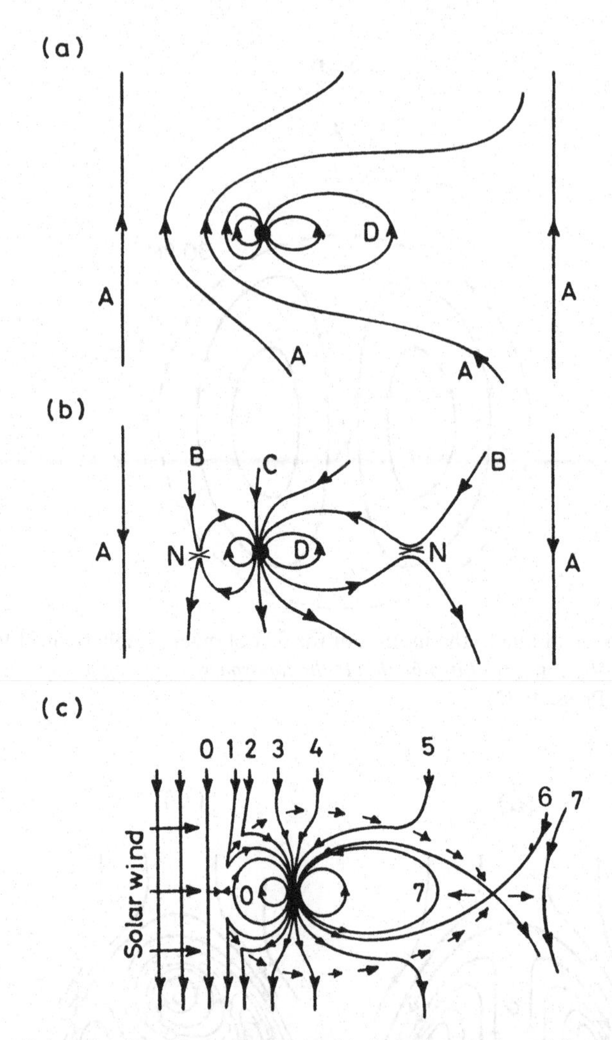

Fig. 5.22 Terrestrial and solar-wind magnetic fields in polar section: (a) northward IMF; (b) southward IMF; (c) circulation due to the flow of the solar wind. (After C. T. Russell, *Critical Problems of Magnetospheric Physics*, 1972, and R. H. Levy *et al.*, *Am. Inst. Aeronaut. Astronaut. J.* **2**, 2065, 1964) A: Inter-planetary field-line. B: Inter-planetary field-line connecting to, or disconnecting from, a geomagnetic field-line. C: Open geomagnetic field-line. D: Closed geomagnetic field-line. N: Neutral point (⪤). 0–7: Successive positions of inter-planetary field-line.

Magnetic linkage

The difficulty with the *viscous interaction* theory was the nature of the frictional force, since friction is usually associated with collisions and the solar wind is so tenuous that collisions are virtually absent, the mean free path being some 10^9 km! Attention therefore moved to an alternative mechanism, based on the work of J. W. Dungey which supposed connections between the interplanetary magnetic field (IMF) and the geomagnetic field. Figure 5.22 depicts a distorted dipole field representing the geomagnetic field in polar section, with the addition of (a) a northward IMF and (b) a southward IMF. In the second case, neutral points are formed in the equatorial plane and some lines of the IMF connect to geomagnetic lines. This is not so in the first case. In fact the IMF tends to lie in the solar–ecliptic plane and be oriented at the 'garden hose' angle, but there is usually a north–south component as well and this, when directed southward, is what may connect to the geomagnetic field. (The basic principle is simply illustrated by the elementary laboratory experiment of plotting field-lines to find the neutral points when a bar magnet is placed in an external field. One orientation gives neutral points on the equator and shows field-lines over the poles connecting to the external field. The other has neutral points over the poles, and there are no inter-connected field-lines.)

The IMF is frozen into the solar wind and is therefore carried along with it. When geomagnetic field-lines are connected to those from the IMF they are dragged over the poles from the sunward neutral point, as in Figure 5.22c, and thereby transported from the day to the night side. While over the polar caps the field-lines are open in the sense that they do not connect back to the other hemisphere in any simple or obvious manner. In the tail these lines reconnect and move back towards the Earth. The above picture is of course a simplified one. Figure 5.23 shows a more sophisticated version which includes a degree of connection when the IMF is northward.

Evidence

Since both the proposed driving mechanisms predict rather similar circulations there has been some debate as to their relative importance. There is direct evidence for the merging of magnetic fields (See Section 5.5.2), and other observations demonstrating that major effects are associated with a southward IMF component are that:

(a) the dayside magnetopause moves inward;
(b) the auroral zone and the dayside cusps are displaced equatorward;
(c) the magnetic flux in the tail lobes is increased;
(d) substorms (Section 5.9) occur more frequently.

All of these argue that a southward IMF increases the magnetospheric circulation. There is also an effect of the east–west IMF component on the symmetry of the polar convection – the Svalgaard–Mansurov effect (Section 8.1.3). However geomagnetic activity does not cease altogether when the IMF has a northward component, indicating that the circulation does not stop. Current (1990) opinion is that both viscous interaction and magnetic merging operate but that merging is considerably the more important during periods of southward IMF. Figure 5.24 indicates how the two mechanisms may co-exist: here, merging drives the flow over the poles involving open field-lines, and 'friction' pushes closed field-lines to a limited depth along the flanks of the magnetosphere. It is thought – and this is still a topic for debate – that the magnetosphere takes up a different and more complex pattern of circulation when the IMF has a northward component.

Despite remaining questions there can be no reasonable doubt about the fact of magnetospheric circulation and its significance in geospace. It is a concept that has made it possible to comprehend many of the phenomena that have been observed and well substantiated in geospace research, and is one of the most basic ideas of modern magnetospheric theory.

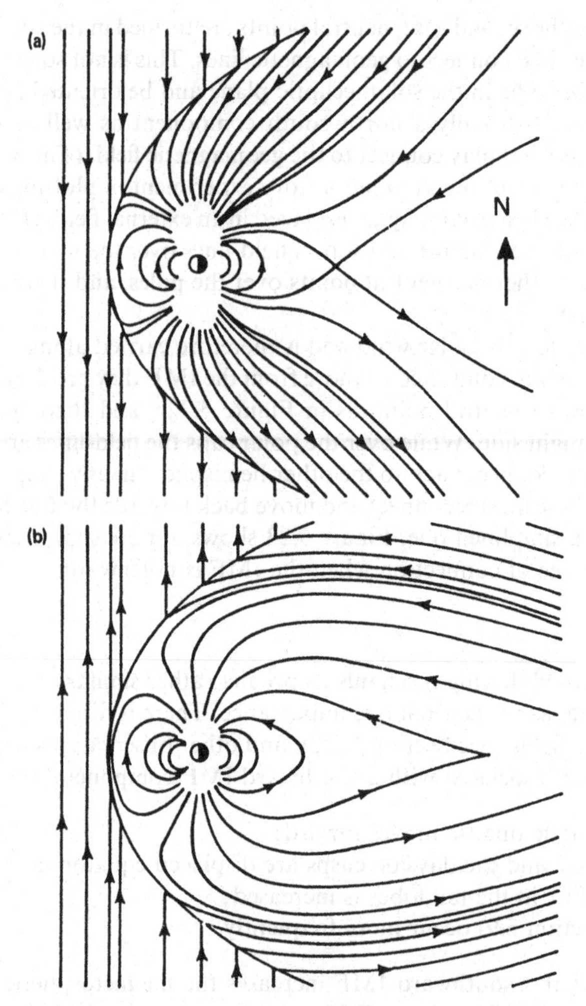

Fig. 5.23 Magnetic field topology with IMF (a) southward, and (b) northward. (T. A. Potemra, *Johns Hopkins APL Tech. Digest* **4**, 276, 1983)

5.5.2 Field merging in the neutral sheet and at the magnetopause

Magnetospheric circulation requires the breaking and reconnection of magnetic field-lines on both the day and night sides of the Earth, and one may well ask what kind of field configurations allow this to happen.

In the centre of the magnetotail open field-lines from the polar regions reconnect and contract towards the Earth. Satellite observations reveal the presence of a small north–south magnetic component across the reconnection region, showing that, as in Figure 5.25, the field-lines are not strictly parallel. This *X-type neutral line* cannot be a static configuration because the tension in the field-lines will produce net forces towards the Earth and into the tail; but there can be dynamic equilibrium, in which the depletion is replaced by other field-lines moving over the poles and then through the lobes towards the neutral sheet. It would also be possible to have a *Y-type neutral line*,

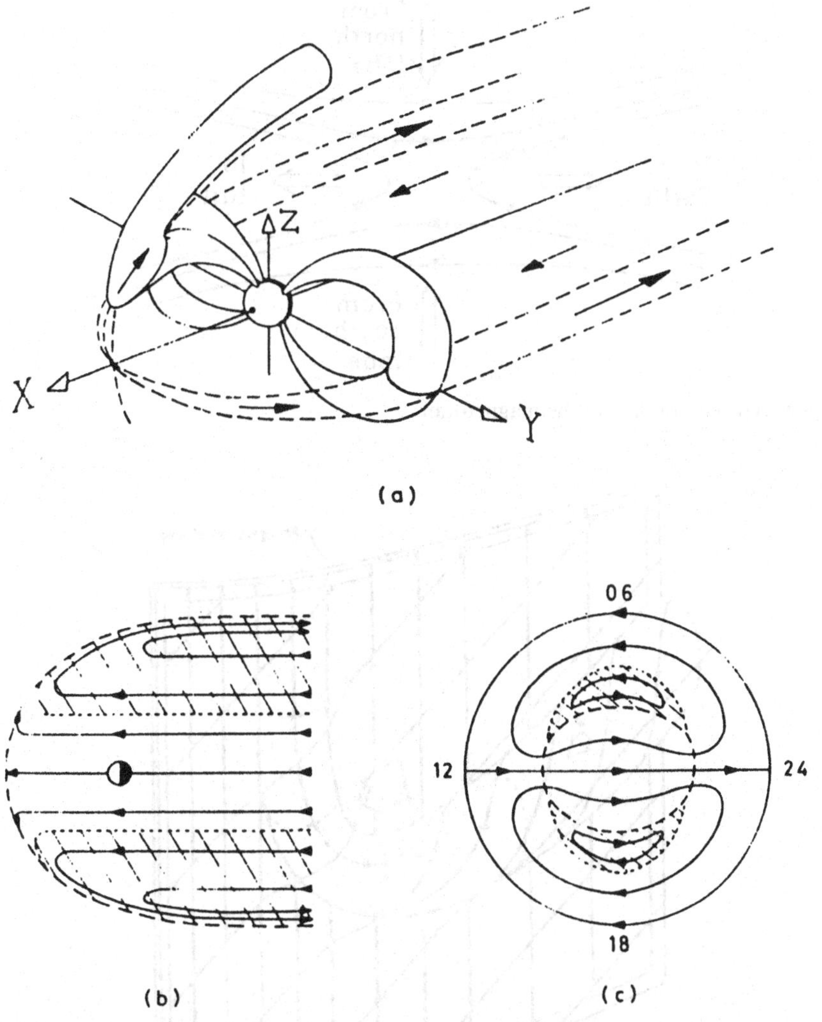

Fig. 5.24 Circulation due to viscous and magnetic drag in combination: (a) flux tube motions; (b) equatorial plane; (c) northern polar region. The hatched region is driven by the boundary layers. (S. W. H. Cowley, *Rev. Geophys. Space Phys.* **20**, 531, 1982, copyright by the American Geophysical Union)

where the field continues to converge on the tailward side, the earthward side being as in Figure 5.25. The velocity of field-lines towards the Earth is estimated as about 100 km/s and the drift towards the neutral sheet as about 10 km/s. It is likely that reconnection in the tail occurs not steadily but intermittently in limited regions, and this will be taken up again in Section 5.9.1.

The principal region of connection between the IMF and magnetospheric field-lines is on the dayside. When connection occurs a geomagnetic flux tube has to break and connect with an IMF tube (Figure 5.26), after which the newly connected tube of plasma moves poleward and joins the general circulation. The characteristic signature

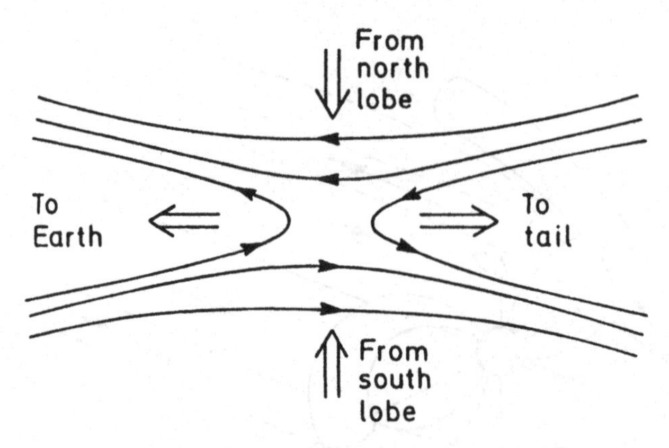

Fig. 5.25 X-type neutral line in the magnetotail.

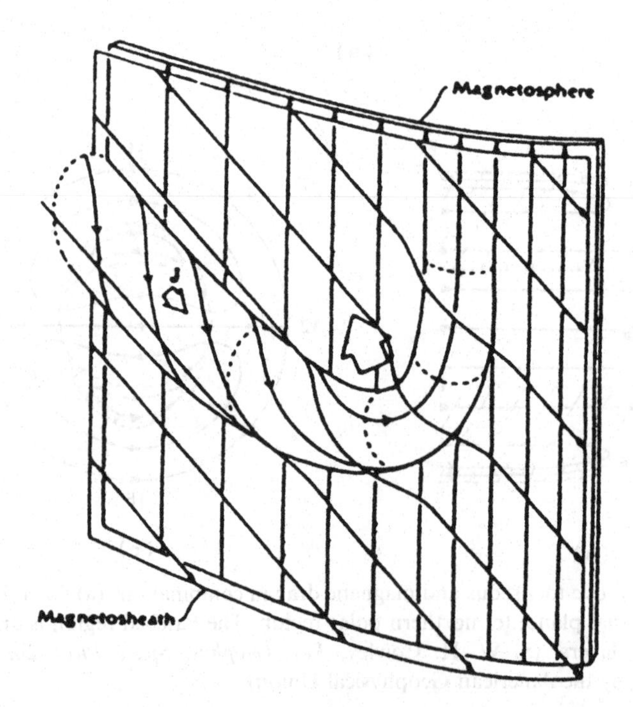

Fig. 5.26 The elbow model of a flux transfer event, connecting the inter-planetary field to the magnetosphere. A field-aligned current causes the field-lines to twist. (After C. T. Russell and R. C. Elphic, *Space Science Reviews* **22**, 681, 1978. By permission of Kluwer Academic Publishers)

which this causes consists of short positive and negative excursions in the magnetic component normal to the magnetopause, amounting to about 10 nT peak-to-peak. The results of such measurements show that during periods of southward IMF dayside connection events (often known as *flux transfer events*) occur frequently (1–2 per hour) but that individually they are of short duration (1–5 min.) and relatively small in spatial extent. The geometry of Figure 5.26 is sometimes called the 'elbow' model. Details of the merging are not, however, settled, and alternative models have been proposed. It is likely that 'quasi-steady' reconnection occurs as well, but again the details are still a topic of investigation. The situation must be different when the IMF has a northward component, and then the merging region may be on the poleward side of the cusp (for example, as in Figure 5.23).

5.5.3 Magnetospheric electric fields

In Section 2.3.7 we set out the equivalence between an electric field and moving magnetoplasma. If an electric field **E** is applied across a magnetoplasma with magnetic flux density **B**, the magnetoplasma moves at velocity

$$\mathbf{v} = \mathbf{E} \times \mathbf{B}/|\mathbf{B}|^2. \quad \text{(Equation 2.40)}$$

(See also Section 6.5.4.) Conversely, it may be shown that if magnetoplasma moves at velocity **v** with respect to a stationary observer, the observer will measure an electric field

$$\mathbf{E} = \mathbf{v} \times \mathbf{B}. \tag{5.19}$$

(This is just equivalent to Equation 2.25, the electric field being the Lorentz force per unit charge.) Electric field and motion of magnetoplasma are equivalent in highly conducting plasmas such as the magnetosphere and the solar wind. (This equivalence can also be used to clarify the idea of a 'moving field-line'. The field-line may be only a physicist's fiction, but as a concept it is very helpful when complicated fields have to be visualized. From the above relation we can say that the motion of the field-line is such that an observer moving with it detects zero electric field.)

An equivalent description of the dynamic magnetosphere can therefore be given in terms of electric fields (as seen by a stationary observer). The general circulation of the outer magnetosphere is equivalent to an electric field directed across the magnetosphere from the dawn to the dusk side. Its magnitude is about 0.3 mV/m, though considerably variable, giving a total potential difference of some 60 kV across the magnetosphere.

The source of this electric field can be appreciated by considering the magnetosphere as a magneto-hydrodynamic generator, in which a jet of plasma is forced through a static magnetic field and an electric potential is developed by dynamo action. If the magnetosphere were static the solar wind would blow across the field-lines and an electric field would be generated. The magnitude of this field is just that required to produce the circulation. The total potential drop is given by

$$V_T = vLB_n, \tag{5.20}$$

where v is the solar wind speed, L is the width of the magnetosphere, and B_n is the magnetic flux density normal to the boundary. B_n is estimated from the magnetic flux leaving the polar caps (approximately the regions within the auroral ovals – Section 8.3.1) which ultimately connects to the IMF. For a polar cap of radius R_p and magnetic flux density B_p, a magnetotail of length S_T and radius $R_T(=L/2)$:

$$B_n = \pi R_p^2 B_p / 2\pi R_T S_T. \tag{5.21}$$

Taking $R_P = 1.7 \times 10^6$ m, $B_p = 5.5 \times 10^{-5}$ Wb/m², $R_T = 20R_E = 1.3 \times 10^8$ m, $S_T = 200R_E = 3.2 \times 10^6$ m, then $B_n = 4.5 \times 10^{-10}$ Wb/m² $= 0.45$ nT.

A solar wind speed of v = 500 km/s gives V_T = 58 kV. Although this calculation assumes values that are not well known, it does give a result of the right magnitude and reinforces the validity of the electric potential approach. It is often more convenient to treat the dynamics of the magnetosphere in terms of electric fields.

The potential distribution across the magnetotail maps across the polar caps, where it is more accessible to direct measurement. Such measurements have related the potential to the solar wind. (See Section 8.1.1.) If the potential difference across the polar cap is 60 kV, the field-line velocity is about 300 m/s.

5.6 Particles in the magnetosphere

5.6.1 Principal particle populations

The geomagnetic field holds within it several distinct populations of charged particles.

(a) Deep within the magnetosphere (in the region often known as the inner magnetosphere) is the *plasmasphere*, closely linked to the mid-latitude ionosphere and comprising electrons, protons and some heavy ions, all having energies in the thermal range.

(b) Also trapped on closed field-lines are the energetic particles generally known as the *Van Allen particles* after their discoverer. Apart from cosmic rays and solar protons, which are merely passing through, the Van Allen particles are the most energetic in the magnetosphere and they make some contribution to the ionization of the upper atmosphere when precipitated out of the trapping region.

(c) The *plasma sheet* is associated with the magnetotail, essentially with the central region where the magnetic field reverses direction. Plasma sheet particles are energized within the magnetotail and they are important in auroral activity and the behaviour of the high-latitude ionosphere. Their energy is intermediate between those of the plasmasphere and the Van Allen belt. The inner edge of the plasma sheet supports the *ring current* that flows in the magnetosphere during magnetic storms.

(d) At the edges of the magnetosphere, and obviously connected with the physics of the magnetopause, are *boundary layers*. Their composition and energy are governed by the solar wind and plasma in the magnetosheath.

The locations of these particle populations are indicated in Figure 5.27. They should not be considered as merely incidental to the magnetosphere; they are in fact essential to its properties and behaviour. In most of the magnetosphere the ratio of the energy density of the particles to that of the magnetic field (β) is less than unity, but there are exceptions.

Originally it was thought that most of the particles in the magnetosphere come from the solar wind, but, on the evidence of heavy ions observed in the magnetosphere, it is now recognized that the ionosphere is also a major source. (See also Section 8.1.4.)

5.6.2 The plasmasphere and its dynamics

Ionized particles in the upper ionosphere (F region and topside) have temperatures up to several thousand degrees Kelvin, and electron energies are therefore several tenths of an electron volt. ($E = \frac{3}{2}kT$; $k = 1.38 \times 10^{-23}$ J K^{-1}; 1 eV $= 1.6 \times 10^{-19}$ J. Hence,

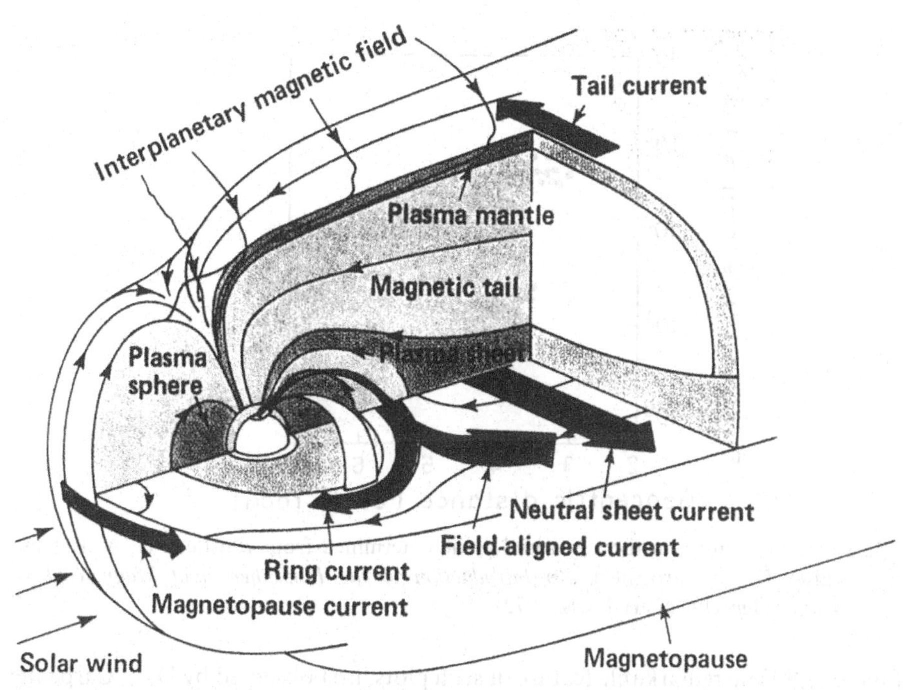

Fig. 5.27 Plasma populations and current systems of the magnetosphere. (T. A. Potemra, *Johns Hopkins APL Tech. Digest* **4**, 276, 1983)

2000 K ≡ 0.26 eV.) The particle density is typically 10^{10} m^{-3} at 1000 km altitude, reducing with increasing height – though not very rapidly because of the large scale height when atomic hydrogen is the principal atom (see Section 4.1.5). The theory of the protonosphere, which is introduced in Sections 6.2.5 and 6.3.3, shows how ionospheric plasma flows up the field-lines to populate the protonosphere as far as the equatorial plane, provided the field-lines are closed. Some of this plasma flows back to lower levels at night where it helps to maintain the ionosphere during hours of darkness, but the plasmasphere nevertheless persists as a permanent feature of the inner magnetosphere. The outer boundary of the plasmasphere is called the *plasmapause*.

Plasmapause

The plasmapause was discovered by the whistler technique. If the travel time of a whistler is displayed against frequency it is seen that there is one frequency where the travel time is a minimum. This is a characteristic of all whistlers, but not all show it clearly and those which do are called *nose whistlers*. The frequency corresponding to minimum travel time indicates which field-line the whistler has travelled along, and the time taken can be interpreted to give the minimum electron density encountered along that field-line. The theory behind the method is outlined in Section 3.5.4.

By means of this technique it is possible, from a ground-based station suitably located for whistler reception (and not all places are equally suitable because the occurrence of whistlers depends on the incidence of lightning in the conjugate hemisphere), to determine the variation of electron density in the equatorial plane, as

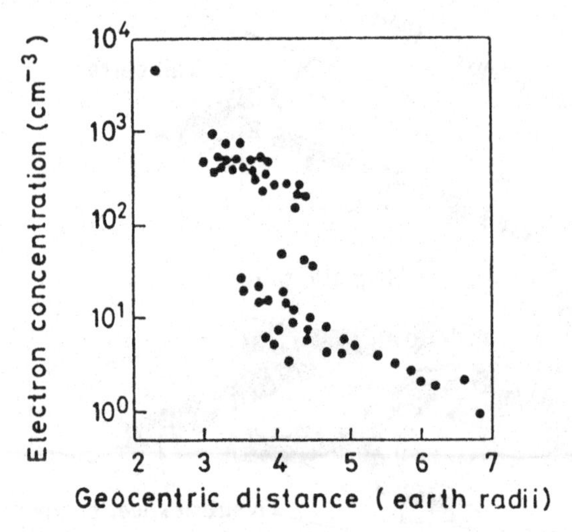

Fig. 5.28 Electron density in the equatorial plane determined from whistlers. (J. A. Ratcliffe (after D. L. Carpenter), *An Introduction to the Ionosphere and Magnetosphere*. Cambridge University Press, 1972)

in Figure 5.28. The remarkable feature of such plots, first observed by D. L. Carpenter in 1963, is that they often show a sudden drop in the electron density near 4 R_E, amounting perhaps to a factor of ten or more within a distance of 0.5 R_E or less. This is the plasmapause, sometimes also known as the *knee*. If traced inward along the geomagnetic field the plasmapause is found to correspond approximately with the ionospheric main trough which effectively marks the poleward extent of the mid-latitude ionosphere (Section 8.1.6). It is clear that the plasmasphere occupies a doughnut-shaped region of the inner magnetosphere where the field-lines are not too distorted from the dipolar form.

Dynamics

Observations show the plasmasphere as a dynamic and variable region. There is a daily variation that includes a bulge in the evening sector (Figure 5.29), and the whole region contracts when geomagnetic activity increases, there being a gradual recovery in the few days following the storm. Figure 5.30 shows measurements of the plasmapause position as a function of the global magnetic activity index K_p. For most of the time it is found between 3 and 6 R_E, though it has been detected as close to the Earth as 2 R_E (i.e. only one Earth radius above the surface). The whistler observations show that the geocentric distance to the plasmapause in Earth-radii (L_{pp}) during the post-midnight hours is related to the greatest value of K_p during the preceding 12 hours ($\hat{K}_p$) by an empirical relation:

$$L_{pp} = 5.7 - 0.47\hat{K}_p. \tag{5.22}$$

Partly, the plasmasphere is explained by ionospheric plasma moving up and down closed geomagnetic field-lines. To explain the dynamics it is also necessary to take account of the circulation of the magnetosphere. As we saw in Section 5.5.1, the inner magnetosphere co-rotates with the Earth while the outer magnetosphere follows its

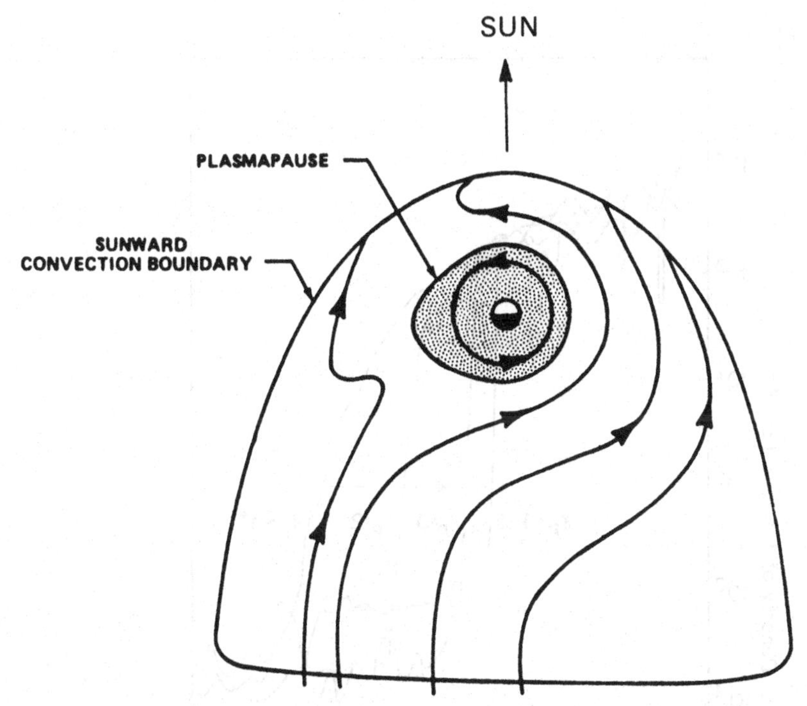

Fig. 5.29 Daily variation of the plasmapause in relation to the plasma convection in the magnetospheric equatorial plane. (After J. L. Burch, in *The Upper Atmosphere and Magnetosphere*. National Academy of Sciences, Washington DC, 1977)

own circulation pattern under the control of the solar wind. Generally speaking, the plasmasphere exists in the co-rotating field, and the plasmapause marks the boundary between the inner and outer regions. In terms of the electric fields of the magnetosphere (Section 5.5.3) the plasmapause is approximately where the cross-tail and co-rotation fields are equal:

$$E_T = \frac{B_E}{L^3} \cdot LR_E\omega, \qquad (5.23)$$

where L is the distance in Earth radii, R_E the radius of the Earth, and B_E the geomagnetic flux density at the surface at the equator. Putting in numerical values gives:

$$E_T = 14.4/L^2 \text{ mV/m}. \qquad (5.24)$$

When the plasmapause is at 4 R_E, the corresponding value of the tail field is about 1 mV/m. A computation of equipotentials about the Earth is shown in Figure 5.31.

Depletion and refilling

The above quoted relation between L_{pp} and $\hat{K}_p$ implies that the effect of a change of activity is delayed. The loss of plasma from the outer part of the plasmasphere when geomagnetic activity increases is probably due to an increase in the cross-tail electric field, representing enhanced magnetospheric circulation that peels off layers of the plasmasphere as in Figure 5.32. The detached plasma is lost in the outer magnetosphere

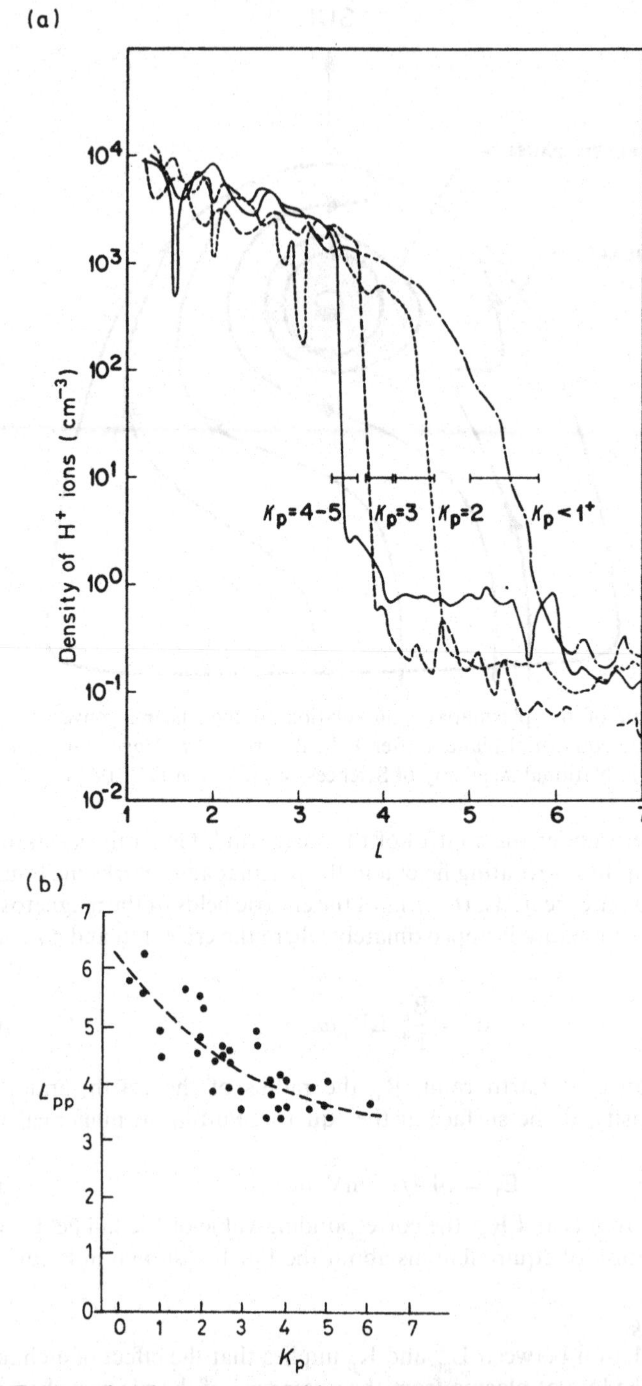

Fig. 5.30 Data on the plasmapause position: (a) satellite observations of ion density, showing the plasmapause at several K_p levels; (b) relation between plasmapause distance, L_{pp}, and K_p. (After C. R. Chappell *et al.*, *J. Geophys. Res.* **75**, 50, 1970, copyright by the American Geophysical Union)

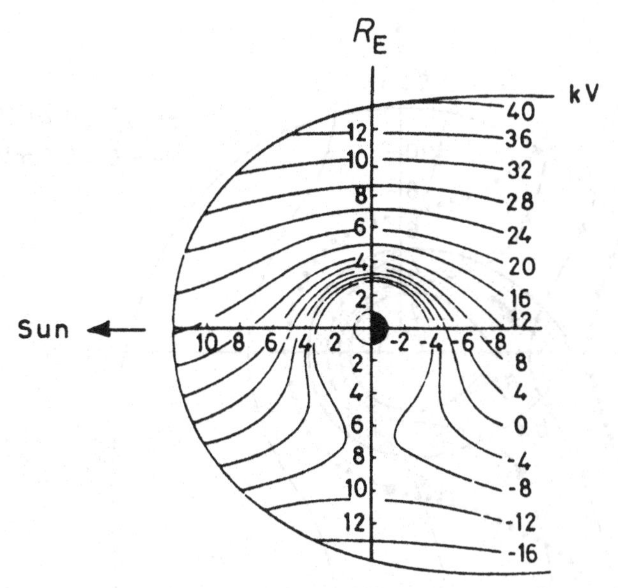

Fig. 5.31 Calculated distribution of equipotentials about the Earth, derived by adding the co-rotation field to a cross-tail field of 2 kV/R_E. (R. J. Walker and M. G. Kivelson, *J. Geophys. Res.* **80**, 2074, 1975, copyright by the American Geophysical Union)

and eventually into the solar wind. (There might also be some draining into the ionosphere during the storm.)

With the decay of activity the magnetospheric circulation and electric fields return to their previous state but now the outer tubes of magnetic flux are devoid of plasma. These gradually refill from the ionosphere over a period of several days. The rate of filling is determined by the diffusion speed of protons (formed in the upper ionosphere by charge exchange between hydrogen atoms and oxygen ions) coming up along the field, and by the volume of the flux tube which varies as L^4. It therefore takes much longer to refill tubes originating at higher latitude. Observations of the filling are shown in Figure 5.33. Since active periods may recur every few days there will be times when the outer tubes are never full and the plasmasphere always has some degree of depletion.

5.6.3 The plasma sheet

Beyond the plasmapause the electron density is much smaller and the temperature is much higher, and this is clearly not the same population of particles which inhabits the plasmasphere. The electron density is only about 0.5 cm^{-3} and the ion density is the same for neutrality. Particle energies are 10^2 to 10^4 eV. The average energy of the electrons is about 0.6 eV and that of the protons about 5 keV. The total energy density of the particles is about 3 keV/cm^3.

The plasma sheet is particularly associated with the central plane of the magnetotail where the magnetic field reverses, and, to a first approximation, the pressure of the particles in the sheet balances the magnetic pressure in the tail lobes. Thus we can put:

$$nkT = B_T^2/2\mu_0, \tag{5.25}$$

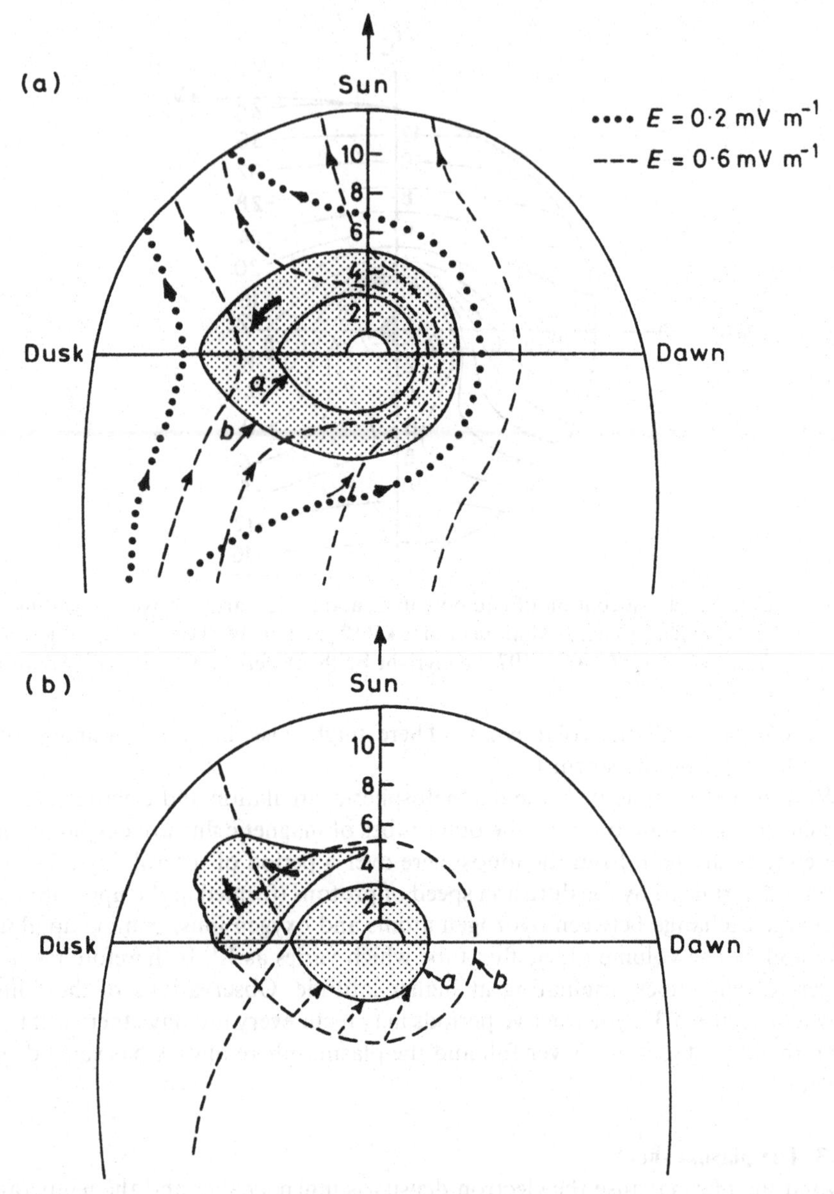

Fig. 5.32 Detaching of plasma due to changing flow patterns during a magnetic storm: (a) flow patterns for cross-tail fields of 0.2 and 0.6 mV/m; (b) resulting detached plasma. (After C. R. Chappell *et al.*, *J. Geophys. Res.* **76**, 7632, 1971, copyright by the American Geophysical Union)

where B_T is the tail magnetic field outside the plasma sheet. As indicated in Figure 5.34 the plasma sheet follows the magnetic field down to lower altitudes in the vicinity of the auroral zone and also continues round to the day side of the Earth. In the equatorial plane there is an identifiable, though variable, inner edge near 7 R_E at

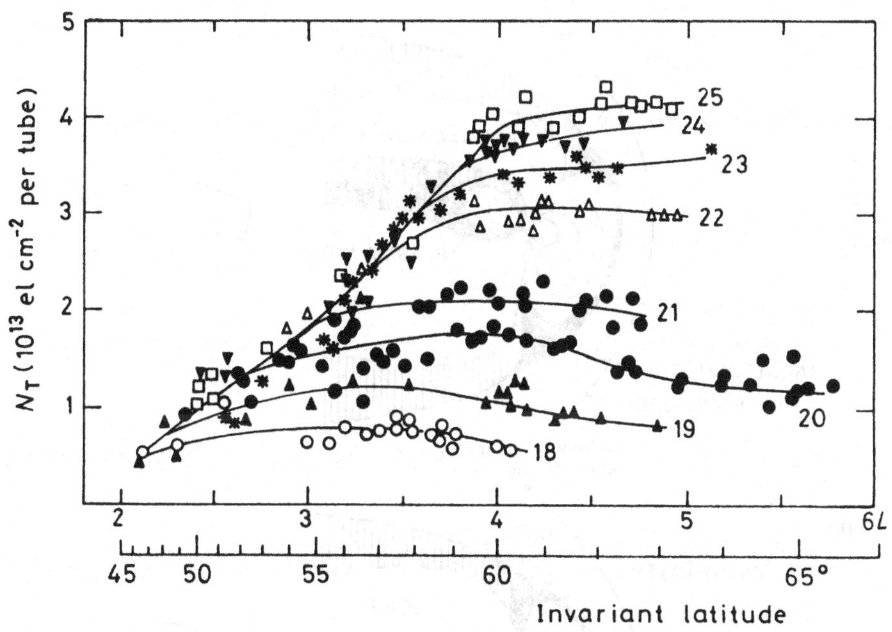

Fig. 5.33 Refilling of the plasmasphere after a storm, 18–19 June 1965. The measurements are of the electron content between conjugate points as a function of L value, by the whistler technique. The content is almost independent of L while the tube is filling, whereas the content of full tubes increases strongly with L. (After C. G. Park, *J. Geophys. Res.* **79**, 165, 1974, copyright by the American Geophysical Union)

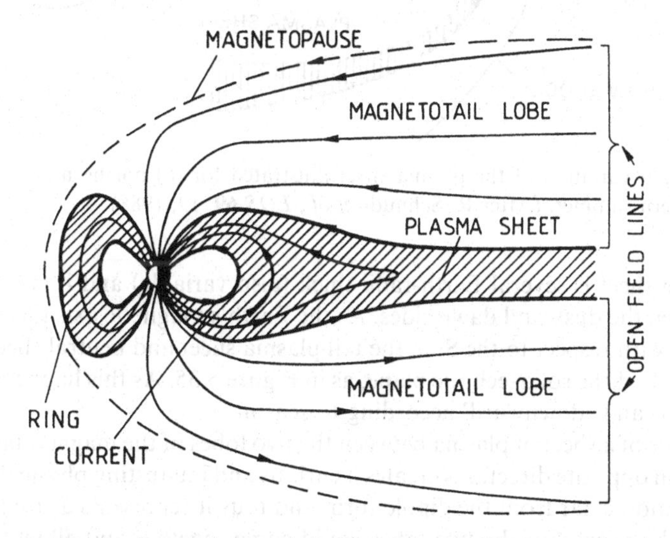

Fig. 5.34 Magnetospheric plasma sheet and ring current. (V. M. Vasyliunas, *Exploration of the Polar Upper Atmosphere*. Reidel, 1981, p. 229. Reprinted by permission of Kluwer Academic Publishers)

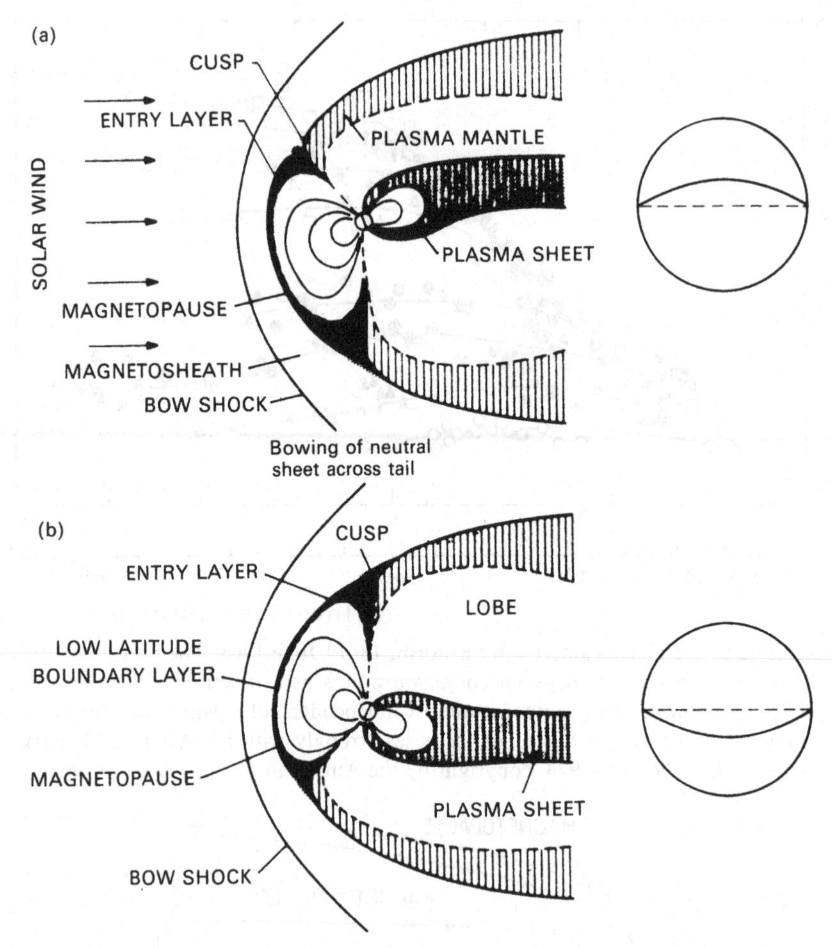

Fig. 5.35 Seasonal variation of the plasma sheet illustrated for (a) northern summer and (b) southern summer. (After R. Schmidt *et al.*, *EOS* **69**, 19, 1988)

midnight. The sheet is several Earth-radii thick (also variable) and it extends across the tail between the dusk and dawn sides. As the Earth's magnetic axis tilts seasonally and diurnally with respect to the Sun, the tail plasma sheet and neutral sheet oscillate north and south of the solar-ecliptic plane, as in Figure 5.35. As this happens the sheet bows 'upward' and 'downward' according to season.

The existence of a sheet of plasma between the two lobes of the magnetotail in which the fields run in opposite directions creates a curious and interesting physical situation. The configuration is far from the dipole form and thus it represents a store of energy which would be released if the two lobes could come together and allow the field to return to a dipolar shape. The magnetic merging in the neutral sheet of the magnetotail, as discussed in Section 5.5.2, involves the mutual annihilation of field from the lobes, and this is the source of the energy that heats the plasma sheet particles. The process is not, in fact, a steady one, as we shall see in Section 5.9.

Figure 5.36 shows the typical profile of ion density with distance from the Earth (i.e.

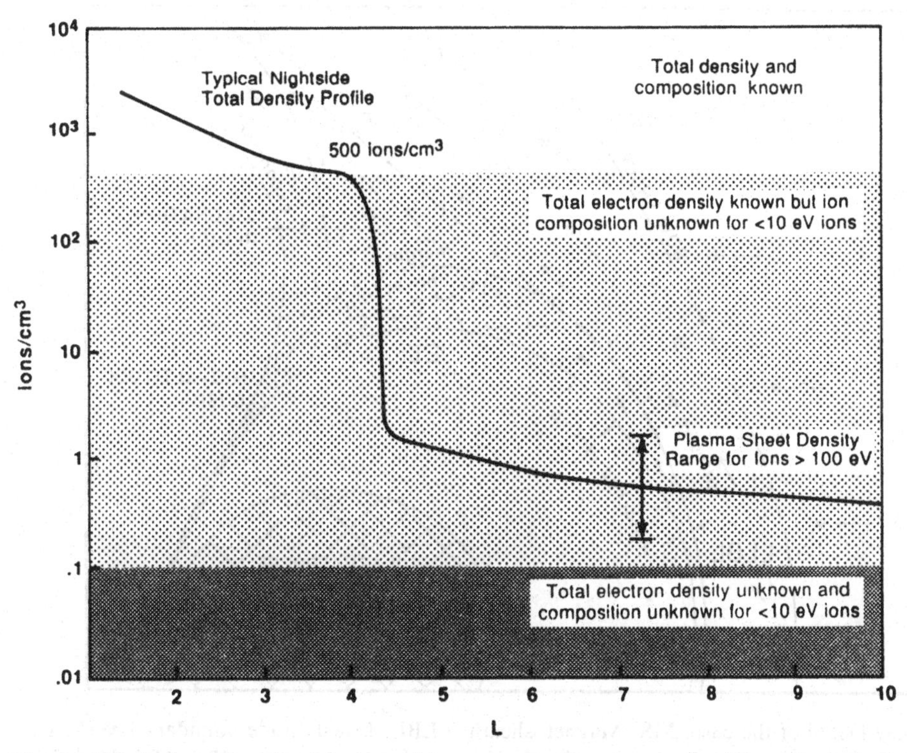

Fig. 5.36 Typical profile of ion density with geocentric distance on the night side of the Earth, with an indication of current (1988) knowledge. (After C. R. Chappell, *Rev. Geophys.* **26**, 229, 1988, copyright by the American Geophysical Union)

geocentric distance in Earth-radii) on the night side. The information is less certain at the greater distances.

5.6.4 Boundary layers

As anticipated in the early magnetospheric theories, the cusps are regions where solar matter from the magnetosheath enters the magnetosphere. There is direct evidence for the entry, since particles of the appropriate energy (< 1 keV) are observed over a limited region centered about 77° magnetic latitude and noon. Further evidence is enhanced electron density and temperature in the ionosphere beneath the cusps.

The structure of the cusp is illustrated in Figure 5.37. The magnetosphere circulation opens field-lines from lower latitudes which move poleward across the cusp and are then swept back over the poles into the tail. Magnetosheath particles flow down the newly opened field-lines, mirror in the stronger field near the Earth (Section 5.7.2), and try to return to the sheath. But because of the convection of the magnetic field-lines some of this plasma finds itself on the open field-lines over the pole. This forms the *plasma mantle*. In the tail the particles are gradually swept from the plasma mantle towards the central region of the tail (see Figure 5.27), but because of their high speed (100–200 km/s) they will not arrive there until beyond 100 R_E. (A component from the

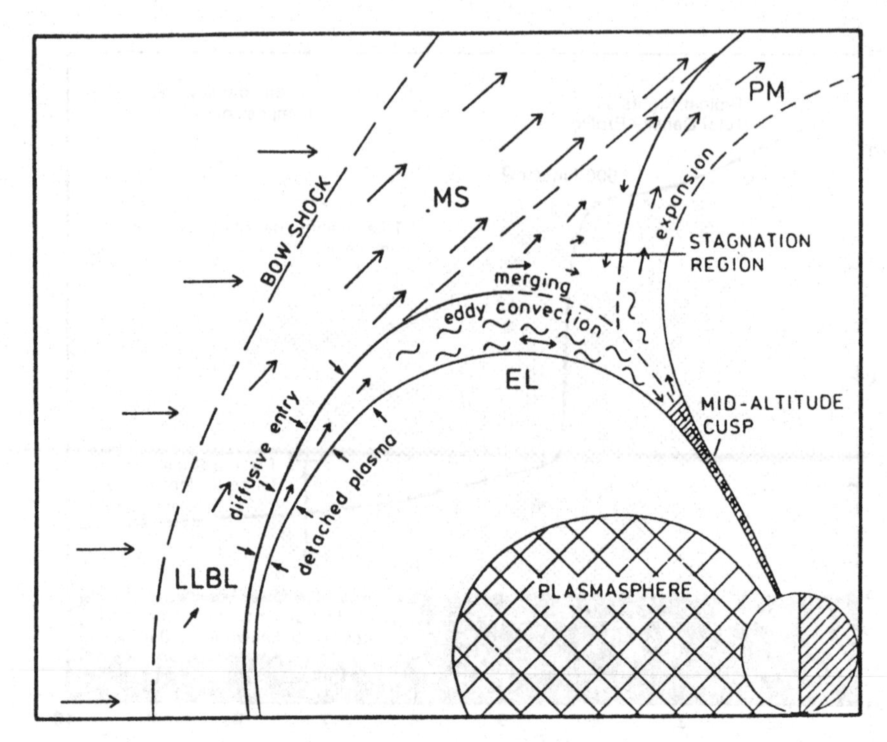

Fig. 5.37 Detail of the cusp. MS: Magnetosheath. LLBL: Low-latitude boundary layer. EL: Entry layer. PM: Plasma mantle. (G. Haerendel *et al.*, *J. Geophys. Res.* **83**, 3216, 1978, copyright by the American Geophysical Union)

ionosphere (the polar wind – Section 8.1.4) arrives at the plasma sheet much closer to the Earth – within 50 R_E.) There are also boundary layers at the equatorward side of the cusp, entrapped by eddy convection or diffusion.

5.7 Van Allen particles

5.7.1 Discovery

The discovery of energetic particles trapped in the magnetosphere was the first major scientific result of the satellite era, and it attracted a great deal of attention at that time. The first reported measurements came from the Explorer 1 satellite, which on 31 January 1958 was the first to be successfully launched by the United States. The key observation was in fact somewhat peculiar. Being interested in cosmic rays, J. A. Van Allen's group at the University of Iowa had built for the satellite a Geiger counter with what should have been the appropriate sensitivity for cosmic ray detection. The measurements found the cosmic ray flux, but there were also times when the counting rate was very large and others when it fell below the expected cosmic ray level. This was interpreted as a particle flux so great as to cause a malfunction of the counter. Similar effects were seen on Explorer 3 in March 1958 and on the USSR's Sputnik 3. These Geiger counters gave little information as to the nature of the particles and they were

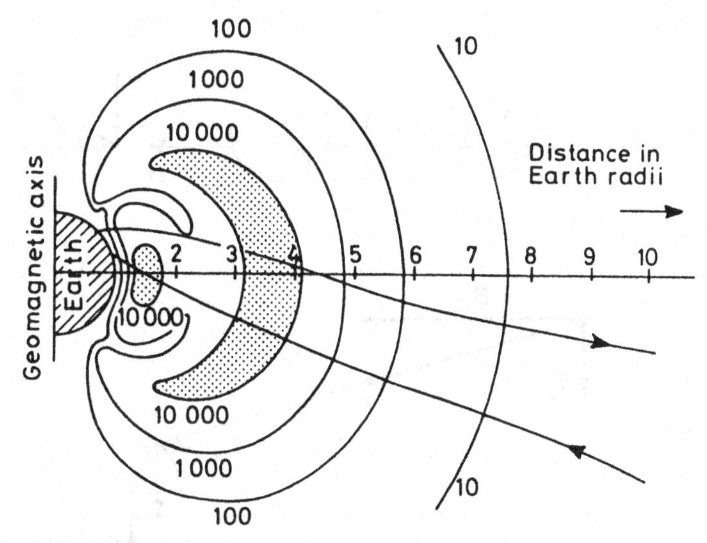

Fig. 5.38 Van Allen's first map of the radiation belts, showing counting rates of the Pioneer-3 Geiger counter. (After J. A. Van Allen and L. A. Frank, *J. Geophys. Res.* **64**, 1683, 1959, copyright by the American Geophysical Union)

at first thought to be electrons of energy 50–150 keV. However Explorer 4, which was more fully instrumented, showed that they were actually protons of energy exceeding 30 MeV. It is of historical interest to note that Sputniks 1 and 2 had also carried Geiger counters, but the crucial observations had been missed for one reason or another.

At the end of 1958, Pioneer 3 went out 107 400 km and found a double structure as in Figure 5.38. These and other early measurements gave rise to the idea that there are two distinct belts of trapped particles, the 'inner' and the 'outer' zones. In fact the total picture is more complicated than this because the distribution depends on the nature and energy of the particles.

5.7.2 Trapping theory

Trapping mechanism

The trapping of energetic charged particles arises from the interaction between a moving electric charge and a static magnetic field. The principles are best expressed in terms of a number of *abiabatic invariants*, which are quantities that do not change provided some other quantity (specifically the magnetic flux density) changes not at all or sufficiently slowly. (Compare the adiabatic expansion of a gas, which applies when the heat loss is slow enough to be negligible during the expansion.)

We saw in Section 2.3.2 that a charged particle of mass m and charge e gyrates in a magnetic field of flux density B at angular frequency, the *gyrofrequency*, $\omega_B = eB/m$ (Equation 2.27). The radius of gyration is $r_B = mv/Be$ (Equation 2.26), where v is the particle velocity. Taking an example, for particles of energy 1 MeV at altitude 2000 km above the equator the periods are 7 μs for electrons and 4 ms for protons. The respective gyration radii are 300 m and 10 km.

If a particle in the magnetosphere has velocity $v_\perp$ normal to the magnetic field and $v_\parallel$ along it, the trajectory is a spiral (Figure 5.39a). Provided no work is done on or by

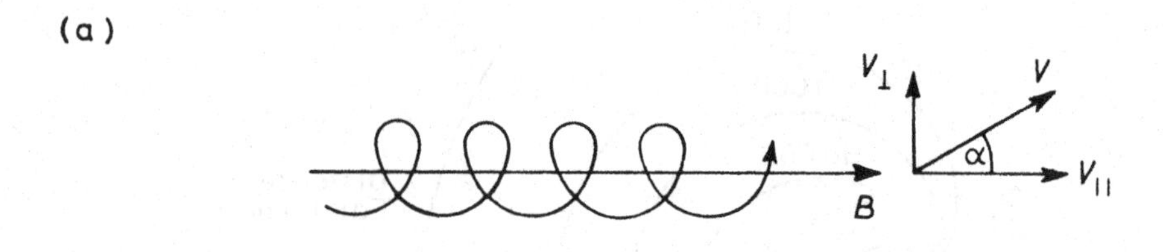

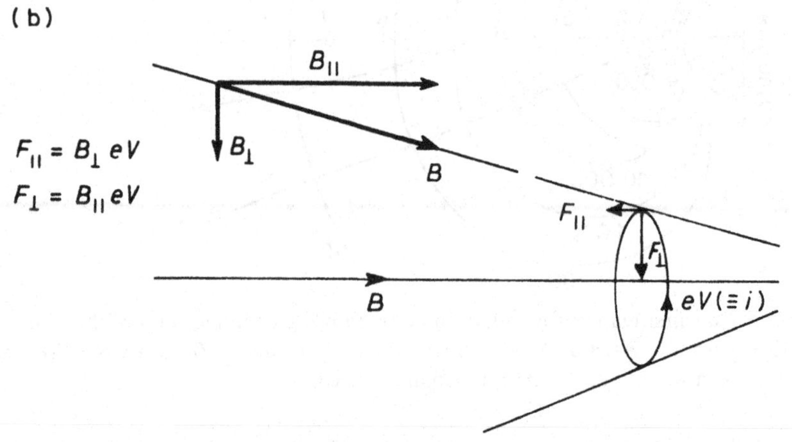

Fig. 5.39 Spiral motion of a trapped particle and forces near a mirror point. (a) Motion of a positive particle along the geomagnetic field. (b) Retarding force on equivalent current loop.

the particle, the magnetic flux through the orbit (Φ_m) is constant (since if $d\Phi_m \neq 0$ an electric field would be present and the particle would be accelerated or retarded). Hence,

$$\Phi_m = B \cdot \pi r_B^2 = 2\pi m E_\perp / e^2 B = \text{constant}, \tag{5.26}$$

where $E_\perp$ is the kinetic energy associated with the transverse velocity component, i.e.

$$E_\perp = m v_\perp^2 / 2. \tag{5.27}$$

Hence, $E_\perp / B$ is constant. But this is the magnetic moment of the current loop represented by the gyrating particle:

$$\mu = \text{current} \times \text{area of loop} = \frac{e v_\perp}{2\pi r_B} \cdot \pi r_B^2 = m v_\perp^2 / 2B = E_\perp / B. \tag{5.28}$$

This gives us the *first adiabatic invariant* of charged particle motion in a magnetic field:

the magnetic moment is constant.

It holds provided the magnetic field does not change significantly during one gyration period.

If the total velocity of the particle is v, then

$$v_\perp = v \sin \alpha \tag{5.29}$$

where α is the angle between the velocity vector and the magnetic field direction, the *pitch angle*. Hence, since the total kinetic energy, E, is constant,

$$E_\perp / B = E \sin^2 \alpha B = \text{constant}.$$

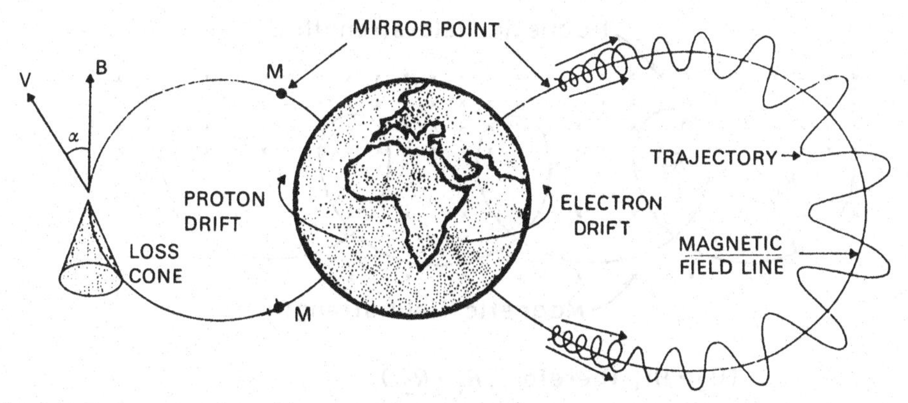

Fig. 5.40 Trajectories of particles trapped on closed field-lines. (Reprinted with permission from J. Lemaire, *Advances in Space Research*, **2**, 3, copyright (1982) Pergamon Press PLC)

Thus,

$$\sin^2 \alpha \propto B. \tag{5.30}$$

As a particle moves from the equator towards higher latitude in a dipole field it encounters increasing B and therefore the pitch angle increases. Eventually, provided the atmosphere is not encountered first, $\alpha = 90°$. Here the forward motion stops and the particle is reflected back along the field towards the equator. The point of reflection is called the *mirror point*. The total energy of the particle does not change during its motion because no acceleration mechanisms are at work; also, $E = E_\perp + E_\parallel$. Hence, for the parallel velocity,

$$v_\parallel^2 = v^2 - v_\perp^2 = v^2 \cos^2 \alpha, \tag{5.31}$$

The parallel energy falls to zero at the mirror point, as would be expected, and the changing pitch angle represents an alternation of kinetic energy between parallel and perpendicular components, the total remaining constant. It is the conservation of the first invariant within a non-uniform magnetic field approximating to a 'magnetic bottle' that makes possible the trapping of energetic charged particles.

The turning at the mirror point can be understood by considering the forces acting on the current loop. As Figure 5.39b illustrates, the convergence of the magnetic field introduces a force component perpendicular to the current loop, which pushes the loop towards weaker field. An alternative model represents the gyrating particle as a small magnetic dipole aligned with the geomagnetic field and oriented so as to oppose that field. In a non-uniform field the net force acts away from the strong field.

For a given particle the position of the mirror point is determined by the pitch angle as the particle crosses the equator (i.e. where the field is weakest), since

$$B_0/B_M = \sin^2 \alpha_0/1.0, \tag{5.32}$$

B_0 and B_M being the flux densities at the equator and at the mirror point respectively. The situation is sketched in Figure 5.40.

Particles will be lost if they encounter the atmosphere before the mirror point. The equatorial pitch angles of particles that will be lost to the atmosphere at the next bounce define the *loss cone*, which will be seen as a depletion within the pitch-angle distribution.

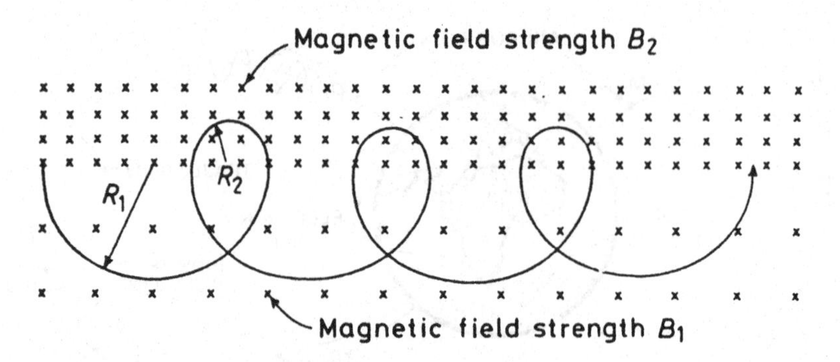

Fig. 5.41 Longitude drift caused by a gradient in geomagnetic field intensity. (After W. N. Hess, *Radiation Belt and Magnetosphere*. Blaisdell, 1968)

The time taken for a particle to travel between mirror points is usually called the *bounce time*, and in the case of 1 MeV particles crossing 2000 km above the equator this time is 0.1 s for an electron and 2 s for a proton.

Longitude drift

In addition to the gyration about the field-line and the oscillation between hemispheres, a trapped particle also drifts around the Earth. This is due to two mechanisms which are about equally important. The first is that the particle finds itself in a slightly weaker field when on that part of its gyration furthest from the Earth. The radius of gyration therefore changes and, as may be seen from the exaggerated situation shown in Figure 5.41, this introduces a lateral shift of the orbit. The second cause is the curvature of the field-line, which produces a centrifugal force to which the particle responds by drifting sideways. (The mechanism is like that by which a neutral wind in the upper atmosphere produces an electric current at right angles to the wind and to the geomagnetic field, as discussed in Section 6.5.3.) The combination of these two effects is called *gradient–curvature drift.*

With reference to Figure 5.41, the respective radii of gyration in fields B_1 and B_2 are $r_1 = mv/B_1e$ and $r_2 = mv/B_2e$. Therefore in one orbit, assuming half within B_1 and half within B_2, the particle steps sideways by

$$2(r_1 - r_2) = \frac{2mv}{e}\left(\frac{1}{B_1} - \frac{1}{B_2}\right).$$

This happens in a time

$$\frac{\pi m}{e}\left(\frac{1}{B_1} + \frac{1}{B_2}\right),$$

giving a drift speed

$$V = \frac{2v}{\pi}\cdot\frac{(B_2 - B_1)}{(B_2 + B_1)}.$$

Now suppose that the magnetic field varies gradually as $\partial B/\partial R$, where R is the radial geocentric distance, and that the amount of variation over an orbit is relatively small. Then

$$(B_2 - B_1) \sim \frac{\partial B}{\partial R}\cdot(r_1 + r_2) = \frac{\partial B}{\partial R}\cdot\frac{mv}{e}\cdot\left(\frac{1}{B_1} + \frac{1}{B_2}\right).$$

Therefore the drift speed

$$V \sim \frac{2v}{\pi}\cdot\frac{(B_2-B_1)}{(B_2+B_1)}$$

$$= \frac{2v}{\pi}\cdot\frac{1}{(B_2+B_1)}\cdot\frac{mv}{e}\cdot\frac{(B_1+B_2)}{B_1B_2}\cdot\frac{\partial B}{\partial R}$$

$$\simeq \frac{1}{2}\cdot\frac{mv^2}{eB^2}\cdot\frac{\partial B}{\partial R} \tag{5.33}$$

(which is the correct formula). We call this V_G, the gradient drift velocity. More generally $\partial B/\partial R$ should be replaced by $\nabla_\perp B$, the vector field gradient normal to B.

The equation holds for non-relativistic and for relativistic particles. For a non-relativistic particle, $mv^2/2$ is the kinetic energy. Thus, the drift speed is proportional to the energy of the particle, irrespective of its mass. The relevant velocity is the component normal to the magnetic field. For a relativistic particle,

$$V_G = \frac{m_0c^2}{2eB^2}\cdot\nabla_\perp B\cdot\frac{\beta^2}{(1-\beta^2)^{\frac{1}{2}}}, \tag{5.34}$$

where m_0 is the rest mass and the relativistic factor is $\beta = v/c$.

Curvature drift arises from the curvature of the geomagnetic field-lines. A particle travelling directly along the field-line at velocity v experiences an outward force $F_c = mv^2/r_c$, where r_c is the radius of curvature of the line. This force produces a particle drift at right angles to itself and to the field-line of magnitude

$$V_C = F_c/Be = mv^2/r_cBe. \tag{5.35}$$

(How this drift arises is discussed in some detail in Section 6.5.2.) It can be shown that $r_c = B/(\partial B/\partial R)$, and therefore

$$V_C = \frac{mv^2}{eB^2}\cdot\frac{\partial B}{\partial R}. \tag{5.36}$$

The curvature drift acts in the same direction as the gradient drift but it depends on the velocity component along the field ($v_\parallel$) instead of the perpendicular component ($v_\perp$); the formulae are also similar. We can therefore combine them into an expression for *gradient–curvature drift*:

$$V_{GC} = \frac{m}{eB^2}\cdot\frac{\partial B}{\partial R}\cdot(\tfrac{1}{2}v_\perp^2 + v_\parallel^2). \tag{5.37}$$

For a particle with velocity v and pitch angle α, $v_\parallel = v\cos\alpha$ and $v_\perp = v\sin\alpha$. Therefore

$$V_{GC} = \tfrac{1}{2}\cdot\frac{mv^2}{eB^2}\cdot\frac{\partial B}{\partial R}(1+\cos^2\alpha); \tag{5.38a}$$

alternatively, in terms of the local field-line radius,

$$V_{GC} = \tfrac{1}{2}\cdot\frac{mv^2}{eBr_c}\cdot(1+\cos^2\alpha). \tag{5.38b}$$

The relative magnitude of the two parts depends on the pitch angle of the trapped particle, but in general they are both important and only in special cases can one or the other be neglected.

Gradient–curvature drift sends electrons to the east and protons to the west. The drift rate depends to some extent on the mirror point (i.e. on the pitch angle). For 1 MeV particles of 'flat' pitch angle, mirroring 2000 km above the equator, an electron completes one circuit of the Earth in about 50 min, and a proton takes about 30 min.

More exactly, the period is given by

$$P = 172.4 \cdot \frac{1+\xi}{\xi(2+\xi)} \cdot \frac{m_e}{m} \cdot \frac{1}{R_0} \cdot \frac{G}{F} \text{ minutes,} \tag{5.39}$$

where $\xi = (1-\beta^2)^{\frac{1}{2}} - 1$, with β being the ratio of the particle speed to the speed of light, m/m_e is the mass of the particle in electron masses, R_0 is the geocentric distance to the particle's equatorial crossing in Earth radii, and G/F is a factor of value 1.0 for a particle mirroring at the equator increasing to 1.5 for one mirroring at the pole. For non-relativistic particles, $\beta^2 \ll 1$ and the formula reduces to

$$P = \frac{733}{E} \cdot \frac{1}{R_0} \cdot \frac{G}{F} \text{ hours,} \tag{5.40}$$

where E is the energy in keV.

Figure 5.42 gives the drift period of trapped electrons as functions of energy and pitch angle.

In a dipole field, gradient–curvature drift would move the particles around at the same distance from the Earth and would serve merely to distribute the particles to all longitudes. However in the actual field, which is not dipolar, the drift paths are not so obvious. The theory of trapped particles shows that the path may be determined from the *second adiabatic invariant*, also known as the *integral invariant*, which states that

the integral of the parallel momentum over one bounce between mirror points is constant:

$$J = 2\int_{l_1}^{l_2} mv_\parallel \, dl = 2mv\int_{l_1}^{l_2} \left(1 - \frac{B}{B_M}\right)^{\frac{1}{2}} dl = \text{constant} \tag{5.41}$$

(since $v_\parallel = v\cos\alpha = v(1-\sin^2\alpha)^{\frac{1}{2}} = v(1-B/B_M)^{\frac{1}{2}}$, from equation 5.30). l is the distance along the field-line, and l_1 and l_2 are the mirror points at which the flux density is B_M. Since mass and total velocity are constant, the invariant can be written

$$I = \int_{l_1}^{l_2} \left(1 - \frac{B}{B_M}\right)^{\frac{1}{2}} dl. \tag{5.42}$$

This is a property of the field configuration and also of the mirror point (or equatorial pitch angle) of the particle, and it defines the surface, or *shell*, on which the particle remains as it drifts around the Earth. It holds provided the field does not change appreciably during one bounce period. Taken together, the first and second invariants define the locus of the mirror points of a bouncing and drifting trapped particle. It has been shown that the flux of particles perpendicular to the magnetic field direction will be the same at all points having the same values of B and I, and thus these variables provide a convenient way of organizing the data when mapping trapped particle distributions.

L-parameter

An important development from the second invariant is the *L parameter* introduced by C. E. McIlwain in 1961. L has the dimension of length taking one Earth radius as the unit. It is analogous to the distance to the equatorial crossing of a field-line (R_0 in Equation 5.16), to which it reduces in a dipole field. Along most closed geomagnetic

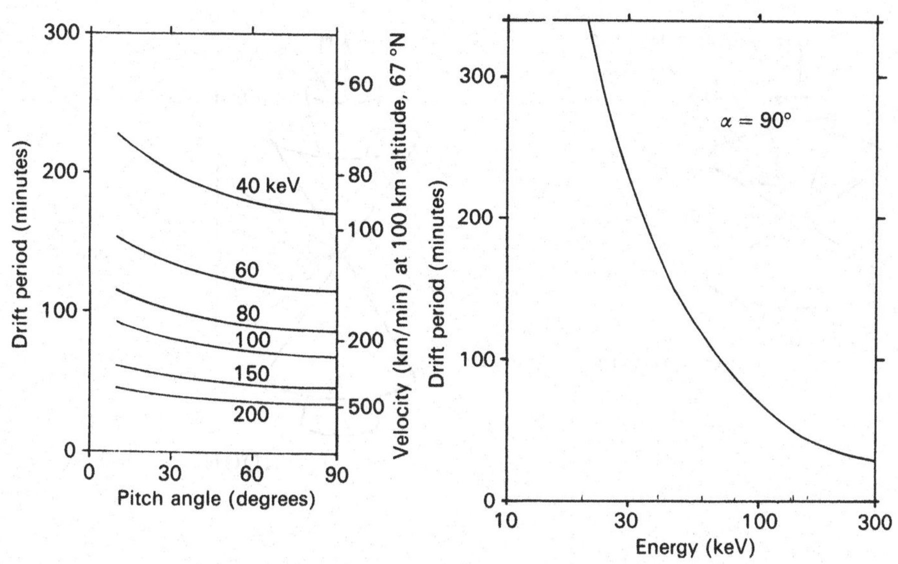

Fig. 5.42 Time for trapped electrons to make one circuit of the Earth at the orbit of a geostationary satellite ($R_0 = 6.6$): (a) against pitch angle for various energies; (b) against energy for $\alpha = 90°$. (a) also gives the velocity of the footprint (67° latitude) 100 km above the Earth's surface. (P. N. Collis, private communication)

field lines L is constant to about 1% and thus it usefully serves to identify field-lines even though they be not strictly dipolar. By analogy with a dipole field, an *invariant latitude* may be defined in terms of L:

$$\Lambda = \cos^{-1}(1/L)^{\frac{1}{2}}. \tag{5.43}$$

This is analogous to Equation 5.17, and through it 'L' is commonly used to label field-lines even at high latitude where L strictly has no meaning because particles cannot be trapped on field-lines that are not closed.

Pseudo-trapping

A complication that arises from the second invariant in an asymmetrical field is the phenomenon of *shell splitting*. This arises because the drift path depends on the pitch angle as well as on the L-value of the field-line. If particles are introduced at midnight, those mirroring at the equator ($\alpha_0 = 90°$) drift so as to remain at constant field intensity. Since the geomagnetic field is squashed in by the solar wind on the day side, increasing the field strength at given distance, these particles of flat pitch angle will move out on the day side. Some will leave the magnetosphere and not return to the midnight meridian. The pitch angle distribution at midnight may therefore include a 'drift loss cone' for pitch angles around 90°, as illustrated in Figure 5.43a. On the other hand, particles mirroring at high latitude (α_0 small) will, in following a shell of constant I, be more influenced by the length of the line, and these particles will not move out so far. The spatial distribution is thereby smeared out. Of particles at the noon meridian, some of those with small pitch angle will be lost in the night sector and not return, and in this case the noon pitch angle distribution includes a drift loss cone for small pitch angle (Figure 5.43a). The normal bounce loss cone, discussed above, is always present as well.

Because of shell splitting and the drift loss cones, there are regions from which the trapped particles will not be able to complete a circuit of the Earth, being lost *en route* in the magnetotail

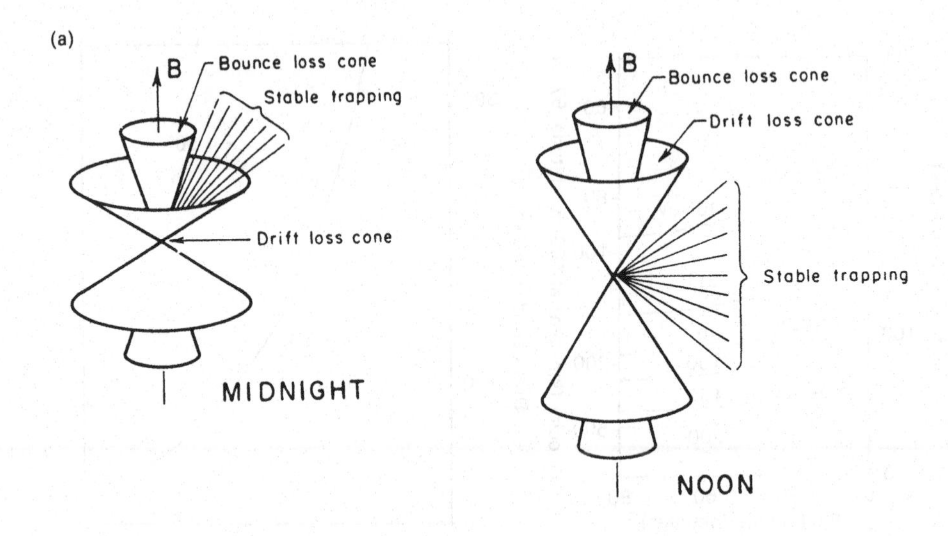

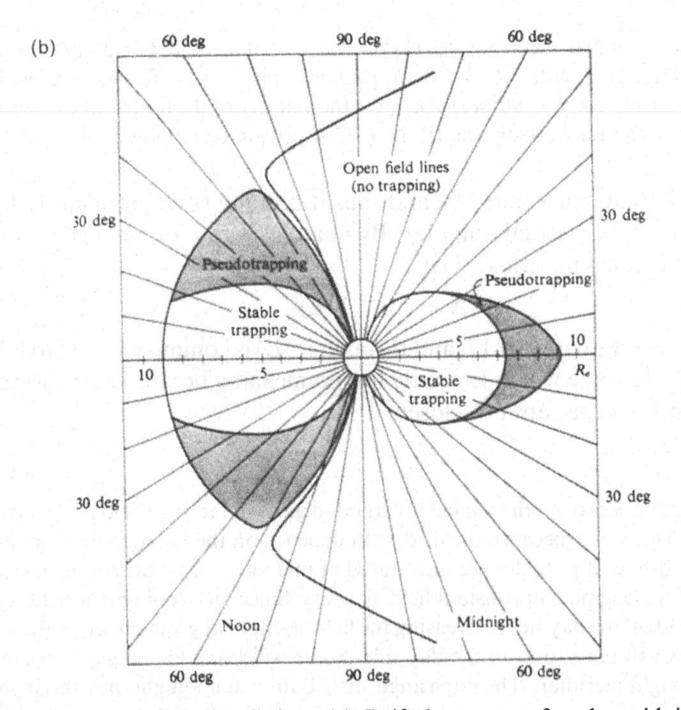

Fig. 5.43 Consequences of shell splitting. (a) Drift loss cones for the midnight and noon meridians. (b) Regions of stable, pseudo-, and no trapping. Particles which mirror in a pseudo-trapping zone do not return from their next circuit of the Earth. (After J. G. Roederer, *Dynamics of Geomagnetically Trapped Radiation*. Springer-Verlag, 1970)

or beyond the magnetopause. These are called *pseudo-trapping regions* and their locations are illustrated in Figure 5.43b.

Flux invariant

The *third adiabatic invariant* or *flux invariant* says that

the total geomagnetic flux enclosed by the drift orbit is constant,

and this enables the effect of very slow changes of magnetospheric structure on particle orbits to be calculated. The third invariant is violated if changes occur in a time less than that taken for the particle to encircle the Earth; such violations are not infrequent, particularly during magnetic storms when conditions tend to change rapidly at the beginning but return to normal more slowly.

Although the theory of particle trapping was originally developed for the Van Allen particles, the ideas and rules derived find more general application within the magnetosphere and to other branches of physics involving the interaction of charged particles and magnetic fields.

5.7.3 Sources, sinks and morphology

In a steady state, if no other processes were operating, particles trapped now would remain trapped for ever. But then one would have to ask how they became trapped in the first place. And if there is indeed a mechanism for trapping charged particles, there must obviously be a loss process or processes too – otherwise the trapped flux would be increasing indefinitely, which is not the case.

The morphology of the radiation belts involves four kinds of process:

(a) injection of charged particles into the trapping region;
(b) acceleration of particles to high energy;
(c) diffusion within the region;
(d) loss processes removing particles from the trapping region.

Inner zone

The source must either create particles *in situ* or introduce them by violating an invariant. In the so called *inner zone*, within about 2.5 R_E, the particles are created *in situ* by the decay of neutrons which themselves come from primary cosmic ray protons of very high energy:

$$\text{5 BeV proton} \rightarrow \text{approx 7 neutrons,}$$
$$\text{neutron} \rightarrow \text{proton} + \text{electron} + \text{antineutrino.}$$

Calculation shows this to be a sufficient source for the inner zone. The loss process involves some kind of interaction with the atmosphere near the mirror points. Three kinds of interaction are possible:

gradual energy loss by ionization;
charge exchange;
nuclear interaction.

The main inner zone loss processes are: retardation and charge exchange, particularly for the less energetic (<100 keV) protons; nuclear collisions, particularly for protons >75 MeV; and scattering into the loss cone, particularly for electrons.

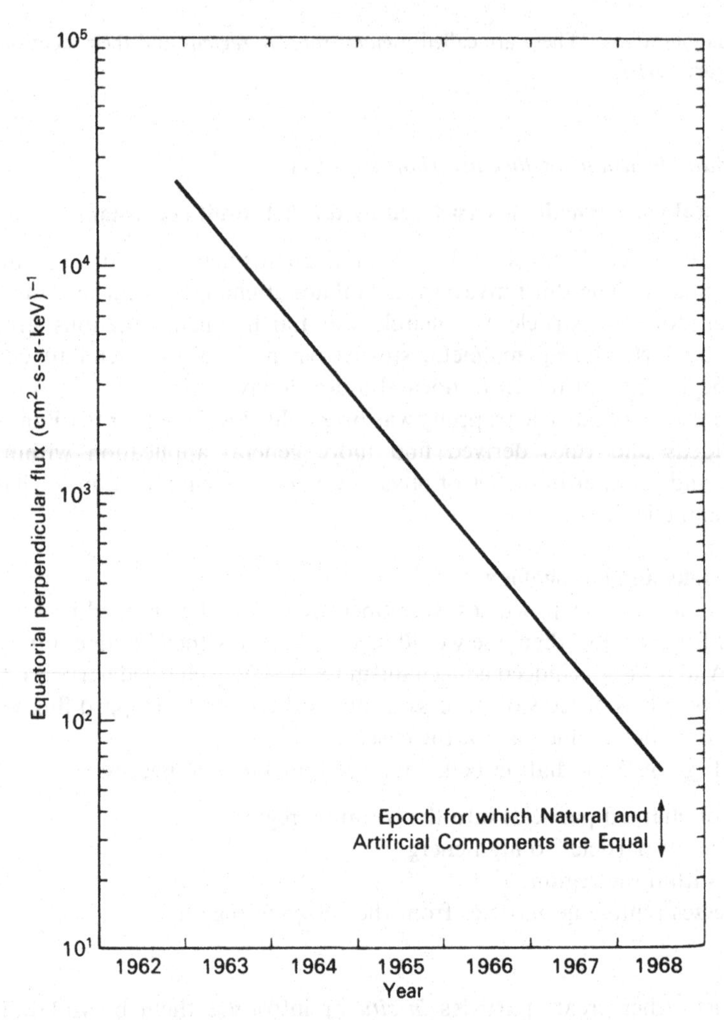

Fig. 5.44 Decay of Starfish electrons of 1.5 MeV energy at L = 1.5 between 1962 and 1968. (After K. W. Chan *et al.*, in *Quantitative Modelling of Magnetospheric Processes* (ed. Olsen). American Geophysical Union, 1979, p. 121)

In 1962 a high-altitude nuclear explosion (Starfish) contaminated the inner zone with electrons, and these did not decay to below the natural level until about 1970 (Figure 5.44). The spectrum of inner zone electrons in Figure 5.45 includes Starfish effects as well as the natural zone. There are also natural variations in the inner zone, but those in the outer zone are much greater.

Outer zone

The question of how the particles of the outer zone (L > 3) are created does not have such a clearcut answer. Some of the more obvious possibilities – neutrons from the Sun; neutral hydrogen from the Sun subsequently ionized; acceleration of plasma originating in the ionosphere or the protonosphere – do not stand up to detailed scrutiny. The charged particles almost certainly come from the outer magnetosphere,

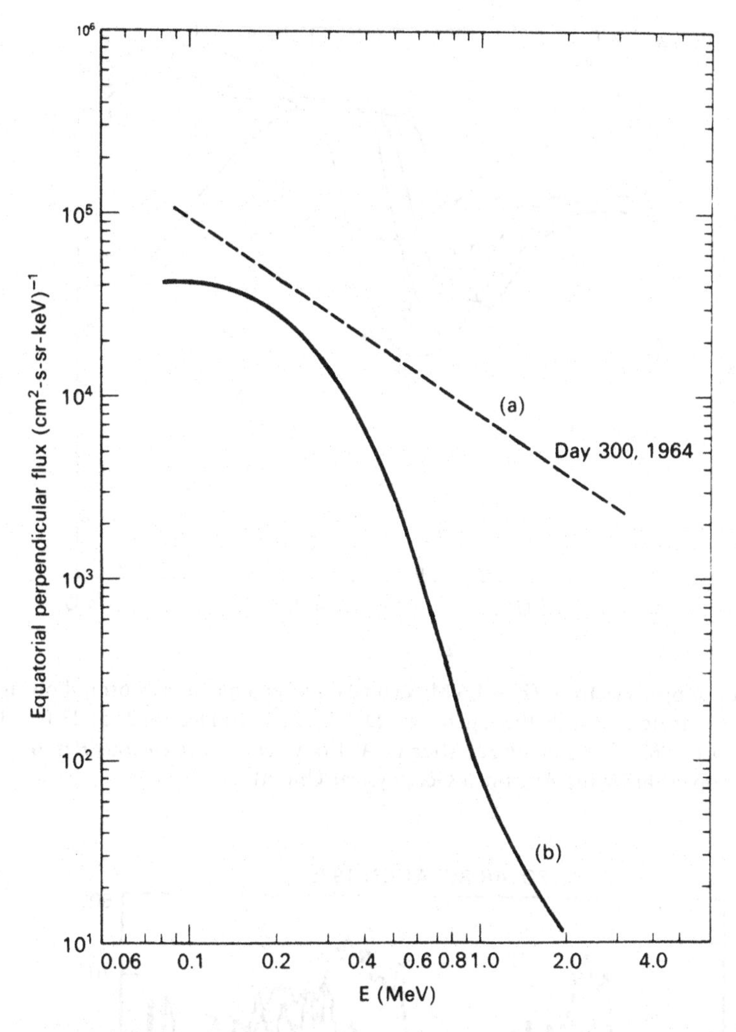

Fig. 5.45 Spectra of inner zone electrons at L = 1.3. (a), which is for day 300 of 1964, includes Starfish electrons. Model (b) is based on 1967 data with the estimated Starfish contribution subtracted. (After K. W. Chan *et al.*, in *Quantitative Modelling of Magnetospheric Processes* (ed. Olsen). American Geophysical Union, 1979, p. 121)

and the main evidence for this is the increases of flux at times of enhanced geophysical activity. Figure 5.46 shows the radiation belt filling and decaying after a magnetic storm, there being also inward drift. At geosynchronous orbit, which is well outside the maximum of the outer zone, the variations are very marked and are related to the velocity of the solar wind. Observations of energetic electrons show that the logarithm of the particle flux correlates with the solar wind velocity, allowance being made for a delay of one or two days depending on the energy (Figure 5.47). Typical energy spectra are shown in Figure 5.48.

The particles probably enter the outer magnetosphere at the sunward cusps. Some will already be of high energy, but local acceleration in the magnetosphere is also

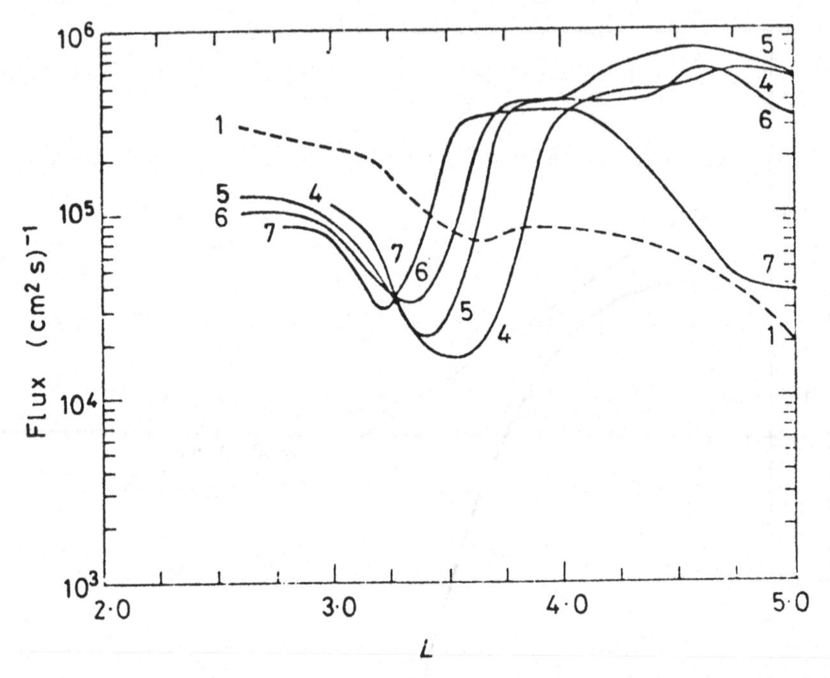

Fig. 5.46 Flux of trapped electrons (E ⩾ 1.6 MeV) on days after a magnetic storm, showing that electrons drifted towards the Earth. 1: 7 Dec 1962; 4: 20 Dec 1962; 5: 23 Dec 1962; 6: 29 Dec 1962; 7: 8 Jan 1963. (After L. A. Frank *et al.*, *J. Geophys. Res.* **69**, 2171, 1964, copyright by the American Geophysical Union)

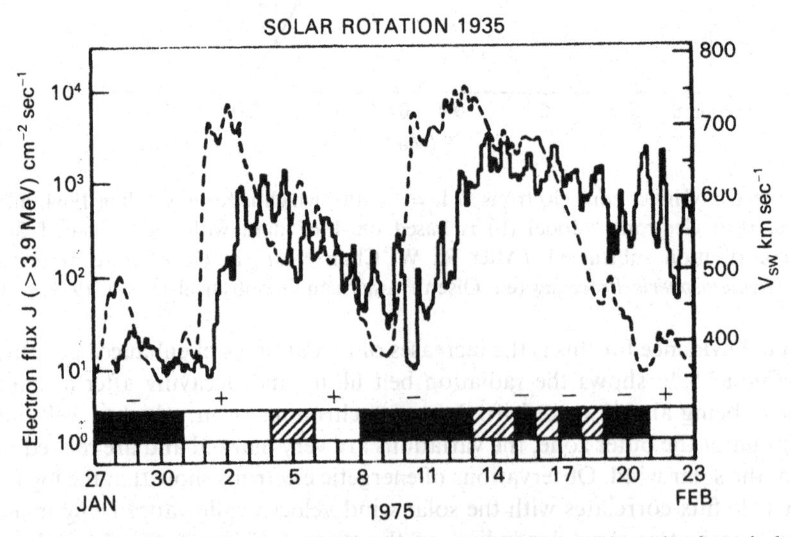

Fig. 5.47 Flux of energetic electrons (solid line) at synchronous orbit and solar wind velocity (dashed line) during one solar rotation. The IMF sectors are indicated. (G. A. Paulikas and J. B. Blake, in *Quantitative Modelling of Magnetospheric Processes* (ed. Olsen). American Geophysical Union, 1979, p. 180)

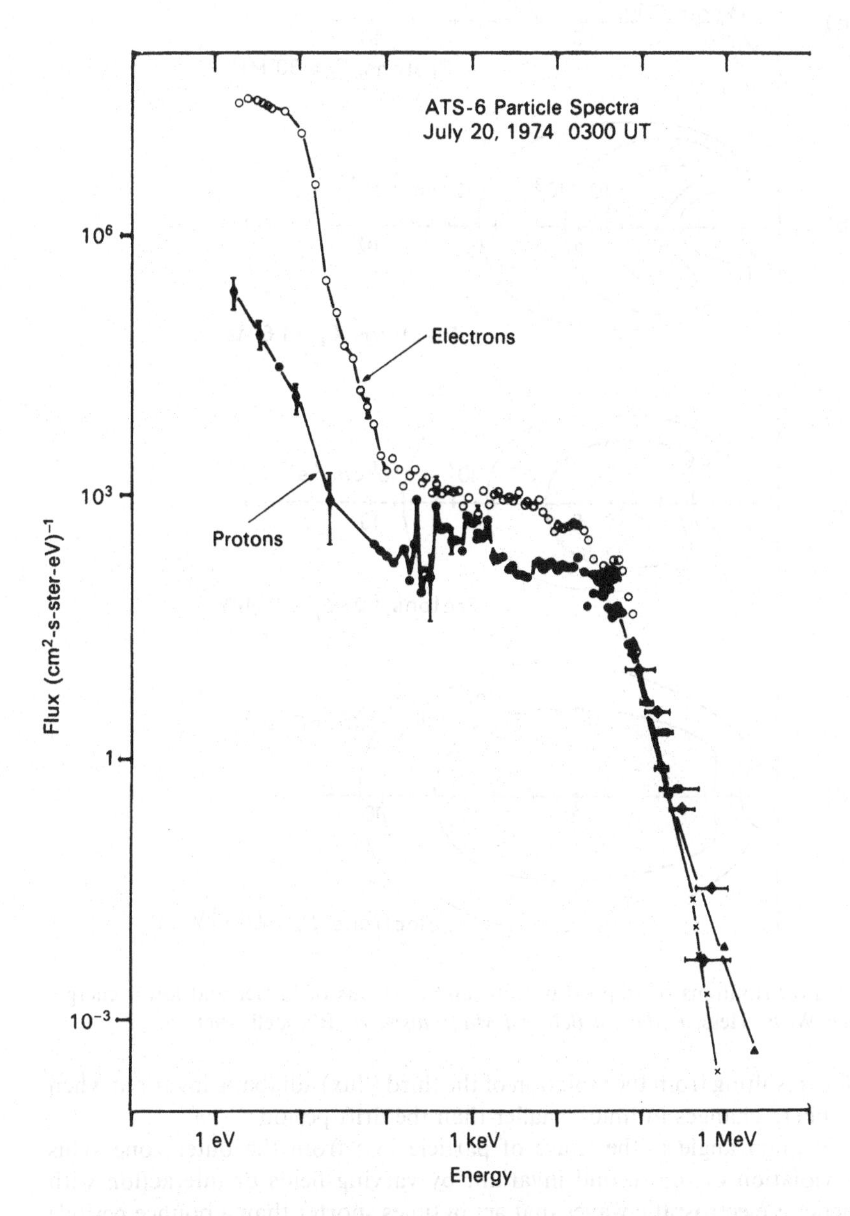

Fig. 5.48 Electron and ion spectra measured at geosynchronous orbit. (G. A. Paulikas and J. B. Blake (after T. A. Fritz *et al.*), in *Quantitiative Modelling of Magnetospheric Processes* (ed. Olsen). American Geophysical Union, 1979, p. 180)

required and it is not entirely clear how this happens. If a particle has to move nearer the Earth it will tend to gain energy because it is moving on to shorter field-lines and into a region of stronger field (Sections 2.3.3 and 2.3.8). There could also be scattering from shock waves in the magnetosphere. The drift to different L-shells is a process of

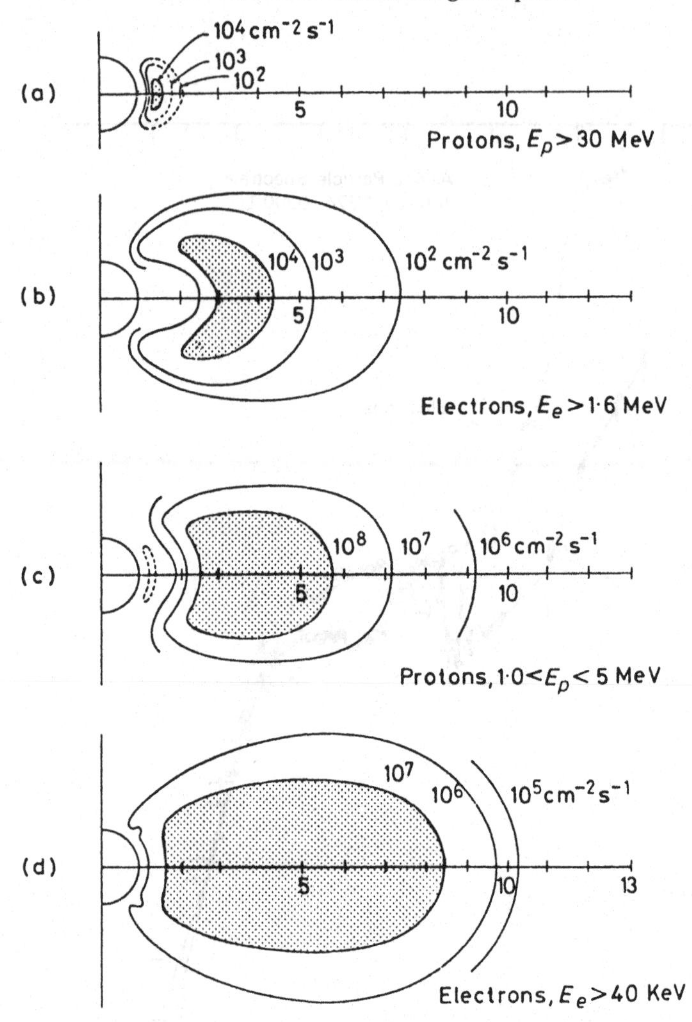

Fig. 5.49 Spatial distributions of trapped protons and electrons of higher and lower energy. (After W. N. Hess, *Radiation Belt and Magnetosphere*. Blaisdell, 1968)

radial diffusion resulting from the violation of the third (flux) adiabatic invariant when the magnetosphere changes in times smaller than the drift period.

Diffusion of pitch angle is the cause of particle loss from the outer zone. This involves the violation of the second invariant by varying fields or interaction with electromagnetic or electrostatic waves that act in times shorter than a bounce period; 'hiss' generated in the plasmasphere (Section 9.3) is believed to be the main cause. When particles diffuse into the loss cone they enter the atmosphere at the next bounce and are lost there. Figure 5.49 illustrates the observed spatial distributions of electrons and protons. The protons are also variable, though not to the same extent as the electrons.

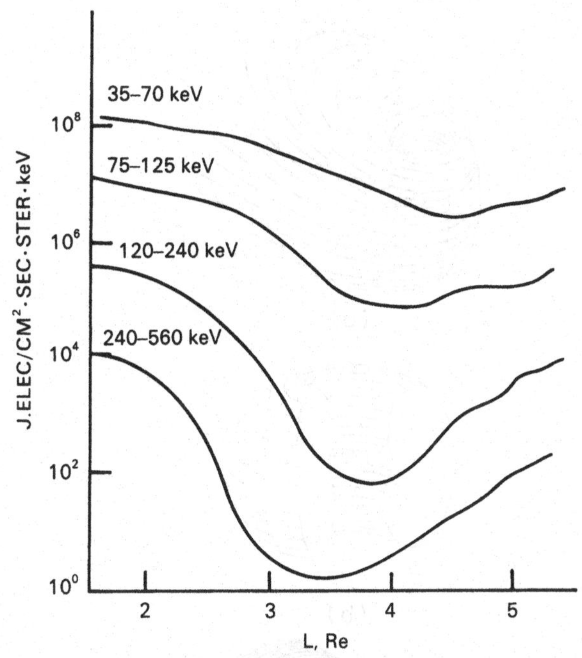

Fig. 5.50 The slot region in quiet conditions: radial profiles of 90° pitch-angle electron flux observed by Explorer 45 near the equatorial plane. (After L. R. Lyons and D. J. Williams, *J. Geophys. Res.* **80**, 3985, 1975, copyright by the American Geophysical Union)

The slot

That the spatial distribution depends on the energy of the particles is clear from Figure 5.49. In general the maximum flux occurs at greater distance for particles of lower energy, and the value of the maximum flux increases. The gap between the inner and outer zones is seen most clearly in the electron distributions, and this feature is called the *slot*. It occurs generally between 3 and 4 R_E and its depth is a factor of 10 to a factor of 10^4 depending on energy (Figure 5.50). The slot tends to fill during periods of enhanced geophysical activity, with a gradual return to normality over the following days and weeks.

5.8 Magnetospheric current systems

The combination of plasma and electric field in the magnetosphere allows electric currents to flow. Several current systems can be identified:

the magnetopause current;
the tail current;
the ring current;
Birkeland (field-aligned) currents.

Figure 5.27 shows them in relation to the main boundaries and plasma populations of the magnetosphere. The various current systems are not independent, and there are connections, some intermittent, between them.

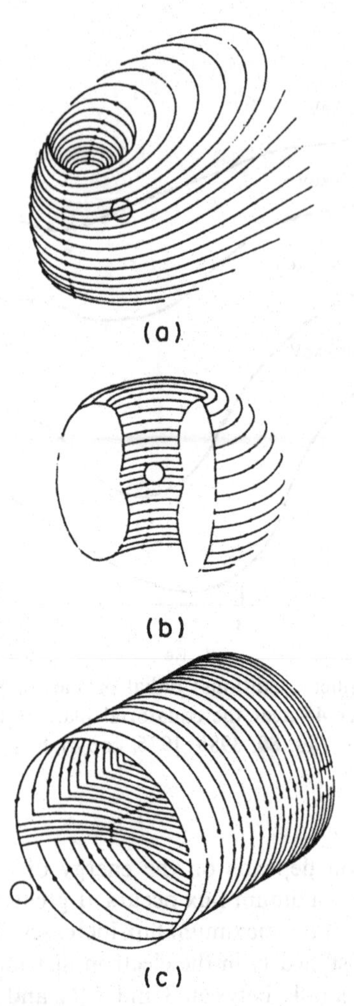

Fig. 5.51 Magnetospheric current systems: (a) magnetopause (Chapman–Ferraro); (b) ring; (c) tail. (Reprinted with permission from W. P. Olsen, *Adv. Space Res.* **2**, 13, copyright (1982) Pergamon Press PLC)

5.8.1 The magnetopause current

Magnetic fields and electric currents are fundamentally related through the Biot–Savart law. The magnetopause is the outer boundary of the geomagnetic field, and the current flowing at the magnetopause is such that its magnetic field cancels the geomagnetic field outside the boundary. In the inner magnetosphere, where the field is almost dipolar, the contribution from the magnetopause current is small compared with that from the Earth's internal dipole. The form of the magnetopause current is shown in Figure 5.51a.

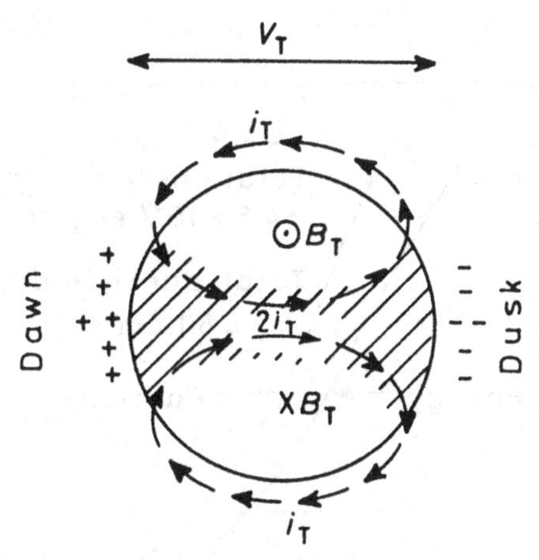

Fig. 5.52 Tail current as seen from the Earth. (After L. Svalgaard, *NASA Report SP*-366, 1975)

5.8.2 The tail current

The down (solar) wind extension of the magnetosphere into a tail indicates the presence of a current flowing across the plasma sheet from the dusk to the dawn side. But the tail is also bounded, and this requires a tail current in the form of a double solenoid, as in Figure 5.51c. Referring to Figure 5.52, if the current density over the surface of the tail is i_T A/m (that in the plasma sheet being $2i_T$), the magnetic flux density in the tail is

$$B_T = \mu_0 i_T. \tag{5.44}$$

The tail current and the tail magnetic field are therefore simply related. Since $B_T \approx 20$ nT, $i_T \approx 1.6 \times 10^{-2}$ A/m. In the sheet the current density is about 3×10^{-2} A/m. Assuming a reasonable length for the tail, the total tail current comes to some 10^8 A, and, taking the cross-tail potential difference as 60 kV, the power extracted from the solar wind by the tail amounts therefore to 6×10^9 kW (6000 GW).

5.8.3 The ring current

The ring current may be detected by ground-based magnetometers at the lower latitudes. It has long been known that under *magnetic storm* conditions (more fully discussed in Section 7.4.2) the magnetic field at the Earth's surface may be depressed for hours or days. At the commencement of the storm the magnetic field can increase, and this is explained by the compression of the magnetosphere with the arrival of increased solar wind, as originally proposed by Chapman and Ferraro (Section 5.3.1). The depression that follows it is due to a *ring current* in the magnetosphere, flowing round the Earth from east to west (clockwise as seen from over the north pole), as in Figure 5.51b. The direction of the current is such that its magnetic field acts to oppose the geomagnetic field. A depression of 30 nT at the surface, which would be a moderate storm, could be explained by 10^6 A at geocentric distance 4.5 R_E. The simple relation between current and field is

$$\Delta B(\text{nT}) = 2\pi I(\text{A})/10r(\text{km}), \tag{5.45}$$

which is based on the standard formula for the flux density at the centre of a current

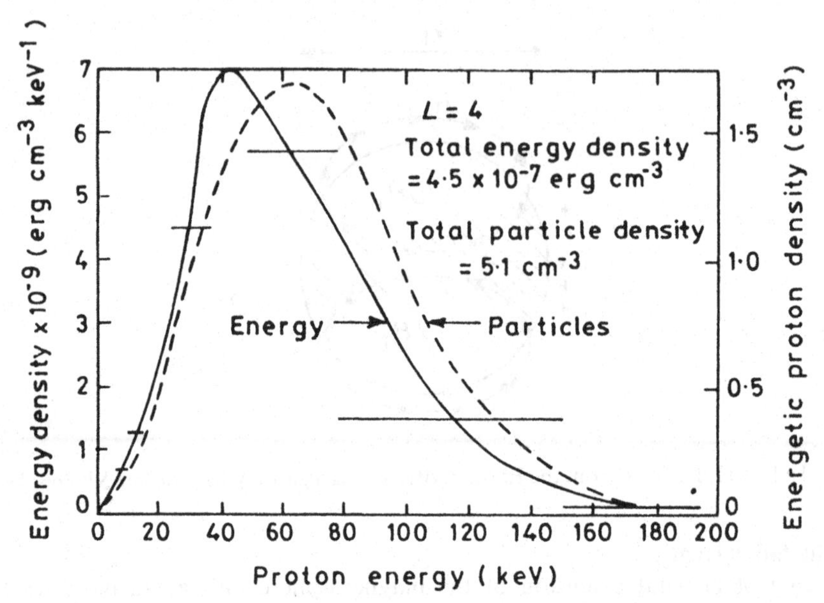

Fig. 5.53 Energy and number spectra of protons in the ring current at L = 4 during a magnetic storm. (M. H. Rees and R. G. Roble, *Rev. Geophys. Space Phys.* **13**, 201, 1975, copyright by the American Geophysical Union)

loop ($B = \mu_0 I/2r$). Currents induced in the ground increase the effect by an amount which depends on the ground conductivity and typically amounts to a factor of about $\frac{3}{2}$. Thus, an approximate relation is just

$$\Delta B \approx 3\pi I/10r \approx Ir. \tag{5.46}$$

The distance, r, cannot of course be determined from ground-based observations alone, though satellite measurements can do so. For example, a satellite-borne magnetometer inside the ring current will observe an effect in the same sense as seen at the ground; if the satellite is outside the ring current the effect will be in the opposite sense.

A clockwise ring current is produced by the drift of trapped particles (Section 5.7.2) since protons drift to the west and electrons to the east. After the discovery of the Van Allen particles in 1958 it was thought that they might account for the ring current, but on further consideration it was seen that the Van Allen belts as then observed did not contain a sufficient flux of particles. The particles reponsible are of somewhat lower energy, and are mainly protons of 10 to 100 keV (Figure 5.53).

Generally the current is located between 4 and 6 R_E, close to the inner edge of the plasma sheet and to the outer edge of the trapping zone. The concentration of particles in that region is a consequence of magnetospheric convection (Section 5.5). The inward convection of the tail field on the Earthward side of the reconnection point accelerates particles inward, but they eventually mirror in the stronger field nearer the Earth and this is where the inner edge of the plasma sheet forms.

According to Equation 5.40 the longitudinal drift rate of a trapped particle is proportional to its energy. Since each proton carries the same charge (+e) the total

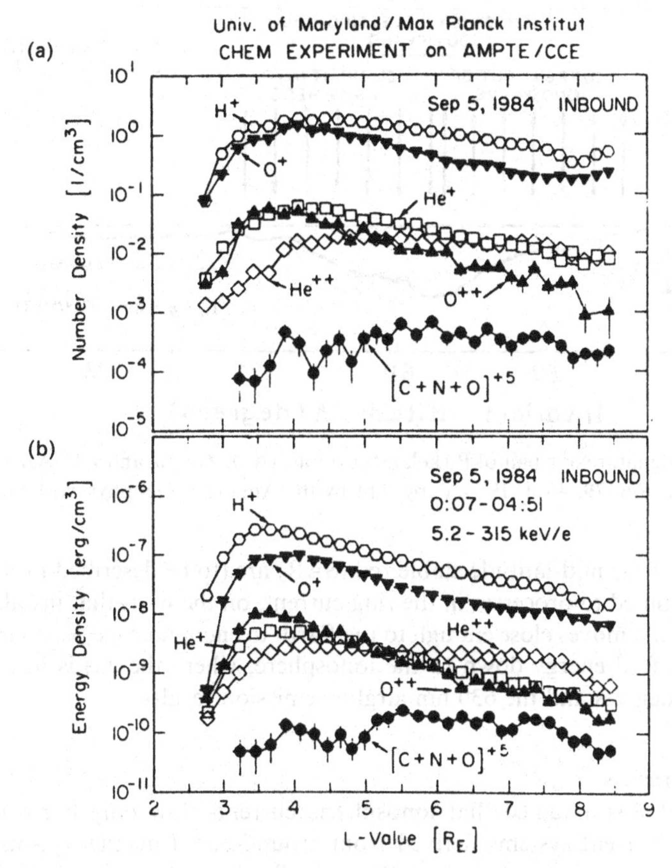

Fig. 5.54 Radial profiles of various heavy ions during an inbound pass of AMPTE on 5 Sept 1984. (a) Number density. (b) Energy density of particles of energy 5–315 keV per unit electronic charge. (D. J. Williams, *Space Sci. Rev.* **42**, 375, 1985. Reprinted by permission of Kluwer Academic Publishers; after G. Gloeckler *et al.*, private communication)

current is proportional to the product of particle density and particle energy integrated over all energies. Therefore the total current is proportional to the total energy of the ring current particles – virtually the total energy of the protons (E). Thus, using Equation 5.46:

$$\Delta B \approx 2.6 \times 10^{14} E(J), \tag{5.47}$$

it being understood that the magnetic field is reduced by the ring current.

Although H^+ is the major ion in the ring current, heavier ions, such as O^+, He^+, He^{++}, and O^{++}, are also present (Figure 5.54).

The energy density of the ring current can exceed that of the geomagnetic field (2×10^{-8} J/m³ at 5 R_E) and then the field becomes distorted. This is called the *inflation* of the geomagnetic field and is an example of the condition $\beta > 1$ (Section 5.6.1). The existence of a stable ring current requires a balance between the centrifugal force on the particles, the magnetic tension along the distorted field-lines, and the gradient of the particle pressure.

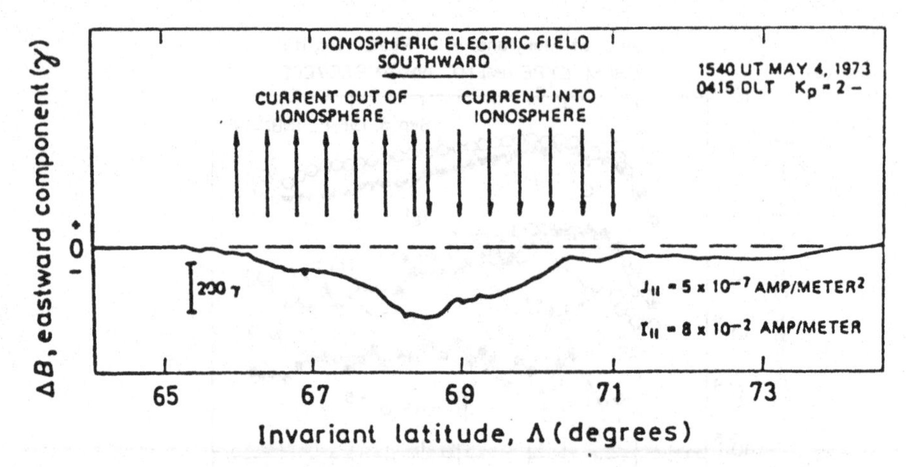

Fig. 5.55 Magnetic signature of a pair of Birkeland currents. (A. J. Zmuda and J. C. Armstrong, *J. Geophys. Res.* **79**, 4611, 1974, copyright by the American Geophysical Union)

The formation of the mid-latitude stable red (SAR) arc (to be described in Section 6.4) has been attributed to processes in the ring current, on the basis that instabilities form within it when it moves close enough to the Earth to encounter the plasmapause. These instabilities feed energy down to the ionosphere, where the gas is heated by several thousand degrees and the 630 nm airglow emission results.

5.8.4 Birkeland currents

For many years it was assumed that ionospheric currents flow only horizontally, which meant that current systems derived from ground-based magnetograms were really only *equivalent current systems* since the possibility of vertical current had been excluded. Indeed, Fukushima's theorem (Section 8.4.2) says that vertical current cannot be detected from the ground. Nevertheless, as we shall see in Section 6.5.5, the conductivity along the geomagnetic field far exceeds that across the field, and thus the idea that some current flows along the field cannot be lightly dismissed. The first suggestion of *field-aligned currents* was put forward by K. Birkeland in 1908, but the idea lay dormant for many years for lack of evidence. It began to return to favour on theoretical grounds in the 1960s, but several more years passed before solid evidence was found.

The convincing observations came from magnetic measurements on satellites (ISIS-2 and TRIAD) in the early 1970s. When the satellites crossed the auroral zone they registered magnetic signatures such as that in Figure 5.55, which shows a westward magnetic field extending over 2–3° of latitude (200–300 km). To explain this it is necessary to suppose that the satellite had passed through a double sheet of electric current, with a magnitude of about 5×10^{-7} A/m^2 and directed downward at the higher latitude and upward at the lower. Other satellite observations since then have plotted the statistical distribution of the currents and filled in many other details.

The currents are nearly always present, with typical distributions as in Figure 5.56; they intensify and move equatorward with increasing disturbance level, and are obviously related to the auroral oval (Section 8.3.2). The components into and out of

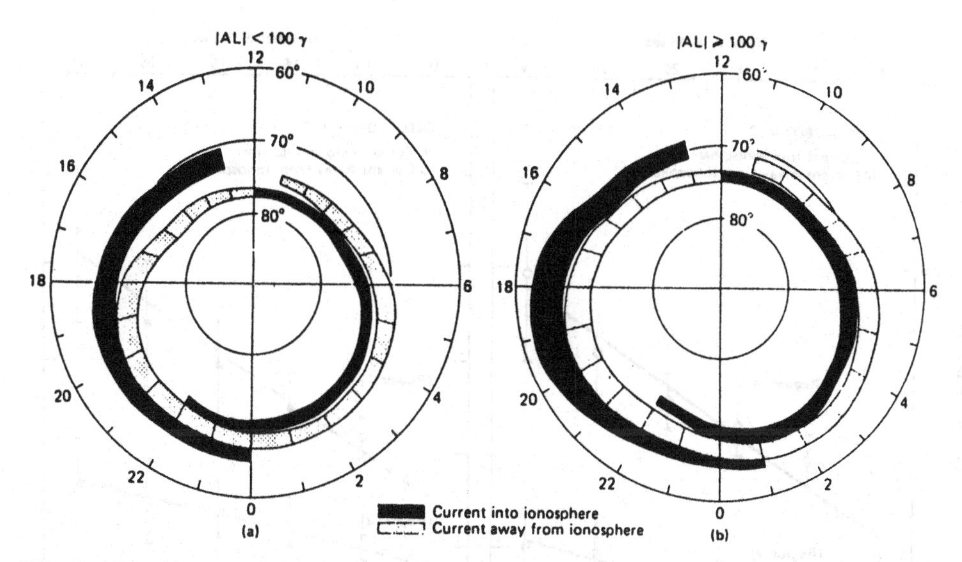

Fig. 5.56 Distribution of Birkeland currents during (a) weak and (b) active disturbances. (T. Iijima and T. A. Potemra, *J. Geophys. Res.* **83**, 599, 1978, copyright by the American Geophysical Union)

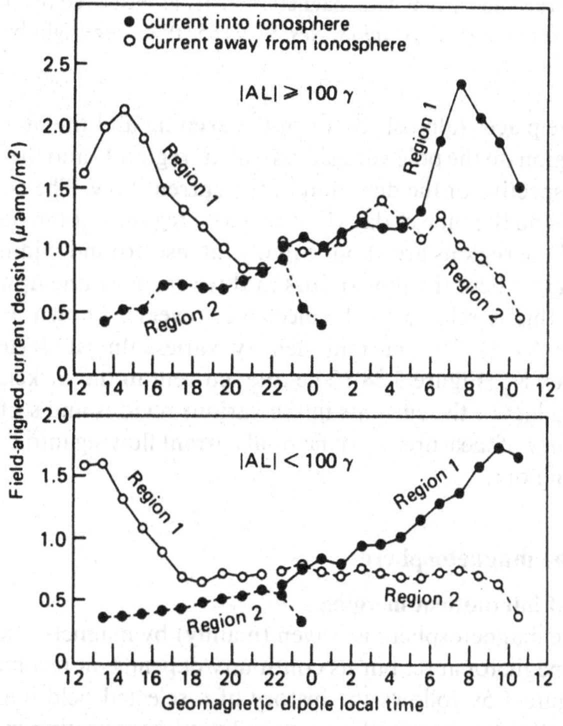

Fig. 5.57 The magnitudes of Birkeland currents in Regions 1 and 2 during active (upper panel) and weakly disturbed (lower panel) periods. (After T. Iijima and T. A. Potemra, *J. Geophys. Res.* **83**, 599, 1978, copyright by the American Geophysical Union)

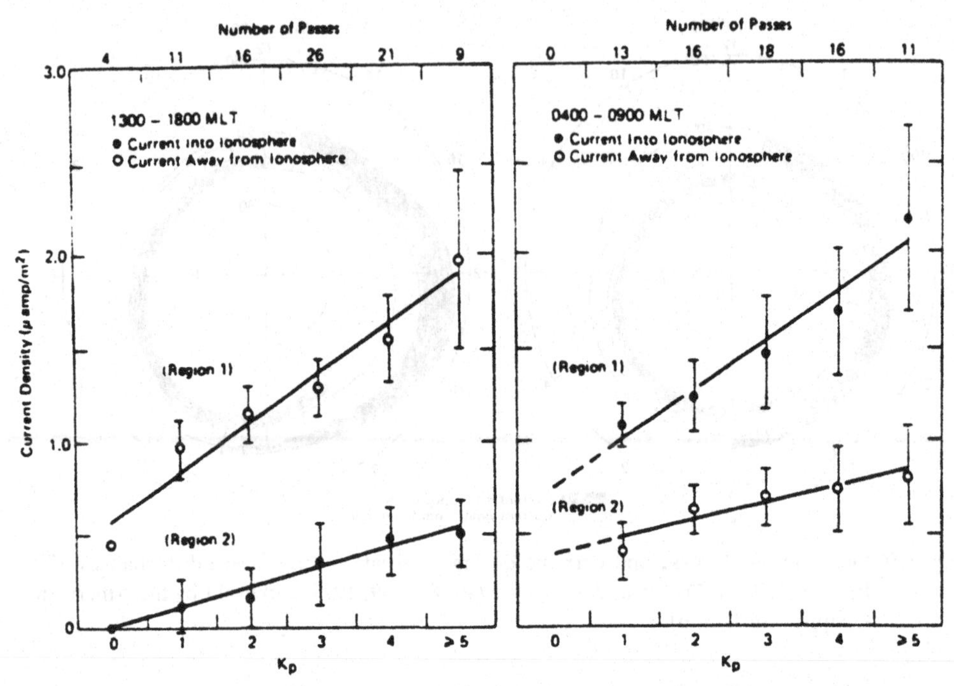

Fig. 5.58 Variation of Birkeland current with magnetic activity as measured by the K_p index. (T. Iijima and T. A. Potemra, *J. Geophys. Res.* **81**, 2165, 1976, copyright by the American Geophysical Union)

the ionosphere change places (almost) between the evening and the morning sectors. By convention, the region on the poleward side is called 'region 1' and the equatorward one is 'region 2', irrespective of the direction of the current flow. The current density in each region depends on the time of day (Figure 5.57); region 1 dominates on the day side of the Earth, and the regions are about equally intense around midnight. There is a transition at midnight where the flow occurs in three regions, one upward between two downward, and this is related to the electrojet reversal known as the *Harang discontinuity* (Section 8.4.1). The current density varies almost linearly with the magnetic activity index K_p (Figure 5.58). The total current in the Birkeland system is 10^6 to 10^7 A, and a budget of the currents in the various regions and sectors indicates that, within the accuracy of measurement, the total current flowing into the ionosphere is equal to that flowing out.

5.9 Substorms in the magnetosphere

5.9.1 Consequences of intermittent merging

The circulation of the magnetosphere is driven (mainly) by magnetic merging on the sunward side of the magnetopause, but its continuity depends on reconnection in the tail plasmasheet. Figure 5.59 follows the history of a selected field-line: connecting with the IMF, convecting over the poles as open lines, reconnecting in the tail, and returning to a nearly dipole form in the earthward return flow. If R_M and R_T are the connection rates at the magnetopause and in the tail respectively, T is the rate of

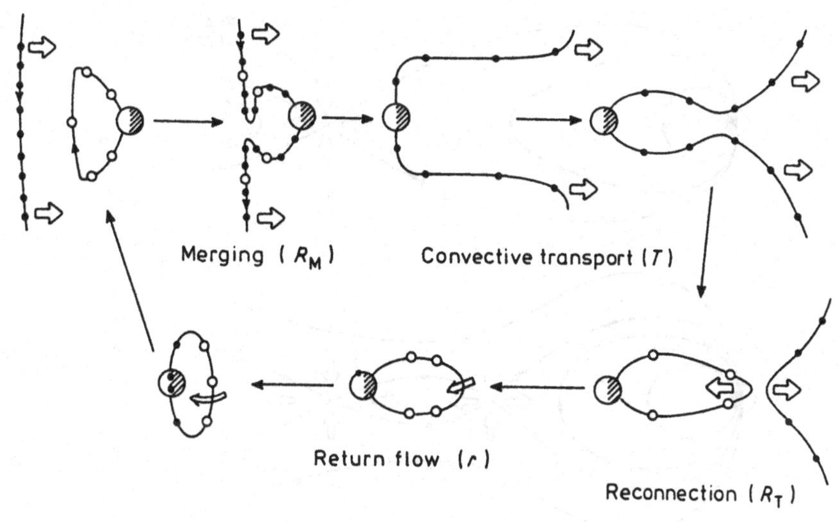

Fig. 5.59 History of a selected field-line. (After S.-I. Akasofu, Chapman Memorial Lecture, 1973)

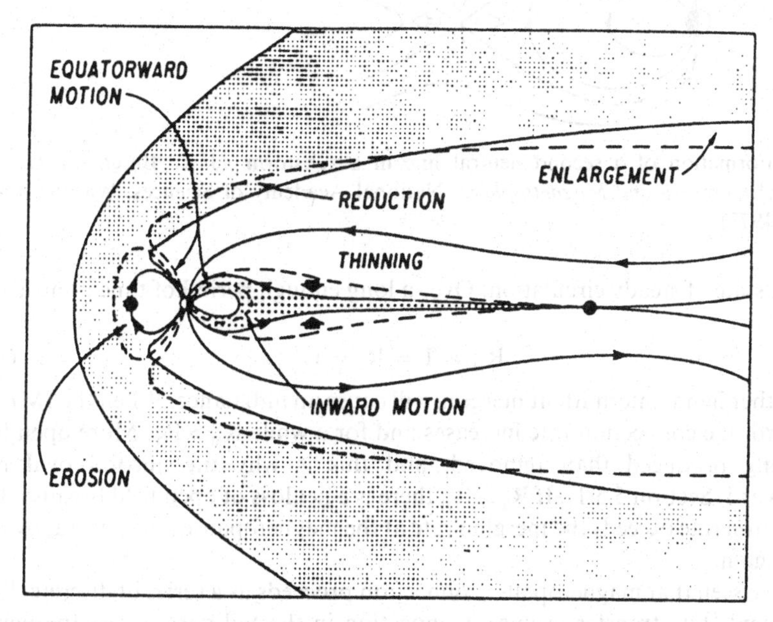

Fig. 5.60 Changes in the magnetosphere during the growth phase of a substorm. (R. L. McPherron *et al.*, *J. Geophys. Res.* **78**, 3131, 1973, copyright by the American Geophysical Union)

transport of open lines into the tail, and r is the rate at which closed flux is returned from the tail to the closed magnetosphere,

$$R_M = T = R_T = r \tag{5.48a}$$

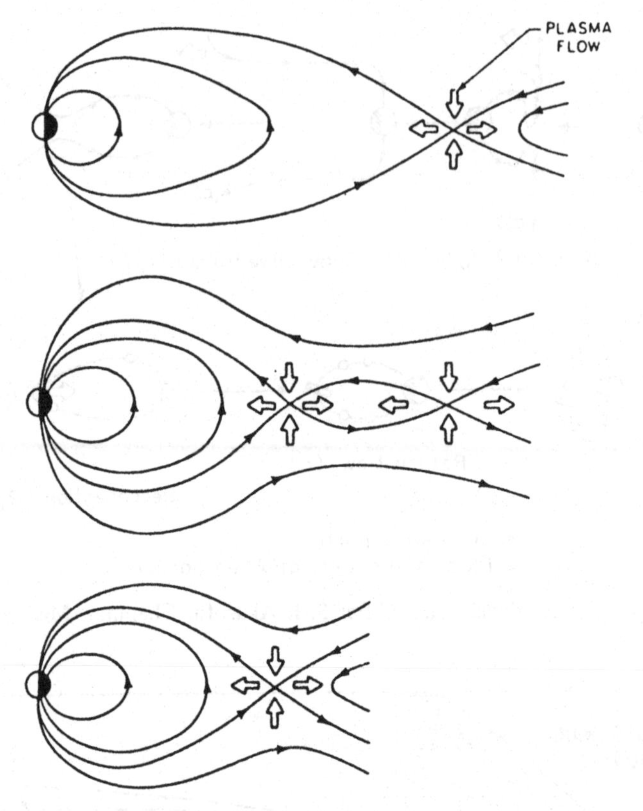

Fig. 5.61 Formation of a second neutral line in a substorm. (J. L. Burch, in *The Upper Atmosphere and Magnetosphere*. National Academy of Sciences, Washington, DC, 1977)

implies a state of steady circulation. Over a long enough period of time it must still be true that

$$\bar{R}_M = \bar{T} = \bar{R}_T = \bar{r}. \tag{5.48b}$$

On the other hand, intermittent merging will cause an imbalance. When the IMF turns southward, the connection rate increases and for a while $R_M > R_T$. More open field is then being produced than removed, and this is why the auroral oval moves equatorward (Section 5.5.1). If $R_T > R_M$ there is a net loss of open field because the tail reconnection rate exceeds the merging rate at the magnetopause; the auroral oval then shrinks again.

We have seen that magnetopause connection proceeds as a series of discrete, limited but frequent 'flux transfer events'. Connection in the tail goes in less frequent but more cataclysmic events, and this is the basis of the *substorm*. The substorm was originally identified from ground-based studies of the aurora (Section 8.4.4), and three phases have been identified:

the growth phase,
the expansion phase, and
the recovery phase.

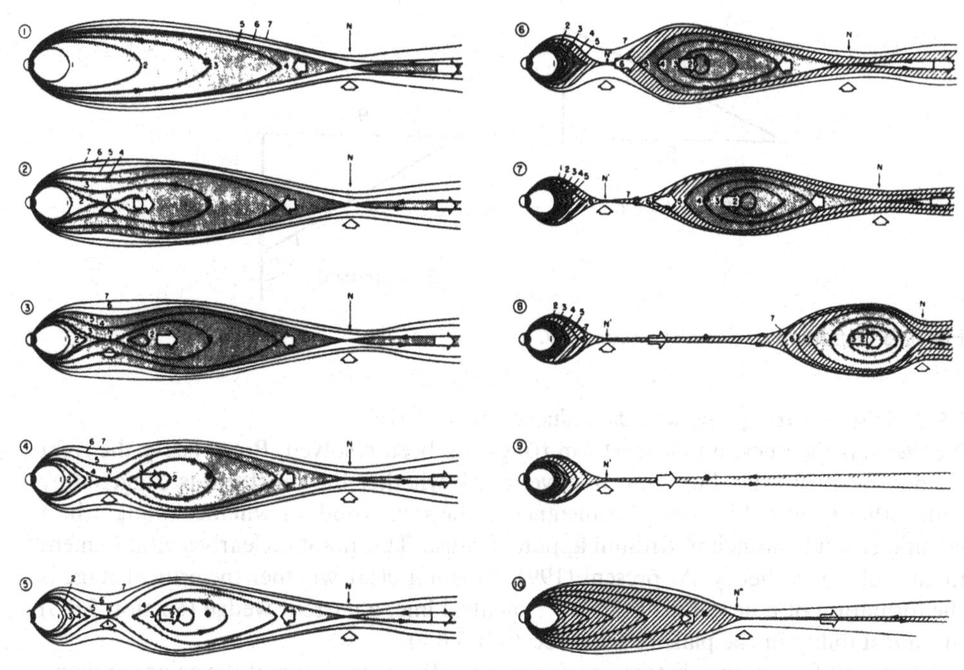

Fig. 5.62 Sequence of events in the magnetotail during a substorm. White arrows indicate plasma flows. The plasma sheet is bounded by field-line 5. N′ is the second neutral line that forms in the substorm, and picture 8 shows the plasmoid being expelled down the tail. (E. W. Hones, in *Magnetic Reconnection* (ed. Hones). A.G.U. Monograph 30, 1984)

In the magnetosphere the growth phase corresponds to increased erosion from the front of the magnetosphere, and at this time the plasma sheet becomes thinner as illustrated in Figure 5.60. The expansion phase begins with the formation of a neutral line nearer to the Earth than during quiet times (Figure 5.61). Satellites at geosynchronous distance observe an increase in the flux of energetic electrons, and the geomagnetic field becomes more dipolar. The likely sequence of events is illustrated in Figure 5.62. Note that the region of the tail between the two neutral lines is ejected along the magnetotail as the recovery phase begins; this is known as a *plasmoid.* The thinning of the plasma sheet in the expansion phase can be detected as a loss of particle flux by a satellite near the neutral sheet. The expulsion of the plasmoid may be detected by satellites 20 to 100 R_E down the tail as a burst of energetic particles moving away from the Earth.

The exact configuation of the magnetotail during the phases of a substorm has not been fully established, and several models – between which the observations have not yet been able to distinguish – have been put forward. Some models involve a local reversal of the tail field, and others include multiple neutral lines to correspond to the multiple arcs seen in aurorae.

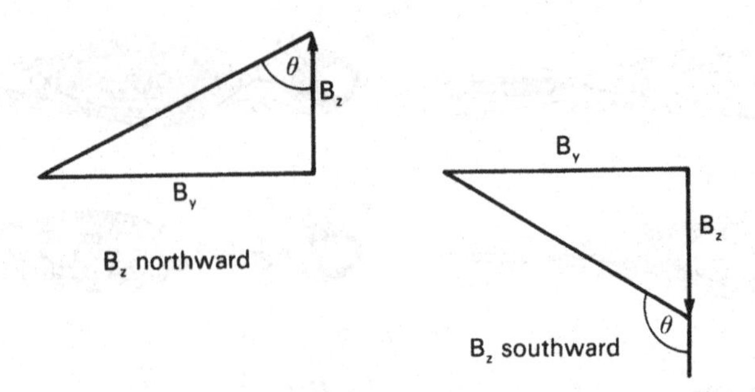

Fig. 5.63 Definition of θ for Equation 5.49.

5.9.2 Substorm triggering and the influence of the IMF

Neither has the question of substorm triggering been resolved. Presumably the right conditions must exist, but then we have to ask whether the substorm is triggered by some other identifiable event, for instance in the solar wind, or whether it might be a spontaneous phenomenon without apparent cause. This point is clearly a vital element in any substorm theory. At present (1991), it is not clear whether the critical stage is the formation of a near-Earth (15 R_E ?) neutral line, a current wedge (Section 5.9.3) or an instability in the plasma sheet (Section 5.6.3).

Important factors in substorm occurrence are the energy flux of the solar wind and the efficiency with which the energy is coupled into the magnetosphere. The index AE, a geomagnetic activity index indicating the activity level of the northern auroral zone, is well correlated with a quantity

$$\varepsilon = vB^2\sin^4(\theta/2)l_0^2, \tag{5.49}$$

where v = solar wind speed, B = IMF magnitude, l_0 = a length related to the cross-section of the magnetosphere (7 R_E), and θ = angle of IMF seen from the Earth (as in Figure 5.63).

If $B_z > 0$ (northward), $\theta = \tan^{-1}|B_y/B_z|$, and if $B_z < 0$ (southward), $\theta = 180° - \tan^{-1}|B_y/B_z|$. The magnetic energy reaching the magnetosphere per unit time is proportional to $vB^2l_0^2$: the magnetic power of the solar wind. The expression $\sin^4(\theta/2)$ represents the fraction of this power coupled into the magnetosphere. It gives a gradual transition between full coupling when the IMF is fully southward ($\sin^4(\theta/2) = 1$) and zero coupling when it is fully northward ($\sin^4(\theta/2) = 0$). If $B_z \ll B_y$, $\theta/2 = 45°$, and the coupling factor is 0.25. Some other expressions based on different combinations of solar wind parameters also correlate with substorm occurrence, though ε is perhaps the best (Figure 5.64).

That substorms occur most frequently when B_z is southward has been known for some time, and it is found that the beginning of the substorm often coincides with a southward turning of the IMF. But there are also cases when the substorm begins as the IMF turns northward, having previously been southward for an hour or two. In such a case it appears that southward IMF puts energy into the magnetosphere and then the shock of the northward turning triggers its release in the substorm.

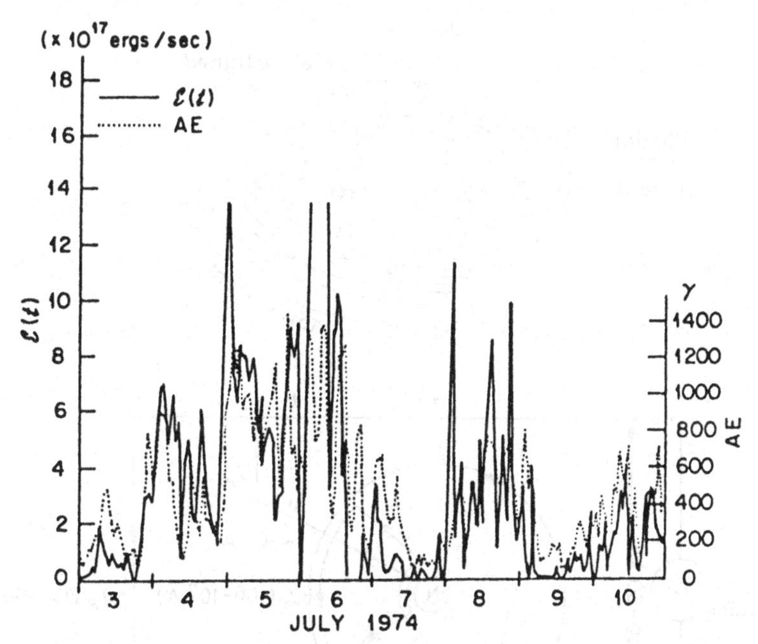

Fig. 5.64 Correlation between the AE index of magnetic activity and the ε parameter during a storm in July 1974. (Reprinted with permission from S.-I. Akasofu, *Planet. Space Sci.* **27**, 425, copyright (1979) Pergamon Press PLC)

5.9.3 Substorm currents

When the tail collapses at a neutral point the cross-tail current must be reduced in that region. Indeed, the collapse would not occur if the current remained at its original value. The current is diverted along the field-lines into the auroral ionosphere, where the circuit is completed by an *electrojet*, as in Figure 5.65a. This E-region current flows in a region where the conductivity is enhanced by particle precipitation. The large-scale diversion of tail current through the auroral E region is sometimes called a *current wedge*.

Some of the particles accelerated by the collapsing field become trapped to enhance the ring current, at least some of which is partial and does not encircle the Earth. As the particles drift away from the injection region in the midnight sector some are lost to the atmosphere, and thus the westward ring current tapers off towards the day side. All the current must be accounted for, and the partial ring current is thought to be completed by Birkeland currents to the ionosphere and an eastward current in the ionosphere. A circuit diagram of the substorm currents is given in Figure 5.65b. Figure 5.66 indicates the likely connections between various current components in a substorm. This is consistent with the observation that during a substorm the Birkeland currents are enhanced most strongly in the night sector.

A consideration of the currents enables one to consider whether the magnetotail contains enough energy to maintain a substorm. If the electrojet amounts to 5×10^5 A and the cross-tail current before the substorm was 3×10^{-2} A/m, the tail current must have been disrupted over a length of about 2×10^4 km (several Earth-radii). Taking the

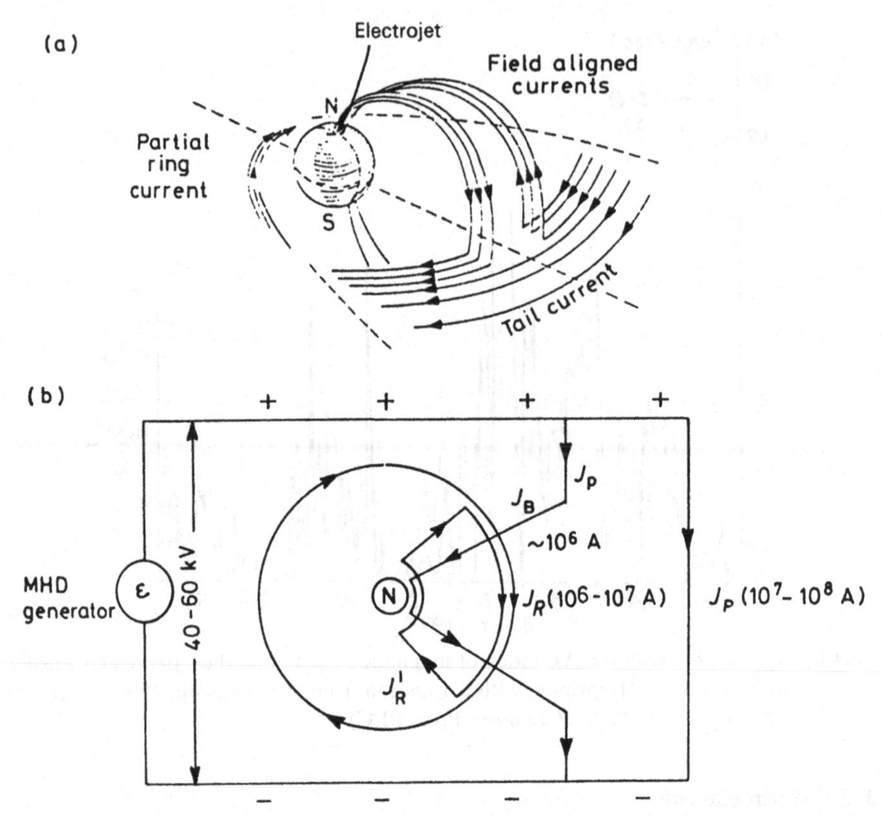

Fig. 5.65 Substorm currents: (a) pictorial; (b) schematic. (After L. Svalgaard, *NASA Report SP*-366, 1975, and W. J. Heikkila, *J. Geophys. Res.* **79**, 2496, 1974, copyright by the American Geophysical Union)

tail field as $B_T = 20$ nT and the radius of the tail as $R_T = 10^5$ km, the magnetic energy in a section 2×10^4 km long is

$$\frac{B_T^2}{2\mu_0} \cdot \pi R_T^2 \cdot 2 \times 10^7 = 10^{14} \text{ J.} \tag{5.50}$$

The ionosphere offers about 0.1 Ω resistance to an electrojet, and thus a current of 5×10^5 A dissipates ($i^2r =$) 2.5×10^{10} W. Auroral precipitation dissipates a similar amount, and both hemispheres have to be included, giving about 10^{11} W in total. A loss of 10^{14} J from the magnetic energy of the tail can therefore support a substorm for 10^3 s – approximately 20 minutes – which is in fact a typical duration for a burst of auroral activity.

5.10 Magnetospheres of other planets

The same principles that govern the formation and properties of the terrestrial magnetosphere apply equally to other planets, and the investigation of planetary magnetospheres in general has been one of the objectives of space missions within the solar system. All the planets lie within the solar wind, whose density decreases with

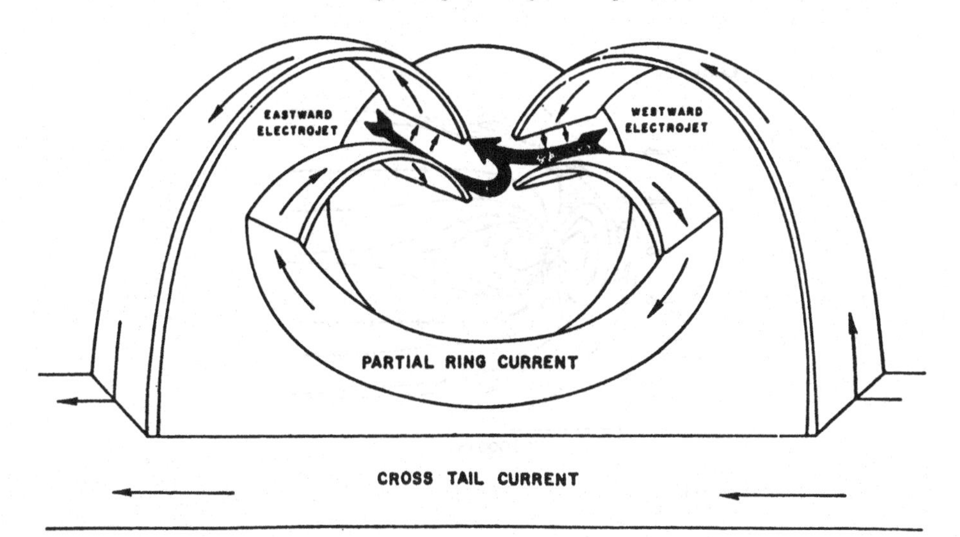

Fig. 5.66 A possible connection between ionospheric and magnetospheric currents. (After D. W. Swift, *Rev. Geophys. Space Phys.* **17**, 681, 1979, copyright by the American Geophysical Union)

Table 5.1 *Magnetic and magnetospheric parameters of the planets*

	Dipole moment (gauss cm²)	Equatorial surface field (gauss)	Polarity (with respect to Earth)	Angle of magnetic axis (degrees to rotation axis)	Plasma sources*	Typical magneto-pause position (planetary radii)
Mercury	$\sim 3 \times 10^{22}$	0.0035	Same	~ 10	W	1.1
Venus	$< 10^{21}$	< 0.0003	—	—	A	1.1
Earth	8×10^{25}	0.31	Same	11.5	W, A	10
Mars	2.5×10^{22}(?)	0.00065(?)	Opposite	—	?	?
Jupiter	1.5×10^{30}	4.1	Opposite	~ 10	W, A, S	60–100
Saturn	1.5×10^{29}	0.4	Opposite	< 1	W, A, S	20–25
Uranus	?	?	—	—	?	?
Neptune	?	?	—	—	?	?

*W = solar wind; A = atmosphere; S = satellites.
(L. J. Lanzerotti and S. M. Krimigis. *Physics Today*, **38** (11), 24, Nov. 1985.)

increasing distance from the Sun. But the various planetary magnetospheres differ mainly because the planets have different magnetic moments. Relative to Earth, the outer planets Jupiter and Saturn have strong dipoles, whereas of the inner planets Mercury and Mars have weak fields and Venus a very weak field if any (Table 5.1). This table also gives in each case the position of the magnetopause, which is the distance where the solar wind pressure is balanced by the combined pressure of the

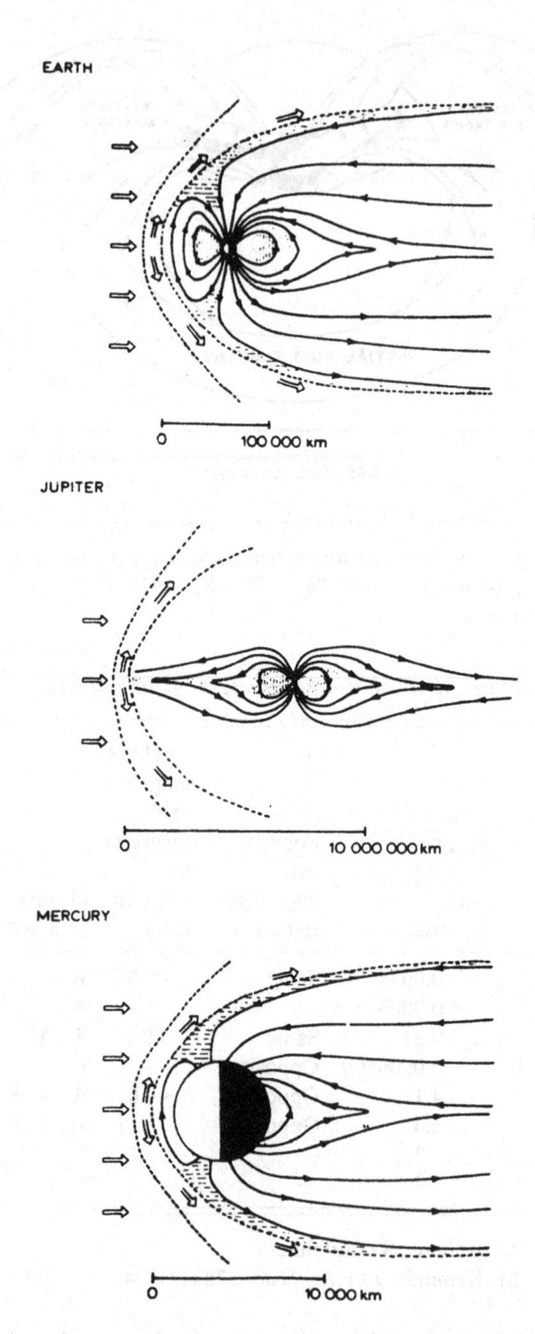

Fig. 5.67 A comparison between the magnetospheres of Earth, Jupiter and Mercury. (After V. M. Vasyliunas, private communication)

planetary magnetic field and any ionosphere. Taking the respective planetary radii as the units of distance, Mercury and Venus have small magnetospheres, terminating close to the surface, whereas those of Jupiter and Saturn are larger than the Earth's. In true dimensions the general sizes of these planetary magnetospheres range over a factor of 1000 in linear size and a factor of 10^9 in volume.

Since Mercury has no atmosphere there is no ionosphere, and the position of the magnetopause is determined by the magnetic field, as is the Earth's. These two magnetospheres are compared in Figure 5.67. The absence of an atmosphere means that particles in the magnetosphere of Mercury can strike the planet's surface directly. Venus has no detectable magnetic field, but it has an atmosphere and an ionosphere, and in this case it is the ionosphere that stops the solar wind; Venus has an *ionopause* rather than a magnetopause. There is a bow shock in front of each of these magnetospheres. Little is known about the magnetosphere of Mars, though there is evidence for a bow shock about half a planetary radius above the surface.

The magnetospheres of the great planets are different. These planets have satellites within the magnetosphere and these serve as a third source of plasma, in addition to the solar wind and the ionosphere as for the Earth. The Jovian magnetosphere is very large, extending 60–100 planetary radii in the sunward direction and at least 6 AU downwind. If this magnetosphere were visible from Earth it would appear larger than the Sun. Because the magnetic field is so strong it rotates with the planet to a considerable distance, and centrifugal force is important in determining the size and shape of the magnetosphere. Saturn's magnetic dipole moment, though still large, is only a tenth of Jupiter's, and the magnetosphere is smaller.

The planets of the solar system display an interesting variety of magnetospheres on which theories can be tested. The Earth's, it seems, is in many respects an 'average' one. In particular, whereas Jupiter's magnetosphere is rotation dominated, and others are driven by the solar wind, the Earth's magnetosphere includes both a co-rotating (inner) region and a circulating (outer) region.

Further reading

S.-I. Akasofu and S. Chapman. *Solar–terrestrial Physics*. Oxford University Press, Oxford, England (1972); particularly Chapter 1 on the Sun and interplanetary space, Chapter 5 on the formation of the magnetosphere, Chapter 6 on energetic particles and plasma in the magnetosphere, and Chapter 7 on solar storms.

P. A. Sturrock (ed.). *Solar Flares*. Colorado Associated University Press, Boulder, Colorado (1980).

E. Tandberg-Hansen and A. G. Emslie. *The Physics of Solar Flares*. Cambridge University Press, Cambridge, England (1988).

Z. Svestka. *Solar Flares*. Reidel, Dordrecht and Boston (1976).

C. Sawyer, J. W. Warwick and J. T. Dennett. *Solar Flare Prediction*. Colorado Associated University Press, Boulder, Colorado (1986).

D. J. Schove (ed.). *Sunspot Cycles*. Hutchinson Ross, Stroudsberg, Pennsylvania (1983).

P. S. McIntosh and M. Dryer (eds.). *Solar Activity Observations and Predictions*. MIT Press, Cambridge, Massachusetts (1972); particularly the articles by Sturrock, Magnetic models of solar flares, and Dryer and Cuperman, The solar wind: a review; also those by Evans, Dodson and Hedeman, and Schatten.

A. S. Jursa (ed.). *Handbook of Geophysics and the Space Environment*. Air Force Geophysics

Laboratory, US Air Force. National Technical Information Service, Springfield, Virginia (1985). Chapters 1–5, covering the Sun, the solar wind, the geomagnetic field, and the radiation belts; Chapter 6 on galactic cosmic radiation and solar energetic particles; Chapter 8 on electrodynamics of the magnetosphere; Chapter 10 on solar radio emissions.

D. P. LeGalley and A. Rosen (eds.). *Space Physics*. Wiley, New York, London and Sydney (1964); particularly Chapter 9 on the geomagnetic field (L. J. Cahill), Chapter 10 on interplanetary magnetic fields (E. J. Smith), Chapter 11 on solar plasma (W. Bernstein), Chapter 12 on the solar wind and its interaction with magnetic fields (F. L. Scarf), and Chapter 14 on trapped-radiation zones (B. J. O'Brien).

A. J. Hundhausen. *Coronal Expansion and the Solar Wind*. Springer-Verlag, Berlin and New York (1972).

W. N. Hess and G. D. Mead (eds.). *Introduction to Space Science*. Gordon and Breach, New York, London and Paris (1968). Chapter 8, The interplanetary medium by N. F. Ness, and Chapter 9, The boundary of the magnetosphere by W. N. Hess and G. D. Mead.

R. L. Carovillano and J. M. Forbes (eds.). *Solar–Terrestrial Physics*. Reidel, Dordrecht, Boston and Lancaster (1983); particularly Solar–wind magnetosphere coupling by T. W. Hill, Magnetic field line merging: basic concepts and Comparative magnetospheres both by V. M. Vasyliunas, and The quasi-static (slow-flow) region of the magnetosphere by R. A. Wolf.

W. N. Hess. *The Radiation Belt and Magnetosphere*. Blaisdell, Waltham, Massachusetts (1968).

L. R. Lyons and D. J. Williams. *Qualitative Aspects of Magnetospheric Physics*. Reidel, Dordrecht (1984); particularly Chapter 2 on charged particle motions, Chapter 3 on trapping regions, and Chapter 4 on electric fields.

J. G. Roederer. *Dynamics of Geomagnetically Trapped Radiation*. Springer-Verlag, Berlin and New York (1974).

S.-I. Akasofu. The magnetospheric currents: an introduction, in *Magnetospheric Currents* (ed. T. A. Potemra), Geophysical Monograph 28, American Geophysical Union, Washington, DC (1983).

A. Nishida. *Geomagnetic Diagnosis of the Magnetosphere*. Springer-Verlag, New York (1978). Chapter III, Implosion in the magnetotail; Chapter IV,Dynamic structure of the inner magnetosphere.

Special issue on solar system plasmas and fields (eds. J. Lemaire and M. J. Rycroft). *Advances in Space Research* **2**(1) (1982).

Special issues on progress in solar-terrestrial physics. *Space Science Reviews* **34** (1–4) (1983).

Special issues on the sun, interplanetary medium, and the magnetosphere. *Space Science Reviews* **17** (1–4) (1975).

D. P. Stern. A brief history of magnetospheric physics before the spacecraft era. *Reviews of Geophysics* **27**, 103 (1989).

Special issues on space plasma simulations. *Space Science Reviews* **42** (1–4) (1985).

Energy Release in Solar Flares. Proceedings of Workshop in Cambridge, Massachusetts, 1979. Report UAG-72, World Data Center A for STP, Boulder, Colorado (1979).

C. de Jager. Solar flares and particle acceleration. *Space Science Reviews* **44**, 43 (1986).

G. L. Siscoe. Evidence in the auroral record for secular solar variability. *Reviews of Geophysics* **18**, 647 (1980).

G. W. Pneuman. Driving mechanisms for the solar wind. *Space Science Reviews* **43**, 105 (1986).

Special issues on the source region of the solar wind. *Space Science Reviews* **33** (1–2), (1982).

M. Neugebauer. Large-scale and solar-cycle variations of the solar wind. *Space Science Reviews* **17**, 221 (1976).

K. H. Schatten. Large-scale properties of the interplanetary magnetic field. *Reviews of Geophysics and Space Physics* **9**, 773 (1971).

A. C. Fraser-Smith. Centered and eccentric geomagnetic dipoles and their poles, 1600–1985. *Reviews of Geophysics* **25**, 1 (1987).

D. P. Stern and I. I. Alexeev. Where do field lines go in the quiet magnetosphere? *Reviews of Geophysics* **26**, 782 (1988).

K. Schindler and J. Birn. Magnetotail theory. *Space Science Reviews* **44**, 307 (1986).

W. I. Axford. Magnetospheric convection. *Reviews of Geophysics* **7**, 421 (1969).

T. G. Forbes and E. R. Priest. A comparison of analytical and numerical models for steadily driven magnetic reconnection. *Reviews of Geophysics* **25**, 1583 (1987),

S. W. H. Cowley. The causes of convection in the Earth's magnetosphere: a review of developments during the IMS. *Reviews of Geophysics* **20**, 531 (1982).

M. I. Pudovkin and V. S. Semenov. Magnetic field reconnection theory and the solar wind–magnetosphere interaction: a review. *Space Science Reviews* **41**, 1 (1985).

D. P. Stern. Energetics of the magnetosphere. *Space Science Reviews* **39**, 193 (1984).

S.-I. Akasofu. Energy coupling between the solar wind and the magnetosphere. *Space Science Reviews* **28**, 121 (1981).

A. Nishida. Interplanetary field effect on the magnetosphere. *Space Science Reviews* **17**, 353 (1976).

M. G. Kivelson. Magnetospheric electric fields and their variation with geomagnetic activity. *Reviews of Geophysics and Space Physics* **14**, 189 (1976).

C. R. Chappell. Recent satellite measurements of the morphology and dynamics of the plasmasphere. *Reviews of Geophysics and Space Physics* **10**, 951 (1972).

S. W. H. Cowley. Plasma populations in a simple open model magnetosphere. *Space Science Reviews* **26**, 217 (1980).

Special issue on the physics of the plasmasphere. *J. Atmos. Terr. Phys.* **38**, 1039 (1976).

M. Schulz. Earth's radiation belts. *Reviews of Geophysics* **20**, 613 (1982).

D. L. Carpenter. Remote sensing of the magnetospheric plasma by means of whistler mode signals. *Reviews of Geophysics* **26**, 535 (1988).

C. R. Chappell. The terrestrial plasma source: a new perspective in solar–terrestrial processes from Dynamics Explorer. *Reviews of Geophysics* **26**, 229 (1988).

T. A. Potemra. Field-aligned (Birkeland) currents. *Space Science Reviews* **42**, 295 (1985).

M. S. Gussenhoven. Low-altitude convection, precipitation, and current patterns in the baseline magnetosphere. *Reviews of Geophysics* **26**, 792 (1988).

G. Rostoker, S.-I. Akasofu, W. Baumjohaum, Y. Kamide and R. L. McPherron: The roles of direct input of energy from the solar wind and unloading of stored magnetotail energy in driving magnetospheric substorms. *Space Science Reviews* **46**, 93 (1987).

C. F. Kennel. Magnetospheres of the planets. *Space Science Reviews* **14**, 511 (1973).

L. J. Lanzerotti and S. M. Krimigis. Comparative magnetospheres. *Physics Today* **38**(11), 24 (Nov 1985).

6

Principles of the ionosphere at middle and low latitudes

There are more things in heaven and earth, Horatio, than are dreamt of in your philosophy. W. Shakespeare, *Hamlet* Act I Scene (v)

6.1 Introduction

The ionized part of the atmosphere, the *ionosphere*, contains significant numbers of free electrons and positive ions. There are also some negative ions at the lower altitudes. The medium as a whole is electrically neutral, there being equal numbers of positive and negative charges within a given volume. Although the charged particles may be only a minority amongst the neutral ones they exert a great influence on the medium's electrical properties, and herein lies their importance.

The first suggestions of electrified layers within the higher levels of the terrestrial atmosphere go back to the 19th century, but interest was regenerated with Marconi's well known experiments to transmit a radio signal from Cornwall in England to Newfoundland in Canada in 1901, and with the subsequent suggestions by Kennelly and by Heaviside (independently) that, because of the Earth's curvature, the waves must have been reflected from an ionized layer. The name *ionosphere* was coined by R. Watson-Watt in 1926, and came into common use about 1932.

Since that time the ionosphere has been extensively studied and most of its principal features, though not all, are now fairly well understood in terms of the physical and chemical processes of the upper atmosphere. Typical vertical structures are as shown in Fig. 6.1. The identification of the regions was much influenced by their signatures on ionograms (Section 3.5.1), which tend to emphasize inflections in the profile, and it should be appreciated in particular that the various layers may not be separated by distinct minima. The main regions are designated D, E, F1 and F2, with the following daytime characteristics:

D region, 60–90 km: 10^8–10^{10} m^{-3} (10^2–10^4 cm^{-3});
E region, 105–160 km: several 10^{11} m^{-3} (10^5 cm^{-3});
F1 region, 160–180 km: several 10^{11}–10^{12} m^{-3} (10^5–10^6 cm^{-3});
F2 region, maximum variable around 300 km: up to several 10^{12} m^{-3} (10^6 cm^{-3}).

The D and F1 regions vanish at night, and the E region becomes much weaker. The F2 region, however, tends to persist though at reduced intensity.

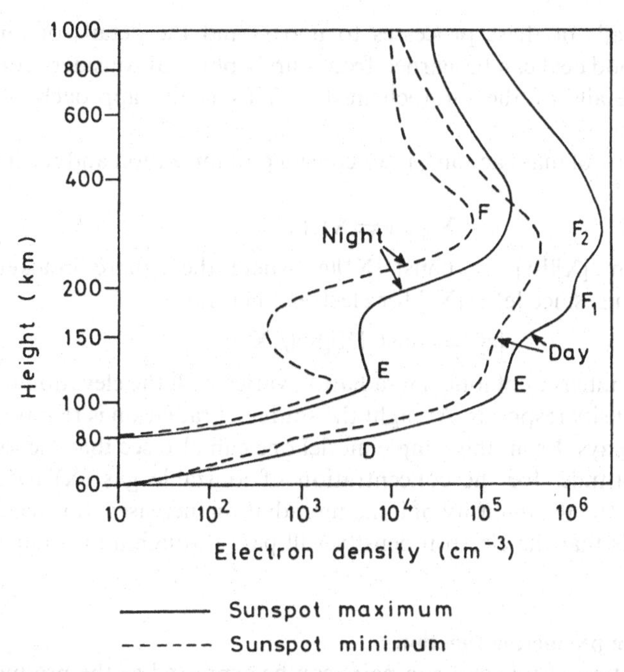

Fig. 6.1 Typical vertical profiles of electron density in the mid-latitude ionosphere. (After W. Swider, Wallchart *Aerospace Environment*, US Air Force Geophysics Laboratory)

The behaviour of the terrestrial ionosphere indicates that it can be considered in two parts, corresponding to the two principal regimes of magnetospheric circulation (Section 5.5). The present chapter discusses the principles governing the ionosphere at middle and low latitudes, in effect latitudes less than about 60° geomagnetic, though many apply to the high latitudes as well. Further considerations affecting particularly the high-latitude ionosphere will be introduced in Chapter 8.

6.2 Physical aeronomy

6.2.1 Principles

The ionosphere is formed by the ionization of atmospheric gases such as N_2, O_2 and O. At middle and low latitude the energy required comes from solar radiation in the extreme ultra-violet (EUV) and X-ray parts of the spectrum. Once formed, the ions and electrons tend to recombine and to react with other gaseous species to produce other ions. Thus there is a dynamic equilibrium in which the net concentration of free electrons (which, following standard practice, we shall call the *electron density*) depends on the relative speed of the production and loss processes. In general terms the rate of change of electron density is expressed by a *continuity equation*:

$$\frac{\partial N}{\partial t} = q - L - \operatorname{div}(N\mathbf{v}), \tag{6.1}$$

where q is the production rate, L the loss rate by recombination, and div ($N\mathbf{v}$) expresses the loss of electrons by movement, $\mathbf{v}$ being their mean drift velocity. Plainly, one has to go into the details of these processes to understand the details of ionospheric formation, but a good deal can be learned from purely physical considerations that do not involve the details of the photochemistry. This is the approach of *physical aeronomy*.

Following the 'law of mass action', if we consider an ionization and recombination reaction,

$$X + h\nu \rightleftharpoons X^+ + e,$$

then at equilibrium $[X][h\nu] = \text{const}\cdot[X^+][e]$ where the square brackets signify concentrations. Thus, since $[e] = [X^+]$ for electrical neutrality,

$$[e]^2 = \text{const}\cdot[X][h\nu]/[X^+].$$

During the day the intensity of ionizing radiation varies with the elevation of the Sun, and the electron density responds. At night the source of radiation is removed and the electron density decays. From this simple model one can also see that the ionosphere must vary with altitude, for the concentration of ionizable gas (X) reduces with increasing height while the intensity of ionizing radiation increases. It is reasonable to anticipate from this that the electron density will pass through a maximum at some altitude.

6.2.2 The Chapman production function

The rate of production of ion–electron pairs can be expressed as the product of four terms:

$$q = \eta\sigma n I. \tag{6.2}$$

Here, I is the intensity of ionizing radiation at some level of the atmosphere and n is the concentration of atoms or molecules capable of being ionized by that radiation. For an atom or molecule to be ionized it must first absorb radiation, and the amount absorbed is expressed by the *absorption cross-section*, σ: if the flux of incident radiation is I ($J/m^2\,s$) then the total energy absorbed per unit volume of the atmosphere per unit time is $\sigma n I$. However, not all this energy will go into the ionization process, and the *ionization efficiency*, η, takes that into account, being the fraction of the absorbed radiation that goes into producing ionization.

From this simple beginning S. Chapman, in 1931, developed a formula which predicts the form of a simple ionospheric layer and how it varies during the day. (It should be pointed out that, although the formula is generally attributed to Chapman's 1931 paper, it had previously appeared in P. O. Pedersen's book, *The Propagation of Radio Waves*, which was published in 1927.) At this stage we deal only with the rate of production of ionization (q), and the formula expressing this is the *Chapman production function*. The derivation makes some assumptions:

(a) The atmosphere is composed of a single species, exponentially distributed with constant scale height;
(b) The atmosphere is plane stratified: there are no variations in the horizontal plane;
(c) Solar radiation is absorbed in proportion to the concentration of gas particles (as in Equation 6.2);
(d) The absorption coefficient is constant: this is equivalent to assuming monochromatic radiation.

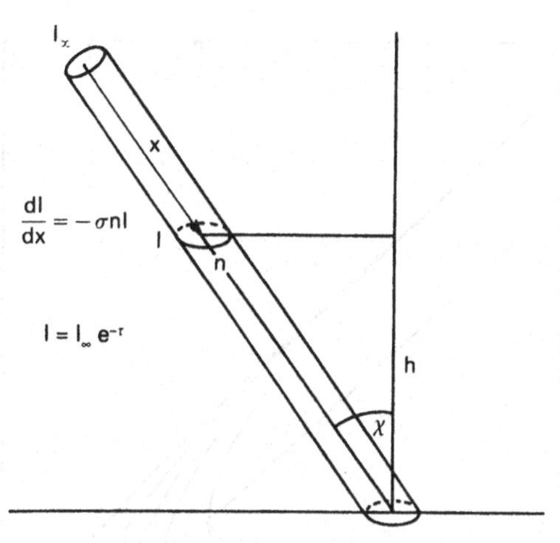

Fig 6.2 Absorption of ionizing radiation in the atmosphere.

The Chapman production function is usually written in a normalized form as

$$q = q_{m0}\exp(1 - z - \sec\chi \cdot e^{-z}). \tag{6.3}$$

Here, z is the reduced height for the neutral gas, $z = (h - h_{m0})/H$, where H is the scale height. χ is the solar zenith angle, h_{m0} is the height of maximum production rate when the Sun is overhead ($\chi = 0$), and q_{m0} is the production rate at h_{m0}, also when the Sun is overhead. Equation (6.3) can also be written

$$q/q_{m0} = e \cdot e^{-z} \cdot e^{(-\sec\chi \cdot \exp(-z))}, \tag{6.4}$$

where the first term is a constant, the second expresses the height variation of the density of ionizable atoms, and the third is proportional to the intensity of the ionizing radiation.

The formula can be proved as follows. Referring to Figure 6.2, if there are n absorbing atoms per unit volume, each with absorption coefficient σ, the intensity of the radiation varies with distance x, as

$$dI/dx = -\sigma nI. \tag{6.5}$$

The intensity may thus be written

$$I = I_\infty e^{-\tau}, \tag{6.6}$$

where τ, the *optical depth*, is the absorption coefficient times the number of absorbing atoms in a unit column down to the level being considered (σN_T). I_∞ is the intensity at a great height, effectively outside the atmosphere. Thus, using Equation (4.9),

$$\tau = \sigma(nH \cdot \sec\chi). \tag{6.7}$$

In an exponential atmosphere $n = n_0 e^{-z}$ (Equation 4.4), n_0 being the particle concentration at the level where $z = 0$. Thus we see that

$$q \propto e^{-z} \cdot e^{-K}\exp(-z),$$

K being a constant.

By differentiating q with respect to z it is readily shown that q passes through a maximum when $Ke^{-z} = 1$; that is,

the production rate is greatest at the level where the optical depth is unity.

This is a useful general result. Further, from Equation (6.7), $n_m = 1/\sigma H \sec\chi$. Putting $\sec\chi = 1$ (for overhead Sun) we obtain $n_{m0} = 1/\sigma H$ for the concentration of ionizable atoms at the level of maximum ion production when $\chi = 0$.

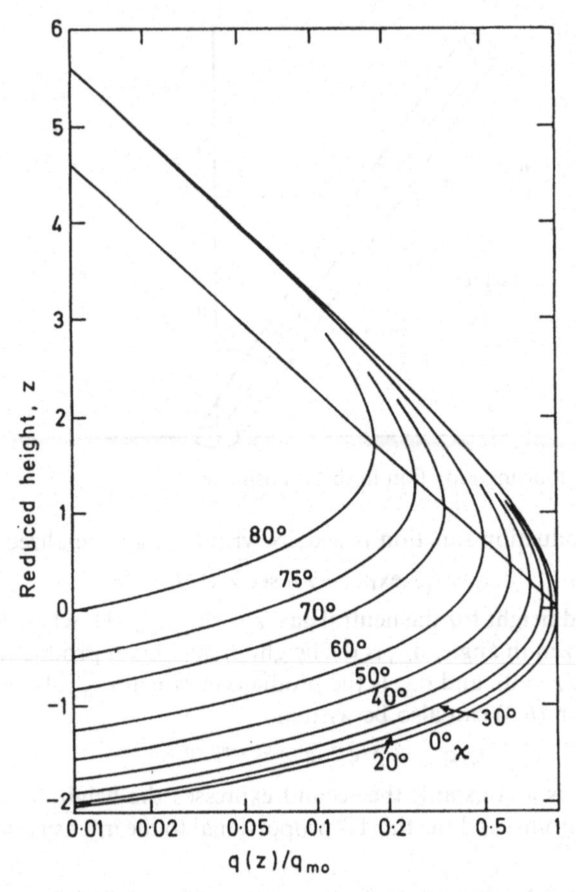

Fig. 6.3 Chapman production function. (After T. E. VanZandt and R. W. Knecht, in *Space Physics* (eds. LeGalley and Rosen). Wiley, 1964)

The production rate at this level is

$$q = \eta\sigma n I = \eta\sigma n_{mo} I_\infty e^{-1}. \tag{6.8}$$

Writing $n = n_{mo}e^{-z}$ in general, we obtain for the production rate at any level that

$$q = \eta\sigma n_{mo} e^{-z} I_\infty e^{-(\sigma n_{mo} H \sec\chi \exp(z))}$$

$$= q_{mo} \cdot e^{+1} \cdot e^{-z} \cdot e^{(\sec\chi \cdot \exp(-z))}. \tag{6.4}$$

The following properties of the Chapman production function are worth noting.

(a) The general shape of the production rate curves plotted in Figure 6.3. At a great height, where z is large and positive, $q \to q_{mo} e \cdot e^{-z}$. Thus the curves merge above the peak, becoming independent of χ and showing an exponential decrease with height due to the reducing density of the neutral atmosphere. In the region well below the peak, when z is large and negative, the shape becomes dominated by the last term of Equation 6.4, producing a rapid cut-off. As anticipated in Section 6.2.1, the production rate is limited by a shortage of ionizable gas at the greater altitudes and by a lack of ionizing radiation low down. On a plot of ln(q) against z all the curves are the same shape, but are displaced upward and to the left as the zenith angle, χ, increases.

(b) As proved above, the production rate maximizes where the optical depth is unity. An expression for the maximum production rate follows from Equations 6.4 and 6.8, giving

$$q_{m0} = \eta I_\infty / eH$$

when $\chi = 0$, or, in the more general case of any solar zenith angle

$$q_m = \eta I_\infty / eH \sec\chi. \tag{6.9}$$

(c) By differentiating Equation 6.3 it is readily proved that

$$z_m = \ln(\sec\chi) \tag{6.10}$$

where z_m is the reduced height of maximum production (the height at $\chi = 0$ being taken as zero). Substituting back into Equation 6.3 gives

$$q_m = q_{m0} \cos\chi \tag{6.11}$$

(essentially like Equation 6.9). These simple results are important in studies of the ionosphere because the maximum of a layer is the part most readily observed. From Equations 6.10 and 6.11 we see that a plot of $\ln(q_m)$ against z_m is effectively a plot of $\ln(\cos\chi)$ against $\ln(\sec\chi)$, obviously a straight line of slope -1. This line is shown in Figure 6.3.

The Chapman production function is important because it expresses fundamentals of ionospheric formation and of radiation absorption in any exponential atmosphere. Although real ionospheres may be more complicated, the Chapman theory provides an invaluable reference point for interpreting observations and a relatively simple starting point for ionospheric theory.

6.2.3 Ionization by energetic particles

The other source of ionization in the terrestrial ionosphere is energetic particles, which are not entirely absent at middle latitudes but are much more important at high latitudes where they frequently provide the dominant source of ionization. As we shall see in Chapter 8, two very significant sources at high latitude are electrons associated with the aurora, and protons (plus some alpha particles) emitted from the Sun during some solar flares.

Electrons

Various methods have been used to calculate the rate of ion production by a stream of energetic electrons arriving from some source above the atmosphere. The most generally useful one relies on laboratory measurements of the range of electrons in air, as in Figure 6.4. The range, r_0, is determined by the loss of energy from an electron as it excites and ionizes neutral particles with which it collides in passing through the neutral atmosphere. The rate of loss obviously depends on the number of gas particles encountered, and in a uniform atmosphere the distance travelled varies in inverse proportion to the gas pressure.

An energetic particle entering the atmosphere from above travels into a medium of increasing density, and the altitude, h_p, to which it penetrates is such that the product of pressure and distance, integrated above h_p, is equal to the range r_0. Obviously, this particle will ionize only above height h_p, and the total number of ion–electron pairs produced will depend on $E/\Delta E$ where E is the initial energy of the particle and ΔE is the energy required for each ionization (35 eV).

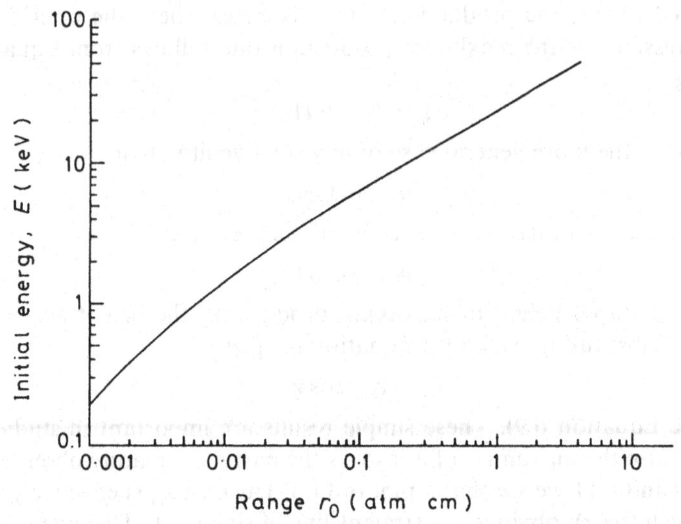

Fig. 6.4 Range of electrons in air, according to various measurements and computations. (After M. H. Rees, *Planet. Space Sci.*, **12**, 722, copyright (1964) Pergamon Press PLC)

We then have to take into account the distribution of ion production along the path, which is a function of the particle velocity. This is again done using results from laboratory experiments, which give the rate of ion production in terms of the atmospheric depth (the total mass of gas in a column of unit cross-section along the path of the particle) in the form of an efficiency λ (which here has nothing to do with wavelength!). λ is written as a function of (s/s_p), where s and s_p are respectively the atmospheric depths at a point on the path and at the point where the particle stops.

Given these definitions we can now express the production rate due to a flux of incident electrons. For an incident flux F ($m^{-2}s^{-1}$) of electrons whose initial energy is E, the total rate of ion–electron pair production over the whole path is $FE/\Delta E$. Beyond the end of the path λ must be zero because there is no further ionization, and the integral of λ over the path (from $(s/s_p) = 0$ to 1) must be unity to account for all the energy. Hence, at some point along the path, the production rate is

$$q = \frac{FE}{\Delta E}\cdot\lambda\left(\frac{s}{s_p}\right) \text{ per unit of } \frac{s}{s_p},$$

$$= \frac{FE}{\Delta E}\cdot\lambda\left(\frac{s}{s_p}\right)\cdot\frac{1}{s_p} \text{ per unit of s},$$

$$= \frac{FE}{\Delta E}\cdot\lambda\left(\frac{s}{s_p}\right)\cdot\frac{\rho}{s_p} \quad \text{per unit distance along the path,} \tag{6.12a}$$

(since $s = \int\rho\,dl$ by definition). This is the number of ion–electron pairs produced per unit time per unit volume at the altitude where the atmospheric depth is s and the density is ρ. Since $s_p = r_0\rho_0$, where r_0 is the range (Figure 6.4) and ρ_0 the density at the stopping point, the formula can also be written

$$q = \frac{FE}{\Delta E}\cdot\lambda\left(\frac{s}{s_p}\right)\cdot\frac{1}{r_0}\cdot\frac{\rho}{\rho_0}. \tag{6.12b}$$

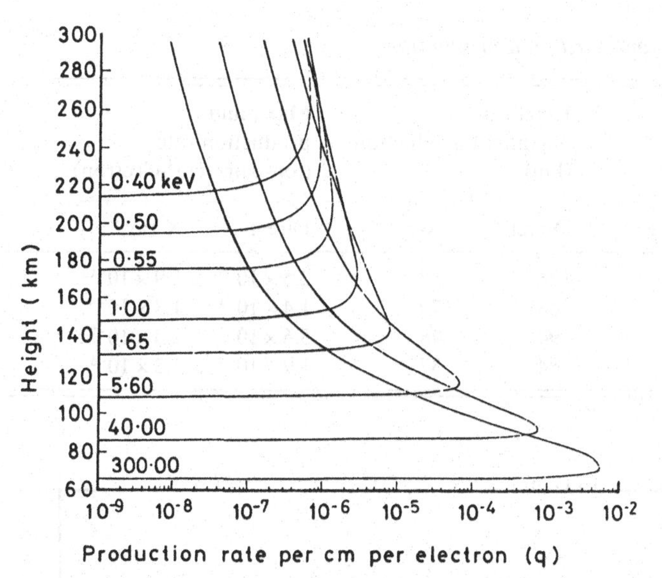

Fig. 6.5 Production rates due to monoenergetic electrons of various initial energy. (After M. H. Rees, *Planet. Space Sci.* **11**, 1209, copyright (1963) Pergamon Press PLC)

The form of λ depends on the angular distribution of the incident particles. If they are mono-directional λ is maximum at $s/s_p = 0.4$. If the electrons arrive over a range of angles, as would be the case in the auroral regions, some particles travel in a spiral path and thus cover a greater distance; in this case λ maximizes at a smaller value of s/s_p such as 0.1 or 0.2.

Figure 6.5 shows calculated production rates in a model atmosphere due to mono-energetic electrons of various initial energies. To get the effect of a more realistic spectrum the production rate must be integrated over energy. The height of maximum production as a function of energy is given in Figure 6.6.

Bremsstrahlung X-rays

When energetic electrons collide with neutral gas particles a small amount of their energy is converted to X-rays through the *bremsstrahlung* process – literally 'braking radiation' – as they are rapidly decelerated. The X-rays penetrate deeper into the atmosphere than the primary electrons and may be observed by balloon-borne detectors at heights of 30 to 40 km. Some X-rays are scattered back out of the atmosphere and can be 'viewed' by means of an X-ray camera carried on a satellite (Section 3.3.2).

The computation of the bremsstrahlung X-ray flux is fairly complicated. An electron of energy E can produce photons of energy E or less, and the X-rays are emitted over a wide range of angles. For non-relativistic electrons ($v \ll c$) the intensity varies according to

$$f(\theta) = \sin^2\theta, \tag{6.13}$$

where θ is the angle between the direction of the primary electron and that of the emitted photon. Inverting an observed X-ray spectrum to give the spectrum of the

Table 6.1 *Direct and X-ray ion production*

	Height of maximum production (km)		Maximum production rate (ion pairs/cm^3-electron)	
E (keV)	Direct	X-ray	Direct	X-ray
3	126	88	2.5×10^{-5}	5.9×10^{-10}
10	108	70	1.4×10^{-4}	1.3×10^{-8}
30	94	48	5.6×10^{-4}	2.3×10^{-7}
100	84	37	1.9×10^{-3}	1.3×10^{-5}

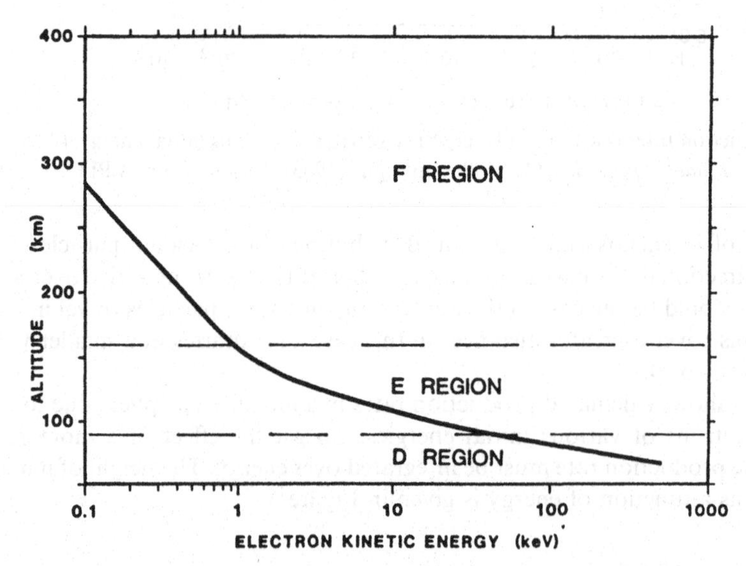

Fig. 6.6 Height of maximum ionization production by incident energetic electrons. (J. A. Whalen (after M. H. Rees), private communication)

primary electrons is even more difficult. The usual practice is to draw on a set of computations giving the bremsstrahlung due to single electrons of specified energy. It is usually necessary to assume a form for the spectrum (e.g. exponential).

When the X-rays are stopped by the atmosphere they create ionization at that level. This occurs both by the process of photoelectron absorption (in which the photon is totally absorbed and its energy transferred to one or more electrons) and by the Compton effect (in which an electron is emitted as well as another photon of longer wavelength). The ionization rate due to bremsstrahlung is several factors of ten smaller than that due to the primary electrons higher up, but at the height concerned, possibly 50 km or below, it is the major ionization source at times of auroral electron precipitation. Table 6.1 compares the heights and maximum production rates due to direct and bremsstrahlung ionization for several initial electron energies.

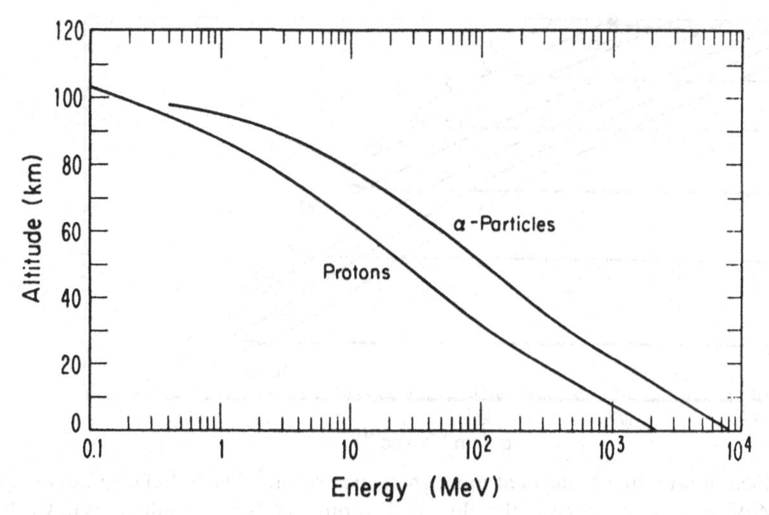

Fig. 6.7 Stopping heights of incident protons and alpha particles. (G. C. Reid, *Fundamentals of Cosmic Physics* **1**, 167, 1974)

Protons

Significant ionization may also be caused by energetic protons, especially at high latitudes during *polar cap events* due to fluxes of protons released from the Sun at the time of a solar flare (Section 8.5). A lesser flux of alpha particles will generally arrive simultaneously. These particles, significantly more energetic than the auroral electrons discussed above, lose energy in colliding with the atmospheric gas and leave ionized trails. The gas concerned is principally that of the mesosphere, having essentially the same familiar composition as the troposphere, and therefore the rate of energy loss is well known from laboratory measurements. A graph showing the rate of energy loss against the distance travelled is called a *Bragg curve*. In the energy range of present concern the loss rate increases as the proton slows down because it then spends a longer time near each air molecule. Over the range 10–200 MeV the loss rate is almost inversely proportional to the energy, a typical value being 0.8 MeV/m of path in air at standard temperature and pressure when the energy is 100 MeV. The energy may be assumed to be used entirely in creating ion–electron pairs, each requiring about 35 eV.

A proton entering the atmosphere from space meets a gas with sharply increasing density, and, combined with the nature of the Bragg curve, this means that the ionization is very concentrated towards the end of the path; for example, a vertically incident 50 MeV proton loses half its energy in the last 2.5 km of the path and the last 10% in only 100 m. One consequence is that the penetration level does not depend strongly on the angle of incidence except near 90°. The altitudes at which protons and alpha particles of various energy stop are shown in Figure 6.7, and the production rate profiles for protons of different initial energy are given in Figure 6.8. For a given spectrum of proton energies the total effect would be calculated by appropriate summing over these curves at each height.

The same curves can be applied to alpha particles because they have almost the same range as protons of the same velocity, but the ionization rate is four times as great.

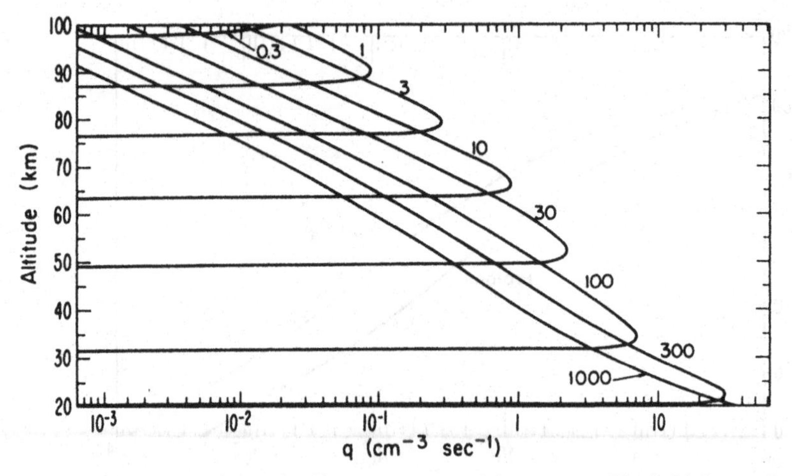

Fig. 6.8 Production rates due to incident monoenergetic protons. The initial energies are given in MeV, and in each case the flux is 1 proton/cm^2-sec-steradian. (G. C. Reid, *Fundamentals of Cosmic Physics* **1**, 167, 1974)

Thus, for an alpha particle of 400 MeV we would use the 100 MeV curve of Figure 6.8 but multiply the production rates by four. This trick may be used down to energies of a few MeV.

6.2.4 Principles of chemical recombination

Working out the rate of electron production is just the first step in calculating the electron density in an ionized layer, and the next step is to reckon the rates at which electrons are removed from the volume under consideration. These are represented in the continuity equation (6.1) by two further terms, one for the recombination between ions and electrons to reform neutral particles, and the other to account for movement of plasma into or out of the volume. We deal first with the principles of chemical recombination. The question of which individual reactions are most important in different parts of the ionosphere will be addressed in Section 6.3.

First we assume that the electrons recombine directly with positive ions and that no negative ions are present: $X^+ + e \rightarrow X$. Then the rate of electron loss is

$$L = \alpha[X^+]N = \alpha N^2, \tag{6.14}$$

where N is the electron density (equal to the ion density $[X^+]$) and α is the *recombination coefficient*. At equilibrium, therefore,

$$q = \alpha N^2. \tag{6.15}$$

Taking the production rate q from the Chapman production function (Equation 6.3), we obtain

$$N = N_{m0}\exp\tfrac{1}{2}(1 - z - \sec\chi \cdot e^{-z}) \tag{6.16}$$

in which $z = (h - h_0)/H$. And from Equation 6.11 it is seen that the electron density at the peak of the layer varies as $\cos^{1/2}\chi$:

$$N_m = N_{m0}\cos^{\frac{1}{2}}\chi. \tag{6.17}$$

A layer with these properties is called an *α-Chapman layer*.

In the lower ionosphere there are also significant numbers of negative ions. Electrical neutrality then requires $N_e + N_- = N_+$, where N_e, N_- and N_+ are respectively the concentrations of electrons, negative ions and positive ions. Since the negative and positive ions may also recombine with each other the overall balance between production and loss is now expressed by

$$q = \alpha_e N_e N_+ + \alpha_i N_- N_+, \tag{6.18}$$

α_e and α_i being recombination coefficients for the reactions of positive ions with electrons and negative ions respectively. The ratio between negative ion and electron concentrations is traditionally represented by λ – which again has nothing to do with wavelength! In terms of λ, $N_- = \lambda N_e$ and $N_+ = (1+\lambda)N_e$, and thus

$$q = (1+\lambda)(\alpha_e + \lambda\alpha_i)N_e^2 \tag{6.19}$$

which, in cases where $\lambda\alpha_i \ll \alpha_e$, becomes

$$q = (1+\lambda)\alpha_e N_e^2. \tag{6.20}$$

In the presence of negative ions the equilibrium electron density is still proportional to the square root of the production rate but its magnitude is changed. The term $(1+\lambda)(\alpha_e + \lambda\alpha_i)$ is often called the *effective recombination coefficient*.

If one is concerned particularly with electron loss, then attachment to neutral particles to form negative ions can itself be regarded as another type of electron loss process. In fact, as we shall see, this becomes the dominant type at somewhat higher levels of the ionosphere. Without at this stage specifying chemical details, we can see that the attachment type of reaction can be written $M + e \rightarrow M^-$, and the rate of electron loss is $L = \beta N$ where β is the *attachment coefficient*. The loss rate is now linear with N because the neutral species M is assumed to be by far the more numerous, in which case removing a few of them has no significant effect on the total remaining and [M] is effectively constant.

At equilibrium,

$$q = \beta N \tag{6.21}$$

and, taking q from the Chapman production function, as before, gives

$$N = N_{m0} \exp(1 - z - \sec\chi \cdot e^{-z}). \tag{6.22}$$

The peak electron density now varies as

$$N_m = N_{m0} \cos\chi. \tag{6.23}$$

Such a layer is a *β-Chapman layer*.

This simple formulation assumes that β does not vary with height, though this restriction does not affect the validity of Equation 6.21 nor of 6.22 at constant z.

In fact β is expected to vary with height because it depends on the concentration of the neutral molecules (M), and this has important consequences for the form of the terrestrial ionosphere. It is known that electron loss in the F region occurs in a two-stage process:

$$X^+ + A_2 \rightarrow AX^+ + A \tag{6.24}$$

$$AX^+ + e \rightarrow A + X \tag{6.25}$$

in which A_2 is one of the common molecular species like O_2 and N_2. The first step moves the positive charge from X to AX, and the second one dissociates the charged molecule through recombination with an electron, a *dissociative recombination* reaction. The rate of Equation 6.24 is $\beta[X^+]$ and that of 6.25 is $\alpha[AX^+]N$. At low altitude β is large, Equation 6.24 goes quickly and all X^+ is rapidly converted to AX^+; the overall rate is then governed by the rate of Equation 6.25, giving an α-type process because $[AX^+] = N$ for neutrality. At a high altitude β is small and Equation 6.24 is slow and controls the overall rate. Then $[X^+] = N$ and the overall process appears to be β-type. As height increases, the reaction type therefore alters from α-type to β-type. The reaction scheme represented by Equations 6.24 and 6.25 leads to equilibrium given by

$$\frac{1}{q} = \frac{1}{\beta(h)N} + \frac{1}{\alpha N^2}, \tag{6.26}$$

where q is the production rate as before. The change from α- to β-behaviour occurs at height h_t where

$$\beta(h_t) = \alpha N. \tag{6.27}$$

6.2.5 Vertical transport

The final term of the continuity equation (6.1) is to account for changes of electron and ion density at a given location due to bulk movement of the plasma. Such movements can have various causes and can occur in the horizontal and the vertical planes in general, but since our present emphasis is on the overall vertical structure of the ionosphere we shall concentrate here on the vertical component of movements in the F region. Assuming that photochemical production and loss are negligible by comparison with the effect of movements, the continuity equation gives (Section 2.2.3)

$$\frac{\partial N}{\partial t} = -(N\mathbf{v}),$$

where $\mathbf{v}$ is the drift velocity of the plasma.

Taking vertical movement only,

$$\frac{dN}{dt} = -\frac{\partial(Nw)}{\partial h}, \tag{6.28}$$

where w is the vertical drift speed and h is the height. Supposing that this drift is due to diffusion we can put (Section 2.2.5)

$$w = -\frac{D}{N}\cdot\frac{\partial N}{\partial h}, \tag{6.29}$$

D being the diffusion coefficient, given in its simplest form by $D = kT/m\nu$. Here k is Boltzmann's constant, T the temperature, m the particle mass and ν the collision frequency.

This expression for D was derived in Section 2.2.5 (Equation 2.21) by equating the driving force due to pressure gradient to the drag force due to collisions as the minority gas diffuses through the stationary majority gas. In the present case the minority gas is the plasma composed of ions and electrons, and the majority gas is the neutral air. Moreover, for drift in the vertical direction the force of gravity also acts on each particle, giving

$$-\frac{dP}{dh} = Nmg + N\nu mw = -kT\frac{dN}{dh}, \tag{6.30}$$

where we have used $P = NkT$. Since $D = kT/m\nu$ and $H_N = kT/mg$ (H_N being the scale height of the minority gas), this may be rearranged to give

$$Nw = -D\left(\frac{dN}{dh}+\frac{N}{H_N}\right),$$

and substitution into the continuity equation gives

$$\frac{dN}{dt} = \frac{\partial}{\partial h}\cdot\left[D\left(\frac{dN}{dh}+\frac{N}{H_N}\right)\right]. \tag{6.31}$$

This is the equation that has to be satisfied by the time and height variations of the upper F region and the protonosphere where ion production and recombination are both sufficiently small.

In this equation the scale height H_N merely represents the value of (kT/mg), and does not necessarily describe the actual height distribution. This is given by the *distribution height*, defined as

$$\delta = \left(-\frac{1}{N}\cdot\frac{dN}{dh}\right)^{-1}.$$

We can easily see that at equilibrium δ is equal to the scale height, since $dN/dt = 0$ gives $(dN/dh + N/H_N) = 0$, and hence $\delta = H_N$.

A plasma is composed of two minority species, ions and electrons. Initially the ions, being heavier, tend to settle away from the electrons, but the resulting separation of opposite charges produces an electric field, E, and a restoring force eE on each charged particle. To find how this affects the drift of the plasma, we write separate equations for each species and include the electrostatic force:

electrons

$$-dp_e/dh = Nm_eg + EeN + Nm_e\nu_ew = -kT_e\,dN/dh,$$

ions

$$-dp_i/dh = Nm_ig - EeN + Nm_i\nu_iw = -kT_i\,dN/dh.$$

Summing,

$$Nm_ig + Nm_i\nu_iw = -k(T_e+T_i)\,dN/dh,$$

it being assumed that $m_e \ll m_i$, $m_e\nu_e \ll m_i\nu_i$, $N_e = N_i(= N)$, and $w_e = w_i(= w)$. This equation is like 6.30 with T replaced by (T_e+T_i). Hence, for a plasma,

$$Nw = -D_p(dN/dh + N/H_p), \tag{6.32}$$

where

$$D_p = k(T_e+T_i)/m_i\nu_i \tag{6.33}$$

and

$$H_p = k(T_e+T_i)/m_ig, \tag{6.34}$$

respectively known as the *ambipolar* or *plasma diffusion coefficient* and the *plasma scale height*.

In that part of the ionosphere where plasma diffusion matters, the electron temperature usually exceeds the ion temperature. But taking $T_e = T_i$ by way of illustration, we see that the plasma diffusion coefficient and scale height are double those of the neutral gas at the same temperature. Effectively, the light electrons have the effect of halving the ion mass since the two species cannot separate very far. Equation 6.31 for a plasma becomes

$$\frac{dN}{dt} = \frac{\partial}{\partial h}\cdot\left[D_p\left(\frac{dN}{dh}+\frac{N}{H_p}\right)\right] \tag{6.35}$$

and at equilibrium $dN/dh = -N/H_p$ with the plasma exponentially distributed as $N/N_0 = \exp(-h/H_p)$ with scale height H_p. Note that this distribution has the same form as a Chapman layer but with (about) twice the scale height.

If the plasma is not in equilibrium the distribution changes with time at a rate depending on the value of the diffusion coefficient, which itself depends on the relevant collision frequency and therefore increases with altitude. If H is the scale height of the neutral gas then the height variation can be written as

$$D = D_0 \exp(h - h_0)/H, \tag{6.36}$$

where D_0 is the value of D at a height h_0. Thus, diffusion becomes ever more important at greater heights as the photochemistry becomes less important.

Another consequence of the height variation of D is that it leads to a second solution of Equation 6.35 in the case $dN/dt = 0$.

Substituting $D = D_0 \exp(h-h_0)/H$ and $N = N_0 \exp[-(h-h_0)/\delta]$ into 6.35,

$$\frac{dN}{dt} = \frac{dD}{dh}\left(\frac{dN}{dh} + \frac{N}{H_p}\right) + D\left(\frac{d^2N}{dh^2} + \frac{1}{H_p}\cdot\frac{dN}{dh}\right) = \frac{D}{H}\left(-\frac{N}{\delta} + \frac{N}{H_p}\right) + D\left(\frac{N}{\delta^2} - \frac{1}{H_p}\cdot\frac{N}{\delta}\right)$$

which rearranges to

$$\frac{dN}{dt} = DN\left(\frac{1}{\delta} - \frac{1}{H_p}\right)\left(\frac{1}{\delta} - \frac{1}{H}\right). \tag{6.37}$$

If $dN/dt = 0$ this has two solutions. The first, $\delta = H_p$ is diffusive equilibrium as already pointed out, and in this case the vertical drift speed (Equation 6.29) $w = -D(-1/\delta + 1/H_p) = 0$.

The second solution is $\delta = H$ (the scale height of the neutral gas, governing the diffusion coefficient). Here, $dN/dt = 0$ as before, but the drift speed is

$$w = D\left(-\frac{1}{\delta} + \frac{1}{H_p}\right) = D\left(-\frac{1}{H} + \frac{1}{H_p}\right),$$

which is not zero since $H_p > H$. The upward flow of plasma is

$$Nw = ND\left(\frac{1}{H} - \frac{1}{H_p}\right), \tag{6.38}$$

and in fact this is independent of height when $\delta = H$ because the height variations of D and of N cancel. Thus this second solution represents an unchanging distribution of electron density and a constant outflow of plasma.

6.3 Chemical aeronomy

6.3.1 Introduction

We have considered the physical principles governing the intensity and form of an ionospheric layer. To work out what the actual ionosphere should be like on Earth or any other planet requires a detailed consideration of many factors. We need to know the composition of the neutral atmosphere and its physical conditions such as density and temperature. We also need full information on the solar spectrum and any energetic particle fluxes able to ionize the constituents of the atmosphere. We would have to determine what gases could be ionized by the radiation incident, and then determine the ionization rate of each species, summing over all wavelengths and all

Table 6.2 *Ionization potentials*

Species	Ionization potential I (eV)	Maximum wavelength λ_{max} (Å)	(nm)
NO	9.25	1340	134.0
O_2	12.08	1027	102.7
H_2O	12.60	985	98.5
O_3	12.80	970	97.0
H	13.59	912	91.2
O	13.61	911	91.1
CO_2	13.79	899	89.9
N	14.54	853	85.3
H_2	15.41	804	80.4
N_2	15.58	796	79.6
Ar	15.75	787	78.7
Ne	21.56	575	57.5
He	24.58	504	50.4

gases to get the total production rate in a given volume. Then the loss processes, both chemical and by transport, have to be taken into account. Fortunately, for some regions major source and loss processes can be identified, though in others minor species are important and these are the regions that have proved the most difficult to comprehend.

The composition of the neutral atmosphere was dealt with in Section 4.1.5. We saw that it is composed mainly of the nitrogen/oxygen mixture generally known as 'air' up to about 100 km, and above that the atomic species O, He, and H progressively enter the scene due to diffusive separation.

To be ionized a species must absorb a quantum of radiation whose energy exceeds the ionization potential. Table 6.2 lists the ionization potential and the corresponding maximum wavelength of radiation able to effect an ionization for various atmospheric gases. For easy reference the wavelengths are given in both ångströms and nanometres.

These values of λ_{max} immediately identify the relevant parts of the solar spectrum as the X-ray (0.1–17 nm, 1–170 Å) and extreme ultra-violet, EUV, (17–175 nm, 170–1750 Å), emissions which come from the solar chromosphere and corona. Some are enhanced during flares. Figure 6.9 illustrates the solar spectrum up to 3000 Å (300 nm), including the major emission lines. This region is well to the shortwave side of the overall solar spectrum (see Figure 5.3).

In the Chapman formulation (Section 6.2.2) the value of the absorption cross-section, σ, generally increases with increasing wavelength up to λ_{max} and then falls rapidly to zero. Figure 6.10 shows how σ depends on wavelength for some common gases. For a value $\sigma(m^2)$, a volume of 1 m^3 containing n particles per m^3 and irradiated by flux I will absorb an amount $I\sigma n$ per unit time.

The efficiency with which the absorbed radiation leads to ionization is expressed by the ionization efficiency, η. With atomic species, all the absorbed energy goes into ion

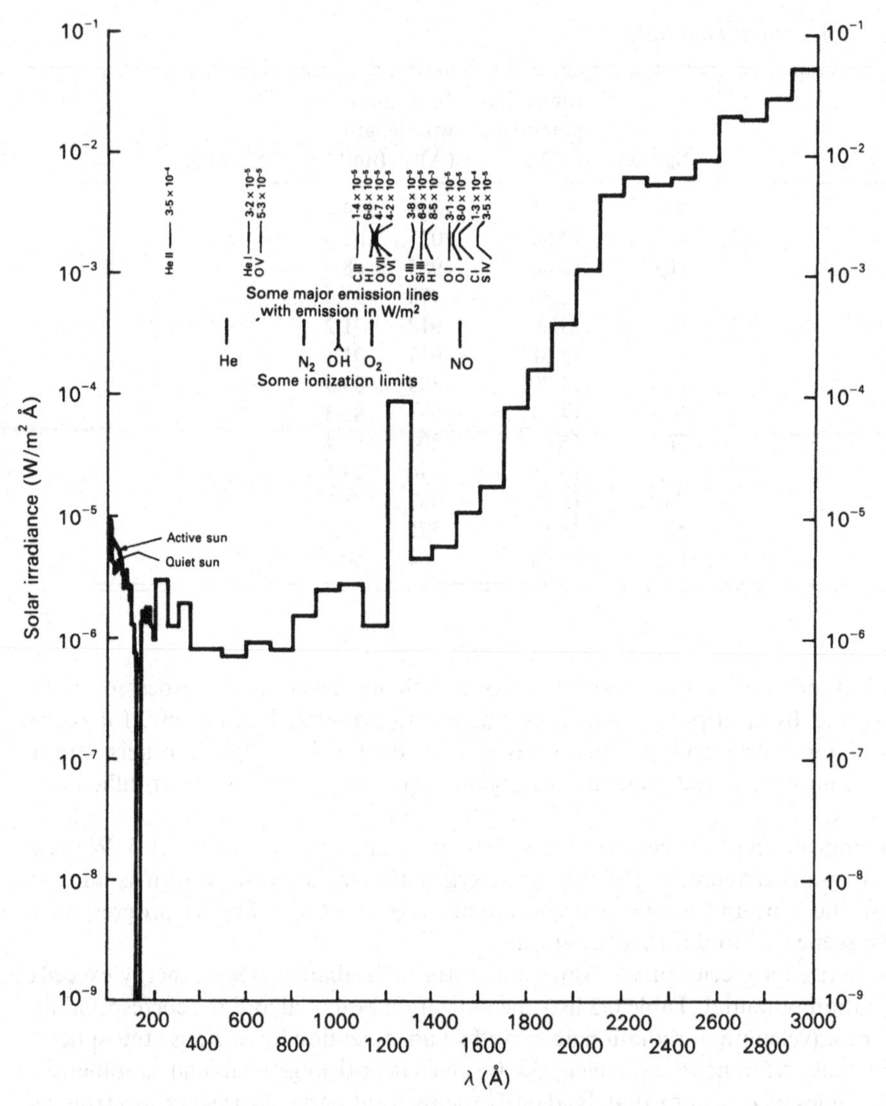

Fig. 6.9 Solar spectrum in the X-ray and UV regions. (Data from E. V. P. Smith and D. M. Gottlieb, *Space Sci.Rev.* **16**, 77, 1974. By permission of Kluwer Academic Publishers)

production at the rate of one ion–electron pair for every 34 eV of energy. The energy is inversely proportional to the wavelength, and a convenient formula in terms of wavelength is

$$\eta = 360/\lambda\,(\text{Å}). \tag{6.39}$$

The Chapman theory shows that the production rate is a maximum at the level where the optical depth, $\sigma n H \sec\chi$, is unity. If the absorption at given wavelength is due to several species, then the condition for maximum production is

$$\sum \sigma_i n_i H_i \sec\chi = 1. \tag{6.40}$$

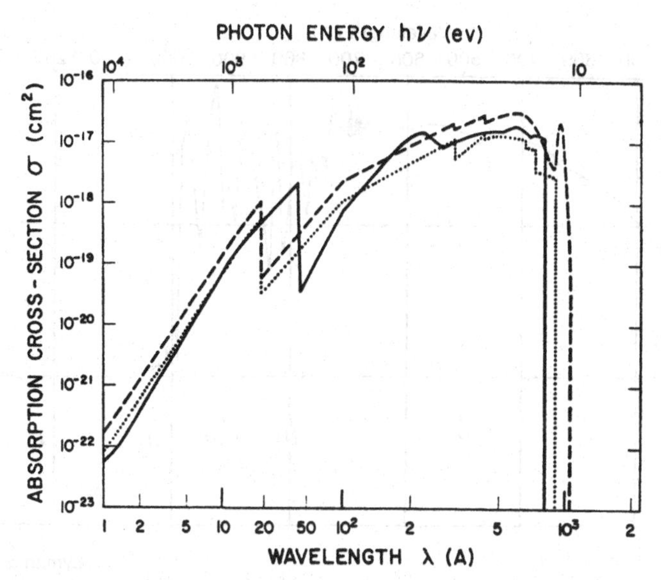

Fig. 6.10 Absorption cross-sections for O_2 (----), O (······) and N_2 (——). There is a change of scale at 100 Å. (T. E. VanZandt and R. W. Knecht, in *Space Physics* (eds. LeGalley and Rosen). Wiley, 1964)

At that height the rate is given by Equation 6.9,

$$q_m = \eta I_\infty \cos\chi / eH$$

summed over several species if necessary.

The height of unit optical depth in a model terrestrial atmosphere is given as a function of wavelength in Figure 6.11 and this, not the intensity of the ionizing radiation, is what determines the height of the ionospheric layers. The simple theory of Section 6.2.2 deals with the shape and intensity of an ionosphere produced by monochromatic radiation acting on a single gas. On a real planet the effect of all gases at a given wavelength has to be considered and then, since the ionosphere is in effect a number of overlapping Chapman layers, the production rate due to all relevant wavelengths has to be summed at each height.

6.3.2 E and F1 regions

Both the E region, which peaks at 105–110 km, and the F1 region at 160–180 km are fairly well understood. The F1 region is attributed to the most heavily absorbed part of the solar spectrum, between about 200 and 900 Å – the ionization limit of atomic oxygen is at 911 Å – for which the optical depth reaches unity from about 140 to 170 km. The band includes an intense solar emission line at 304 Å. The primary reaction products are O_2^+, N_2^+, O^+, He^+, and N^+, but subsequent reactions leave NO^+ and O_2^+ as the most abundant.

The E region is formed by the less strongly absorbed, and therefore more penetrating, parts of the spectrum. EUV radiation between 800 and 1027 Å (the ionization limit of O_2) is absorbed by molecular oxygen to form O_2^+. As can be seen from Figure 6.9, the band includes several important emission lines. At the short

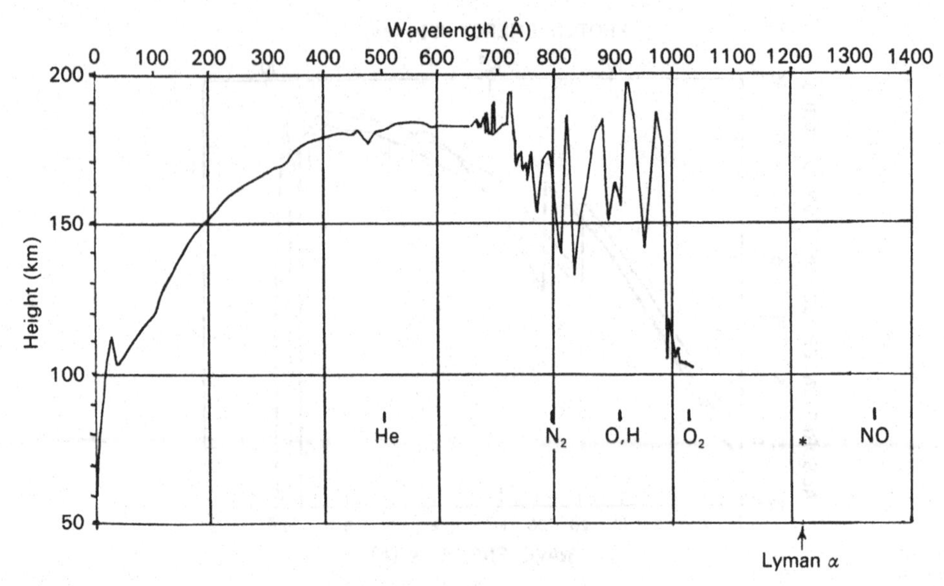

Fig. 6.11 Height where the optical depth reaches unity for radiation vertically incident on a model atmosphere. Ionization limits for common gases are marked. The wavelength range is that involved in producing the ionosphere. (J. D. Mathews, private communication)

wavelength end X-rays of 10–100 Å (1–10 nm) ionize all the atmospheric constituents. The main primary ions are N_2^+, O_2^+, and O^+, but the most numerous are again observed to be NO^+ and O_2^+. The intensity of solar X-rays varies over the solar cycle and they probably make little contribution to the E region at solar minimum.

Direct *radiative recombination* of the type

$$e + X^+ \rightarrow X + h\nu \tag{6.41}$$

is slow relative to other reactions and is not significant in the E and F regions. *Dissociative recombination*, as

$$e + XY^+ \rightarrow X + Y, \tag{6.42}$$

is 10^5 times faster (with a reaction coefficient of 10^{-13} m^3/s) and in both E and F regions the electron and ion loss proceeds via molecular ions. The main recombination reactions of the E region are therefore

$$e + O_2^+ \rightarrow O + O$$
$$e + N_2^+ \rightarrow N + N$$
$$e + NO^+ \rightarrow N + O. \tag{6.43}$$

Metallic ions of meteoric origin, such as Fe^+, Mg^+, Ca^+ and Si^+, are also present in the E region. They cannot recombine dissociatively and therefore have recombination coefficients typical of the radiative process (10^{-18} m^3/s), which gives them relatively long lifetimes (see Section 7.1.1).

In the F region the principal primary ion is O^+, which is first converted to a molecular ion by a *charge exchange* reaction

$$\left.\begin{array}{l} O^+ + O_2 \rightarrow O_2^+ + O \\ \text{or} \\ O^+ + N_2 \rightarrow NO^+ + N. \end{array}\right\} \tag{6.44}$$

The molecular ion then reacts with an electron as above, to give as the net result

$$\left.\begin{array}{l} e + O^+ + O_2 \rightarrow O + O + O \\ \text{or} \\ e + O^+ + N_2 \rightarrow O + N + N. \end{array}\right\} \tag{6.45}$$

In the F1 region the overall reaction is controlled by the rate of the dissociative recombination.

6.3.3 F2 region and protonosphere

Bradbury layer

At first sight the F2 region presents a puzzle because it peaks at 200–400 km whereas Figure 6.11 shows no band of radiation which has its maximum ionization rate at any height above about 180 km. The answer is found in the height variation of the recombination rate, which forms the F2 region as an upward extension of F1 even though the production rate is now decreasing with height.

Taking O^+ as the major ion, the two stage recombination process is

$$O^+ + N_2 \rightarrow NO^+ + N \quad \text{with rate} \propto \beta[O^+]$$

$$NO^+ + e \rightarrow N + O \quad \text{with rate} \propto \alpha[NO^+]N_e.$$

As discussed in Section 6.2.4 the second reaction controls the overall rate at low altitude and the first is the rate determining step at high levels, the transition being where $\alpha N_e = \beta(h_t)$. Here h_t is the transition height, generally between 160 and 200 km. The appearance of the F1-ledge depends on h_t being above the height of maximum production rate, h_m: that is, on there being a production maximum within an α-type region.

To explain the F2 region we consider the upper part where the recombination is β type and β depends on the concentration of N_2. On the other hand, the production rate depends on the O concentration. Thus, at equilibrium

$$N_e = q/\beta \propto [O]/[N_2]$$

$$\propto \exp\left[-\frac{h}{H(O)} + \frac{h}{H(N_2)}\right],$$

where $H(O)$ and $H(N_2)$ are the scale heights for O and N_2. Since the masses of N_2 and O are in the ratio 1.75:1, this rearranges to give

$$N_e \propto \exp\left[-\frac{h}{H(O)}\left(1 - \frac{H(O)}{H(N_2)}\right)\right] = \exp\left[+\frac{0.75h}{H(O)}\right]. \tag{6.46}$$

This is a layer whose electron density increases with height because the loss rate falls off more quickly than the production rate. It is often called a *Bradbury layer*.

F2 peak

Since the Bradbury layer increases with height indefinitely it cannot explain the peak of the F2 layer. On the other hand, diffusion produces an electron density distribution that decreases with height. With increasing height, *in-situ* production and loss become less important but diffusion becomes more important because of the reducing air density. The F2 layer peaks where these two kinds of process are equally important. To decide the level at which this will occur we regard the two loss processes – β-type recombination and transport – as being in competition, and compare their time constants for electron loss on the principle that the more rapid will be in effective control.

For recombination this is straightforward. If the ionization source is removed the electron density decays as

$$\frac{dN_e}{dt} = -\beta N_e = -\frac{N_e}{\tau_\beta}.$$

Therefore the characteristic time for recombination is

$$\tau_\beta = 1/\beta. \tag{6.47}$$

In diffusion the rate of change of electron density at a given height is given (Equation 6.37) by

$$\frac{dN_e}{dt} = DN\left(\frac{1}{\delta} - \frac{1}{H_p}\right)\left(\frac{1}{\delta} - \frac{1}{H}\right).$$

Equating this to $-N_e/\tau_D$ gives

$$\tau_D = \left[-D\left(\frac{1}{\delta} - \frac{1}{H_p}\right)\left(\frac{1}{\delta} - \frac{1}{H}\right)\right]^{-1}, \tag{6.48}$$

where H and H_p are the neutral and plasma scale heights respectively. Since δ is the distribution height (Section 6.2.5) whose value changes continuously in the approach to equilibrium, τ_D does not have a constant value. We therefore take approximate values. If we start with a plasma distribution such that $\delta \gg H, H_p$, the time constant is $\tau_D = HH_p/D$. Starting from the other extreme, $\delta \ll H, H_p$, gives $\tau_D = \delta^2/D$. The numerator always has the dimensions of a characteristic distance squared, and, as an approximation, we therefore take

$$\tau_D = H_1^2/D, \tag{6.49}$$

where H_1 is a typical scale height for the F2 region. Comparing Equations 6.47 and 6.49 then places the F2 peak at the level where

$$\beta \sim D/H_1^2. \tag{6.50}$$

The electron density at the peak is given by

$$N_m \sim q_m/\beta_m. \tag{6.51}$$

Protonosphere

At some level in the topside the ionosphere dominated by O^+ gives way to the protonosphere dominated by H^+. The ionization potential is almost the same for these two ions (Table 6.2), and therefore the reaction

$$H + O^+ \rightleftharpoons H^+ + O \tag{6.52}$$

is rapid, and around the transition level there is equilibrium in which

$$[H^+][O] = \tfrac{9}{8}[H][O^+]. \tag{6.53}$$

(The factor $\frac{9}{8}$ arises for statistical reasons, and there is also a temperature dependence proportional to $(T_n/T_i)^{\frac{1}{2}}$.) This reaction enables ionization to move readily between ionosphere and protonosphere, an important aspect of topside behaviour.

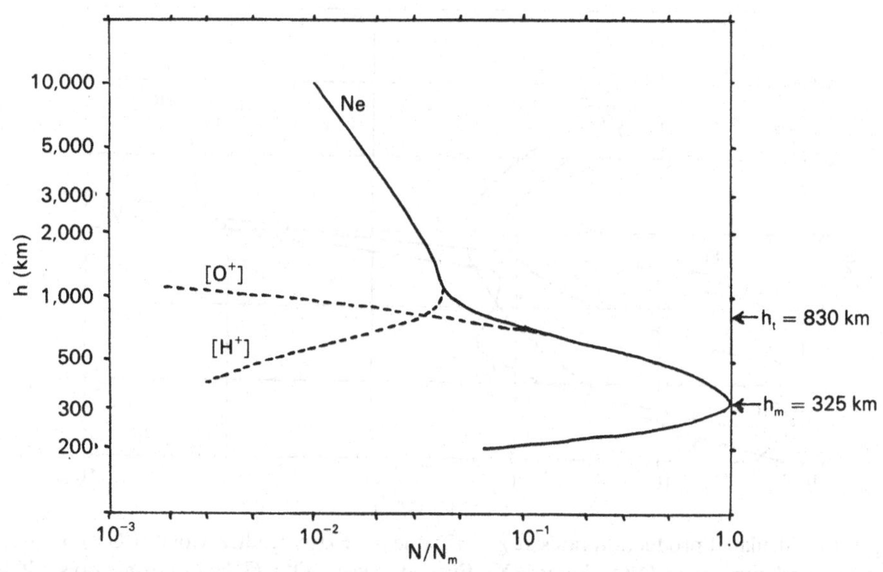

Fig. 6.12 Model ionospheric profile, showing transition between O^+ and H^+, for equinoctial night at Boulder, Colorado.

The transition effectively defines the *base of the protonosphere*. Below that level the H^+ distribution is determined by 6.53, and it is tied to the O^+ distribution by

$$[H^+] \propto [H][O^+]/[O] = \frac{\exp\left[-\frac{h}{H(H)}\right] \cdot \exp\left[-\frac{h}{H(O^+)}\right]}{\exp\left[-\frac{h}{H(O)}\right]} = \exp\left[+\frac{7h}{H(H)}\right]. \quad (6.54)$$

The strong upward gradient of H^+ concentration is apparent in the computed ionosphere/protonosphere profile shown in Figure 6.12.

Above the transition the concentration of O^+ decreases rapidly, and the protonosphere when in equilibrium assumes an exponential profile with plasma scale height according to Equation 6.34. The corresponding change of gradient of the electron density profile is evident in Figure 6.12.

As for the F2 peak, the transition level between ionosphere and protonosphere can be estimated by comparing time constants. If the rate constant of the reaction $H^+ + O \rightarrow H + O^+$ is k, then the lifetime of a proton is $(k[O])^{-1}$. Taking the time constant for diffusion in the protonosphere as H_2^2/D, the boundary occurs where $k[O] \sim D/H_2^2$. This occurs at 700 km or above, which is always well above the peak of the F2 layer.

6.3.4 D region

The D region of the ionosphere does not include a maximum but is that part below about 95 km that is not accounted for by the processes of the E region. The D region is also the most complex part of the ionosphere from the chemical point of view. This is due, first, to the relatively high pressure, which causes minor as well as major species to be important in the photochemical reactions, and, second, because several different sources of ionization contribute significantly to ion production.

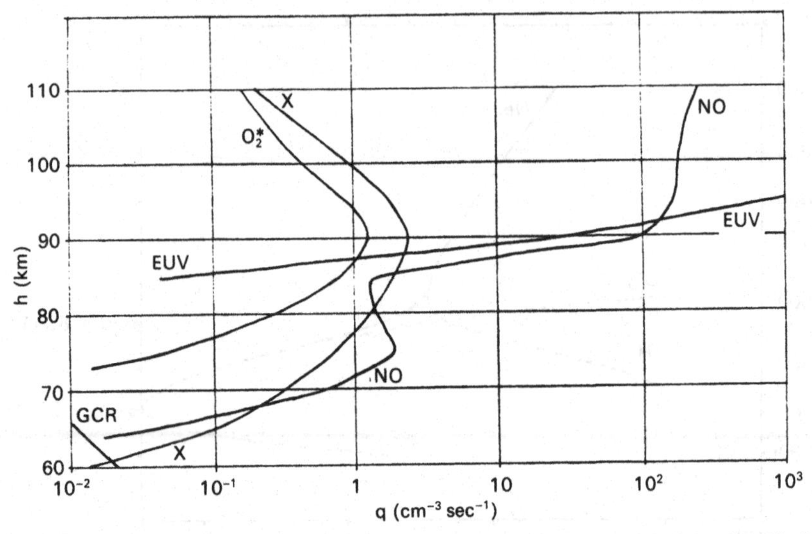

Fig. 6.13 Calculated production rates at $\chi = 42°$ due to: Extreme ultra-violet (EUV), Lyman-α and nitric oxide (NO), X-rays (X), Excited oxygen (O_2^*), Galactic cosmic rays (GCR). (J. D. Mathews, private communication)

Sources

To treat the whole of the D region from 60 to 95 km over all latitudes one has to reckon with 6 sources of ion production:

(a) The Lyman-α line of the solar spectrum at 1215 Å penetrates below 95 km and ionizes the minor species nitric oxide (NO) whose ionization limit is at 1340 Å.
(b) The EUV spectrum between 1027 and 1118 Å ionizes another minor constituent, excited oxygen in the state $O_2(^1\Delta_g)$.
(c) EUV radiation also ionizes O_2 and N_2, as in the E region.
(d) Hard X-rays of 2–8 Å ionize all constituents, most effect being therefore from the major species O_2 and N_2.
(e) Galactic cosmic rays, which affect the whole atmosphere down to the ground, become a major ionization source in the lower D region. At this level the production rate increases downward in proportion to the total air density.
(f) Energetic particles from the Sun or of auroral origin ionize the D region at high latitudes, where at times they form the main source. However we shall not be concerned with this source in this section.

The relative contributions of these different sources vary with latitude, time of day, and level of solar activity. Theoretical profiles of production rate (for solar zenith angle 42° and a 10-cm solar flux of 165 units) are given in Figure 6.13. These show EUV as the dominant source down to 92 km (E region), Lyman-α important from 92 to 68 km (this depending on the NO model assumed), X-rays as the main source below 68 km and also at 81–85 km, and galactic cosmic rays assuming the major role below 61 km. $O_2(^1\Delta_g)$ contributes between 80 and 105 km but is never the major source. At greater solar zenith angles the contributions from Lyman-α and X-rays are reduced, cosmic rays becoming relatively more important below about 70 km. The X-ray flux varies strongly with solar activity (by a factor of a hundred to a thousand) and is

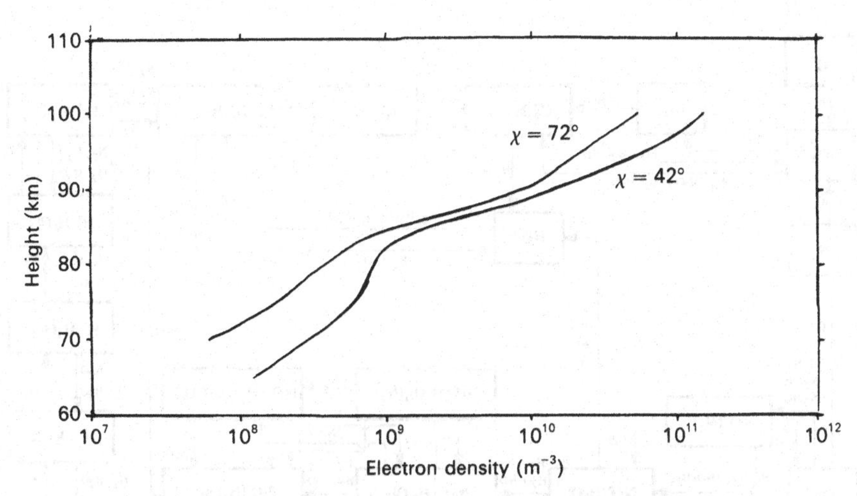

Fig. 6.14 Electron density profiles observed at Arecibo, Puerto Rico (18° N, 67°W) for two solar zenith angles. (J. D. Mathews, private communication)

probably not significant in the D region at sunspot minimum. These production rate profiles are consistent with D-region electron density measurements by incoherent scatter at Arecibo (Figure 6.14).

Positive ions

A lot of chemistry lies between the production rates of Figure 6.13 and the electron densities of Figure 6.14. The primary ions are NO^+, O_2^+ and N_2^+, but the latter are rapidly converted to O_2^+ by the charge-exchange reaction

$$N_2^+ + O_2 \rightarrow O_2^+ + N_2,$$

leaving NO^+ and O_2^+ as the major ions. However, below 80 or 85 km, apparently the level of the mesopause, rocket-borne mass spectrometers find heavier ions which are hydrated species such as $H^+ \cdot H_2O$, $H_3O^+ \cdot H_2O$, and hydrates of NO^+. These hydrates occur when the water vapour concentration exceeds about 10^{15} m^{-3}.

Where simple ions dominate, the loss process is dissociative recombination as in the E region, with a recombination coefficient of about 5×10^{-13} m^3s, the reaction of NO^+ being somewhat faster than that of O_2^+. In total the situation is much more complex, as illustrated in Figure 6.15. This scheme includes O_2^+, NO^+, O_4^+, hydrates and others, and has to be solved by means of a computer program. The significance of the hydrated ions is that, being larger molecules, they have greater recombination rates than the simple ions, of the order of 10^{-12} to 10^{-11} m^3s, depending on their size. Thus the electron density is relatively smaller in regions where hydrates dominate.

Negative ions

Below about 70 km by day or 80 km by night much of the negative charge is in the form of negative ions. The process begins with the attachment of an electron to an oxygen molecule, forming O_2^-. This is a three-body reaction involving any other molecule, M, whose function is to remove excess kinetic energy from the reactants:

$$e + O_2 + M \rightarrow O_2^- + M. \tag{6.55}$$

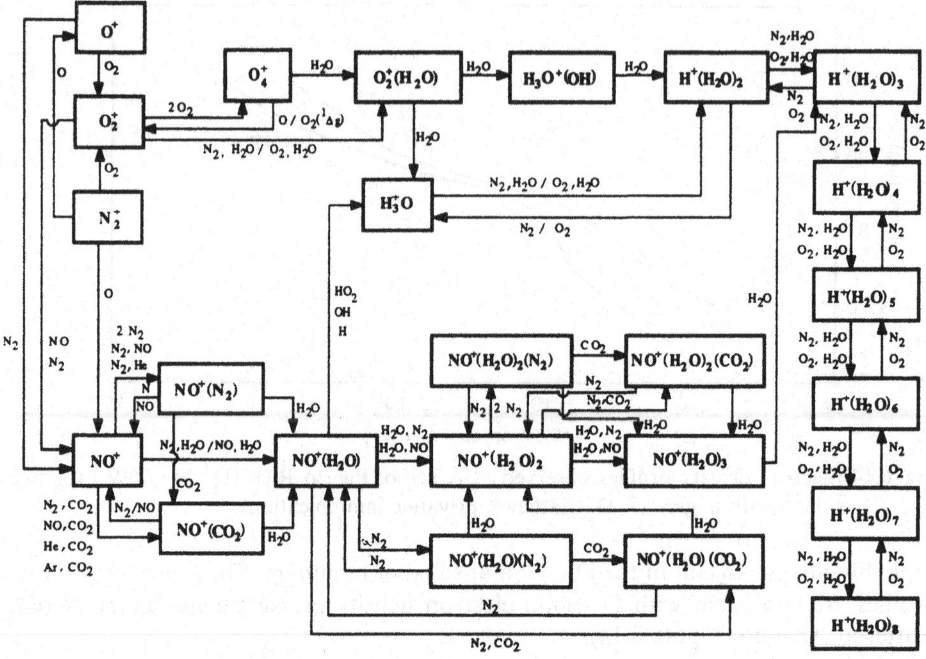

Fig. 6.15 A scheme of positive-ion chemistry for the D region. (From a model developed at Sodankylä Geophysical Observatory, Finland. E. Turunen, private communication)

This is followed by further reactions forming other and more complex negative ions including CO_3^-, NO_2^-, NO_3^- (the most abundant negative ion in the D region) and clusters such as $O_2^- \cdot O_2$, $O_2^- \cdot CO_2$ and $O_2^- \cdot H_2O$. A scheme of negative ion chemistry for the D region is shown in Figure 6.16.

Because the electron affinity of O_2 is small (0.45 eV) the electron may be removed by a photon of visible light or near infra-red:

$$O_2^- + h\nu \rightarrow O_2 + e. \qquad (6.56)$$

It is also detached in reactions with atomic oxygen to form ozone, and with the excited $O_2(^1\Delta_g)$:

$$O_2^- + O \rightarrow O_3 + e,$$

$$O_2^- + O_2(^1\Delta_g) \rightarrow 2O_2 + e.$$

These detaching reactions are included in the scheme of Figure 6.16. However some of the negative ion species are very stable against detachment and these are called *terminal ions*.

The effect of negative ions on the balance between electron production and loss was included in Equations 6.18 to 6.20. Variations of electron density in the D region can be due to changes in the negative-ion/electron ratio, λ, as well as to production rate changes.

Because of the complexity and uncertainty of D-region photochemistry it is common practice, when relating electron production rates to electron densities, to work with an 'effective recombination coefficient' (Equation 6.20) which may be either empirically or experimentally determined.

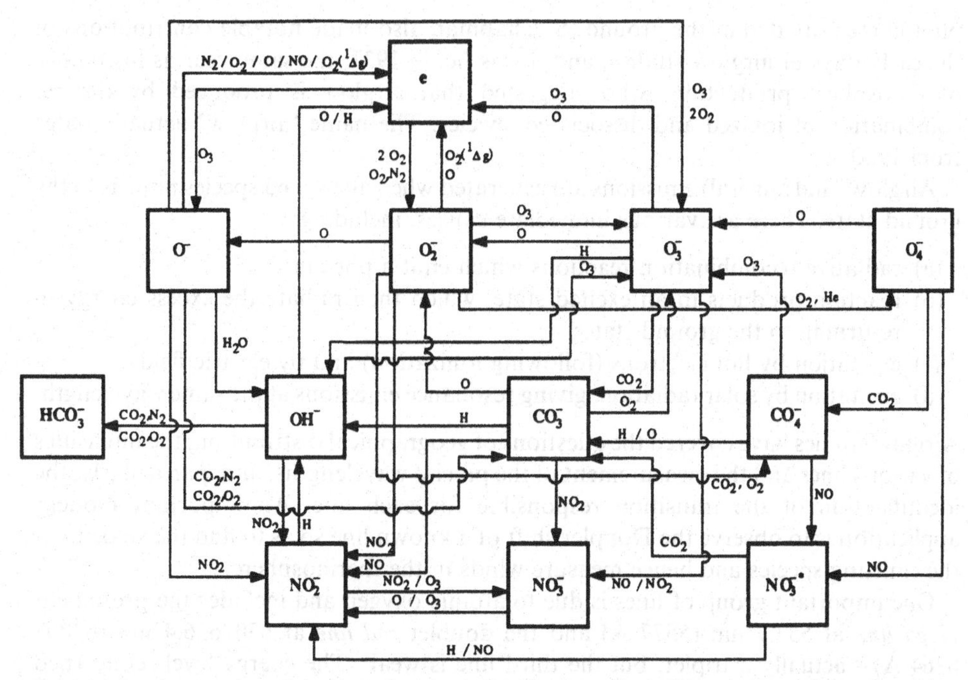

Fig. 6.16 A scheme of negative-ion chemistry for the D region. (From a model developed at Sodankylä Geophysical Observatory. E. Turunen, private communication)

6.4 Principles of airglow

Many of the photochemical reactions within the upper atmosphere are accompanied by the emission of radiation in the visible, infra-red or ultra-violet parts of the electromagnetic spectrum. These emissions comprise the *airglow*, which is a highly developed scientific topic having its own extensive literature. While a detailed discussion of airglow would be out of place here, an introduction is included to show its connection with aeronomy.

Aurora and airglow are often treated together, but there are some important differences. They are, for instance, energized from different sources. The aurora is a high-latitude phenomenon, occurring most intensely during and after solar disturbances, which is energized by charged particles entering the atmosphere from the magnetosphere. It shows much structure in both space and time. Airglow, in contrast, is always present, covers all latitudes, and is virtually unstructured. On a moonless night the airglow contributes the major part of the light arriving from the sky, exceeding starlight in total intensity though its presence is not generally appreciated because of its uniform distribution across the sky.

Being a less obvious phenomenon than aurora, it is perhaps not surprising that airglow is also younger as a scientific topic. R. J. Strutt, fourth Baron Rayleigh and son of the great Lord Rayleigh (the third Baron), pioneered airglow studies in the 1920s, and in his honour the unit of airglow emission rate is called the rayleigh:

$$1\,\text{R} = 10^6\ \text{photons/cm}^2\,\text{s}. \tag{6.57}$$

This measures the height-integrated emission rate, such as would be observed by a photometer situated at the ground. S. Chapman also made notable contributions in the early days of airglow studies, and it was he, in 1931, the same year as his theory of ionosphere production, who suggested that airglow is produced by the recombination of ionized and dissociated species. The name 'airglow' actually dates from 1950.

Airglow (and auroral) emissions are generated when an excited species returns to the ground state. There are various immediate causes, including :

(a) radiative recombination reactions which emit a photon;
(b) reaction products in an excited state, which then radiate the excess energy in returning to the ground state;
(c) excitation by hot electrons (following ionization) and by electric fields;
(d) excitation by solar radiation, giving resonance emissions at the same wavelength.

Airglow studies have covered the questions of geographical distribution, the intensities of various lines and the measurement of the precise wavelengths, and, particularly, the identification of the transition responsible for each line. An important modern application is to observe the Doppler shift of a known line so as to find the velocity of the emitting species and hence measure winds in the thermosphere.

One important group of lines is due to atomic oxygen and includes the prominent *green line* at 557.7 nm (5577 Å) and the doublet *red line* at 630/636.4 nm (6300/6364 Å) – actually a triplet, but the third line is weak. The energy levels concerned are shown in Figure 6.17. These lines are emitted when atomic oxygen in the state ^1S reverts to the ^3P ground state in two steps via ^1D. The first transition, lifetime 0.74 s, gives the green line, and the second, lifetime 110 s, produces the red line. (The UV line at 297.2 nm, corresponding to direct transition to the ground state, is too weak to be of interest in airglow.)

In the F region the excited O atoms come from the dissociative recombination

$$O_2^+ + e \rightarrow O^* + O^*, \quad \text{(Equation 6.43)}$$

where the * indicates the electronically excited species. Since 7 eV of energy is released from each such reaction there are atoms in both the ^1S and the ^1D states, and both the red and the green lines are emitted. In the E region the main source of excited O is the reaction first suggested by Chapman:

$$O + O + O \rightarrow O_2 + O^* \tag{6.58}$$

and the O* is in the ^1S state. However at this lower height the green line is observed but the red one is missing. At 110 s, the lifetime of the intermediate ^1D state is so long that the energy is removed through collisions with other molecules before the red photon can be emitted. This process is called *quenching*.

By virtue of its mechanism, the green line is suitable for dynamical studies, and its Doppler shift can be interpreted as the velocity of the neutral air at that level. The intensity of the red line from the F region is found to be closely related to F-region parameters, which might be expected since it arises from part of the O^+ recombination process. However the Chapman reaction at lower height appears to have no correlation with the ionosphere.

Ultra-violet emissions around 130 nm (1300 Å), again from excited O atoms, have been much used for observing aurora from space (Section 8.3.2). This emission and other significant O lines are included in Figure 6.17.

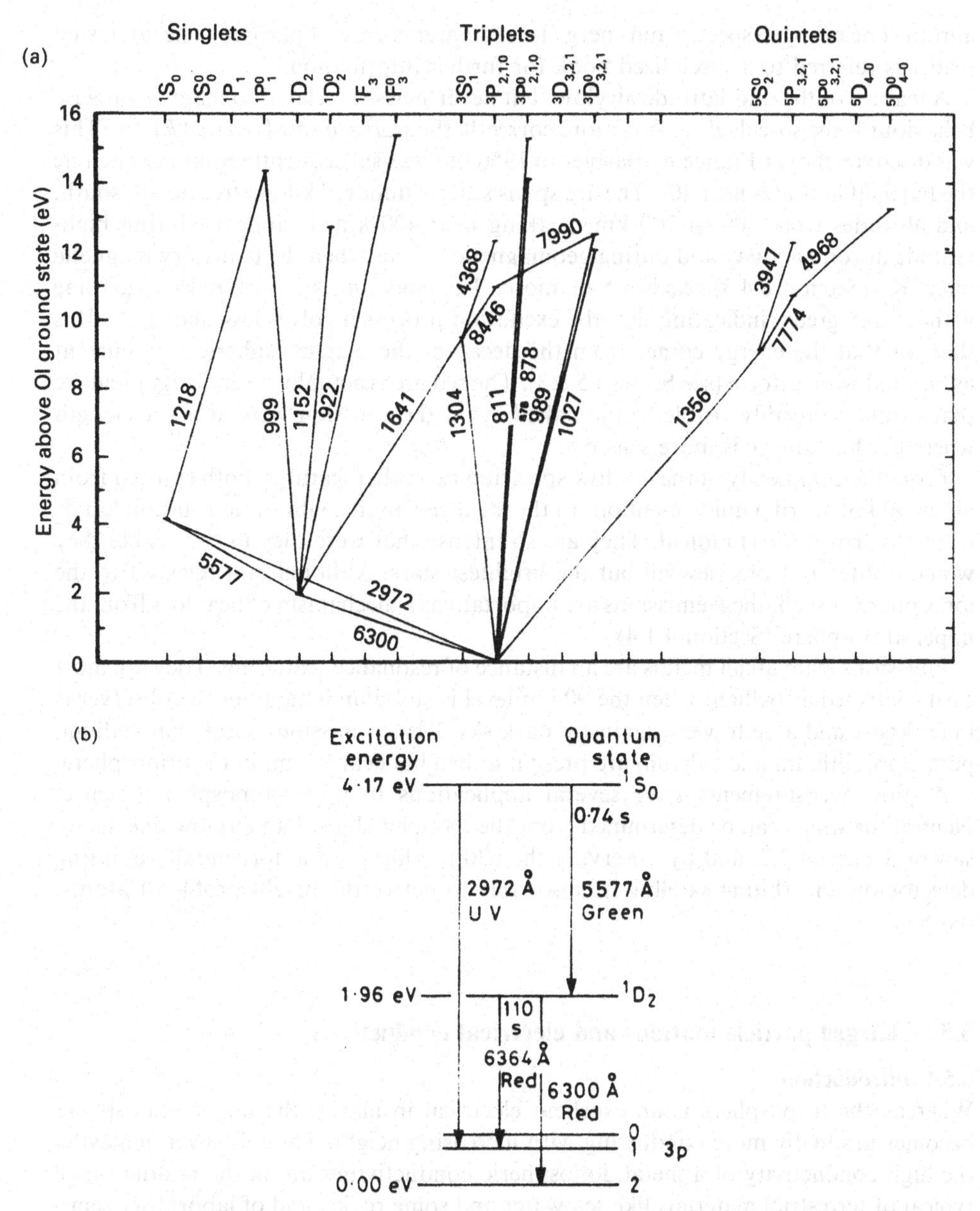

Fig. 6.17 Energy levels and transitions in atomic oxygen. (a) Transitions that have been observed in airglow or aurora. (M. H. Rees, *Physics and Chemistry of the Upper Atmosphere*. Cambridge University Press, 1989) (b) Details of the most important lines. (After S. J. Bauer, *Physics of Planetary Ionospheres*. Springer-Verlag, 1973) In each case the unit of wavelength is the ångström.

Nitrogen lines are also observed, amongst them 391.4 nm, which is due to the excitation of N_2 molecules. The intensity of this line is directly proportional to the ionization rate and is useful as a measure of the energy input to the atmosphere in an aurora. The nitrogen spectra and energy level diagrams are complicated. The interested reader is referred to a specialized book for further information.

A feature of the mid-latitude airglow that nevertheless is related to magnetospheric behaviour is the so-called *red arc*, more correctly the *stable auroral red* (*SAR*) *arc*. This was discovered over France by Barbier in 1956 but was subsequently found to encircle the Earth at latitudes near 40°. The arc spans several hundred kilometres north–south, and altitudes from 300 to 700 km, peaking near 400 km. It appears during high-latitude auroral activity and during geomagnetic storms when the planetary magnetic index K_p (Section 8.4.3) reaches 5 or more. The emission is a pure red oxygen line without any green, indicating that the excitation process involves less than 4 eV. It is thought that the energy comes from the decay of the magnetospheric ring current associated with a storm (see Section 5.8.3). The red arc is actually a very bright feature that would be readily visible to the naked eye if the emission were at a wavelength where the human eye is more sensitive.

From the complexity of the airglow spectrum two other features, both emitted from about 90 km, merit a quick mention. In the infra-red are the vibration–rotation bands of the hydroxyl (OH) radical. They are so intense that were they in the visible they would obliterate from view all but the brightest stars. Although not relevant to the ionosphere as such, these emissions are important as a mechanism of heat loss from the upper atmosphere (Section 4.1.4).

Emissions from alkali metals are an instance of resonance scattering. They are most easily detected at twilight when the 90 km level is still illuminated but the observer is in darkness and able to view against a dark sky. These emissions show that sodium, potassium, lithium and calcium are present at heights near 90 km in the atmosphere.

Airglow measurements have several applications to upper-atmosphere science. Neutral-air winds can be determined from the Doppler shift of an airglow line, as we saw in Section 4.2.2, and by observing the 630 nm line with a horizontally pointing detector on an orbiting satellite it is possible to derive the height profile of atomic oxygen.

6.5 Charged particle motions and electrical conductivity

6.5.1 Introduction

Whereas the troposphere is an excellent electrical insulator, the upper atmosphere becomes gradually more conducting with increasing height. Though never achieving the high conductivity of a metal, ionospheric conductivities are in the middle range typical of terrestrial materials like sea water and some rocks, and of laboratory semiconductors. The conductivity of the upper atmosphere arises, of course, from the presence of free electrons and ions which can move readily under impressed forces and thereby act as efficient charge carriers.

We shall see that the ions and the electrons may respond to forces in different ways, so that ion and electron currents are not necessarily the same. If the positive and the negative particles move in opposite directions the ion and electron currents are simply added together. The other extreme case is when the ions and the electrons move exactly together; now the two currents cancel and there is no net current, but the plasma as a whole is in motion. This is called a *plasma drift*. The general situation is sketched in

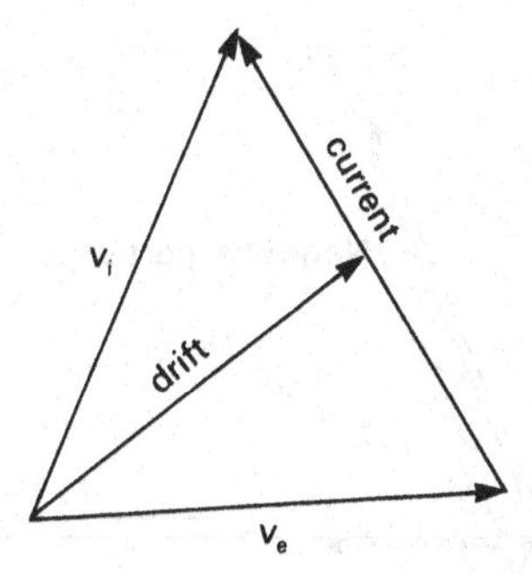

Fig. 6.18 Resolution of electron (v_e) and ion (v_i) velocities to give electric current and plasma drift.

Figure 6.18: the drift is given by the vector sum of the ion and electron velocities, and the current by their vector difference, with suitable scaling in each case. Clearly the conductivity varies with altitude in direction as well as in magnitude.

Two different driving forces occur naturally in the upper atmosphere, and these act in somewhat different ways upon the medium. When an electric field is present the positive and the negative particles tend to be driven in opposite directions and one expects an electric current to be the first result. The other driving force is a wind in the neutral air. Through collisions some of the motion is transferred to the ionized species, initially driving them along in the same direction as a plasma drift.

There are, however, two complicating factors. Collisions between the charged and the neutral particles is one. We can think of the collision frequency between two species as the number of times a selected particle of one kind collides with any particle of the other kind, though a more correct definition is based on momentum transfer, since particle mass m, relative velocity v, and collision frequency ν causes a drag force $mv\nu$ (Section 2.2.4). The collision frequency is a key parameter in conductivity considerations.

The other complicating factor is the geomagnetic field characterized by the gyrofrequency, the frequency at which a charged particle gyrates in the field (Section 2.3.2), given by

$$\omega_e = Be/m_e \quad \text{for an electron,}$$

and

$$\omega_i = Be/m_i \quad \text{for an ion.}$$

Although the direction of gyration depends on the sign of the charge, our treatment takes the convention that all frequencies are considered positive. The gyrofrequency is the other key parameter, and an important consequence of the geomagnetic field is that all upper atmosphere motions and currents are anisotropic.

6.5.2 Particle motion in a magnetic field in the presence of collisions

To understand the conductivity structure of the upper atmosphere we must first appreciate how a charged particle in a magnetic field responds to a driving force when collisions with neutral particles also occur. The simplest case is when the force F acts along the magnetic field. The particle, mass m, initially accelerates at F/m, but if the collision frequency is ν it travels for a time of only $1/\nu$ and a distance of $F/2m\nu^2$ before colliding with a neutral particle. If we imagine that the particle stops for an instant at

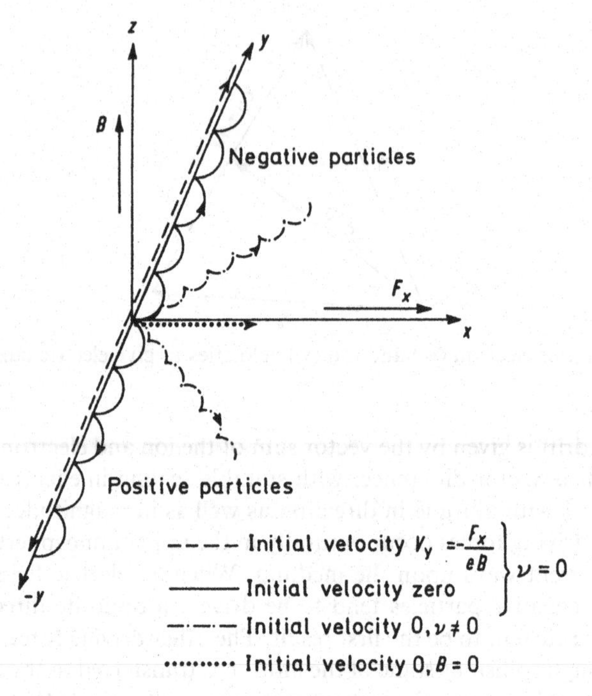

Fig. 6.19 Charged particle motions in a magnetic field.

each collision, after which it is again accelerated by the force F, the overall speed will be $v = F/2m\nu$. However this is inexact because the particle may not stop on collision. The correct result is obtained, also more economically, by equating the driving force, F, to the drag force due to collisions, $mv\nu$, giving

$$v = F/m\nu. \tag{6.59}$$

The magnetic field has no effect in this case since it is parallel to the direction of motion.

Referring to Figure 6.19, consider now a force F_x directed along the x axis and acting on a positively charged particle at the origin. Initially the particle moves in the x direction, but as soon it moves, with velocity v, it experiences a Lorentz force of magnitude Bev deviating it to the right, the magnetic field, B, being directed along the z axis. In vector notation, $\mathbf{F} = e\mathbf{v} \times \mathbf{B}$. If no collision occurs the particle will move in a loop, being retarded by F_x during the second part and coming to rest at a point further along the $-y$ axis. The average velocity in the $-y$ direction is approximately $F_xP/4m$ where P is the gyroperiod ($2\pi m/Be$). Thus an approximate expression for the drift speed is $v = \pi F_x/2Be$.

The exact expression is $v = -F_x/Be$, which may be verified as follows. The equations for the velocity components in the x and y directions are

$$\frac{dv_x}{dt} = \frac{F_x}{m} + \frac{eB}{mv_y},$$

$$\frac{dv_y}{dt} = -\frac{eB}{mv_x}.$$

(Note that $eB/m = \omega_B$, the gyrofrequency.) The solution of these equations depends on what initial velocity is assumed. It is easily seen that they are satisfied by $v_x = 0$, $v_y = -F_x/eB$; if this is the initial motion the particle continues with the same velocity. If the initial velocity is zero, the solution is

$$v_x = \frac{F_x}{eB} \sin \omega_B t,$$

$$v_y = \frac{F_x}{eB} \cos \omega_B t - \frac{F_x}{eB},$$

representing a circular motion with a superimposed drift F_x/eB in the $-y$ direction. The foregoing applies to a particle with positive charge. A negatively charged particle drifts in the $+y$ direction since the Lorentz force acts in the opposite direction. There would be appropriate changes of sign in the above equations.

If collisions occur we can visualize their effect as to 'stop' the particle every $1/\nu$ instead of every $P/2$ (provided $1/\nu < P/2$), and, as Figure 6.19 illustrates, this introduces a drift component in the x direction at the expense of motion along the y axis. If ν is large enough $v_x = F/m\nu$, as though the magnetic field were not present. To derive a general formula including the effects of both the magnetic field (introducing a Lorentz force) and collisions (introducing a drag force) we consider the vector equation balancing the various forces:

$$\mathbf{F} \pm e\mathbf{v} \times \mathbf{B} - m\nu\mathbf{v} = 0. \tag{6.60}$$

Here the first term is the driving force, the second the Lorentz force (+ for a positive ion, − for an electron) and the third is the drag force due to collisions. Because of the cross product $\mathbf{v} \times \mathbf{B}$ the second term always acts at right angles to the third, and the vector diagram is a right angled triangle as sketched in Figure 6.20. Plainly, $\tan\theta = \pm eB/m\nu$ and $\cos\theta = m\nu v/F$, from which may be derived the expressions for the velocity components for electrons and positive ions respectively:

Electron:

$$\left.\begin{aligned} v_y &= \frac{F_x}{eB} \cdot \frac{\omega_e^2}{(\nu_e^2 + \omega_e^2)}, \\ v_x &= \frac{F_x}{eB} \cdot \frac{\nu_e \omega_e}{(\nu_e{}^2 + \omega_e{}^2)} \\ \text{Ion:} \quad v_y &= -\frac{F_x}{eB} \cdot \frac{\omega_\cdot{}^2}{(\nu_i^2 + \omega_i^2)} \\ v_x &= \frac{F_x}{eB} \cdot \frac{\omega_\cdot \nu_\cdot}{(\nu_i^2 + \omega_i^2)}. \end{aligned}\right\} \tag{6.61}$$

In these expressions ω_e and ω_i are the electron and ion gyrofrequencies and ν_e and ν_i are the respective collision frequencies. The electron charge, e, is considered positive. Since the fractions appear frequently it is convenient to express them by

$$\left.\begin{aligned} f_1(e) &= \frac{\omega_e \nu_e}{(\nu_e^2 + \omega_e^2)}; \\ f_2(e) &= \frac{\omega_e^2}{(\nu_e^2 + \omega_e^2)}, \end{aligned}\right\} \tag{6.62}$$

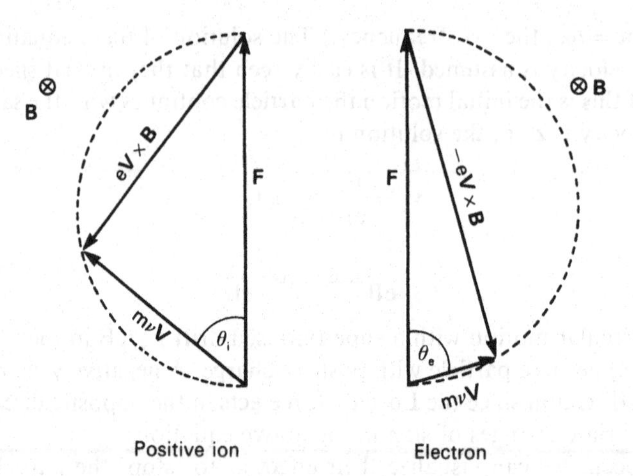

Fig. 6.20 Forces on electron and positive ion in the presence of a magnetic field.

with similar expressions for the ions. For motion along B or in the absence of a magnetic field

$$v_z = \frac{F_z}{m\nu_e} = \frac{F_z}{eB}\cdot\frac{\omega_e}{\nu_e}$$

and we define

$$f_0(e) = \frac{\omega_e}{\nu_e}, \tag{6.63}$$

with a similar expression for ions. Recall that the angle between the force and the resulting velocity, θ is given by

$$\tan\theta = \omega/\nu = f_0.$$

This ratio varies with height, being very small at low height and very large at a great height: thus θ varies between 0 and 90°, as illustrated in Figure 6.21.

6.5.3 Responses to a neutral-air wind

Motion across the magnetic field

The force due to a wind in the neutral air, U, is $F = m\nu U$ where ν is the relevant collision frequency. First we consider that the wind blows perpendicular to the geomagnetic field. Substituting this value for F_x in Equations 6.61 then gives the resulting velocity components in the x and y directions (Figure 6.19).

Two limiting cases are of interest. At a sufficiently low height $\nu \gg \omega$, f_0 is small, $v_y \ll v_x$, and $v_x = (m\nu U/eB)f_0 = U$. The plasma is carried along with the wind as expected. At a great height $\nu \ll \omega$, f_0 is large, $v_x \ll v_y$; then

$$v_y = m_e\nu_e U/eB = U/f_0(e)$$

for electrons, and

$$v_y = -U/f_0(i)$$

for ions. The ions and electrons now move in opposite directions, comprising a current at right angles to the wind direction, though its value is relatively small because f_0 is large.

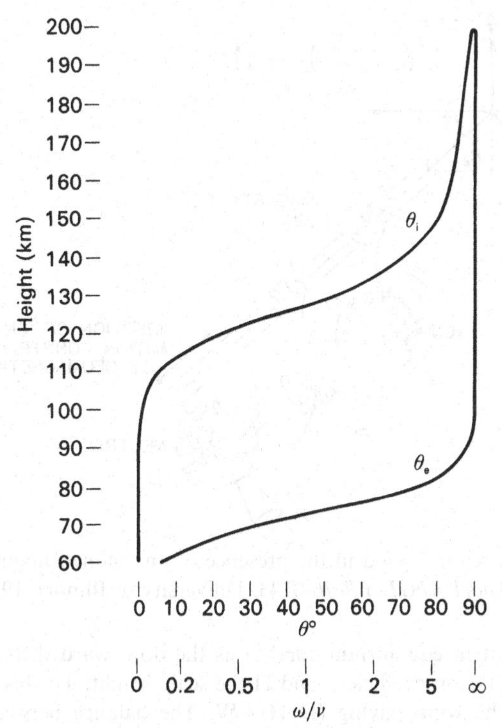

Fig. 6.21 Angle between force and velocity for an electron (θ_e) and a positive ion (θ_i). If $\theta = 0$ the particle moves with the neutral air, and if $\theta = 90°$ it moves with the magnetic field (**E** × **B**).

If $\nu = \omega$ the motion is at 45° to the wind. This occurs at about 75 km for electrons and 120 km for ions. Between these levels the ions move more or less with the wind whereas the electrons move nearly across it. This region, where the wind is most effective in producing an electric current, is the *dynamo region*. The principal reason for the existence of such a region is the mass ratio between the ion and the electron:

$$\omega_e/\omega_i = m_i/m_e = 1836M_i,$$

where M_i is the atomic weight of the ion.

Motion along the magnetic field

Suppose now that there is a wind component along the geomagnetic field. In general this will be so, because the neutral-air wind blows in the horizontal plane but the geomagnetic field is not vertical except at the magnetic poles. For simplicity assume that the wind, U, is in the magnetic meridian. If the dip angle is I, the neutral air component along the field is $U_{\parallel} = U \cos I$ – See Figure 6.22 – and the plasma moves in the same way. Compared to its vertical structure the ionosphere is relatively uniform in the horizontal plane. The principal effect of the field aligned drift, $U_{\parallel}$, is therefore a lifting or depression of the layer depending on the vertical component

$$W = U_{\parallel} \sin I = \tfrac{1}{2} \cdot U \sin 2I. \tag{6.64}$$

The effect is greatest where the magnetic dip is 45°.

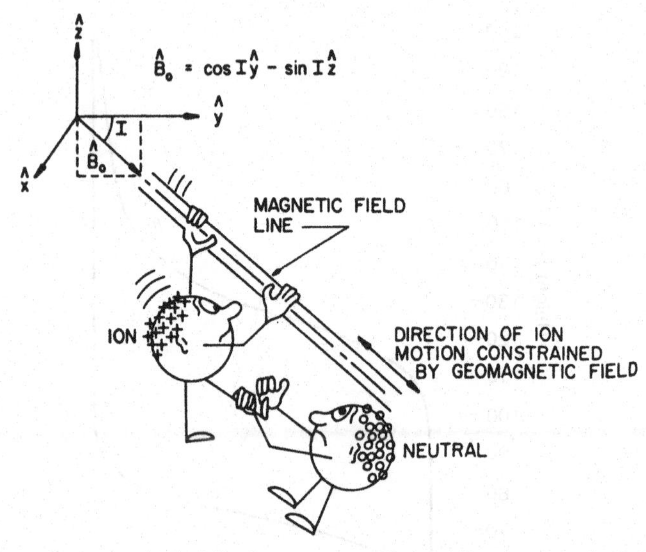

Fig. 6.22 Lifting effect of a neutral wind in the presence of an inclined magnetic field. (R. E. DuBroff *et al.*, *Report AFGL-TR*-76-0143, University of Illinois, 1976)

Equation 6.29 shows that in equilibrium conditions the downward drift due to diffusion $\approx D/H$, where D is the diffusion coefficient and H the scale height. To this we now add the vertical drift, W, due to the wind, giving $(D/H)+W$. The balance between diffusion and recombination which determines the height of maximum is expressed by $\beta \sim D/H^2$ in the absence of wind. With the wind present this becomes

$$\beta \sim \left(\frac{D}{H^2}+\frac{W}{H}\right),$$

where β is the recombination coefficient at the height of maximum. Since β varies with height as $\exp(-h_m/H)$, we can write

$$\frac{(D/H+W)}{D/H} = 1+\frac{WH}{D} \sim \exp\left(\frac{\Delta h_m}{H}\right) \sim 1+\frac{\Delta h_m}{H},$$

for a small increment, Δh_m, in the height of the maximum. Hence the effect of the neutral wind is to change the height of the layer maximum by

$$\Delta h_m \sim \frac{WH^2}{D} \sim \frac{W}{\beta}. \tag{6.65}$$

6.5.4 Response to an electric field

An electric field, E_x, exerts force $E_x e$ on a positive ion and $-E_x e$ on an electron. Replacing F_x by $E_x e$ and $-E_x e$ respectively in Equations 6.61 gives for a positive ion,

$$v_x = \frac{E_x}{B} f_1(i);\ v_y = -\frac{E_x}{B} f_2(i);$$

and for an electron,

$$v_x = -\frac{E_x}{B} f_1(e);\ v_y = -\frac{E_x}{B} f_2(e).$$

Table 6.3 *Effects of wind and electric field*

Altitude	Wind (U)	Electric field (E)
High ($f_0 \gg 1$)	Current $\perp$U	Plasma drift $\perp$E
Intermediate ($f_0(e) > 1$ but $f_0(i) < 1$)	Ion current $\Vert$U Small electron current $\perp$U	Electron current $\perp$E Small ion current $\Vert$E
Low ($f_0 \ll 1$)	Plasma drift $\Vert$U	Current $\Vert$E

In each case, if $\nu \ll \omega$ then $v_x \ll v_y$ and $v_y = -E_x/B$. Therefore at high altitude the plasma as a whole drifts at velocity E_x/B in the $-y$ direction. This is the same result as that derived in Section 2.3.7 for $E \times B$ drift, given in vector form by Equation 2.40.

At a low level $\nu \gg \omega$ and $v_y \ll v_x$. Then for positive ions,

$$v_x = \frac{E_x}{B} \cdot f_0(i)$$

and for electrons,

$$v_x = -\frac{E_x}{B} \cdot f_0(e).$$

The electrons and ions move in opposite directions, representing currents in the same direction and parallel to the applied electric field. However the currents are relatively small because f_0 is small at low altitude.

As before, the drifts are at 45° to the electric field when $\nu = \omega$ (i.e. $f_0 = 1$), and there is again a range of heights where the electrons tend to respond to the electric field but the ions remain more under the control of the neutral air.

Table 6.3 summarizes the responses to wind and electric field in the three height regions.

6.5.5 Conductivity

Electrical conductivity is the ratio between current density (i in A/m) and electric field (E in V/m): $\sigma = i/E = Nev/E$, where v is the velocity of the relevant particles, N their concentration and e the charge on each.

In the absence of a magnetic field, or if the electric field is along the magnetic field, $v = F/m\nu = Ee/m\nu$, from Equation 6.59. Thus,

$$\sigma = \frac{Ne^2}{m\nu}. \qquad (6.66)$$

Since the ion and electron currents are simply added, the conductivity due to both is

$$\sigma_0 = \left(\frac{N_e}{m_e\nu_e} + \frac{N_i}{m_i\nu_i}\right)e^2, \qquad (6.67)$$

This is called the *direct* or *longitudinal* conductivity.

For electric fields (or electric field components) perpendicular to the magnetic field

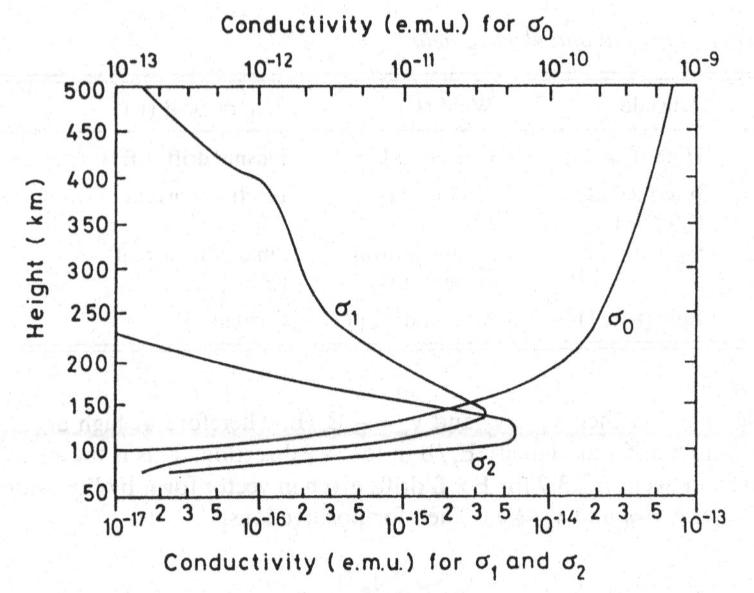

Fig. 6.23 Conductivity profiles calculated for middle latitude at noon. (S.-I. Akasofu and S. Chapman (after K. Maeda and H. Matsumoto), *Solar–Terrestrial Physics*. Oxford University Press, 1972. By permission of Oxford University Press)

we define two conductivities to deal with currents flowing respectively parallel and perpendicular to the electric field. These are represented by the symbols σ_1 and σ_2, called the *Pedersen* and *Hall* conductivities respectively

$$\sigma_1 = (N_e f_1(e) + N_i f_1(i))\frac{e}{B} = \left[\frac{N_e}{m_e \nu_e}\cdot\frac{\nu_e^2}{(\nu_e^2+\omega_e^2)} + \frac{N_i}{m_i \nu_i}\cdot\frac{\nu_i^2}{(\nu_i^2+\omega_i^2)}\right]e^2; \quad (6.68)$$

$$\sigma_2 = (N_e f_2(e) - N_i f_2(i))\frac{e}{B} = \left[\frac{N_e}{m_e \nu_e}\cdot\frac{\omega_e \nu_e}{(\nu_e^2+\omega_e^2)} - \frac{N_i}{m_i \nu_i}\cdot\frac{\omega_i \nu_i}{(\nu_i^2+\omega_i^2)}\right]e^2. \quad (6.69)$$

When deriving and manipulating these expressions it should be born in mind that $\omega_B = eB/m$.

If the electric field is expressed as a vector

$$\mathbf{E} = \mathbf{i}E_x + \mathbf{j}E_y + \mathbf{k}E_z,$$

where E_x, E_y and E_z are components and **i**, **j** and **k** are unit vectors, the z axis being along the magnetic field (as in Figure 6.19), then the current vector is given by

$$\mathbf{J} = \mathbf{i}J_x + \mathbf{j}J_y + \mathbf{k}J_z = \begin{pmatrix} \sigma_1 & \sigma_2 & 0 \\ -\sigma_2 & \sigma_1 & 0 \\ 0 & 0 & \sigma_0 \end{pmatrix}\begin{pmatrix} E_x \\ E_y \\ E_z \end{pmatrix}, \quad (6.70)$$

where the elements of the tensor are the conductivities derived above. This form allows any current component to be derived if the three conductivities and the vector electric field are known.

A calculated height distribution of conductivity in the mid-latitude ionosphere at noon is shown in Figure 6.23. Note that the Pedersen and Hall conductivities peak in the E region whereas the direct conductivity continues to increase with altitude and is

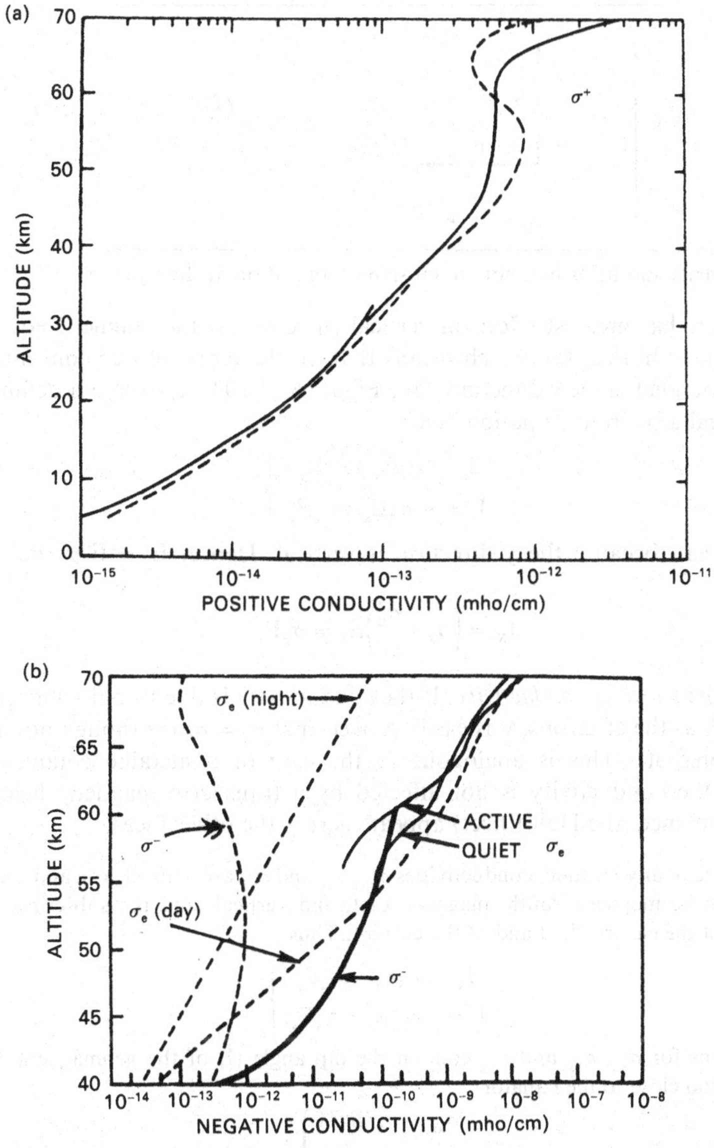

Fig 6.24 Conductivities, 5–70 km, due to (a) positive and (b) negative particles. (——) Measured and (-----) theoretical values, due to electrons (σ_e), negative ions (σ^-) and positive ions (σ^+). The negative ion conductivity is equal to the positive ion conductivity below 35 km. (After W. Swider, *Radio Science* **23**, 389, 1988, copyright by the American Geophysical Union)

of much greater magnitude. Figure 6.24 shows estimates of the conductivity profiles due to positive and negative particles between 5 and 70 km.

A complication of the real ionosphere is its spatial non-uniformity, which affects the currents through the continuity requirement – i.e. all the current has to go somewhere. Latitudinal and time-of-day variations have some effect, but the horizontal layering of

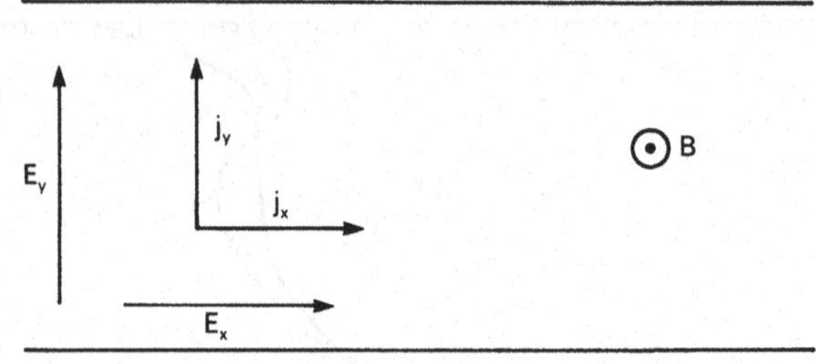

Fig. 6.25 Currents and fields in a horizontal slab of ionization at the equator.

the ionosphere has greatest effect on vertical currents. At the magnetic equator the magnetic field is horizontal, which means that, in the terms of our convention, the medium is bounded in the y direction. (See Figure 6.25.) Then, from our definitions of σ_1 and σ_2 (and also from Equation 6.70),

$$\left.\begin{aligned} J_x &= \sigma_1 E_x + \sigma_2 E_y, \\ J_y &= -\sigma_2 E_x + \sigma_1 E_y. \end{aligned}\right\} \tag{6.71}$$

J_y must be zero because the y direction is vertical. Hence, $E_y = E_x\sigma_2/\sigma_1$, and, by substitution,

$$J_X = \left(\sigma_1 + \frac{\sigma_2^2}{\sigma_1}\right)E_x = \sigma_3 E_x, \tag{6.72}$$

where σ_3 is the *Cowling conductivity*. If the conductivity is due to only one species of particle, such as the electrons, it is easily proved that $\sigma_3 = \sigma_0$ (as though no magnetic field were present). This is analogous to the case of a metallic conductor in a laboratory; the conductivity is not affected by a transverse magnetic field, but a potential difference (the Hall Effect) appears across the other faces.

To generalize to any latitude, conductivities σ_{xx}, σ_{xy} and σ_{yy} are defined. x, y and z are taken respectively to be magnetic south, magnetic east, and vertical, and the subscripts give the components of the electric field and of the current. Thus,

$$\left.\begin{aligned} J_x &= \sigma_{xx} E_x - \sigma_{xy} E_y \\ J_y &= -\sigma_{xy} E_x + \sigma_{yy} E_y. \end{aligned}\right\} \tag{6.73}$$

The expressions for σ_{xx}, σ_{xy} and σ_{yy} contain the dip angle (I) of the geomagnetic field. At latitudes not too close to the equator

$$\left.\begin{aligned} \sigma_{xx} &\sim \sigma_1/\sin^2 I \\ \sigma_{xy} &\sim \sigma_2/\sin I \\ \sigma_{yy} &\sim \sigma_1. \end{aligned}\right\} \tag{6.74}$$

If the ionosphere is regarded as a layer carrying horizontal currents only it is useful to integrate the conductivities with respect to height:

$$\left.\begin{aligned} \Sigma_{xx} &= \int \sigma_{xx}\,dh \\ \Sigma_{xy} &= \int \sigma_{xy}\,dh \\ \Sigma_{yy} &= \int \sigma_{yy}\,dh. \end{aligned}\right\} \tag{6.75}$$

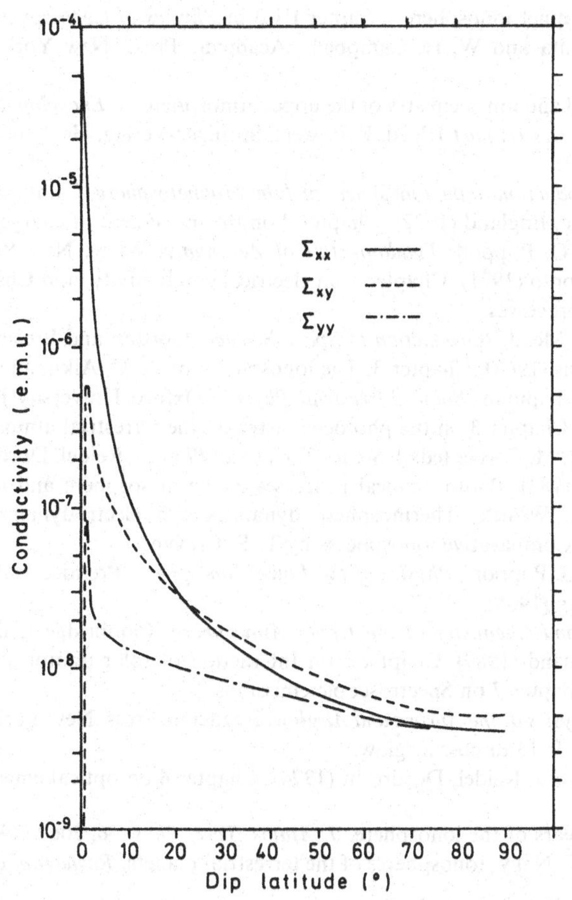

Fig. 6.26 Layer conductivities for local noon. (S.-I. Akasofu and S. Chapman (after K. Maeda and H. Matsumoto, *Solar–Terrestrial Physics*. Oxford University Press, 1972. By permission of Oxford University Press)

The variations of these *layer conductivities* are shown against dip latitude in Figure 6.26. These values could be used with equations like 6.70 to predict the currents due to an applied electric field.

Further reading

S. J. Bauer. *Physics of Planetary Ionospheres*. Springer-Verlag, Berlin, Heidelberg and New York (1973), particularly Chapters 2–5.

J. A. Ratcliffe and K. Weeks. The ionosphere, Chapter 9 in *Physics of the Upper Atmosphere* (J. A. Ratcliffe, ed.). Academic Press, New York and London (1960).

H. Rishbeth and O. K. Garriott. *Introduction to Ionospheric Physics*. Academic Press, New York and London (1969); particularly Chapter 3 on photochemical processes and Chapter 4 on transport processes in the ionosphere.

G. Schmidtke. Modelling of the solar extreme ultraviolet irradiance for aeronomic applications, in *Encyclopaedia of Physics XLIX/7* (*Geophysics III part VII*, ed. K. Rawer). Springer-Verlag, Heidelberg, New York and Tokyo (1984).

T. E. Van Zandt. The quiet ionosphere, Chapter III-3 in *Physics of Geomagnetic Phenomena* (eds. S. Matsushita and W. H. Campbell). Academic Press, New York and London (1967).

L. Thomas. The neutral and ion chemistry of the upper atmosphere, in *Encyclopaedia of Physics XLIX/6* (*Geophysics III part VI*, ed. K.Rawer), Springer-Verlag, Heidelberg, New York and Tokyo (1982).

J. A. Ratcliffe. *An Introduction to the Ionosphere and the Magnetosphere*. Cambridge University Press, Cambridge, England (1972). Chapter 7 on the movement of charged particles.

R. C. Whitten and I. G. Poppoff. *Fundamentals of Aeronomy*. Wiley, New York, London, Sydney and Toronto (1971). Chapter 7 on electrical conductivity, and Chapters 8 and 9 on ionospheric processes.

W. N. Hess and G. D. Mead. *Introduction to Space Science*. Gordon and Breach, New York, London and Paris (1968). Chapter 3, The ionosphere, by A. C. Aikin and S. J. Bauer.

S.-I. Akasofu and S. Chapman. *Solar–Terrestrial Physics*. Oxford University Press, Oxford, England (1972). Chapter 3 on the photochemistry on the terrestrial atmosphere.

R. L. Carovillano and J. M. Forbes (eds.). *Solar– Terrestrial Physics*. Reidel, Dordrecht, Boston and Lancaster (1983). Photochemical processes in the mesosphere and lower thermosphere by J. R. Winick, Thermosphere dynamics and electrodynamics by A. D. Richmond, and Comparative ionospheres by T. E. Cravens.

R. C. Whitten and I. G. Poppoff. *Physics of the Lower Ionosphere*. Prentice-Hall, Englewood Cliffs, New Jersey (1965).

M. H. Rees. *Physics and Chemistry of the Upper Atmosphere*. Cambridge University Press, Cambridge, England (1989). Chapter 2 on Interaction of solar photons, Chapter 5 on Composition, Chapter 7 on Spectroscopic emissions.

J. W. Chamberlain. *Physics of the Aurora and Airglow*. Academic Press, New York and London (1961). Chapters 9–13 discuss airglow.

A. Vallance Jones. *Aurora*. Reidel, Dordrecht (1974). Chapter 4 on optical emissions.

Special issue on fifty years of the ionosphere. *J. Atmos. Terr. Phys.* **36**, 2069 (1974).

R. W. Schunk and A. F. Nagy. Ionospheres of the terrestrial planets. *Reviews of Geophysics* **18**, 813 (1980).

D. G. Torr. Neutral and ion composition of the thermosphere. *Reviews of Geophysics* **21**, 245 (1983).

D. R. Bates. Reactions in the ionosphere. *Contemp. Phys.* **11**, 105 (1970).

E. V. P. Smith and D. M. Gottlieb. Solar flux and its variations. *Space Science Reviews* **16**, 771 (1974).

Special issue on recent advances in the physics and chemistry of the E region. *Radio Science* **10**, 229 (1975).

E. E. Ferguson. D-region ion chemistry. *Reviews of Geophysics and Space Physics* **9**, 997 (1971).

M. H. Rees and R. G. Roble. Observations and theory of the formation of stable auroral red arcs. *Reviews of Geophysics and Space Physics* **13**, 201 (1975).

7

Ionospheric phenomena at middle and low latitudes

There is no excellent beauty that hath not some strangeness in the proportion.
Francis Bacon (1561–1626), Essay 43, *Of Beauty*

The geospace environment is complex and subtle. The concepts of the previous chapter lay down the basis of ionospheric behaviour, but in the real world additional factors conspire to complicate matters. To a large extent the ionosphere varies in a regular and predictable manner, but these regularities may not always accord with simple theory. In addition, major perturbations called storms occur from time to time, and the spatial structure includes irregularities of various sizes. Indeed, it appears that the structure of the ionosphere includes all the scales of space and time that are accessible to observation.

In this chapter we will first consider the regular behaviour of the ionosphere at middle and low latitudes, including ionospheric electric currents, and will then discuss perturbations and irregularities.

7.1 Observed behaviour of the mid-latitude ionosphere

We will begin with the regularities of the ionosphere – those variations with altitude, time of day, latitude, season, and solar cycle which are repeatable and therefore predictable (at least to an extent), and also with the ionosphere's response to solar flares and eclipses. All of these relate to the larger scales of variation: vertical distances measured in tens and hundreds of kilometres, horizontal distances in hundreds and thousands of kilometres, times in hours to years.

It should not be assumed, though, that because behaviour is regular it is necessarily understood. Historically, ionospheric observations were compared with the Chapman theory of the production of an ionospheric layer (Sections 6.2.2 and 6.2.4), and major departures from the theory were christened 'anomalies'. Some of these anomalies have now been explained (for instance by taking account of ionization movements, which Chapman theory excludes) though others remain.

Today, the ionosphere at middle and low latitudes is no longer a great mystery. It is quite well understood in general terms and there are good explanations for most of its behaviour – indeed, sometimes too many explanations! However we shall encounter some phenomena, including some that have been known for many years, that still defy an adequate scientific explanation.

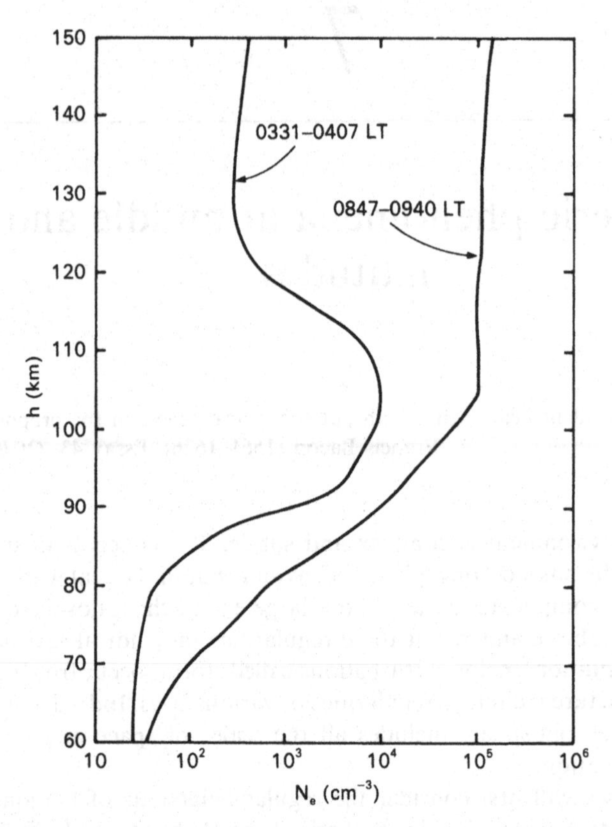

Fig. 7.1 Night and day electron density profiles through the E region, measured with the incoherent scatter radar at Arecibo, Puerto Rico (18° N, 67° W), January 1981. (J. D. Mathews, private communication)

7.1.1 E region and sporadic-E

The behaviour of the E region is close to that of an α-Chapman layer (Equation 6.16). On average its critical frequency, f_0E, varies with the solar zenith angle, χ, as $(\cos\chi)^{\frac{1}{4}}$, which means that the peak electron density, $N_m(E)$, varies as $(\cos\chi)^{\frac{1}{2}}$. Writing $f_0E \propto (\cos\chi)^n$, the value of n ranges between about 0.1 and 0.4 in practice.

Given that the E region is an α-Chapman layer, the Chapman theory can be applied to determine the recombination coefficient (α) from observations. There are three approaches:

(a) To use the relation $\alpha = q/N^2$, with an observed electron density and an observed or computed production rate;
(b) To observe the rate of decay of the layer after sunset, assuming $q = 0$;
(c) To measure the asymmetry of the diurnal variation about local noon, an effect sometimes called the *sluggishness* of the ionosphere, the time delay being given by

$$\tau = 1/2\alpha N. \qquad (7.1)$$

Such methods give values of α in the range 10^{-13} to 10^{-14} m³/s (10^{-7} to 10^{-8} cm³/s).

The E layer does not quite vanish at night, but a weakly ionized layer remains with

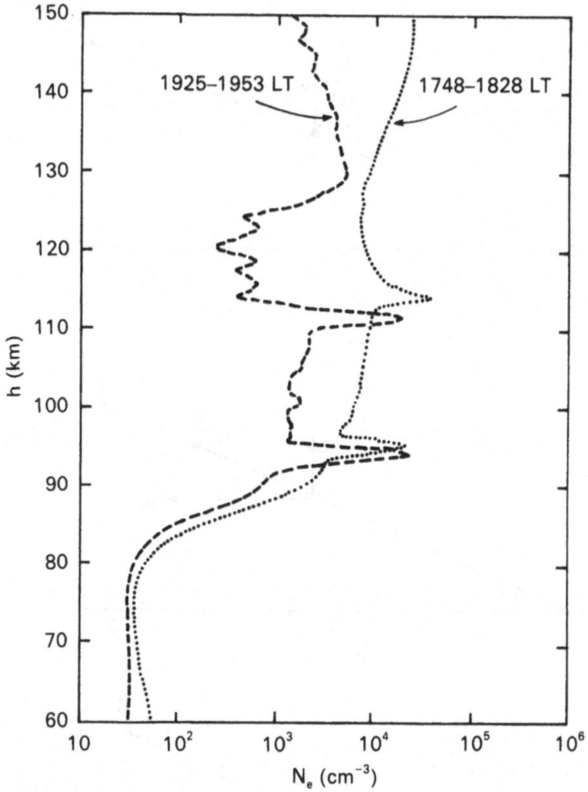

Fig. 7.2 Some sporadic-E layers observed with the Arecibo incoherent scatter radar, January 1981. (J. D. Mathews, private communication)

electron density about 5×10^9 m^{-3} (against 10^{11} m^{-3} by day). One possible cause is meteoric ionization, though other weak sources might also contribute. Figure 7.1 shows speciman electron density profiles of the E region for day and night, measured by incoherent scatter radar (Section 3.6.4–5).

The most remarkable anomaly of the E region is *sporadic-E*. On ionograms sporadic-E is seen as an echo at constant height which extends to a higher frequency than is usual for the E layer; for example to above 5 MHz. Rocket measurements, and more recently incoherent scatter radar, show that at mid-latitude these layers are very thin, perhaps less than a kilometre across. Examples are shown in Figure 7.2.

The principal cause of sporadic-E at middle latitude is a variation of wind speed with height, a *wind shear*, which, in the presence of the geomagnetic field, acts to compress the ionization. The mechanism is similar to that by which a neutral-air wind may raise or lower the F region (Section 6.5.3).

If electric fields are negligible, the motions of the ions and the neutrals are related by $e\mathbf{v} \times \mathbf{B} = m\nu_i(\mathbf{v} - \mathbf{U})$, where $\mathbf{v}$ and $\mathbf{U}$ are the ion and neutral velocities, ν_i is the ion-neutral collision frequency, m is the ion mass and $\mathbf{B}$ is the magnetic flux density. Suppose the neutral wind is horizontal, and its magnetically southward and eastward velocity components are U_S and U_E. The dip angle is I. Let the ionosphere be horizontally uniform so that we are concerned only with the resulting vertical motion of the ions, w. We also deal with the northern hemisphere.

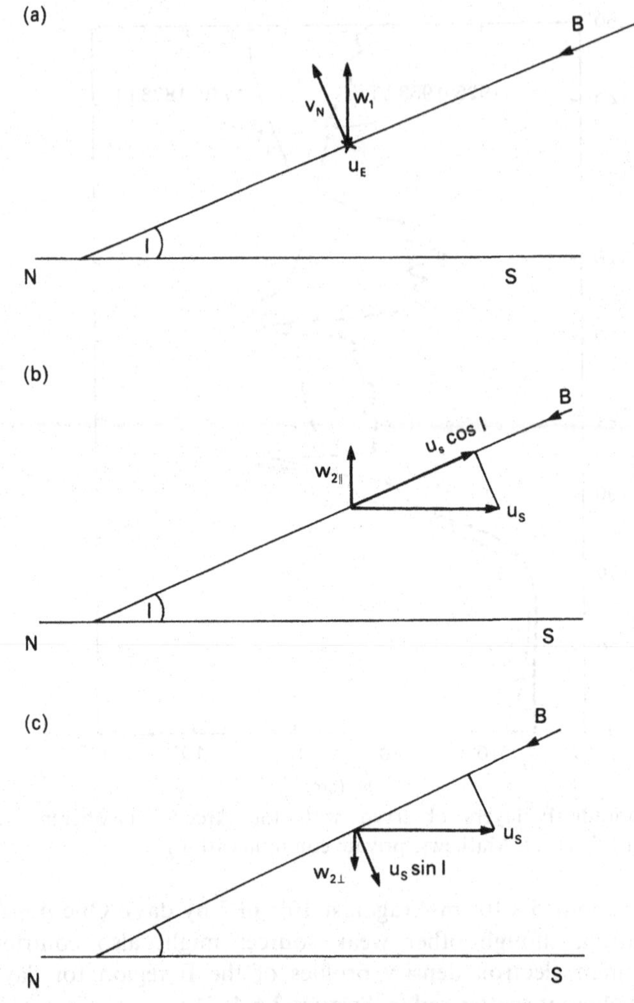

Fig. 7.3 Vertical components of ion velocity due to a neutral-air wind. (a) w_1, due to eastward wind (U_E). (b) $w_{2\parallel}$, due to component of southward wind ($U_S \cos I$) parallel to magnetic field (**B**). (c) $w_{2\perp}$., due to component of southward wind ($U_S \sin I$) perpendicular to **B**.

Consider, first, the effect of U_E. Referring to Figure 7.3(a), the force due to U_E is $m\nu_i U_E$. We are not concerned with the ion motion parallel to U_E because it is horizontal. From Equations 6.61, the ion velocity normal to both U_E and **B** is

$$v_N = \frac{m\nu_i U_E}{eB} \cdot \frac{\omega_i^2}{\nu_i^2+\omega_i^2} = \frac{U_E(\nu_i/\omega_i)}{1+(\nu_i/\omega_i)^2},$$

since $\omega_i = eB/m$. Hence the vertical component of velocity due to U_E is

$$w_1 = \frac{U_E \cos I(\nu_i/\omega_i)}{1+(\nu_i/\omega_i)^2}.$$

Now consider the effect of the southward component, U_S. This may be further resolved into components along **B** ($U_S \cos I$) and perpendicular to it ($U_S \sin I$). The first is not affected by **B**, and it gives the ions a vertical component

$$w_{2\parallel} = U_s \cos I \sin I.$$

The second component produces ion motion in the horizontal plane, with which we are not concerned, and a downward component given (again using Equations 6.61) by

$$w_{2\perp} = \frac{U_S \sin I \cos I\,(\nu_i/\omega_i)^2}{1+(\nu_i/\omega_i)^2}.$$

Adding up the three contributions to vertical ion motion,

$$w = \frac{U_E \cos I\,(\nu_i/\omega_i) + U_S \cos I \sin I}{1+(\nu_i/\omega_i)^2}. \tag{7.2}$$

Note that the characteristic frequencies ν_i and ω_i appear as a ratio. For values of $\nu_i \ll \omega_i$, $w = U_S \cos I \sin I$, as for the effect of wind in the F region (Equation 6.64). Lower down, the eastward wind also matters, and it has most influence if $\nu_i \gg \omega_i$.

If the wind varies with height, w also varies with height and this means that a wind shear acts to alter the height distribution of the ionization. Under the right conditions a thin layer can be formed.

The time scale of the process needs ions of relatively long life, and it is thought that these are metallic ions of meteoric origin (Section 6.3.2).

Figure 7.4 indicates the occurrence probability of sporadic-E against time of day and season in three latitude zones. Sporadic-E tends to be particularly severe at low latitude; some is due to instabilities in the equatorial electrojet (Section 7.3), but layers of metallic ions compressed by a wind-shear mechanism are thought to be responsible for others. At high latitude, sporadic-E is attributed to ionization by incoming energetic particles (Section 6.2.3).

Sporadic-E is significant in radio propagation since it may reflect signals that would otherwise penetrate to the F region. The irregularities within a sporadic-E layer can scatter radio waves if their dimensions are comparable to half a radio wavelength, and at times they may produce scintillation of trans-ionospheric signals (Section 7.5.1), though F-layer irregularities are the more usual cause of this phenomenon.

7.1.2 F1 region

The F1 region is another well behaved layer as viewed from the Chapman theory, though it seldom exists as a distinct peak and is more correctly called the *F1 ledge*. Its critical frequency, f_0F1, varies as $(\cos\chi)^{\frac{1}{4}}$, indicating α-Chapman behaviour. However the ledge is not always present, being more pronounced in summer and at sunspot minimum. (F1 is never seen in winter at sunspot maximum.) The explanation is to be found by comparing h_t, the transition height between α-type and β-type recombination – as discussed in Section 6.2.4 – and h_m, the height of maximum electron production rate. The F1 ledge appears only if $h_t > h_m$, and since h_t depends on the electron density (Equation 6.27) the ledge vanishes when the electron density is greatest.

7.1.3 F2 region and its anomalies

Since the F2 region has the greatest concentration of electrons, it is also the region of greatest interest in radio propagation; how unfortunate, then, that it is also the most variable, the most anomalous, and the most difficult to predict! If considered only in

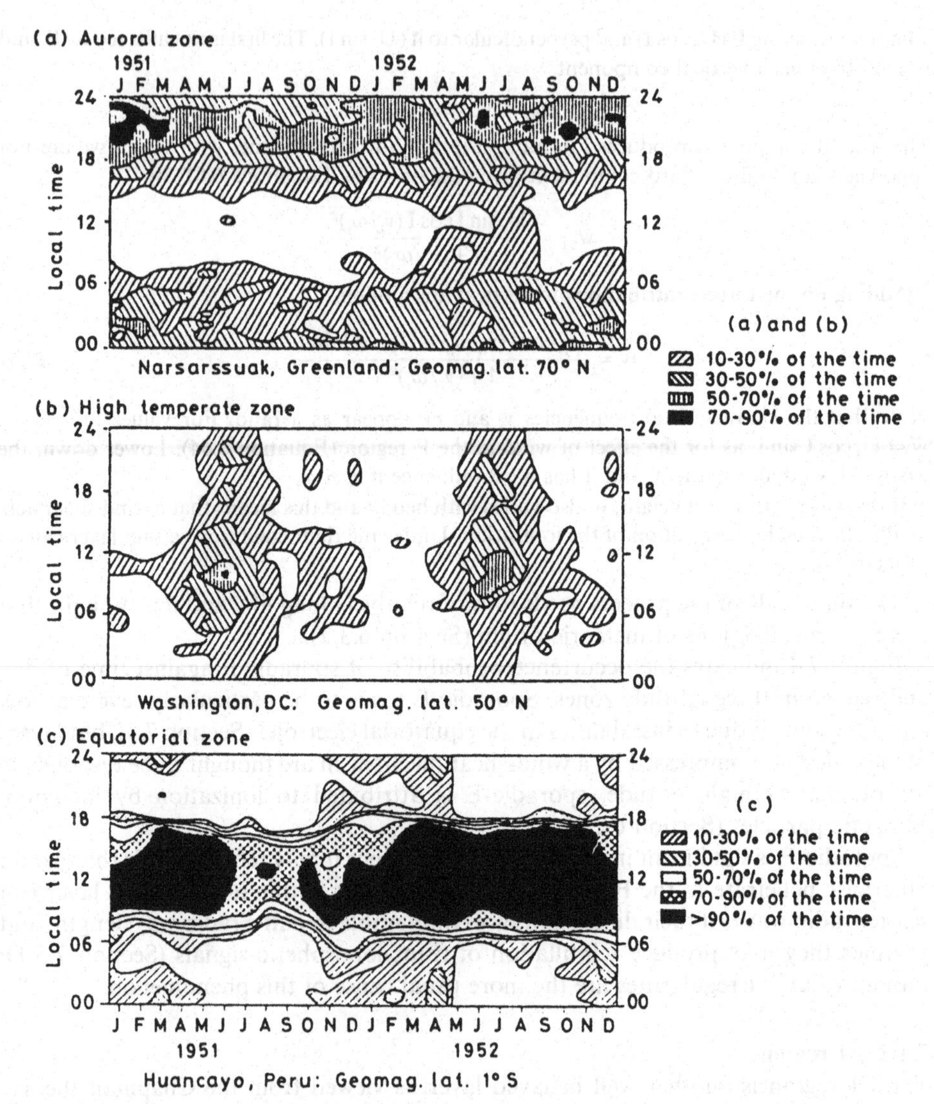

Fig. 7.4 Diurnal and seasonal occurrence of three kinds of sporadic-E. (a) The auroral kind maximizes at night but shows no seasonal variation. (b) The temperate kind peaks near noon in summer. (c) The equatorial kind occurs mainly by day but shows no seasonal preference. (After E. K. Smith, *NBS Circular* 582, 1957)

terms of the Chapman theory the F2 region's behaviour is anomalous in many ways – and this is hardly surprising since it is only by including diffusion, which the Chapman theory omits, that the existence of the F2 peak can be explained at all (Section 6.3.3). Much more is known now about the so-called *classical anomalies* of the F2 layer, and the term may be considered somewhat outdated, but we will continue to use it here because of its historical interest and because some uncertainties do remain. The anomalies may be summarized as follows:

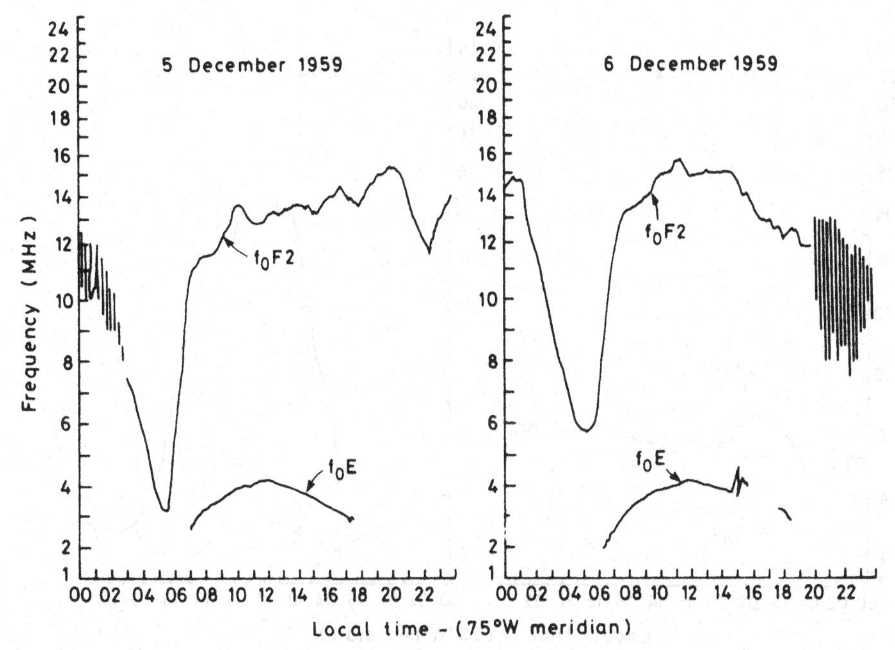

Fig. 7.5 Diurnal change of f_0F2 on successive days in December 1959 at a low-latitude station, Talara, Peru. Note, by contrast, the regularity of the E layer. (T. E. VanZandt and R. W. Knecht, in *Space Physics* (eds. LeGalley and Rosen). Wiley, 1964)

(a) The diurnal variation may be asymmetrical about noon. There can be rapid changes at sunrise but perhaps little or no change at sunset. The daily peak may occur either before or after local noon in the summer, though it is likely to be near noon in the winter. On some days a secondary minimum appears near noon between the morning and evening maxima.

(b) The daily pattern of variation may not repeat from day to day. (If it did, the next day could at least be predicted from the previous one.) Figure 7.5 illustrates points (a) and (b).

(c) There are several anomalous features in the seasonal variation. Noon values of f_0F2 and of electron content are usually greater in winter than in summer, whereas the Chapman theory leads us to expect the opposite. This is the *seasonal anomaly*, which is shown in Figure 7.6. There is also an *annual anomaly*, such that, if averaged between hemispheres, the electron densities are 20% greater in December than in June. The changing distance between Sun and Earth would account for only a 6% difference. Also, the electron content is abnormally large at the equinoxes, giving the *semi-annual anomaly*.

(d) The mid-latitude F2 region does not vanish at night, but remains through to the next sunrise at a substantial level.

A full discussion of the ways and means of the F layer would require a book to itself. However, we can summarize the situation by saying that there now appear to be four main causes of seemingly anomalous behaviour:

(a) reaction rates are sensitive to temperature;

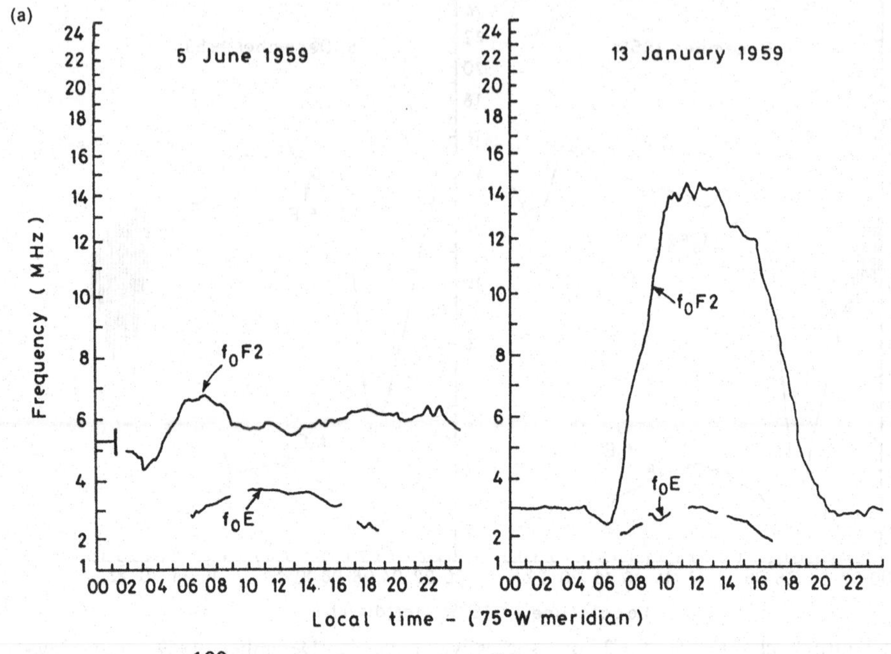

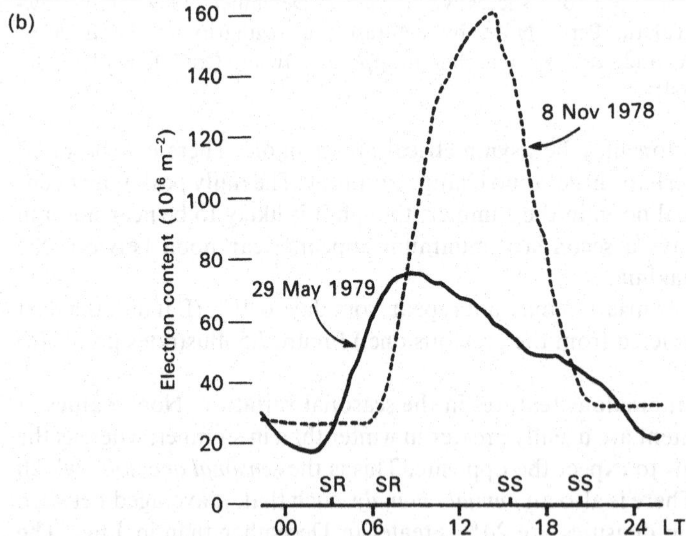

Fig. 7.6 (a) Diurnal change of f_oF2 in summer and winter at a high-latitude station in the northern hemisphere, Adak, Alaska. The F region is anomalous while the E layer behaves as expected. (T. E. VanZandt and R. W. Knecht, in *Space Physics*, (eds. LeGalley and Rosen). Wiley, 1964) (b) Summer and winter electron content measured at Fairbanks, Alaska. (R. D. Hunsucker and J. K. Hargreaves, private communication)

(b) the chemical composition varies;
(c) there are winds in the neutral air which lift or depress the layer by the mechanism given in Section 6.5.3;
(d) the ionosphere is influenced by the protonosphere and the conjugate hemisphere.

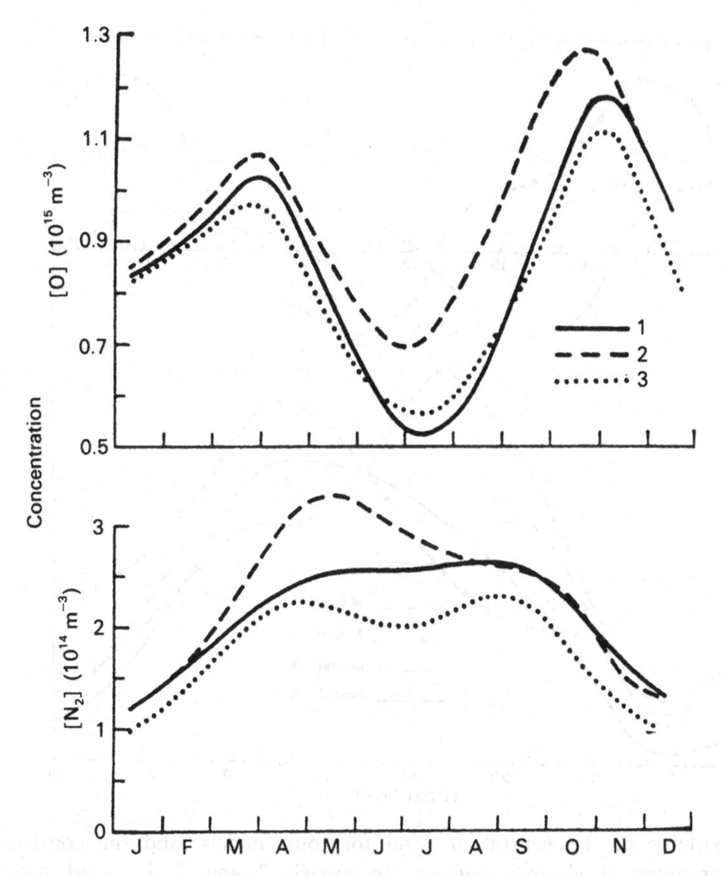

Fig. 7.7 Variations of O and N_2 concentration at 300 km at latitude 45° and 1400 LT, according to three models, of which model 2 is MSIS (see Section 4.4). (After G. S. Ivanov-Khalodny and A. V. Mikhailov, *The Prediction of Ionospheric Conditions*. Reidel, 1986. By permission of Kluwer Academic Publishers)

Reaction rates are generally temperature sensitive. The rate for the reaction

$$O^+ + N_2 \rightarrow NO^+ + N,$$

the first step in an important two-stage loss process (Equation 6.44), varies strongly with the temperature (T) of neutral N_2 and increases by a factor of 16 between 1000 K and 4000 K. The reaction rate constant k is given by

$$k = 4.5 \times 10^{-20}(T/300)^2\,\mathrm{m^3/s} \quad \text{if } T > 1000\ \mathrm{K},$$

and

$$k = 5 \times 10^{-19}\,\mathrm{m^3/s} \quad \text{if } T < 1000\ \mathrm{K}. \tag{7.3}$$

This property obviously contributes to both the persistence of the night F region and to the seasonal anomaly.

Section 6.3.3 showed that the electron production rate depends on the concentration of atomic oxygen, O, whereas the loss rate is controlled by the molecular species N_2 and O_2. Thus, increases in the ratios [O]/[O_2] and [O]/[N_2] will increase the equilibrium electron density. This idea was first proposed in the early 1960s, and satellite measurements (Fig.7.7) since then have shown that such variations do occur. The ratio

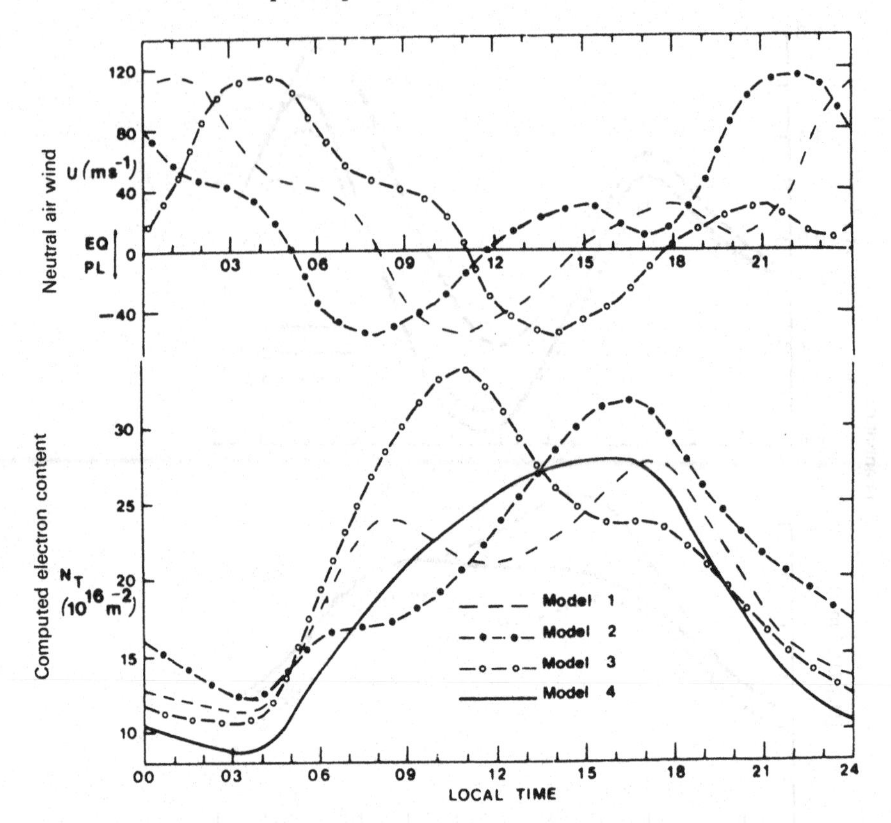

Fig. 7.8 Daily patterns of the neutral-air wind for four models, and the consequential daily variations of electron content. In models 2 and 3 the wind pattern is shifted by 3 hours with respect to model 1. The wind is zero in model 4. (Reprinted with permission from G. C. Sethia *et al.*, *Planet. Space Sci.*, **31**, 377, copyright (1983) Pergamon Press PLC)

[O]/[N_2] at 250–300 km is measured as about 6 in winter and about 2 in summer, a seasonal change amounting to a factor of three. The change of composition is attributed to the pattern of global circulation in the thermosphere (Section 4.2.4).

More than a change of composition must be involved, though, because the pattern of daily variation of the ionosphere also changes with the season. In winter months representative quantities such as N_m (the peak electron density in the layer) and N_T (the electron content) show a single daily maximum at or near to local noon. In summer such quantities vary more gradually during the middle of the day when there might even be a local minimum or 'biteout'; peaks are more likely to appear in the morning or in the evening, or sometimes in both. The meridional component of the thermospheric neutral wind, which acts to depress the ionosphere when flowing poleward and elevates it when flowing equatorward (Section 6.5.3), can exert a major influence both on electron densities and on electron content. At 300 km the neutral wind flows poleward by day and equatorward by night (Section 4.2.4) at speeds ranging between tens and hundreds of metres per second. Thus its effect is usually to depress the ionosphere and thereby increase the rate of loss by day, but to lift the region and reduce its rate of decay at night. These effects can be modelled mathematically (Figure 7.8) and they are found to be major.

The amount by which the layer is raised or lowered may be estimated by applying Equation 6.65: $\Delta h_m \sim WH^2/D$. Taking H = 60 km for the neutral scale height, $D = 2 \times 10^6$ m²/s for the diffusion coefficient, and W = 30 m/s as a typical vertical drift due to the daytime poleward wind, the layer peak is lowered by about 50 km.

It is not possible to understand the F2 region without taking account of its temperature variations. The photoelectrons produced by the ionization processes are hotter than the neutral atoms from which they were formed. This excess energy is gradually shared with the positive ions, though transfer to the neutrals is less efficient. Consequently the plasma is hotter than the neutral air, and within the plasma the electrons are hotter than the ions ($T_e > T_i$). The electron temperature can be two or three times the ion temperature by day, though by night their temperatures are more nearly equal. Diurnal temperature variations, measured by incoherent scatter radar, are illustrated in Figure 7.9. These temperature changes strongly affect the distribution of F2-region plasma. When hotter, the plasma has a greater scale height (Equation 6.34) and so spreads to greater altitudes, where it tends to persist for longer because the loss rate is smaller.

At the greater altitudes the positive ions are protons, and, as discussed in Section 6.3.3, the ionosphere and the protonosphere are strongly coupled through the charge exchange reaction between protons and atomic oxygen ions (Equation 6.52). As the F region builds up and is also heated during the hours after sunrise, plasma moves to higher altitudes where protons are created, and these then flow up along the field lines to populate the protonosphere. In the evening the proton population flows back to lower levels, where it charge exchanges back to oxygen ions and so helps to maintain the F region at night.

Via the protonosphere the magnetically conjugate ionosphere may also have an effect, since protonospheric plasma, coming mainly from the summer ionosphere, is equally available to replenish the winter ionosphere. Computations show that this is a significant source. Indeed, it is useful to treat the mid-latitude plasmasphere as consisting of winter and summer ionospheres linked by a common protonosphere; the ionospheres act as sources to the protonosphere, which in turn serves as reservoir to the ionospheres. Overall, the winter ionosphere benefits from the conjugate region in the summer hemisphere. The inter-hemispheric coupling provided by the geomagnetic field has another consequence since photoelectrons arriving from the conjugate region may help to heat the local ionosphere. This is most noticeable at sunrise when electron densities are low and thus a given amount of heat causes the greatest temperature rise. A marked increase in the slab thickness (Section 3.5.2) is observed at sunrise, which is plainly due to the arrival of photoelectrons.

One of the most remarkable things about the F region is its variability from one day to the next (Figure 7.5). In the polar regions this might not be surprising because of the sporadic nature of solar and auroral activity. But these are not dominant influences inside the plasmasphere, and the magnitude of the day-to-day changes poses a riddle whose solution so far lacks even good experimental clues. Presumably the origin must be a source in the terrestrial atmosphere or in the solar wind. Although correlations

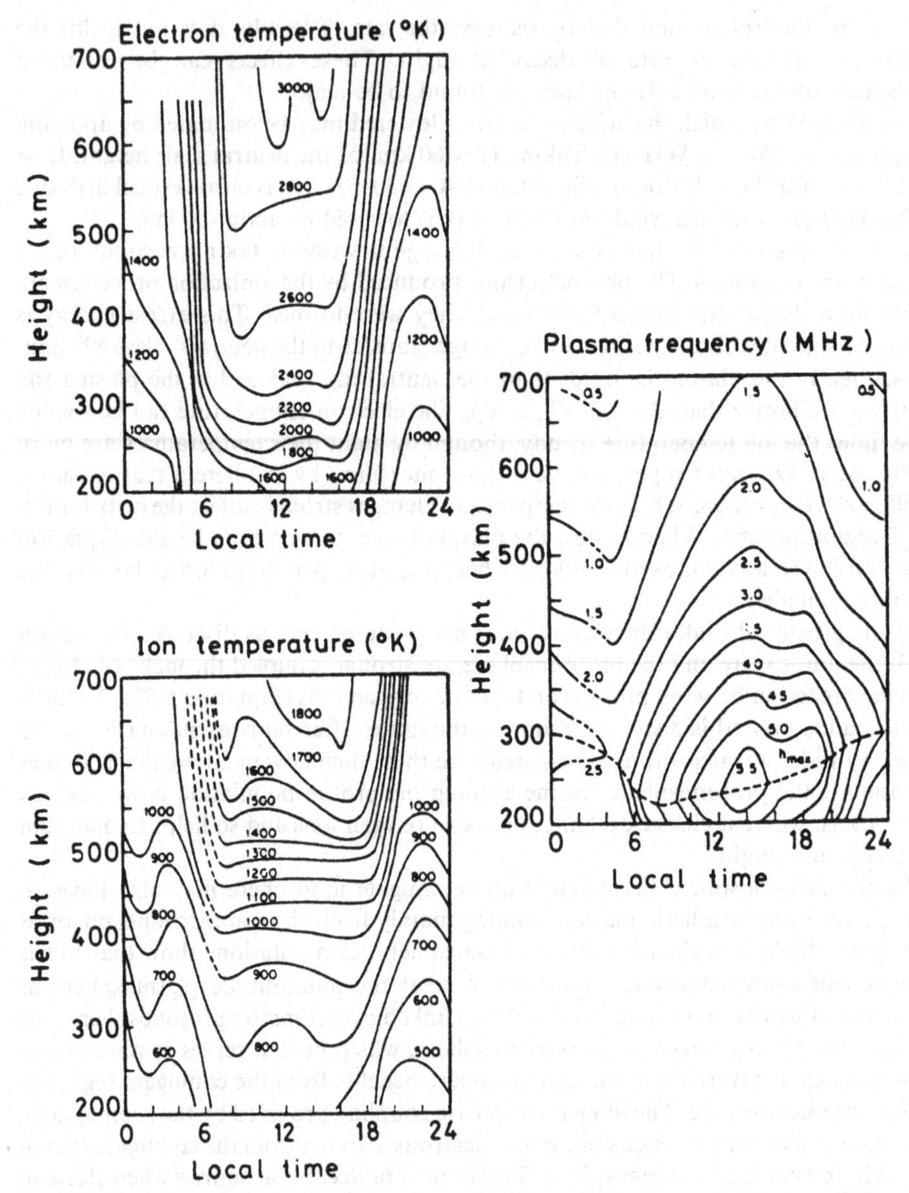

Fig. 7.9 Diurnal variation of electron temperature (T_e), ion temperature (T_i), and plasma frequency (f_N) between 200 and 700 km, measured by incoherent scatter at a mid-latitude station, April 1964. (After J. V. Evans, *Planet. Space Sci.* **15**, 1387, copyright (1967) Pergamon Press PLC)

between properties of the F region and sunspot numbers are significant over the long term, from day to day the correlation is generally poor and, by elimination, it seems most likely that the origin must lie in the neutral-air wind in the thermosphere. Even if thermospheric wind variations are accepted as a plausible hypothesis, another lurking problem is that the source of that variation is itself not known.

It appears likely that the various classical anomalies of the F2 region arise from combinations of the factors outlined above, though the details may not be clear in any particular case. With the mechanisms established in essence, further progress depends on acquiring better data on the thermosphere and on more precise modelling of their effects using computer techniques.

7.1.4 D region

The mid-latitude D region is complex chemically (Section 6.3.4), but observationally its behaviour may be deceptively simple. The region is under strong solar control and it vanishes at night. We know that, to a first approximation, VLF ($f < 30$ kHz) radio waves are reflected as at a sharp boundary in the D region because the refractive index changes markedly within one wavelength (Section 2.5.3). For VLF waves incident on the ionosphere at steep incidence, the reflection height, h, appears to vary as

$$h = h_0 + H \ln \sec \chi, \qquad (7.4)$$

where χ is the solar zenith angle. h_0 comes to about 72 km, and H to about 5 km which happens to be the scale height of the neutral gas in the mesosphere. This form of height variation is just what is predicted for a level of constant electron density in the underside of a Chapman layer and it is consistent with NO ionization by solar Lyman-α radiation.

At oblique incidence, when transmitter and receiver are more than about 300 km apart, the height variation follows a quite different pattern. The reflection level falls sharply before ground sunrise, remains almost constant during the day, and then recovers fairly rapidly following ground sunset. The reason has to do with the formation and detachment of negative ions at sunset and sunrise, coupled with electron production by cosmic ray ionization – a source with no diurnal variation. This lower part of the D region is sometimes called a *C layer*. These patterns of height variation are illustrated in Figure 7.10.

The D region is the principal seat of radio absorption, and absorption measurements (Section 3.5.1) are one way of monitoring the region. The absorption per unit height depends on both the electron density and the electron–neutral collision frequency (Equation 3.12), and the measurement gives the integrated absorption up to the reflection level. Multi-frequency absorption measurements can provide some information about the height distribution.

Generally, the absorption varies with solar zenith angle as $(\cos \chi)^n$ with n in the range 0.7–1.0. However, the seasonal variation contains an intriguing anomaly, which is that during the winter months the absorption exceeds by a factor of two or three the amount that would be expected by extrapolation from the summer. Moreover, the absorption is much more variable from day to day in the winter. This phenomenon is the *winter anomaly of ionospheric radio absorption*; it will be considered in more detail in Section 7.4.5.

7.1.5 Effects of solar flares

The solar flare serves as a useful indicator of solar activity since more occur when the Sun is more active (Section 5.2.4). It also has direct consequences in the ionosphere. In 1937, J. H. Dellinger recognized that *fadeouts* in high-frequency radio propagation

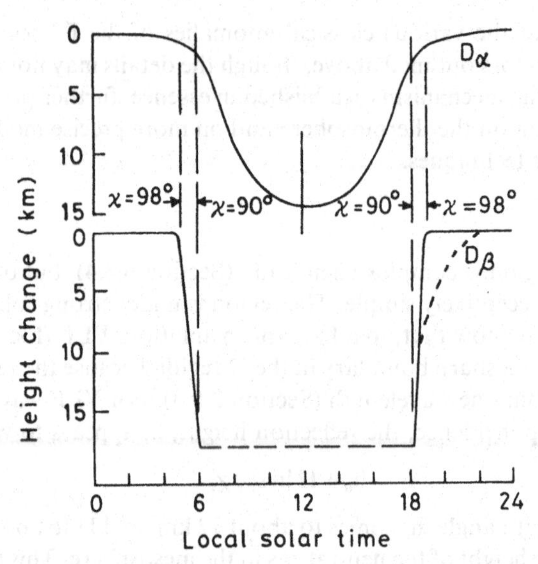

Fig. 7.10 Two kinds of diurnal behaviour of the D region inferred from VLF radio propagation at vertical and oblique incidence. The regions originally called D_α and D_β are now more usually known as D and C. The evening recovery at oblique incidence tends to be more gradual than in a simple D_β pattern and similar to the dashed curve. (After R. N. Bracewell and W. C. Bain, *J. Atmos. Terr. Phys.* **2**, 216, copyright (1952) Pergamon Press PLC)

were the result of abnormally strong absorption in the ionosphere occurring at the same time as a solar flare. The fadeout had a rapid onset and a typical duration of tens of minutes, like the visible flare. Because they begin suddenly, all the immediate effects of a solar flare are known as *sudden ionospheric disturbances* (*SID*). The absorption effect discovered by Dellinger was for long called the *Dellinger fade* but is now generally termed a *shortwave fadeout* (*SWF*).

The nature and timing of the SWF immediately provide two clues to its nature. The simultaneity between the fadeout and the visible flare shows that the cause is electromagnetic; and the occurrence of radio absorption indicates that the electron density in the D region has been increased. Thus (Figure 6.11), the enhancement is most likely to be in the Lyman-α line or in the X-ray flux. Lyman-α is enhanced by a few per cent during a flare, and for many years this was thought to be the cause of the SWF. But when it became possible to measure hard X-rays from rockets it was observed that they intensified by several powers of ten during a flare; thus the SWF is now attributed to the X-rays emitted from the flare.

There is in fact a wide range of ionospheric disturbances, as summarized in Table 7.1. Though the effects are most marked in the D region, E and F region effects can also be detected. The electron content, governed mainly by the F region, is increased by a few per cent. The range of effects shows that a considerable band of the spectrum is enhanced. Moreover, there is some difference of timing: the EUV radiation and the hardest X-rays tend to be enhanced early in the flare (at the same time as the 10 cm radio noise), whereas the softer X-rays last longer and correspond more closely with the optical flare. These phases are shown schematically in Figure 7.11.

Table 7.1 *SID phenomena*

		Technique	Effect	Region	Radiation
SWF	Shortwave fadeout	HF radio propagation	Absorption	D	Hard X-rays 0.5–8 Å
SCNA	Sudden cosmic noise absorption	Riometer	Absorption	D	Hard X-rays 0.5–8 Å
SPA	Sudden phase anomaly	VLF radio propagation	Reflection height reduced	D	Hard X-rays 0.5–8 Å
SEA	Sudden enhance-ment of atmospherics	VLF atmospherics	Intensity enhanced	D	Hard X-rays 0.5–8 Å
SFE	(Magnetic) solar flare effect	Magnetometer	Enhanced ionospheric conductivity	E	EUV and soft X-rays
SFD	Sudden frequency deviation	HF Doppler	Reflection height reduced	E+F	EUV
—	Electron content enhancement	Faraday effect	Content enhanced	F	EUV

Comparing the predicted and observed effects of flares is a useful way to verify models of the ionosphere and its production and loss mechanisms. The computation would take X-ray and EUV fluxes measured directly on a satellite, include an atmospheric model and assume the photochemistry. Figure 7.12 shows increases of electron density calculated from the enhancement of flux in six bands. Note that most of the F-region increase comes from the middle bands, 260–796 Å, whereas radiation at longer and shorter wavelength contributes more at the lower altitudes. The D-region effect is due to the hardest X-rays. (Compare this with Figure 6.11.) Figure 7.13 shows how the electron density profile changed during an actual flare. The F-region effect was by no means negligible, though it was smaller than the D- and E-region effects in percentage terms.

All SID effects cover the whole of the Earth's sunlit hemisphere and are essentially uniform except for a dependence on the solar zenith angle. Even flares that are not observed visually can be detected by their ionospheric effects and some properties of the flare deduced. One good reason for studying the effects of solar flares is that nuclear explosions also create dramatic effects in the ionosphere and it is important not to confuse the two!

7.1.6 Variations with sunspot cycle

The numbers of sunspots on the solar disc, the rate at which flares occur and the intensity of the 10 cm radio flux are all part of the evidence that the activity of the Sun goes through an 11-year cycle. These signs are accompanied by variations in the

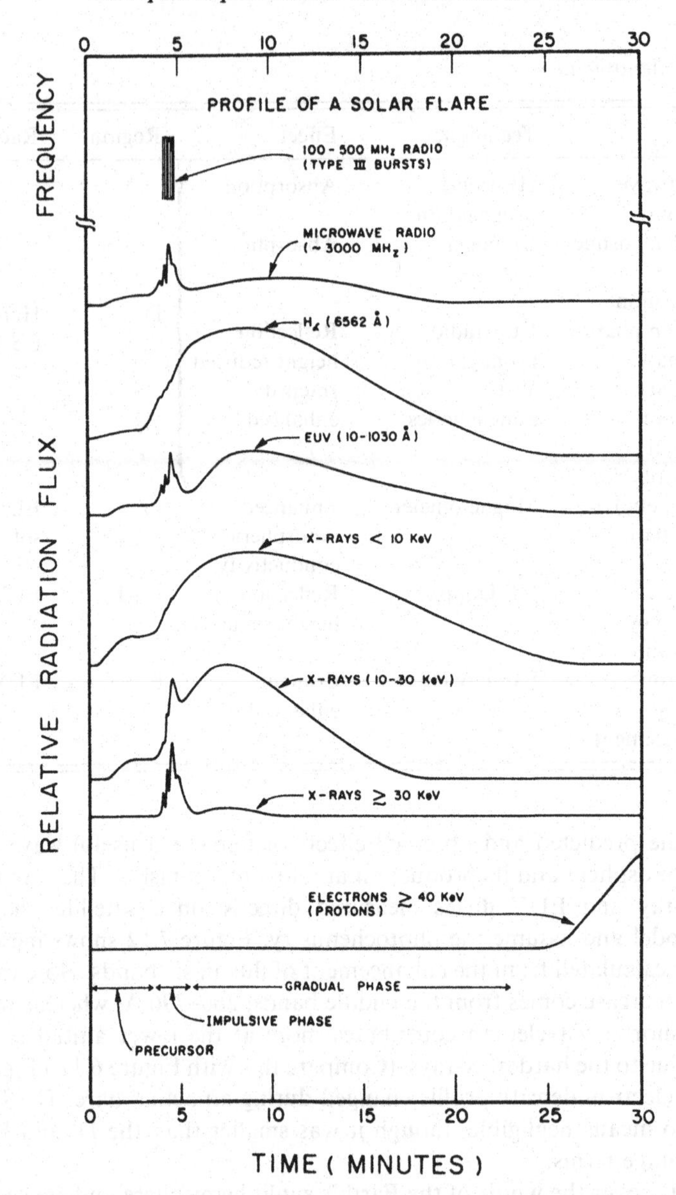

Fig. 7.11 Typical time scale of emissions from a flare. (D. M. Rust, in *Solar System Plasma Physics* (eds. Parker *et al.*). North-Holland, 1979. Elsevier Science Publishers)

ionizing radiations in the X-ray and EUV bands, which produce an 11-year cycle in the ionosphere. The temperature of the upper atmosphere also varies, by a factor of two between sunspot minimum and maximum; this leads, in turn, to large density variations at given height.

The ionospheric critical frequencies, f_oE, f_oF1 and f_oF2, all depend on the sunspot number, R (defined in Section 5.2.4). This influence can be seen in Figure 7.14. We have seen that the E and the F1 layers both behave as Chapman layers, varying with

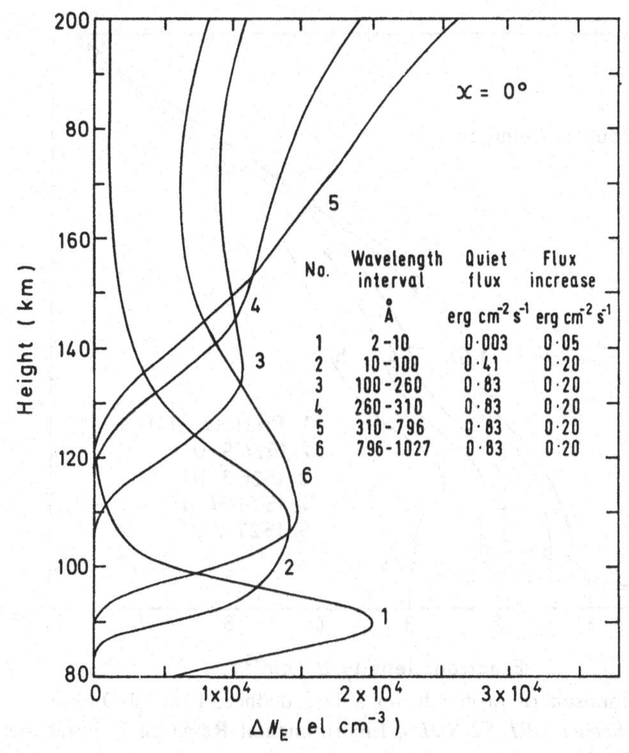

Fig. 7.12 The computed increase of electron density due to flux increases in the stated wavelength intervals. (A. D. Richmond, private communication)

solar zenith angle (χ) as $(\cos\chi)^{\frac{1}{4}}$. Taking the sunspot number into account as well gives the empirical laws:

$$f_oE = 3.3[(1+0.008R)\cos\chi]^{\frac{1}{4}}\ \text{MHz}, \qquad (7.5)$$

$$f_oF1 = 4.25[1+0.015R)\cos\chi]^{\frac{1}{4}}\ \text{MHz}. \qquad (7.6)$$

The F1 layer is nearly twice as sensitive as the E layer to the sunspot number.

From the status of E and F1 as Chapman layers it follows that the ratios $(f_oE)^4/\cos\chi$ and $(f_oF1)^4/\cos\chi$ are proportional to the ionization rates (q) in the E and F1 layers respectively. These ratios are called *character figures*. Taking $R = 10$ for a typical solar minimum and $R = 150$ for a maximum, we see from Equation 7.5 that the E-region production rate varies a factor of 2 over a typical sunspot cycle.

The F2 layer is not a Chapman layer but the dependence on R can be seen in the noon values of f_oF2. If these are smoothed over 12 months to remove the seasonal anomalies, a dependence such as

$$f_oF2 \propto (1+0.02R)^{\frac{1}{2}}\ \text{MHz} \qquad (7.7)$$

can be recognized.

The height of the maximum and the thickness of the F2 layer also vary with the sunspot cycle. Figure 7.15 shows measured noon and midnight electron density profiles for winter, equinox, and summer at solar maximum and solar minimum. Note the seasonal anomaly and the effect of sunspot cycle on N_{max}, and also the variations of h_{max} with the solar cycle and the time of day.

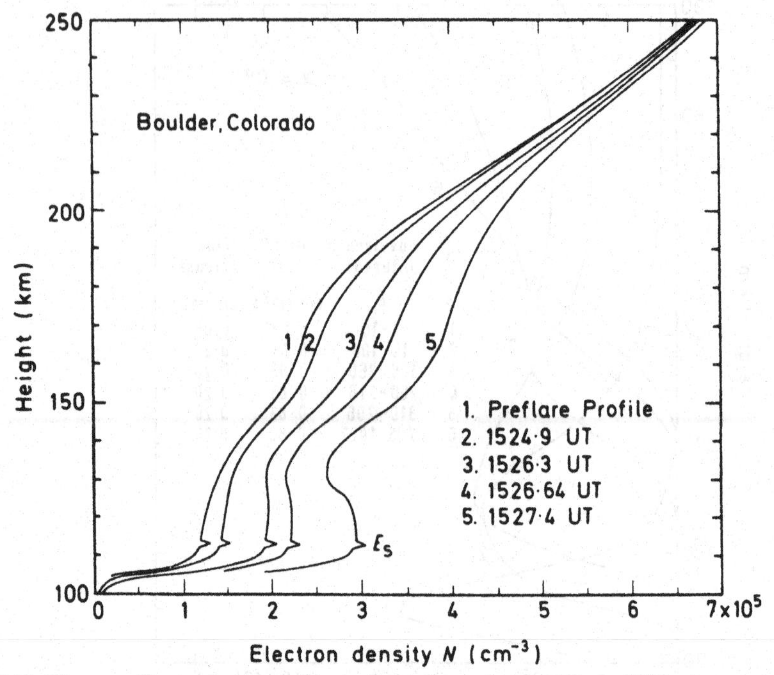

Fig. 7.13 Change of ionospheric profile during a flare, deduced from SFD observations. (R. F. Donnelly, *Report ERL* 92-*SDL*6, Environmental Research Laboratories, Boulder, Colorado, 1968)

The radio absorption measured, for example, by pulsed sounding is often taken to indicate the strength of the D region. Other parameters being constant, the absorption is observed to increase about 1% for each unit of sunspot number:

$$A(\text{dB}) \propto (1+0.01R). \tag{7.8}$$

At mid-latitude the absorption is expected to vary by about a factor of two over a sunspot cycle.

7.1.7 Eclipse effects

During a solar eclipse there is a gradual variation of ionization rate as the Moon passes in front of the Sun. In the D, E and F1 regions, where electron densities are controlled by the balance between photoionization and chemical recombination, observing the effects of the eclipse provides a means to determine the loss rate. The accuracy of the method depends on how well the distribution of ionizing radiation across the solar disc is known. In fact some EUV and X-ray emission comes from the corona, which is never totally obscured by the Moon, and X-ray emissions may come from localized regions of the disc, particularly when the Sun is active. Direct monitoring of the emissions from a satellite or rocket helps the interpretation.

F2-region effects are more difficult to interpret, and were generally considered very puzzling when only the bottom side of the layer could be observed. According to

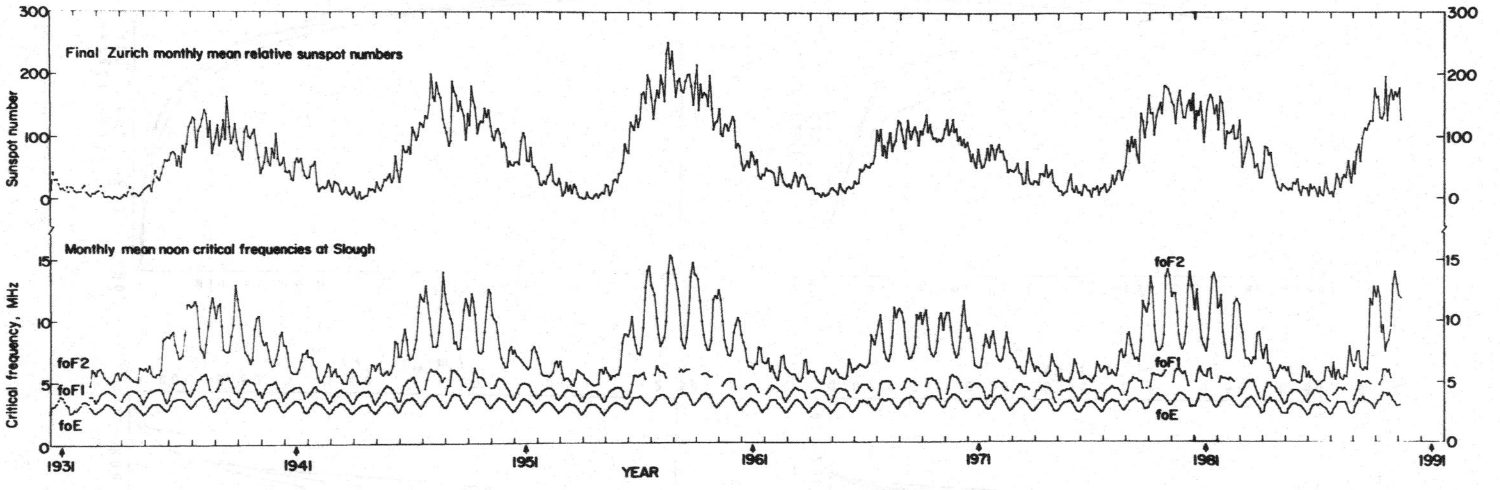

Fig. 7.14 Variation of critical frequencies over several sunspot cycles. Note, also, the seasonal modulation. The E and F1 regions peak in the summer, but F2 peaks in the winter. (Rutherford Appleton Laboratory, Chilton, England.)

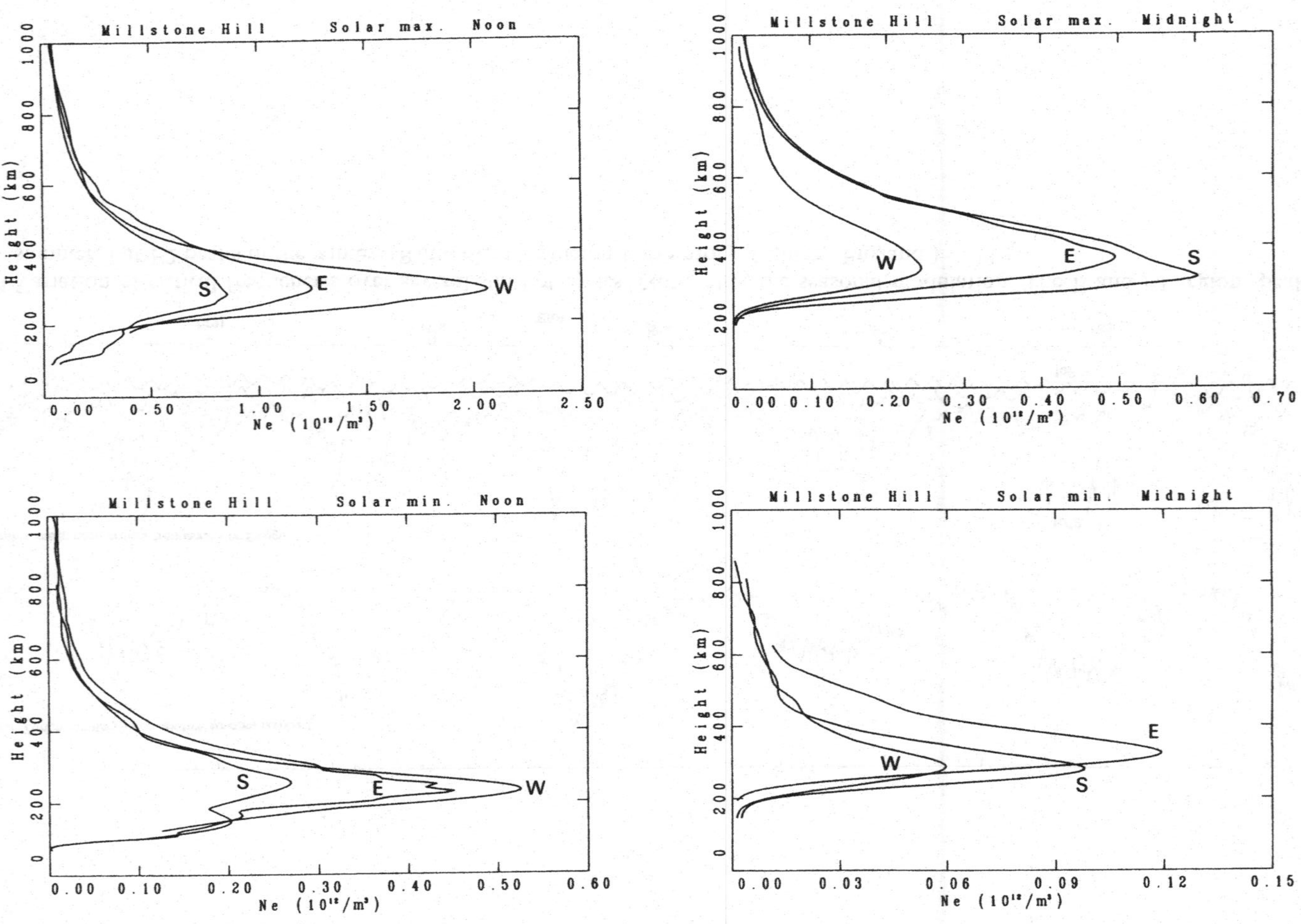

Fig. 7.15 Typical electron density profiles under magnetically quiet conditions at Millstone Hill, USA (42° N, 72° W), at noon and midnight, solar maximum and solar minimum. W: winter. E: equinox. S: summer. (M. J. Buonsanto, private communication)

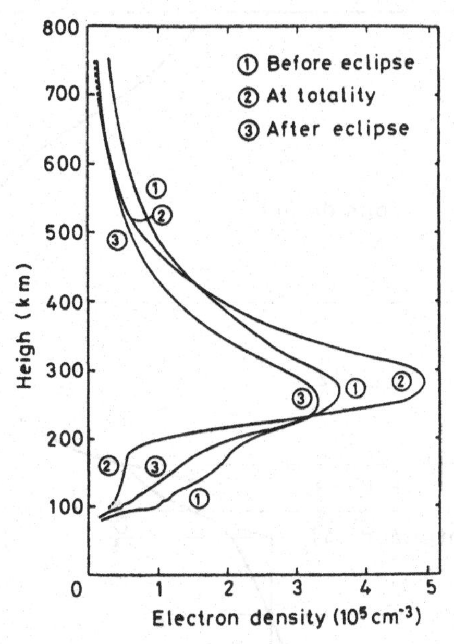

Fig. 7.16 F-region changes during the eclipse on 20 July 1963. The curves are composites from incoherent scatter data and bottomside soundings. (After J. V. Evans, *J. Geophys. Res.* **70**, 131, 1965), copyright by the American Geophysical Union

ionosondes, the quantity f_0F2 can decrease, increase, or remain unchanged during the course of an eclipse. Beacon satellite measurements, giving the electron content, and incoherent scatter radar, giving the complete profile, show that the electron content always decreases during the eclipse, and that the strange behaviour around the peak of the layer results from redistribution of ionization due to cooling. The example shown in Figure 7.16 demonstrates an increase of peak electron density coupled with a decrease of electron content.

7.1.8 Ionospheric models

Section 4.4 introduced the idea of atmospheric models. Both the empirical and the mathematical varieties are as important in ionospheric work as they are in studies of the neutral atmosphere.

The International Reference Ionosphere

The internationally recommended empirical model of the ionosphere is the *International Reference Ionosphere* , or *IRI*, which has been built up by an international working group since the late 1960s. One of the model's vital parts is a 'map' of critical frequencies. Using the international network of ionosondes, a spherical harmonic analysis was performed on each set of monthly mean critical frequencies over 24 hours of the day and for a range of sunspot numbers. The set of maps was thereby reduced to a (large) set of coefficients, from which values of the critical frequencies for any points on the Earth's surface for specified conditions could be retrieved as required. (Note that the fact that the model may reside in a computer does not alter its empirical status.) The distribution of ionosondes over the Earth's surface was, and still is, very

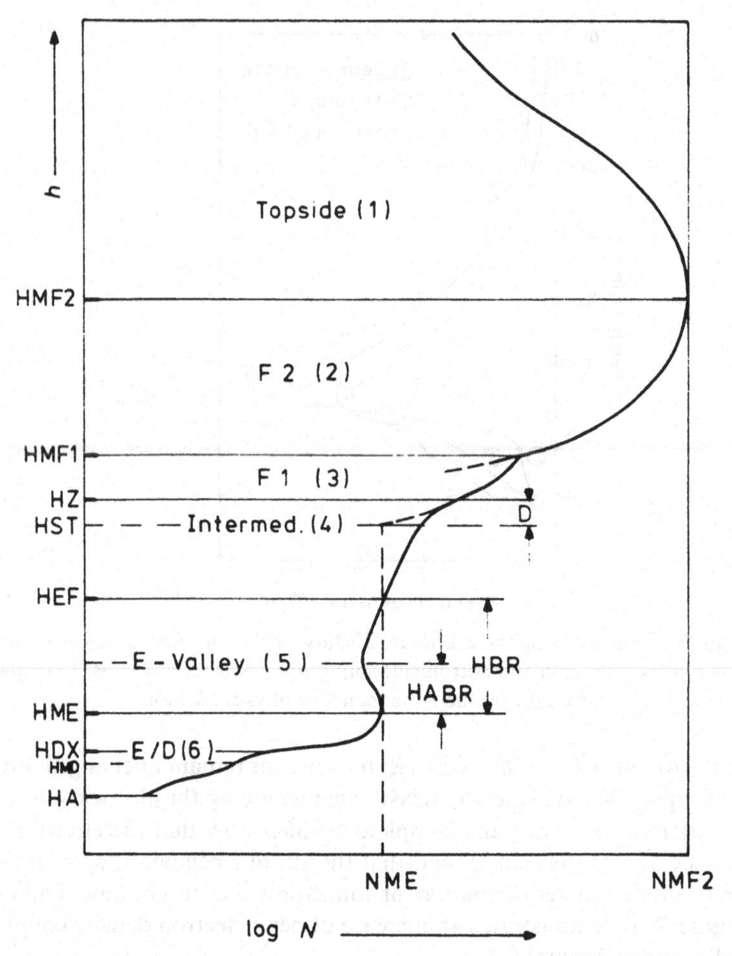

Fig. 7.17 The sections comprising the electron density profile in the International Reference Ionosphere: (1) Topside; (2) F2 region; (3) F1 region; (4) Intermediate region; (5) Valley; (6) E and D regions. (After K. Rawer, *Report UAG*-82, World Data Center A for Solar–Terrestrial Physics, Boulder, Colorado, 1981)

uneven, and some form of interpolation is needed to fill the gaps. These maps have been adopted by the CCIR (International Consultative Radio Committee) as the basis of predictions for HF radio propagation. (See Section 10.2.1.)

The IRI, however, aims to be more comprehensive. It includes the vertical profile of electron density made up of 6 sections (Figure 7.17), electron temperature, ion temperature and ion composition. Ion drift and electron collision frequency are to be added. The required input parameters are altitude, location, time of day, month, and solar activity. In addition to the ionosonde data (taken from the CCIR maps), other data have been incorporated from incoherent scatter radars, radio absorption measurements, and *in-situ* measurements by satellites carrying topside sounders and probes, and from rockets.

The IRI is subject to several restrictions. The values which it provides are 'mean' ones; they do not include any day-to-day variation ($\pm 25\,\%$), though there is an option to put in a measured critical frequency instead of that from the CCIR map. The range of solar activity is slightly restricted at the top end, and irregularities of ionospheric behaviour, such as spread-F, sporadic-E, and magnetic storms, are not included. Neither are auroral latitudes included. While earlier versions were restricted to 1000 km altitude, later versions will include a protonsphere.

The IRI is an important contribution to ionospheric science, but its predictions should not be accepted uncritically. It is still being developed, and comparisons with new data sometimes show marked discrepancies.

Electron content modelling

In matters that involve trans-ionospheric propagation, the electron content is the important quantity. Empirical electron content models are also based on a data compilation, generally known as the *Bent model*, which was developed in the early 1970s from topside-sounder measurements. The same tabular model, in fact, provides the topside in the IRI, illustrating some tendency to inbreeding amongst models. The usual approach in electron content modelling is to integrate the Bent model or the IRI along the propagation path. Again, new observations sometimes show discrepancies. The empirical electron content models have the same difficulty as the critical frequency models when it comes to day-to-day variability, and we could hardly expect some of the more pronounced structuring such as that of the high-latitude ionosphere (Section 8.2) to be included.

Mathematical models

The approach of mathematical modelling is entirely different from the empirical. For the F region and above it is based on time-dependent solutions of the continuity and momentum equations for the relevant ions – usually O^+ and H^+, with possibly He^+ as well. Ion production and loss are based on the known chemistry, and may be varied to take account of seasonal changes in the neutral atmosphere. Electron and ion temperature may be put in from an empirical temperature model or may be computed from the ionization rate. The ion motion is an essential part of an F-region model. The ion flow is constrained along the geomagnetic field, and the wind in the neutral air may contribute through air drag. The transition between O^+ and H^+ at the base of the protonosphere enables the protonosphere to build up during the day and to act as reservoir to the heavy-ion ionosphere during the night. At low and middle latitudes it is possible to include both hemispheres and thus demonstrate their interaction via the protonosphere. The dynamics of the high-latitude ionosphere pose additional problems and require an even more sophisticated approach. In the lower ionosphere the dynamics are different in nature and the ion chemistry more complex (Section 6.3.4.)

At present the principal application of mathematical models is studying the behaviour of the ionosphere to discover the important physical processes. Computation times may be too long for the model to be used in routine prediction work, and its accuracy in absolute terms is unlikely to match that of an empirical model. However, some results are being incorporated in empirical models to fill data gaps, and there is probably a future for mixed models which combine the precision and scope of a mathematical approach with the experience of a data set based on observations.

7.2 Ionospheric electric currents

7.2.1 Generation of global ionospheric currents

The basic mechanism for generating electric fields and currents in the upper atmosphere is the dynamo action of the horizontal wind system. As discussed in Section 6.5.3, the molecules of the neutral air collide with electrons and ions, forcing them initially in the direction of the wind, but their motion depends also on the geomagnetic field. In the dynamo region the gyrofrequency is smaller than the collision frequency for ions but larger for electrons. Consequently the ions are carried along with the wind and the electrons move at right angles (and more slowly). The relative movement constitutes an electric current and the separation of charge produces an electric field, which in turn also affects the current.

In general the ion current **J**, the electric field **E**, the ion velocity **v**, and the magnetic flux density **B** are related by

$$\mathbf{J} = \sigma(\mathbf{v} \times \mathbf{B} + \mathbf{E}) \qquad (7.9)$$

where σ is the tensor conductivity (Equation 6.70). Equation (7.9) reduces to Ohm's law if $\mathbf{v} = 0$. The first term represents the electric field induced by the ion motion across the magnetic field. If some of these quantities are known it is possible, in principle, to solve for others. An additional piece of information is that the current must be continuous from place to place ($\text{div}\,\mathbf{J} = 0$), though in a sense this also complicates the matter because it means that the whole of the global system has to be taken into account.

There are two approaches. Comprehensive data on **J** exist from magnetometers, whereas **v** and σ are not so well known. A distribution of **J** may therefore be taken as the starting point for computing the winds. Alternatively, a computed wind pattern may be used to evaluate a pattern of currents and winds, and the first is then verified against the measured current system.

7.2.2 S_q current system

The current system generated by the tidal motion is called S_q, meaning the variation related to the solar day under quiet geomagnetic conditions. This current system has been known and studied for many years. Indeed, the first suggestions that there was an ionized layer at some high level of the atmosphere originated with the evidence for ionospheric currents, and with Balfour Stewart's suggestion of 1882 that they were the cause of the small daily variations of magnetic field at the Earth's surface. It was also he who proposed that these currents were produced by a dynamo action due to neutral winds blowing ionization across magnetic field-lines.

Some observed S_q variations are illustrated in Figure 7.18. These are readily interpreted as overhead electric currents. The magnetic flux density due to a uniform current sheet J of infinite extent is

$$B = \mu_0 J/2, \qquad (7.10)$$

where B is in webers/m² (= tesla) and J is in amperes/m. Including a factor of $\frac{3}{2}$ for ground effects (as in Section 5.8.3), and altering the units, gives $\Delta B = 3\pi J/10$, where Δ indicates that this is a perturbation, ΔB is in nT, and J in A/km. Approximately, therefore,

$$\Delta B(\text{nT}) \sim J(\text{A/km}). \qquad (7.11)$$

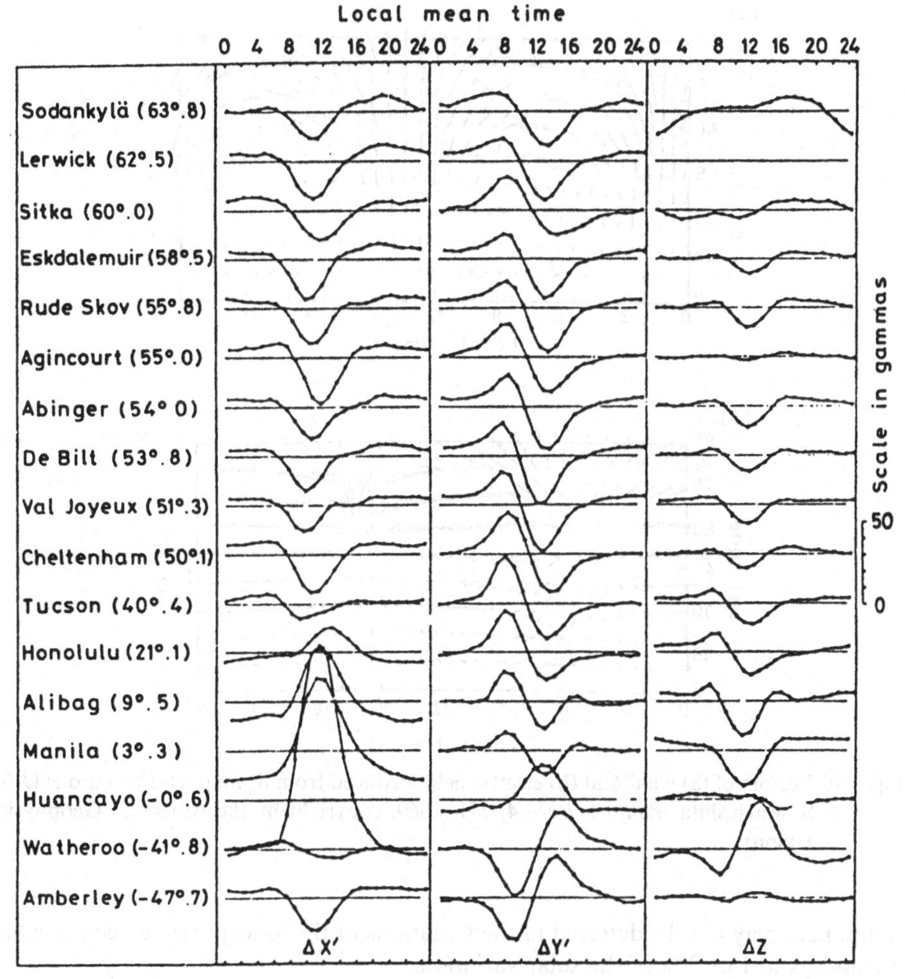

Fig. 7.18 S_q magnetic variations at several geomagnetic latitudes under equinoctial conditions. $\Delta X'$, $\Delta Y'$ and ΔZ are three normal components of the magnetogram, respectively geomagnetic north, geomagnetic east and vertical (downward). Note the large excursion of $\Delta X'$ at the geomagnetic equator, which is due to the equatorial electrojet. (E. H. Vestine, in *Physics of the Upper Atmosphere* (ed. Ratcliffe). Academic Press, 1960)

Wind and electric field variations deduced from S_q observations (Equation 7.9) are shown in Figure 7.19. The winds amount to tens of metres per second and the electric fields to a few millivolts per metre.

There is also a current system related to the phase of the moon, L, which is due to the lunar tide in the atmosphere. It is much smaller than S_q and is difficult to determine, the lunar day being 24 h 50 min on average. L depends on the magnetic activity, its amplitude increasing with disturbance level, and on the sunspot number. Over a range of sunspot numbers from 6 to 96, S (the solar component) increases 68 % and L increases only 22 %. The ratio L/S varies from 0.08 to 0.06 over this range. A lunar

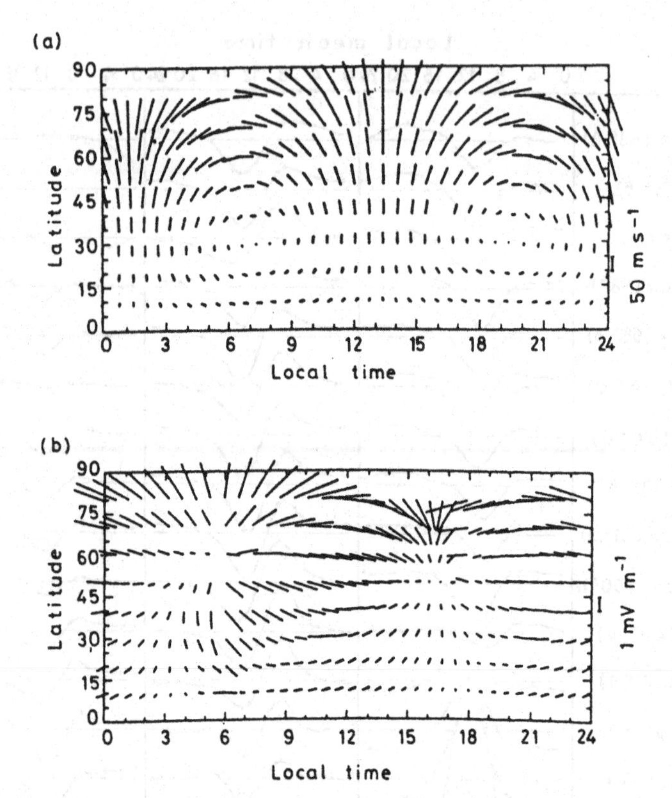

Fig. 7.19 Vectors of (a) wind and (b) electric field, deduced from S_q magnetic variations (After S. Matsushita, *Radio Science* **4**, 771, 1969, copyright by the American Geophysical Union)

component may also be detected in the variations of the ionospheric F region, where it amounts to 1 or 2% of the solar variation.

7.2.3 F-region drifts

The conductivity along the field-lines, σ_0, becomes large in the F region (Figure 6.23), and in both the E and the F regions it is larger than the transverse conductivities, σ_1 and σ_2. If we envisage an electric circuit linking the E and F regions, the generator being in the E region, it becomes obvious that, because of the high conductivity along the field-lines, most of the electric field in the E region will appear in the F region also. The consequence (Section 6.5.4) will be a drift of the F-region plasma. If the E region acts as a dynamo, the F region behaves as a motor.

F-region drifts can be measured by the spaced-receiver method (Section 7.5.2), by the drift of ionized clouds such as barium, and by incoherent scatter radar. Over the equator the drift is eastward by day and westward by night, the amplitude of the oscillation being between 50 and 200 m/s. This east–west component decreases in amplitude with increasing latitude. The north–south component shows a 12-hour component during daylight hours.

The drift and the electric field in the F region are simply related by Equation 2.40. If $B = 5 \times 10^{-5}$ Wb/m^2 (0.5 G), and v = 100 m/s, then E = 5 mV/m. Typical F-region electric fields are 1–3 mV/m in the S_q region, 20 mV/m in the polar cap, and > 100 mV/m, but variable, in the auroral zone. These last two, however, arise from magnetospheric effects rather than neutral winds, and they will be encountered again in Chapter 8.

7.2.4 Ion drag effects

The neutral air and the ionization are driven by different forces in the first instance. However they do interact through collisions, so the motion of one tends to be communicated to the other. Section 2.2.4 considered how collisions between different species of the atmospheric gas transfers momentum between them and so causes a drag force if the species are in relative motion. The force on a unit volume of particles of type 1, concentration n_1, due to collisions with particles of type 2 is

$$F = n_1 m \nu_{12}(v_2 - v_1), \tag{7.12}$$

where v_1 and v_2 are the velocities of types 1 and 2 (here assumed to be in the same direction for simplicity), ν_{12} is the collision frequency of a given particle of type 1 with particles of type 2, and m is the mass of the particles, here assumed the same for both types.

The effect of ions on neutral particles is called *ion drag*, and the ion drag force per unit volume is

$$F_{ni} = n_n m \nu_{ni}(v_i - u), \tag{7.13}$$

where v_i and u are respectively the ion and neutral-air velocities, and the subscripts n and i refer to the neutrals and the ions. The converse effect, of neutral particles on ions, is called *air drag*, and the drag force per unit volume is

$$F_{in} = n_i m \nu_{in}(u - v_i). \tag{7.14}$$

Since these forces are equal and opposite,

$$n_n \nu_{ni} = n_i \nu_{in}. \tag{7.15}$$

Below the E region the neutral-air density and ν_{in} are both relatively large, and thus the ions are carried along in the wind. In the F region the drag force is smaller, though still significant and with a difference between day and night. The neutral wind effectively blows the ionization along the geomagnetic field and thereby it alters the height of the F-layer maximum by some tens of kilometres (Section 7.1.3). In the thermospheric tide the ion drag force exceeds the Coriolis force, so that the air flows across the isobars instead of along them (Section 4.2.4), unlike the situation at ground level. The tidal wind in the thermosphere is larger by night than by day because the ion density, and therefore the ion drag, is smaller at night.

Ion drag is also able to induce motion in the neutral air, communicating to it plasma motion due to an electric field. However, because the ion density is much less than the neutral density, it takes longer for the ions to move the air than it would for the air to move the ions. (Compare Equations 7.13 and 7.14.) The time constant for the acceleration of neutrals due to collisions with the ions is approximately

$$\frac{1}{\nu_{ni}} = \frac{n_n}{n_i \nu_{in}}. \tag{7.16}$$

Table 7.2 *Values of density and collision frequency for neutrals and ions at 300 km*

	Day	Night
$n_n (m^{-3})$	9.8×10^{26}	6.6×10^{26}
$n_i (m^{-3})$	1.0×10^{12}	3.5×10^{11}
$\nu_{in} (s^{-1})$	0.6	0.4
$\nu_{ni} (s^{-1})$	6.5×10^{-5}	2.0×10^{-5}

At 300 km, the typical values of n_n, n_i, ν_{in} and ν_{ni} by day and night are as in Table 7.2. The time constant $(1/\nu_{ni})$ may be as short as half an hour by day, but is likely to be several hours by night. Thus the plasma drift can be communicated to the neutral air, but the process acts more efficiently during the day. The effect is most important in the polar ionosphere, where electric fields are greatest (Section 8.1.2).

7.3 Peculiarities of the low latitude ionosphere

Equatorial electrojet

We saw in Section 7.2.2 that the wind in the neutral air drives the S_q current system in the E region. Over the magnetic equator the conductivity is exceptionally large because of the vertical limitation of the conducting layer, and it takes the Cowling value across the geomagnetic field – which is comparable in value to the direct conductivity along the field (Section 6.5.5 and Figure 6.25). The current is therefore also larger, as evidenced by the enhanced magnetic variations at equatorial stations. This is clearly seen in Figure 7.18. This enhanced current is the *equatorial electrojet*. It is centred near 100 km altitude and is a few degrees of latitude wide. The east–west electric field is about 0.5 mV/m and the vertical field 10 mV/m. The current flows eastward by day and westward by night, but the nighttime section is less evident because the electron density is smaller by night. The westward electron velocity (corresponding to the eastward current) is several 100 metres per second. There can, however, be reversals from the normal direction by day or by night.

The high velocities of the charged particles that comprise the electrojet create instabilities and irregularities that may be detected by radar as well as by rocket observations. These E-region irregularities are discussed in Section 7.5.4.

Electrodynamic lifting

Just away from the magnetic equator, where the geomagnetic field is not quite horizontal, the electric field associated with the equatorial electrojet is communicated along the magnetic field to the F region. Here , in the motor region, it drives a plasma drift which is directed upward by day and downward by night. For an electric field of a few 100 microvolts per metre and a magnetic field of 5×10^{-5} Wb/m² (0.5 G), the vertical drift is about 10 m/s. This is called *electrodynamic lifting*.

Since the lifting operates across the field-lines rather than along them, the plasma is able to descend again along the magnetic field. This motion brings it away from the equator, as shown in Figure 7.20. This is sometimes called the *fountain effect*.

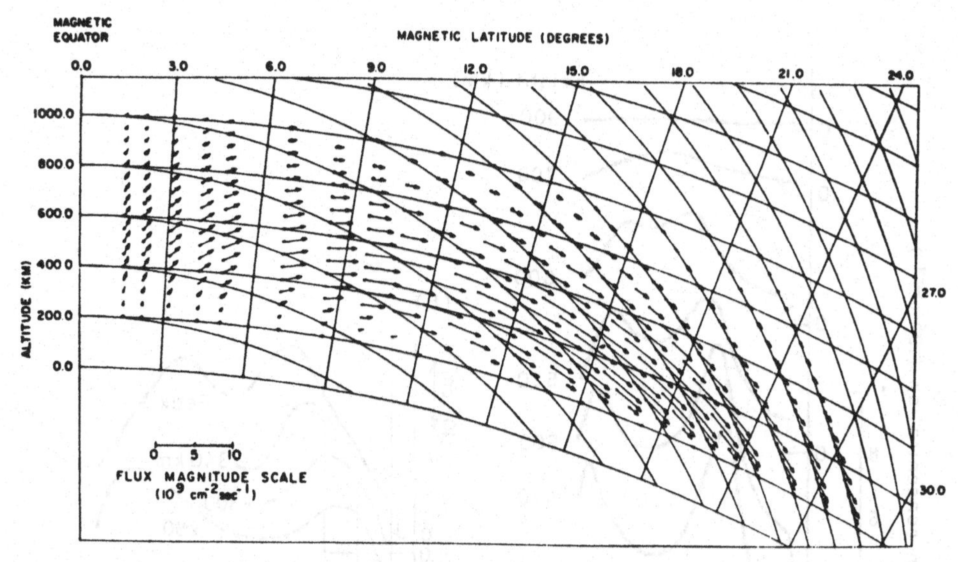

Fig. 7.20 Electrodynamic lifting and fountain effect in the equatorial F region. The upward drift over the magnetic equator is due to an eastward electric field communicated from the E region. The arrows show the flux of ionization in direction and magnitude. The computation is for noon conditions. (W. B. Hanson and R. J. Moffett, *J. Geophys. Res.* **71**, 5559, 1966, copyright by the American Geophysical Union)

Equatorial ionosphere

The fountain effect is responsible for the equatorial ionosphere's best known anomaly, the *Appleton anomaly*, or simply the *equatorial anomaly*, illustrated in Figure 7.21. Instead of the electron density coming to a maximum over the equator, as we might expect, there is actually a minimum over the magnetic equator, and two maxima 10° to 20° north and south of it. Topside and bottomside ionograms show that the separation between the peaks depends on the height and they tend to merge above 500–600 km. The theory has been well substantiated by detailed computations. Note (Figure 7.20) that whereas the plasma descends almost along the field-lines at magnetic latitudes more than about 25°, the flow is more horizontal in the region where the anomaly is strongest.

The equatorial anomaly varies during the day. There is a maximum about 1400 LT, and a second, often larger, peak occurs in the late evening during years of sunspot maximum. The northern and southern peaks are not always equal, and this is thought to be due to inter-hemispheric wind in the neutral air.

One type of 'sporadic-E' observed up to about 30° latitude is attributed to layers of metallic ions (Section 7.1.1) and is characterized by a slow descent from the F to the E region and a drift from middle latitude towards the equator. Electrodynamic lifting and tidal effects operate in the dynamics of these layers, effecting a large scale circulation of the metallic atoms which are meteoric in origin. Since they are not confined to the E region, layers of this kind are better referred to as *intermediate layers*.

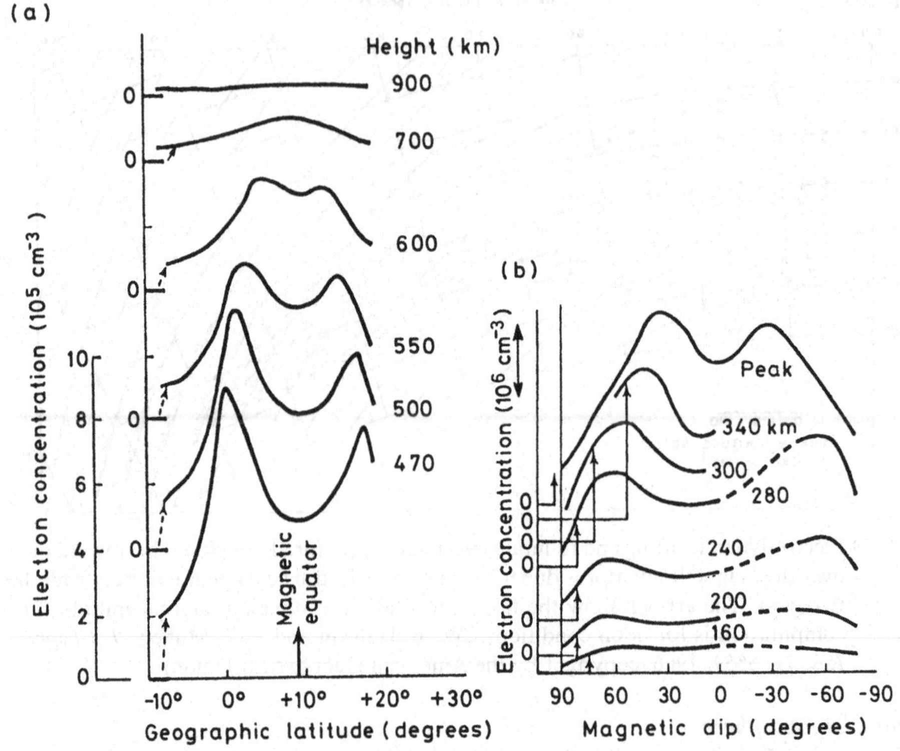

Fig. 7.21 Latitudinal variation of electron density across the equatorial anomaly at several altitudes: (a) above h_{max} (from topside ionograms); (b) below h_{max} (from bottomside ionograms). (J. A. Ratcliffe, *An Introduction to the Ionosphere and Magnetosphere* Cambridge University Press, 1972; after D. Eccles and J. W. King, *Proc. IEEE* **57**, 1012, 1969, and S. Croom *et al.*, *Nature* **184**, 2003, 1959)

7.4 Storms

7.4.1 Introduction

The upper atmosphere is prone to disturbances which, by analogy with conventional weather, are called *storms*. In general terms, a storm is a severe departure from normal behaviour lasting, usually, from one to several days. There are several phenomena which fit the definition, and we shall consider four – the magnetic storm, ionospheric storms of the F region and the D region, and the winter anomaly of the D region. Storms are not a recent discovery – the magnetic storm has been known, if not by that name, since the 18th century – but it is only in recent years that there has been much progress in explaining them. However there are remaining questions, and it cannot yet be said that ionospheric storms are fully understood.

There are, for example, clear phenomenological connections between some of the storms (compare the magnetic and F-region varieties, for instance) but the physical connections are less obvious. Part of the difficulty is that a chain of events is involved; we may have the solar wind as the initial cause, consequences in the magnetosphere, and then more than one kind of effect in the ionosphere, with perhaps a contribution coming up from the neutral troposphere or stratosphere. Altogether, storms have been

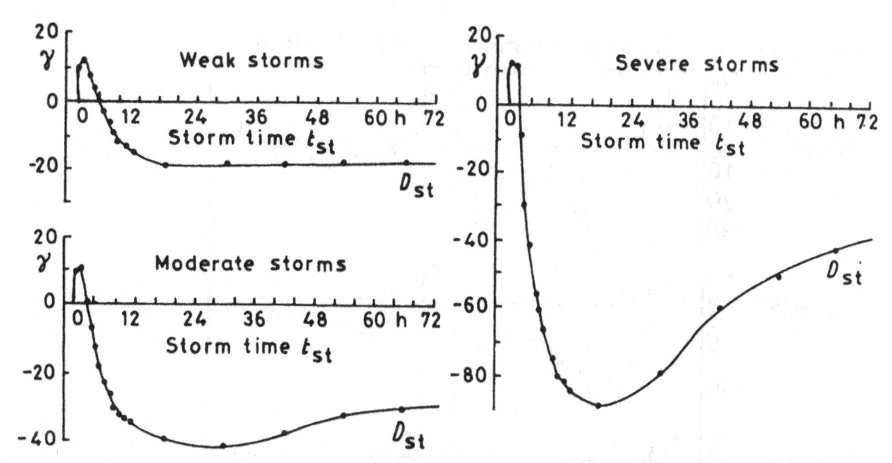

Fig. 7.22 Representative magnetic deviations during magnetic storms. (After M. Sugiura and S. Chapman, *Abh. Akad. Wiss. Göttingen Math. Phys.* **Kl.** Spec. Issue **4**, 53, 1960)

difficult to get to grips with and some major problems remain. This is not necessarily due to any shortage of theories, but may, rather, represent a difficulty in deciding between a number of good candidates!

7.4.2 Magnetic storm and the D_{st} index

During a period of high geophysical activity magnetometers at middle and low latitude tend to show characteristic patterns like those in Figure 7.22. We may call this the *classical magnetic storm* and it consists of three phases:

(a) an increase of magnetic field lasting a few hours only;
(b) a large decrease in the H component building up to a maximum in about a day;
(c) a slow recovery to normal over a few days.

The first part, the *initial phase*, is caused by the compression of the magnetosphere on the arrival of a burst of solar plasma (as in the Chapman and Ferraro theory – Section 5.3.1). The second part, the *main phase*, is due to the ring current which was described in Section 5.8.3. Equation 5.46 relates the magnitude of the ring current to the size of the magnetic depression at the ground.

The D_{st} index of magnetic storms is derived from low-latitude magnetograms. In units of nT ($= \gamma$), it simply expresses the reduction of the magnetic H component at the equator due to the ring current, and it serves as a useful indicator of the intensity and duration of individual storms.

7.4.3 The F-region ionospheric storm

Phenomena

The F-region ionospheric storm resembles the magnetic storm superficially, though its mechanisms can hardly be the same in detail. Its characteristics have been established from ionosonde data as variations of the maximum electron density, N_m(F2), and the height of the maximum, h_m(F2), and from the variations of electron content, I, measured by trans-ionospheric propagation techniques (Section 3.5.2). As in the magnetic storm there is an initial *positive phase* lasting a few hours, when the electron

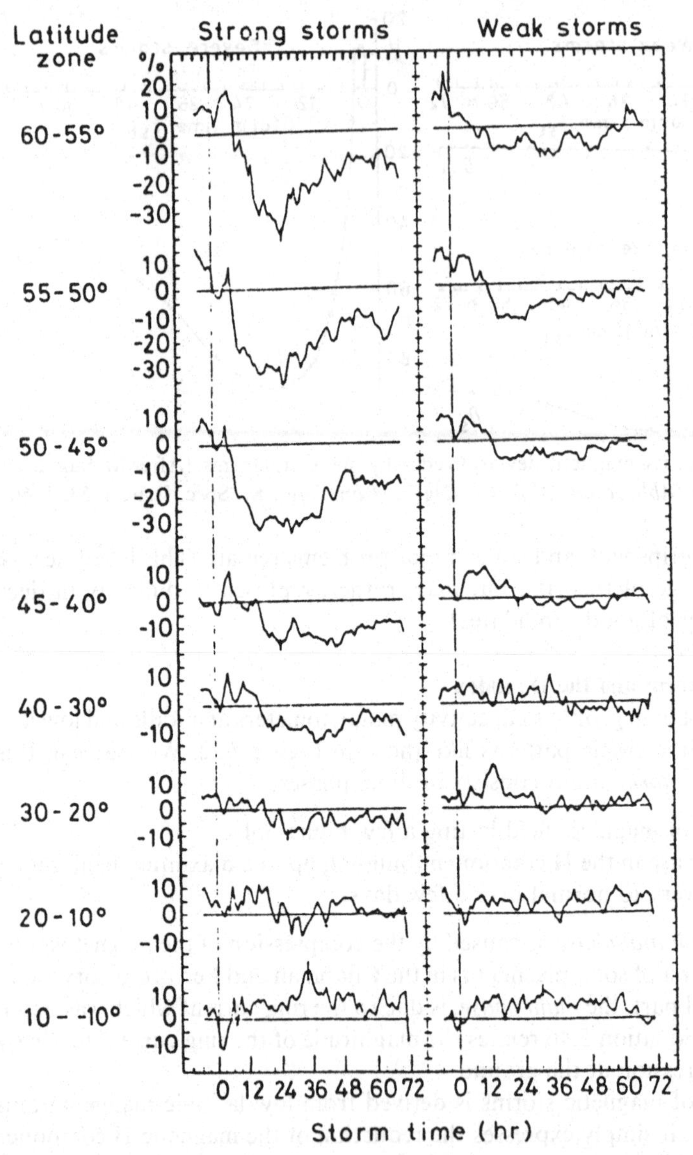

Fig. 7.23 Storm-time (D_{st}) variations of maximum electron density (N_mF2), in eight zones of geomagnetic latitude. The ordinate is the percentage deviation from quiet-day behaviour. (After S. Matsushita, *J. Geophys. Res.* **64**, 305, 1959, copyright by the American Geophysical Union)

density and the electron content are greater than normal, and then a *main* or *negative phase* when these quantities are reduced below normal values. The ionosphere gradually returns to normal over a period of one to several days in the *recovery phase*. Figure 7.23 shows variations of maximum electron density at several magnetic latitudes. The storm has greatest effect at middle and high latitude. The beginning can be sudden or gradual, the term *sudden commencement* being used, as for magnetic

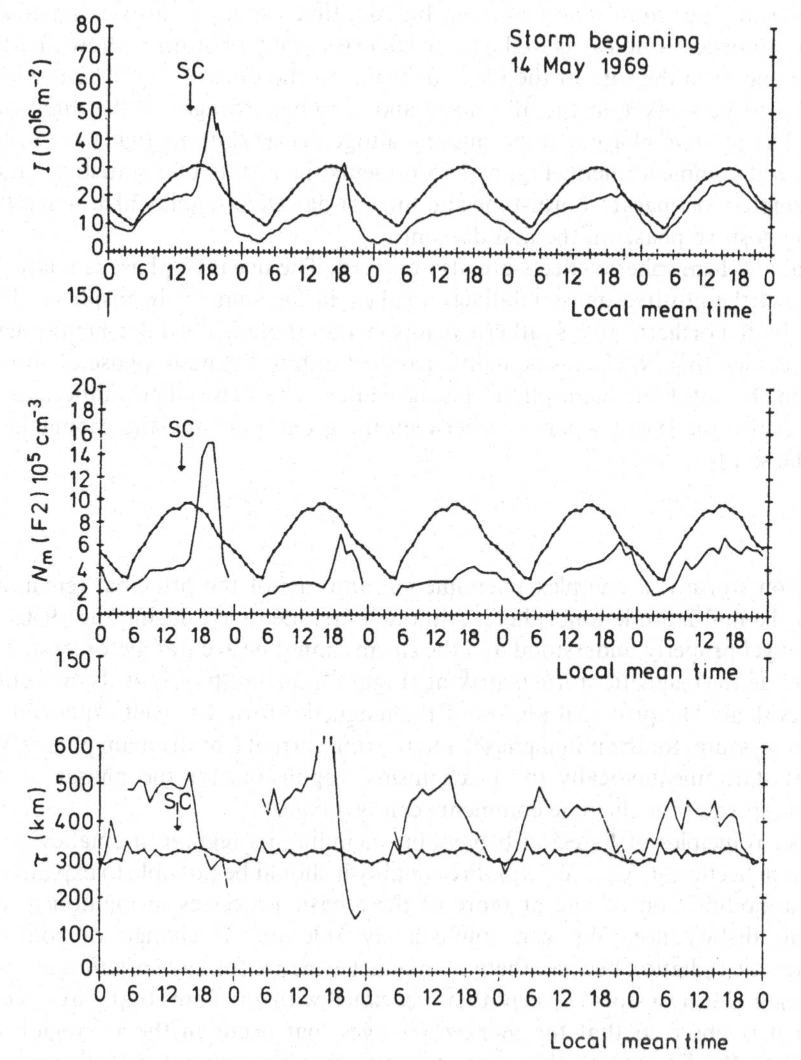

Fig. 7.24 Electron content, electron density and slab thickness at a mid-latitude station during an F-region storm. SC marks the time of the sudden commencement. The 7-day mean is also shown to indicate normal behaviour. (M. Mendillo and J. A. Klobuchar, *Report AFGRL-TR*-74-0065, US Air Force, 1974)

storms, to describe the former. Ionosondes at middle latitudes show an increased height of apparent maximium, h′(F2), though real-height analysis attributes this mainly to greater group retardation (Section 3.5.1) below the peak rather than a genuine lifting of the region. The slab thickness ($\tau = I/N_m$) does increase, however, showing that the F region broadens during the negative phase; this is consistent with increased group retardation of a radio pulse. Figure 7.24 compares electron content, electron density, and slab thickness in a typical mid-latitude storm.

Obviously, the progress of the storm can be related to the time since its commencement, the *storm-time variation*, but the time of day is also a significant parameter. Statistical studies, as well as case histories of major storms, show that the magnitude and even the sign of the effect depends on the time of day. The negative phase tends to be weaker in the afternoon and evening, stronger in the night and morning. The positive phase is often missing altogether at stations that were in the night sector at commencement. Figure 7.25 presents the results of a statistical study which attempted to separate storm-time and time-of-day effects, and which suggests a co-rotating positive phase on the first day only.

Seasonal and hemispheric effects are also marked. The negative phase is relatively stronger, and the positive phase relatively weaker, in the summer hemisphere. This holds for both northern and southern hemispheres, though the inter-hemispheric difference is such that N_m(F2) is actually increased during the main phase of storms occuring in the southern hemisphere during winter. The hemispheric difference is probably due to the larger separation between the geographic and the geomagnetic poles in the south.

Theories

The F-region storm is a complex phenomenon and one of the principal remaining puzzles of the mid-latitude ionosphere. Although first identified during the 1930s its cause is not yet properly understood and the storm cannot be accurately forecast. The resemblence to the magnetic storm is striking (Figure 7.26) but the physical connection is not so evident. The principal phases of the magnetic storm are well explained by solar-wind pressure (for the initial phase) and the ring current (for the main phase). We cannot yet state unequivocally the mechanisms responsible for the phases of the ionospheric storm, though some comments can be made.

The basic principles of F-region behaviour, including its regular anomalies, were introduced in Sections 6.3.3 and 7.1.3. Presumably it should be possible to explain the storm as a modification of one or more of these basic processes in relation to the geophysical disturbance. We can immediately rule out a change of ionizing electromagnetic radiation because there is no direct evidence for any variation of solar EUV emission during a storm (except that associated with any flares that may occur). Moreover it is observed that the marked changes that occur in the F2 region do not extend to the F1 region; this, also, suggests that the cause is not a variation of the ionizing flux. It is also clear that the immediate cause(s) of the electron density and content changes must be 'local' (where this term includes the magnetic flux tube into the protonosphere), though that change may itself be a response to far away events. Because the ionospheric storm tends to occur at the same time as a magnetic storm, there must be some sort of connection with the state of the magnetosphere, or with some disturbance of the high-latitude atmosphere and ionosphere due to the storm.

The major contenders come down to the neutral wind, conditions in the protonosphere, and heating from the magnetosphere. A change in the thermospheric wind will lift or depress the ionosphere and alter the atom/molecule ratio in the neutral air, in both cases changing the relation between ion production and loss. If the protonosphere vanishes or is severely reduced it cannot act as reservoir to the ionosphere. If the medium is heated the recombination rate will be increased.

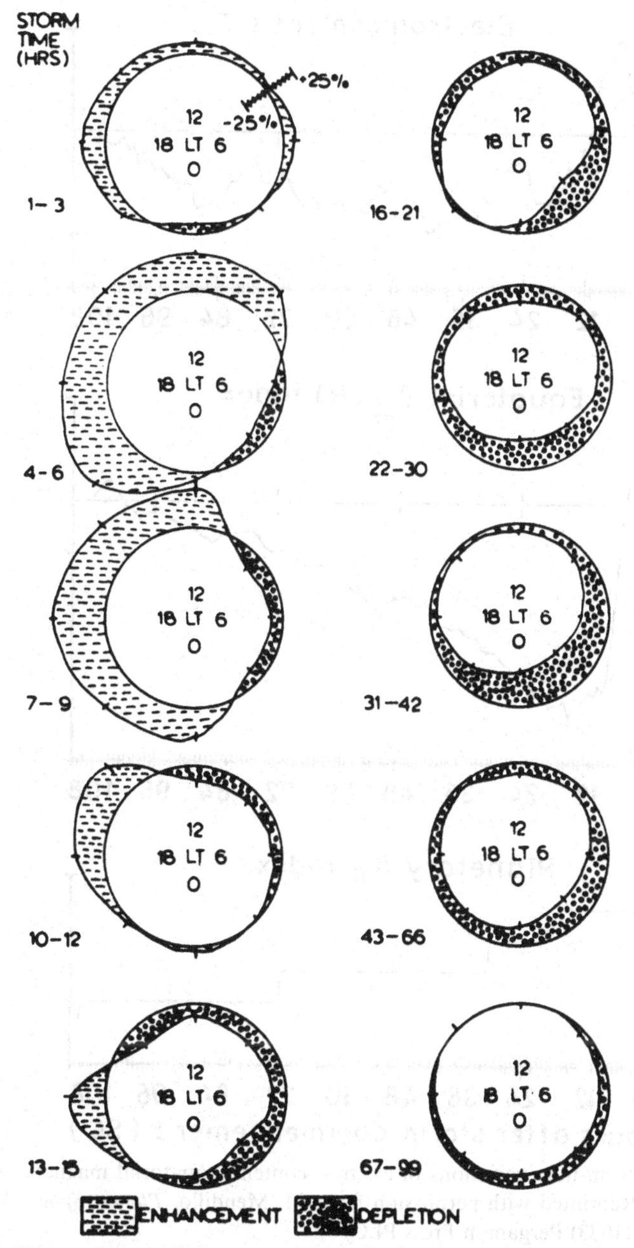

Fig. 7.25 Median variation of electron content (from a normal day) with respect to local time and storm time. The transition between positive and negative phases co-rotates with the Earth during the first 18 hours. (J. K. Hargreaves and F. Bagenal, *J. Geophys. Res.* **82**, 731, 1977, copyright by the American Geophysical Union)

It is known that the mid-latitude trough moves to lower latitude during a magnetic storm, which is consistent with increasing depletion of the protonosphere. The protonosphere normally serves to replenish the ionosphere by night, and this function

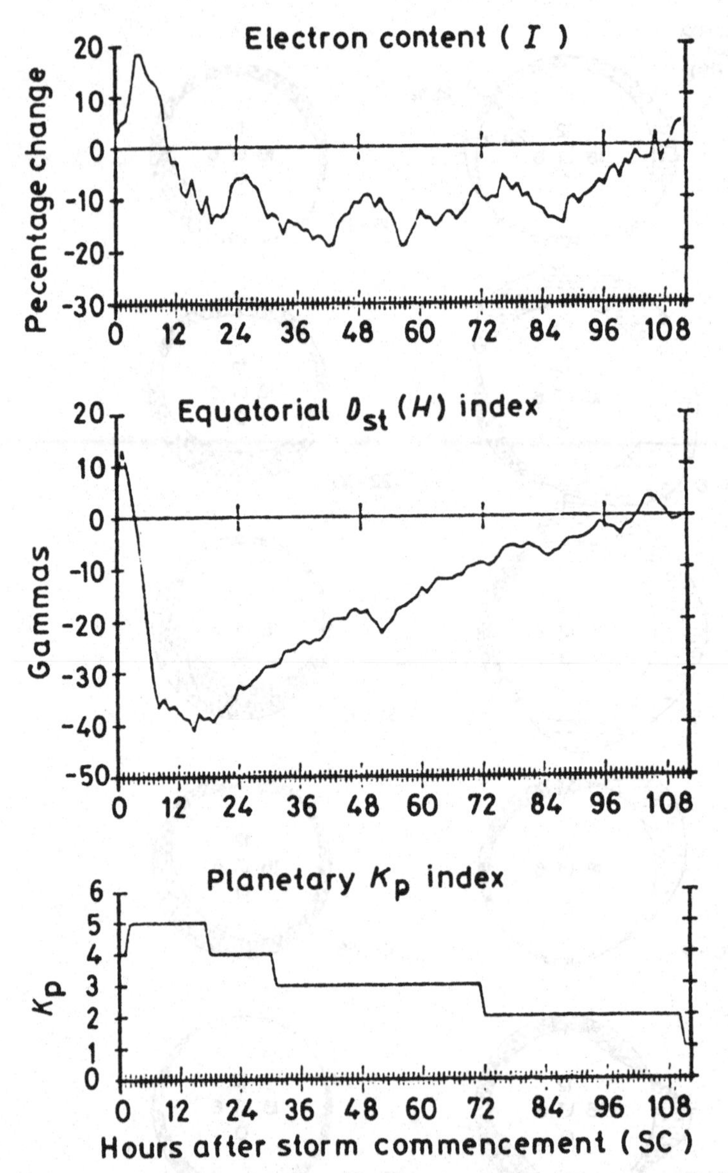

Fig. 7.26 Average storm-time variations of electron content, equatorial magnetic index (D_{st}), and K_p. (Reprinted with permission from M. Mendillo, *Planet. Space Sci.* **21**, 349, copyright (1973) Pergamon Press PLC)

is diminished if the protonosphere is depleted. The effect would be to reduce the ionospheric content between L values 6 and 3.

The magnetosphere can also be a source of energy and heating for the mid-latitude ionosphere through the precipitation of O^+ ions from the ring current; this is an attractive mechanism because it links the progress of the ionospheric storm with that of the magnetic storm.

Mathematical modelling of the thermosphere wind system has shown that the global circulation (Section 4.2.4) is modified by the heating of the auroral zone, which increases during storms due to electrojets and the precipitation of energetic particles (Sections 8.4.2 and 8.3.4). These sources are relatively strong, and the daytime tidal flow, normally poleward, can be reversed to become an equatorward flow up to 500 m/s. The dynamic effect of an equatorward wind is to lift the F2 region, reducing the loss rate and thus increasing the electron density. This does not help with the negative phase though it could contribute to the positive one.

A change of composition is almost certainly one important factor in the negative phase. Satellite observations show that the N_2 concentration is increased and the O concentration is reduced, with their relative concentration changing by a factor of 10 to 20 during a storm, and such a change is consistent with the upward expansion of the atmosphere due to heating of the lower ionosphere as in Figure 4.20. By the mechanism of Section 6.3.3 the ratio q/β would then be reduced and the ion density with it. The molecularly enriched air could then be conveyed from high to middle latitude in the equatorward flow, but it is not at present clear whether or not this happens to a sufficient extent to explain the storm effect. It has been argued, also, that a local source of heat is required at middle latitudes, and it is thought that this may be provided by energetic ions (particularly O^+) coming from the ring current (Section 5.8.3).

Over the years a number of mechanisms, each logically sound, have been put forward to account for the F-region ionospheric storm. The task of finding the relevant one(s) has been largely a process of elimination as more observations and better modelling have come along. Present thinking favours a combination of effects, notably the precipitation of ring current ions to heat the ionosphere, and changes to the thermospheric wind to raise or depress the layer and alter the composition. But there remains more work to be done on this important, long standing, problem of ionospheric behaviour.

7.4.4 D-region storms

In the next chapter we shall see that disturbances of the D region at high latitude are caused directly by the precipitation of energetic particles from the magnetosphere. The mid-latitude D region also experiences changes; these are longer in duration, are not so well defined and do not have a clear explanation.

Most studies of the D-region storm in the ionosphere have been based on radio propagation, particularly of LF (30–300 kHz) and VLF (3–30 kHz) waves. The effect on VLF propagation is illustrated by Figure 7.27. As we saw in Section 7.1.4, VLF waves are reflected as at a sharp boundary whose height is very sensitive to the electron density in the D region. During a storm the reflection height is reduced, indicating that the electron density has been increased at heights around 70–80 km. This has been called a *storm after-effect* or a *post-storm effect* (*PSE*), since it tends to follow a magnetic storm and continues for several days after the magnetic elements have returned to normal. During a VLF after-effect the absorption of LF waves is increased, which again suggests that the lower D region has been enhanced. Effects at LF have been observed for as long as eight days after a magnetic storm.

Figure 7.28 shows some key features of the storm after-effect. The affected region is bounded on both equatorward and poleward sides, and the whole shifts poleward with

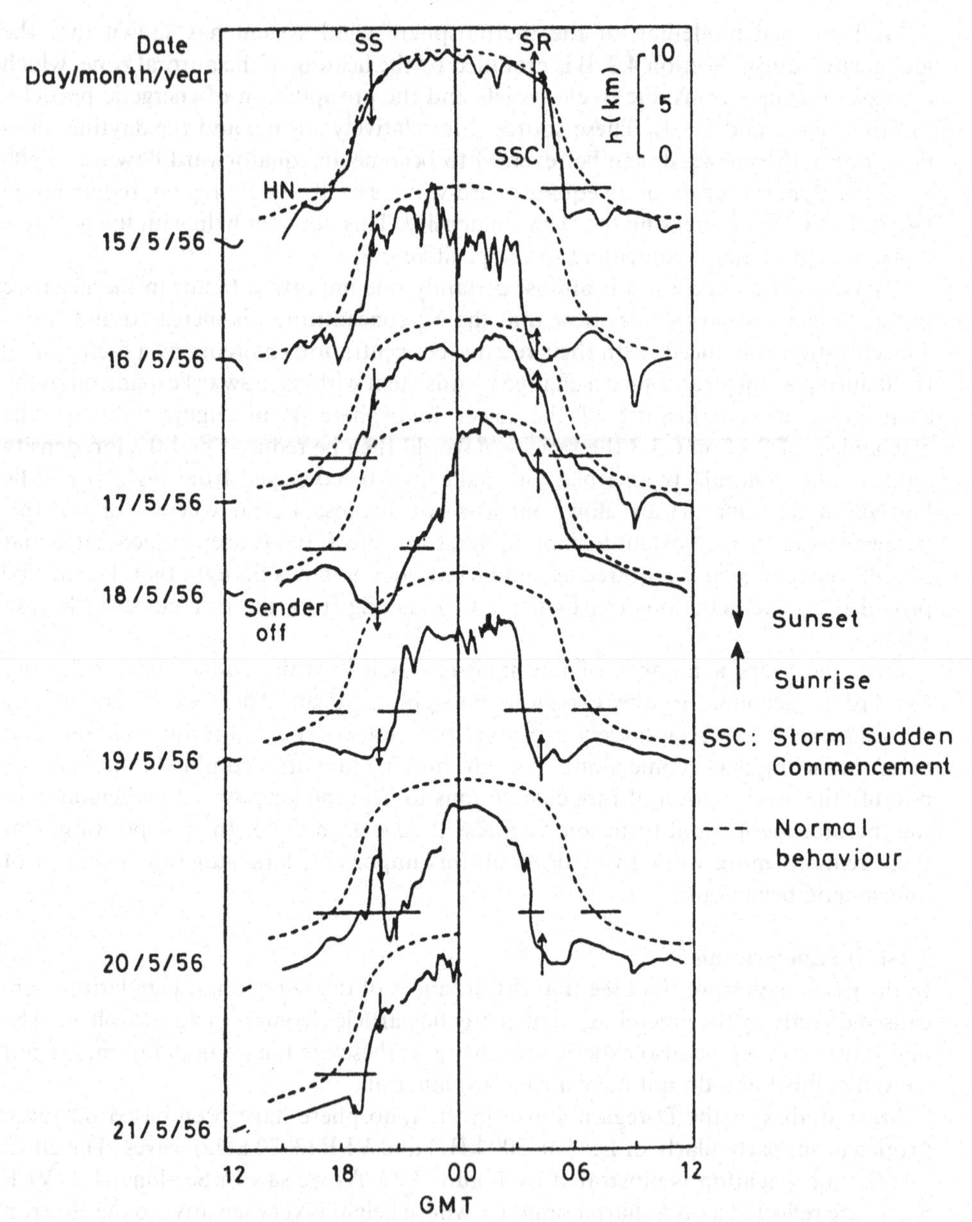

Fig. 7.27 A D-region storm as seen in VLF radio propagation: reflection height of 16 kHz waves over the path Rugby–Cambridge, UK, during a magnetic storm in 1956. (J. S. Belrose, *AGARD Report* 29, 1968)

time. This shift is well correlated with the D_{st} index, whereas the high-latitude disturbances vary with the magnetic index A_p (Section 8.4.3). The poleward edge of the affected region is related to the position of the plasmapause, and it has been observed that the occurrence, duration and magnitude of the effect depend on the direction (i.e. inward or outward) of the inter-planetary magnetic field.

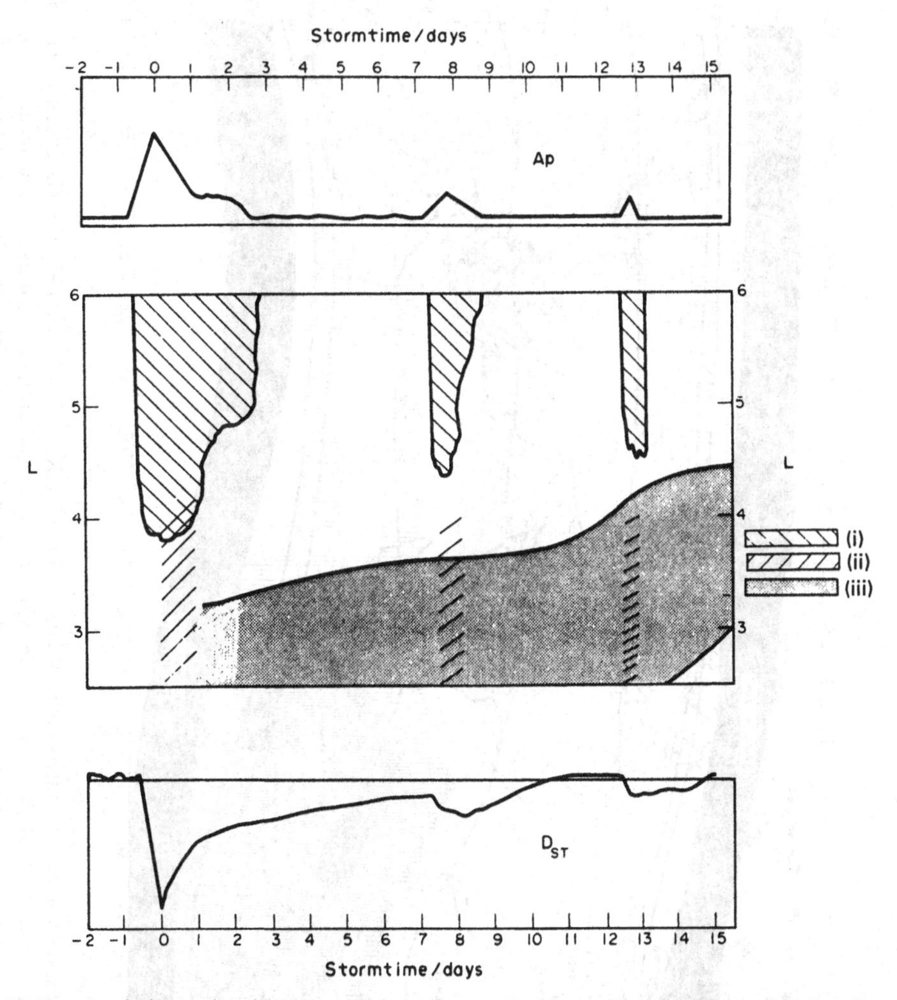

Fig. 7.28 Sketch of absorption increases with respect to L-value and storm time, showing (i) auroral absorption, (ii) enhanced smooth absorption during magnetic disturbances, and (iii) PSE absorption. (After C.-U. Wagner and H. Ranta, *J. Atmos. Terr. Phys.* **45**, 811, copyright (1983) Pergamon Press PLC)

The most probable explanation of the storm after-effect is that it is due to electron precipitation from the outer Van Allen belt. The existence of a mid-latitude precipitation region is revealed in the data from low orbit satellites (Figure 7.29). It is well known from direct observations that in the 'slot' region of the magnetosphere (Section 5.7.3) the trapped particle flux is increased during a storm. Subsequent refilling of the plasmasphere with cold ionospheric plasma may then provide the necessary precipitation mechanism (as described in Section 9.4.5, for example), consistent with the poleward movement of the PSE region. A quantitative treatment is still lacking, however.

The principal alternative explanation is based on a change of chemistry. Particle precipitation, which is greatest at high latitude, is able to produce nitric oxide (Section

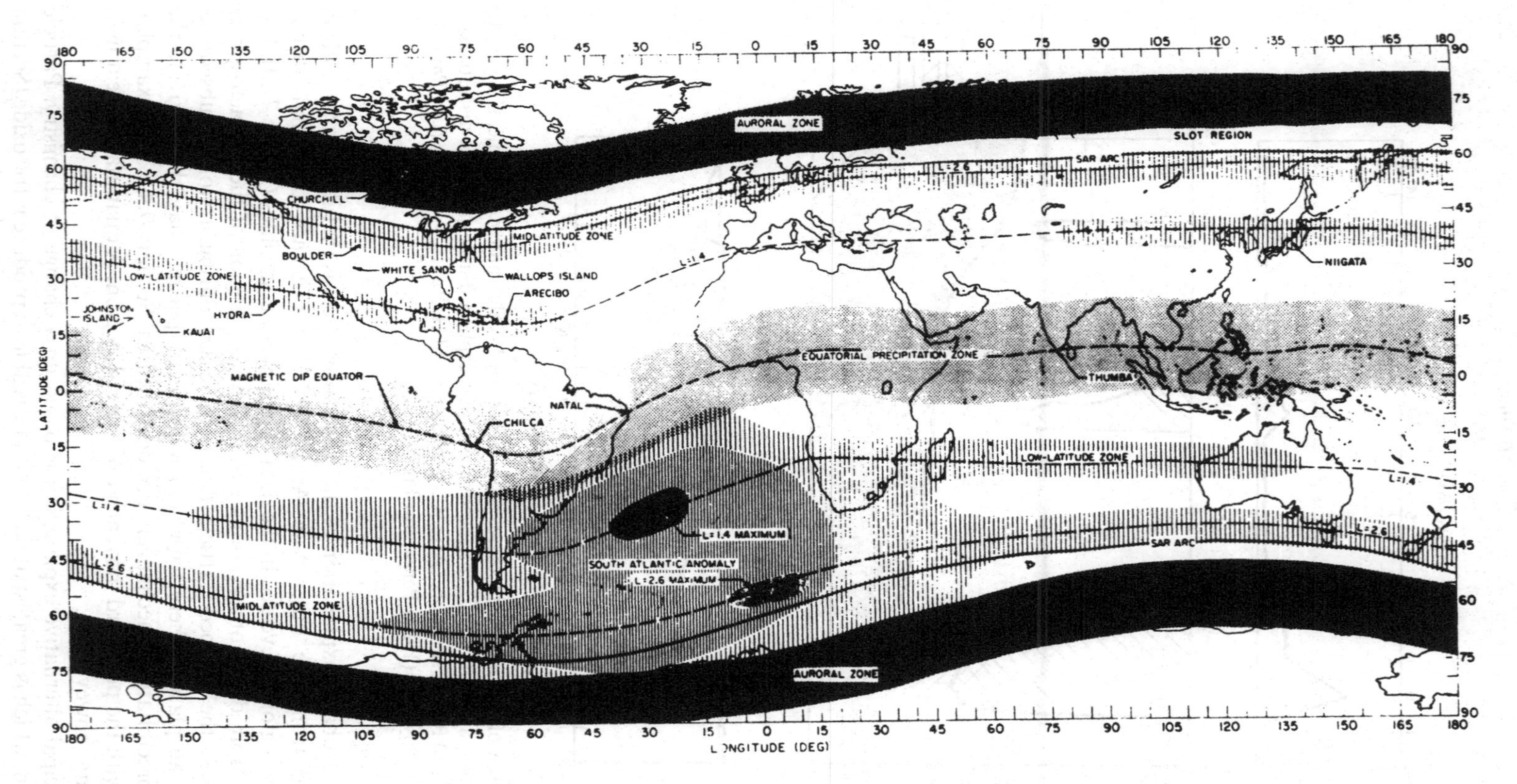

Fig. 7.29 Global zones of night-time particle precipitation for moderately disturbed conditions ($3 < K_p < 7$; $30 < |D_{st}| < 150$). The enhancement in the vicinity of the South Atlantic magnetic anomaly is clearly seen. Some sites for rocket and ground-based measurements are marked. (Reprinted with permission from H. D. Voss and J. G. Smith, *J. Atmos. Terr. Phys.* **42**, 227, copyright (1980) Pergamon Press PLC)

4.1.5) which would lead to an enhancement of electron production by solar Lyman-α radiation in the D region. In principle this NO could be carried by winds to middle latitudes, but it is not known whether this happens in practice.

7.4.5 Winter anomaly of radio absorption

Observations

The *winter anomaly* of ionospheric radio absorption is one of the classical anomalies of the ionosphere. It has been intensively studied and the observational facts are beyond question. Yet it has proved difficult to explain. It has become clear though – and this is part of the difficulty – that the phenomenon cannot be treated in isolation; there is good evidence that it involves relationships between the mesosphere and the stratosphere, including their dynamics.

The winter anomaly shows in absorption measurements, particularly those made by the pulse-reflection method and by the reception of continuous low-frequency transmissions after reflection from the ionosphere, and it has two facets:

(a) on the whole, radio absorption is greater in winter (by about a factor or two) than would be predicted by a simple extrapolation from summer measurements;
(b) the absorption is much more variable in winter than in summer, and groups of days show levels of absorption that are abnormally high even compared to the already enhanced winter level; however, the absorption on some days is relatively low and more typical of summer.

The examples of Figure 7.30 illustrate both aspects of the anomaly. These results show the absorption for one particular value of the solar zenith angle ($\cos\chi = 0.1$). According to the simple theory, the absorption in decibels should vary with zenith angle as $\cos^n\chi$. Whatever the value of n, one would expect to find the same absorption at the same value of χ at all times of year (dashed lines in Figure 7.30). In the northern hemisphere this holds for much of the year, typically from March to October, but the behaviour tends to be anomalous from November to February. The winter anomaly also occurs in the southern hemisphere though with some significant differences. In the south the effects are smaller and the major events occur very late in the winter – usually in November, which would be equivalent to May in the north. In neither hemisphere does the pattern of anomalous days repeat in detail from one year to the next. The magnitude of the anomaly in a given year varies with the mean sunspot number, but there is no day-to-day correlation between absorption and daily sunspot number.

The winter anomaly as observed is essentially a mid-latitude phenomenon, though it may well extend poleward into the auroral zone where its presence would tend to be masked by the phenomenon of auroral absorption (Section 8.3.6). Generally the magnitude and duration of periods of high absorption decrease with decreasing latitude, and in the northern hemisphere the anomaly does not occur south of 40° N. The steady part of the anomaly (a) covers the whole of the hemisphere within the occurrence region, but the variable part (b) is of limited extent on any one day, and the correlation coefficient worked out between pairs of spaced stations falls off considerably over a separation of 200 km.

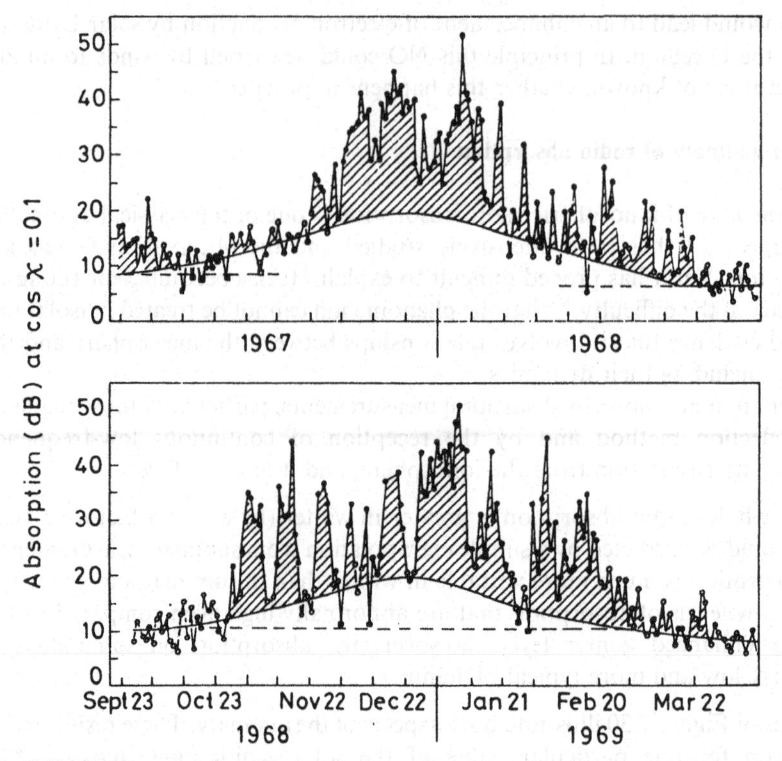

Fig. 7.30 Winter anomaly of ionospheric radio absorption, in terms of the absorption at a constant solar zenith angle during 1967–69. (Reprinted with permission from H. Schwentek, *J. Atmos. Terr. Phys.* **33**, 1647, copyright (1971) Pergamon Press PLC.) The dashed lines extrapolate the trends from summer.

Explanations

There is no doubt about the immediate cause of the anomaly in the ionosphere; it is an enhancement of electron density at heights between about 60 and 90 km. Where the increase is largest, between 70 and 80 km, the enhancement can amount to as much as a factor of ten. This is shown by rocket measurements (Figure 7.31) and it is a well substantiated result. If there is no change of ionizing radiation the cause must be sought in a chemical change. The three favourite mechanisms are:

(a) an increased ion production rate due to an increase in NO concentration;
(b) a decreased loss rate due to the removal of water-cluster ions;
(c) effects of temperature change on the chemical reaction rates.

Possibilities (a) and (b) require large-scale transport of species from the auroral zone. In addition, evidence for an association between the variable part of the anomaly and the incidence of magnetic storms has led to suggestions that

(d) electron precipitation from the trapping zone may play some part.

The ledge in the mid-latitude electron density profile between 80 and 90 km is ascribed, at least in part, to the hydration of ions below the ledge (Section 6.3.4). The

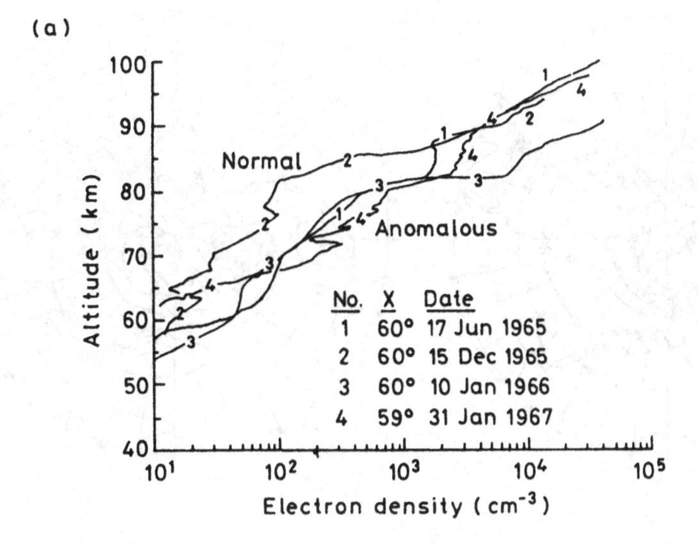

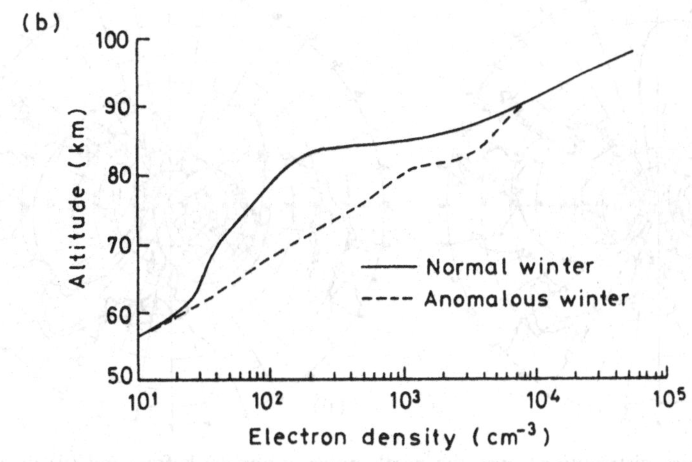

Fig. 7.31 Electron density profiles measured from rockets: (a) individual measurements; (b) average profiles for normal and anomalous winter conditions. (Reprinted with permission from M. A. Geller and C. F. Sechrist, *J. Atmos. Terr. Phys.* **33**, 1027, copyright (1971) Pergamon Press PLC)

disappearance of the ledge during the anomaly (Figure 7.31) supports mechanism (b). There is also direct evidence for an increase of NO concentration (explanation (a)).

Relation to the stratwarm

An important clue to the nature of the winter anomaly came to light in the mid-1960s. It has been known for many years that at the 10 mbar level (about 30 km altitude) the polar atmosphere is warm in summer and cold for most of the winter. When the atmosphere is cold over the pole the stratospheric wind circulates as a vortex around the pole. However, on certain winter days the polar atmosphere becomes warmer than

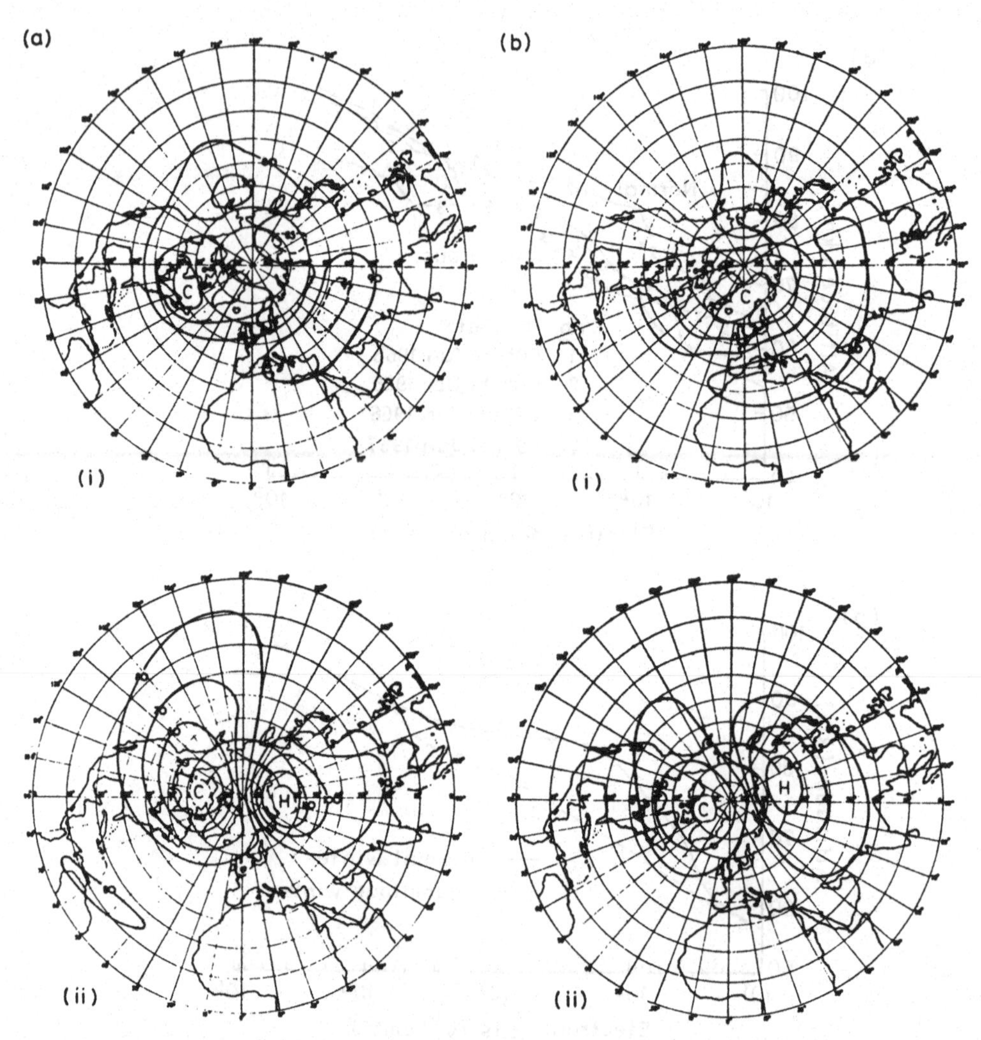

Fig. 7.32 Temperature distributions over the north polar region (i) before and (ii) during a stratwarm in early 1971, at the (a) 2-mbar and (b) 20-mbar levels. C: cold; H: hot. (Reprinted by permission from J. J. Barnett *et al.*, *Nature* **230**, 47, copyright © 1971 Macmillan Magazines Ltd)

usual by 10–30 K, and the circulation pattern alters. (In 1971 a change of nearly 100 K was observed at 45 km.) Such events, generally called *stratospheric warmings* or *stratwarms* for short, were first observed by rocket measurements in the 1950s. Since then, satellite remote-sensing methods, using the infra-red emission from atmospheric carbon dioxide, have delineated and mapped the temperature distributions on the global scale. (The principle of this method was described in Section 3.3.2.)

It is observed that during a stratwarn a hot region develops to one side of the pole, as illustrated in Figure 7.32. Taking readings around the pole at constant latitude, the temperature then shows one maximum and one minimum, and this pattern is said to be 'wave number 1'. Wave number 2 and higher orders appear at times in the

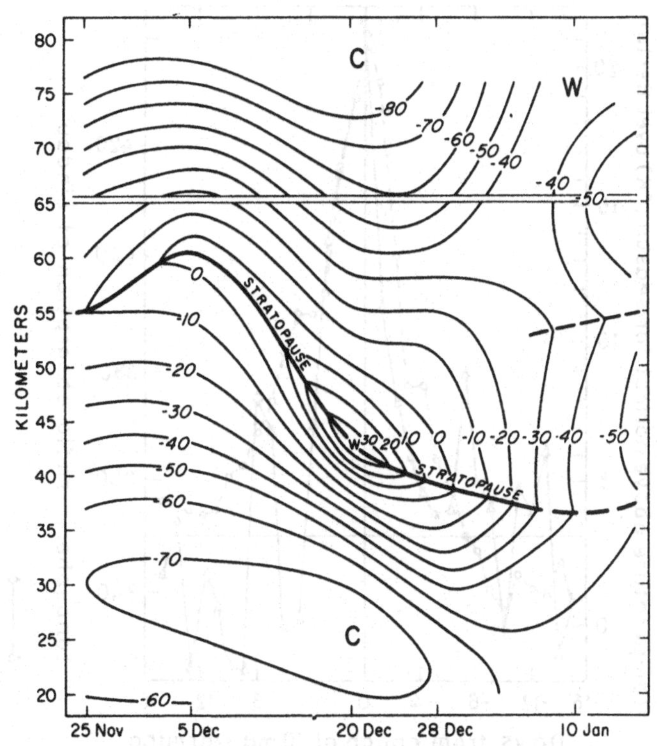

Fig. 7.33 Typical temperature variation (in °C) with height and time during a major mid-winter warming at about 60° N. The Warmer and Cooler regions are marked. (After K. Labitzke, *J. Geophys. Res.* **86**, 9665, 1981, copyright by the American Geophysical Union)

stratosphere, and in the troposphere the polar front typically folds back and forth with wave number 5 or 6. In the stratosphere, however, the characteristic of a major stratwarm, occurring only in winter, is that the temperature distribution becomes dominated by wave 1. As may be noticed in Figure 7.32, the longitudes of the hot and cold spots vary with height; indeed, a warming at one height may be a cooling at another (Figure 7.33). Major stratwarms occur once or twice during most winters in the northern hemisphere, but their incidence and timing are variable from year to year. In the southern hemisphere there are fewer stratwarms, occurring late in the winter season but more repeatably from one year to the next. A distinction is often made between warmings that occur in the middle of the winter and the 'final warming' that marks the return to summer conditions. In addition, there are some 'minor warmings' of the stratosphere that are not associated with a breakdown of the polar vortex.

The relevence of all this to the winter anomaly of ionospheric radio absorption is that, as discovered by A. H. Shapley and W. J. G. Beynon (Figure 7.34), a major stratwarm is always accompanied by a period of anomalous radio absorption. From several points of view this is a fascinating aspect of the phenomenon. The stratwarm is part of the dynamics of the stratosphere and mesosphere, marking a change in the wind pattern – the wind flows along the isothermals, as it does along isobars in the

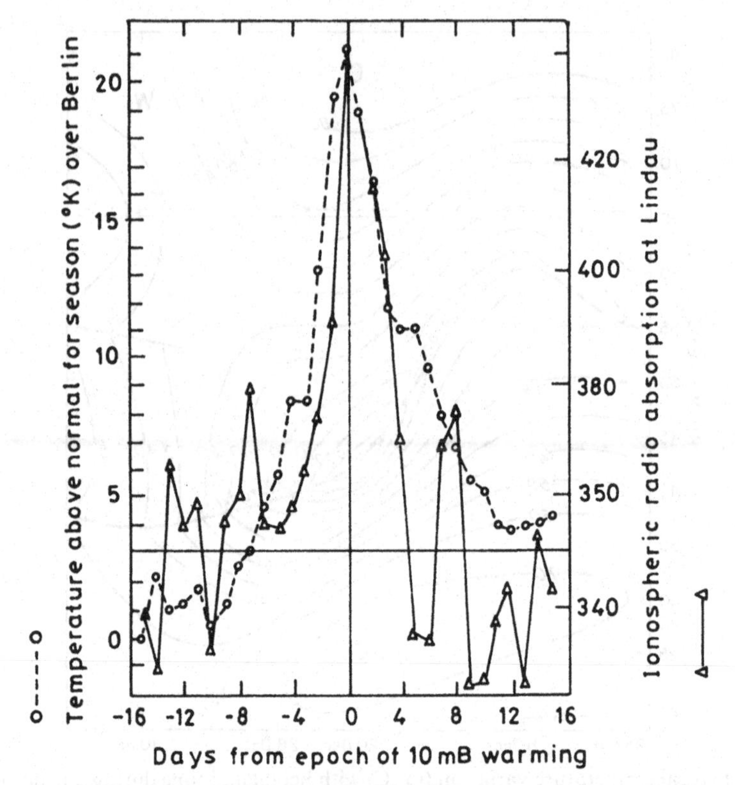

Fig. 7.34 Superimposed epoch analysis comparing 10-mbar temperatures over Berlin, and radio absorption over Lindau. (After A. H. Shapley and W. J. G. Beynon, *Nature* **206**, 1242, copyright © 1965 Macmillan Magazines Ltd)

troposphere – as a different planetary wave becomes established, and it has been argued that this change of wind direction is what brings nitric oxide from the auroral zone to middle latitude to cause the absorption anomaly. There is evidence that during a minor warming the electron density is reduced at 90–110 km, the radio absorption is reduced, and the wind near 95 km reverses in direction. It is supposed that the supply of NO is reduced in these conditions. Since planetary waves propagate upward from the troposphere it appears that the anomaly originates in the lower atmosphere. Certainly this is a phenomenon to be studied in depth!

During the 1980s, special campaigns invoking several techniques (rockets, meteor winds, radio absorption) were mounted to study the winter anomaly, and their results have supported the theory which attributes the anomaly to the arrival of additional nitric oxide from the auroral zone. However, there remains evidence for a magnetospheric connection. Whether or not auroral and magnetic activity had any connection with the winter anomaly was debatable for some years, but by the mid-1960s it had been established to everyone's satisfaction that they were not associated. Then a correlation was found with energetic particle precipitation detected on an orbiting satellite. Considering that energetic electron fluxes are incident in the auroral zone, this essentially interpreted the winter anomaly as an equatorward extension of the auroral precipitation. The amount of the precipitaion seemed insufficient to explain the magnitude of the absorption in a major anomaly, however, and so this could not be the whole explanation. But with attention now directed to the matter, further studies were made, using superimposed-epoch analysis, and these showed a tendency for enhanced mid-latitude absorption to follow high-latitude disturbances by 3–5 days. This is reminiscent of the storm after-effect in the D region, where the production of NO by auroral precipitation and subsequent transport to somewhat lower latitudes could be a factor. It may be, indeed, that there is a relation between the winter anomaly and the PSE that has yet to be clarified.

7.5 Irregularities

The behaviour described in this chapter so far applies to the large scale in distance and to slow changes in time: vertical structure measured in tens and hundreds of kilometres, variations in longitude and latitude involving hundreds and thousands of kilometres, temporal change reckoned mainly over days,months and years.

We turn now to structures on smaller scales. The distances involved now go down to kilometres and metres, small enough to produce spatial variations within the field of view of a single observing instrument. Because it would be impractical to describe all the details of a large number of irregularities, these small-scale structures are usually treated statistically, principal interest being in quantities such as average size, shape and strength, the spread of properties within the population, and velocities.

There appear to be irregularities at all levels of the atmosphere, and one remarkable fact about ionospheric irregularities is that they cover a wide range of spatial size. Consequently the total picture can only be realized by employing a number of observing techniques. Further, since there are good reasons to believe that structures of different size are related to each other through physical mechanisms, it is important to investigate the overall picture as well as the details, which makes irregularity studies an ideal topic for special observation campaigns in which a range of techniques – radar, ionosonde, beacon satellite, airglow photometer, *in-situ* measurement from satellite or rocket – are brought together at one place.

7.5.1 Scintillations

Small irregularities in the ionosphere are detected and studied by two main techniques: direct measurement with satellite probes, and through their effect on radio signals. The mechanism by which irregularities in the ionized medium produce scintillation in radio waves was discussed in Section 2.6. In essence, the irregularity of the medium produces phase irregularity along the emerging wavefront, and the process of diffraction between this 'phase screen' and the observing point converts some of the phase variation to amplitude variation. The term *scintillation*, which is used to describe the irregular variations in the amplitude or the phase of a radio signal received after passing through, or being reflected in, the ionosphere, is given by analogy with the intensity variation of luminous stars when seen through a turbulent atmosphere. The mechanisms, indeed, are very similar.

Small irregularities in the E region can be investigated by means of reflected radio waves of a few megahertz frequency. The fading period is typically tens of seconds, and spatial correlation distances, determined by the spaced receiver technique described in the next section, are a few hundred metres. Irregularities can be observed as low as

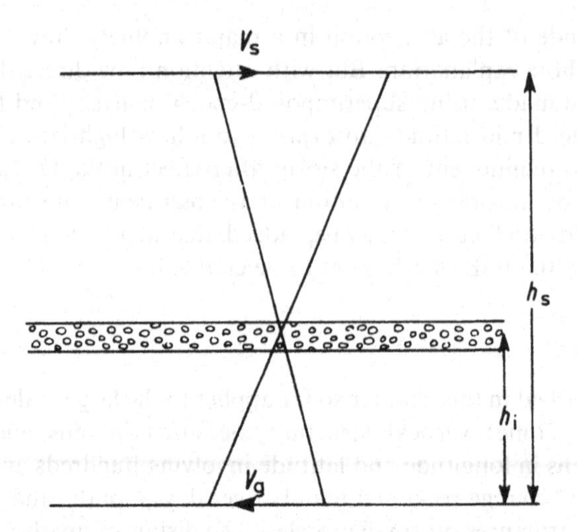

Fig. 7.35 Determination of the height of irregularities from the apparent motion of a fading pattern over the ground.

70 km by the partial reflection technique (Section 3.5.5). Velocity measurements give information about the neutral winds (Section 4.2.2). VLF signals, also reflected in the D region, may fade at night, but in this case the correlation distances, no doubt because of the longer wavelength, appear much larger – about 20 km.

Most scintillation, though, arises in the F region. The irregularities probably exist throughout the whole layer, although there appears to be some bias towards the topside. Figure 7.35 illustrates one method of estimating the height of the irregularities, based on spaced-receiver observations of a radio beacon carried on an orbiting satellite. Since the satellite moves more quickly than the irregularities, the apparent speed of the irregularities over the ground (V_g) depends on the velocity of the satellite (V_s) and on the height of the irregularities (h_i) as a fraction of the satellite altitude (h_s):

$$V_g = V_s h_i/(h_s - h_i).$$

Hence,

$$h_i = h_s/(1 + V_s/V_g). \tag{7.17}$$

Most height determinations fall in the range 200–600 km, with the maximum between 300 and 400 km.

At medium latitude, F-region scintillations tend to occur in patches about 1000 km across. These patches have been identified with the phenomenon of *spread-F*, a well known feature of ionograms which is illustrated in Figure 7.36. There is more than one kind of spread-F and the scintillation regions are associated with the 'range-spread' rather than the 'frequency-spread' variety. Thus, spread-F is not necessarily a guide to scintillation occurrence, but there are general similarities in that both occur most frequently at night, and there are occurrence maxima at low and high latitudes (Figure 7.37).

As determined from amplitude scintillation, F-region irregularities are several hundred metres across, but they show considerable elongation (by a factor as large as 60) along the direction of the geomagnetic field. This, of course, is a consequence of the efficiency of field-aligned plasma diffusion in the F region (Section 6.3.3). Further

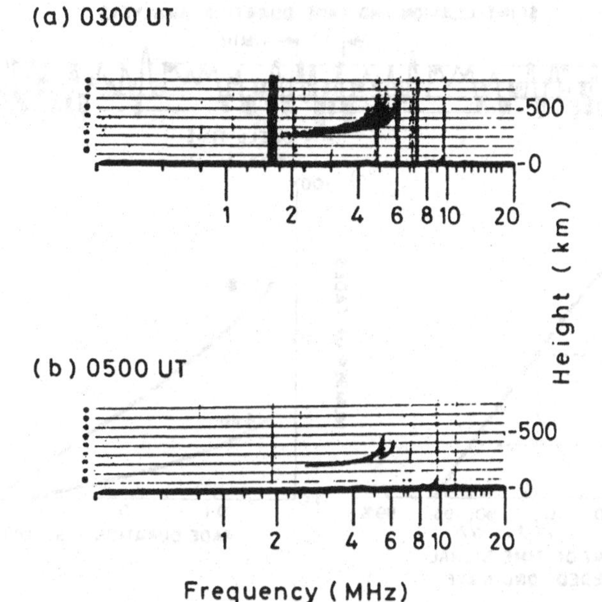

Fig. 7.36 Ionograms (a) with and (b) without spread-F: vertical soundings at Esrange, Sweden (68° N, 21° E), 15 March 1970. (*Kiruna Geophysical Data*, 70/1–3, Kiruna Geophysical Institute, 1970)

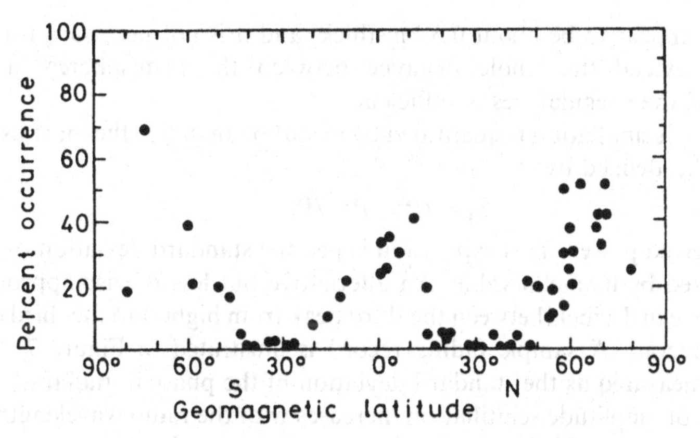

Fig. 7.37 Occurrence of spread-F as a function of geomagnetic latitude during August–September 1957. (T. Shimazaki, *J. Radio Res. Lab. Japan* **6**, 669, 1959)

evidence for the existence of field-aligned irregularities comes from the occurrence of *ducted echoes* on topside ionograms. When the F region is irregular a topside sounder will observe spread-F below, but also some echoes may arrive with such long delays that they can only be reflections from the conjugate ionosphere. The mechanism is that some of the radio energy becomes trapped between columns or sheets of enhanced or depleted ionization, which guide it to the other hemisphere where it is reflected in the topside ionosphere and returned by the same mechanism. With the radio signal obliquely incident on the duct, the variation of electron density need be no more than

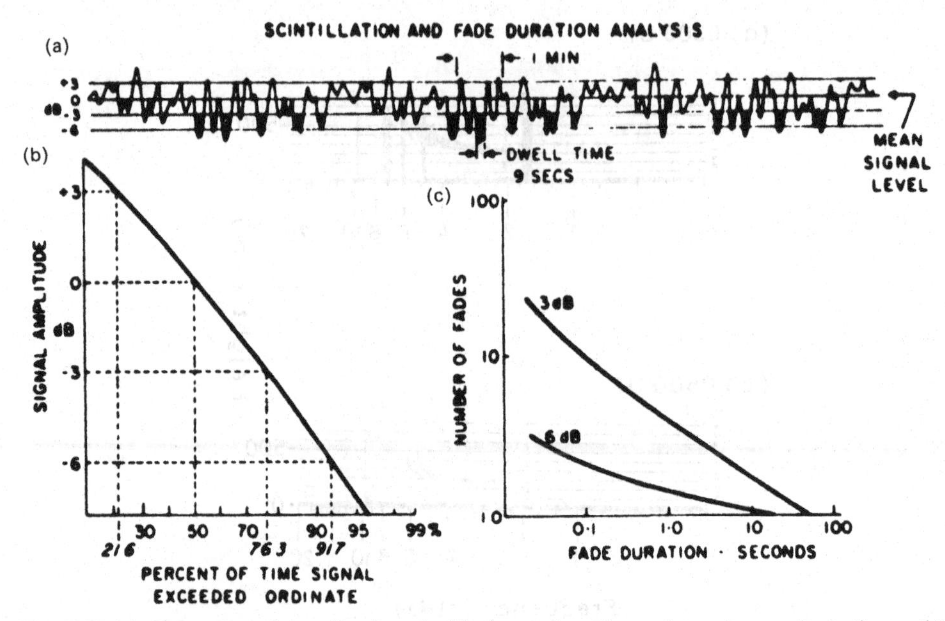

Fig. 7.38 (a) Example of amplitude scintillation, plotted as intensity against time. (b) Cumulative probability distribution. (c) Distribution of fade duration. (After *Handbook of Geophysics and Space Environment* (ed. A. S. Jursa). US Air Force Geophysics Laboratory, 1985)

1%. The ducts appear to be about 0.5 km thick, and it is not necessary for a single irregularity to extend the whole distance between the hemispheres; a general elongation of all the irregularities is sufficient.

The intensity of scintillation is quantified by means of an index, the most used being that known as S_4, defined by

$$S_4 = (\overline{P^2} - \overline{P}^2)^{\frac{1}{2}}/\overline{P}, \tag{7.18}$$

P being the received power. This expression is just the standard deviation of received power normalized by its mean value. An alternative but less precise approach is to measure the range in decibels between the third peak from highest to the third dip from lowest on the record. A sample fading record is illustrated in Figure 7.38. Phase scintillation is measured as the standard deviation of the phase in radians.

The intensity of amplitude scintillation increases with the radio wavelength. If S_4 is not too large it varies as $\lambda^{1.5}$, but the law becomes less steep for stronger scintillation ($S_4 > 0.6$, or peak–peak scintillation exceeding 10 dB), and at extreme levels ($S_4 \sim 1$) the wavelength dependence vanishes. The index of phase scintillation varies in proportion to the wavelength except under the most extreme conditions.

The spectrum of the fading is also of interest. The irregularities in the ionosphere generally show a power law spectrum of form κ^{-p}, κ being the wave number $(2\pi/\lambda_I)$ in which λ_I is the spatial wavelength of the irregularities. Referring to Section 2.6, we may generally suppose that the phase screen in the ionosphere produces a pattern of amplitude and phase fluctuation over the ground which is related to the spectrum of the irregularities themselves, and that scintillations are observed because the pattern is moving across the observing point. By this means, the distance variation is converted

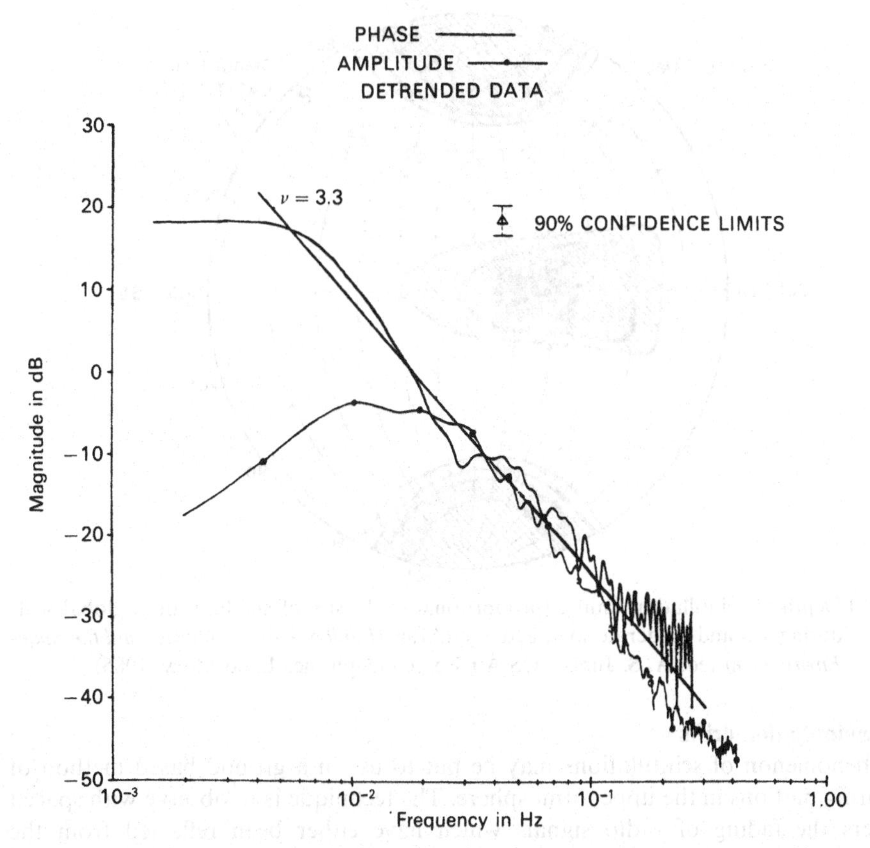

Fig. 7.39 Comparison between amplitude and phase spectra recorded at 40 MHz from a geosynchronous satellite transmission. Power spectra are each plotted on a log scale of relative values marked in decibels. The amplitude spectrum levels off due to data detrending (at 3×10^{-3} Hz), but the turn in the phase spectrum marks the Fresnel frequency. (After W. J. Myers *et al.*, *J. Geophys. Res.* **84**, 2039, 1979, copyright by the American Geophysical Union)

to a time variation. Also, the conversion of phase to amplitude scintillation depends on the size of the irregularity (Equation 2.75). Therefore, the low-frequency (arising from the large-scale) end of the spectrum is attenuated. The attenuation operates at frequencies less than $u/(2\lambda D)^{\frac{1}{2}}$, where u is the velocity (it being assumed that all the irregularities move together) and $(\lambda D)^{\frac{1}{2}}$ is the radius of the first Fresnel zone. Since λ is known and D may be assumed ($\sim$ 350 km), u can be determined by this means. This effect is seen in the spectra of Figure 7.39.

The geographical distribution of scintillation occurrence (Figure 7.40) identifies maxima in the equatorial and the polar regions. High-latitude irregularities are considered in Section 8.2.

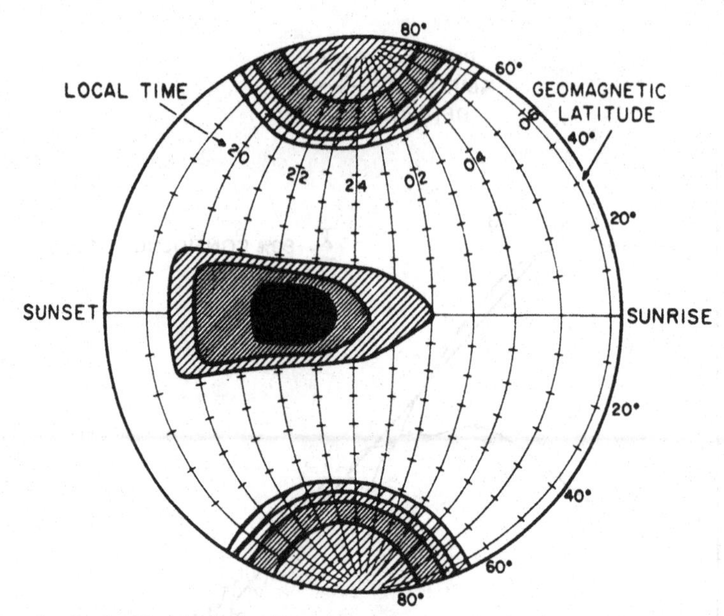

Fig. 7.40 Depth of scintillation fading (proportional to density of shading) on a global scale during low and moderate solar activity. (After *Handbook of Geophysics and the Space Environment* (ed. A. S. Jursa). US Air Force Geophysics Laboratory, 1985)

7.5.2 Scintillation drifts

The phenomenon of scintillations may be put to use in a ground-based method of measuring motions in the upper atmosphere. The technique is to observe with spaced receivers the fading of radio signals which have either been reflected from the ionosphere or transmitted through it. If the spacing is smaller than the correlation distance of the fading it is likely that each receiver will record a similar fading pattern but with a small time displacement superimposed, as in Figure 7.41(a). The simple interpretation is that a region of irregularity (i.e. a diffraction screen – Section 2.6.2) is moving bodily across the observation sites. Knowing the geometry of the receiving network, it is then not difficult to deduce the speed and direction of the ionospheric motion.

To take a simple case, let the stations form a right-angled triangle as in Figure 7.41(b). Then the time differences t_1 and t_2 give 'apparent velocities' V'_x and V'_y in the x and y directions:

$$V'_x = R_1R_3/t_2;$$
$$V'_y = R_1R_2/t_1. \tag{7.19}$$

From Figure 7.41 it is seen that

$$\frac{1}{V^2} = \frac{1}{(V'_x)^2} + \frac{1}{(V'_y)^2},$$

and hence

$$V = \frac{V'_x V'_y}{[(V'_x)^2 + (V'_y)^2]^{\frac{1}{2}}}. \tag{7.20}$$

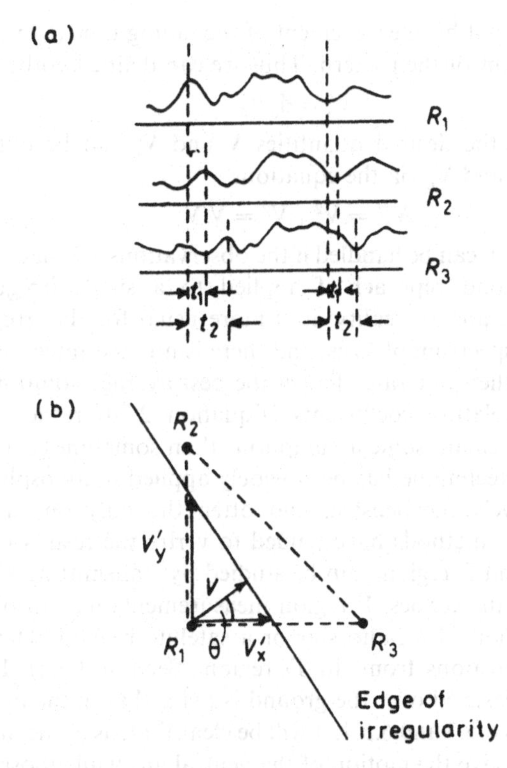

Fig. 7.41 The spaced receiver method for observing the movement of a pattern of irregularities over the ground. (a) Typical fading at stations R_1, R_2 and R_3. (b) Relation between true and apparent velocities.

Also,

$$\theta = \tan^{-1}\left(\frac{V'_x}{V'_y}\right). \tag{7.21}$$

Note that V'_x and V'_y are not velocity components in the usual sense.

The treatment is readily generalized for stations grouped other than in a right-angled triangle, but this is the least of the complications that can arise in practice. What often happens when the stations are far enough separated to give a measurable time difference is that the fading patterns are no longer identical. This means that the pattern is changing as it moves, and in this case the above treatment is not adequate. To take an extreme example, if the pattern is changing but not moving, the average time difference between spaced stations is zero. The true velocity, however, is not infinity but indeterminate. The effect of changes in a moving pattern is to reduce the measured time differences and to increase the apparent velocity. This is handled by introducing a conceptual *characteristic velocity* defined by

$$V_c = d_0/t_0, \tag{7.22}$$

where d_0 is a characteristic distance (i.e. the size of the irregularity) and t_0 is a characteristic fading time, both as would be measured by an observer moving with the pattern. The true velocity, V, is the velocity of this observer. Suppose that a stationary observer measures the speed in the direction of the vector **V**. In principle he can determine d_0 correctly, but his measurement of the fading time (t'_0) will be less than t_0 because of the movement of the pattern. Thus we can define another velocity

$$V'_c = d_0/t'_0. \quad (7.23)$$

It can be proved that the desired quantities V and V_c can be obtained from the observed quantities V′ and V'_c via the equations

$$V'^2_c = V^2_c + V^2 = V'V. \quad (7.24)$$

Anisotropy in the pattern can be handled if the observations are made at four stations.

The method is beyond reproach if applied to a single irregularity and the measurements are accurate. In reality, it is more usual for the irregularities in the ionosphere to cover a spectrum of sizes, and there is no guarantee that they all move together or change at the same rate. To get the best average solution the analysis is usually based on correlation coefficients (Equation 2.70) rather than individual features, though there remain some assumptions than sometimes cannot be justified.

The spaced-receiver technique has been widely applied in ionosphere studies. It is convenient and relatively inexpensive, and often the only one available! Direct comparisons with other methods have tended to verify the results in general terms. Movements in the E and F regions can be studied by transmitting a signal from the ground and observing the echoes. F-region measurements may also be achieved by observing the scintillation of a radio star or a satellite beacon. The technique also works with partial reflections from the D region (Section 3.5.5). In the reflection version the velocity measured over the ground is twice that in the ionosphere.

Referring to Sections 6.5.3 and 6.5.4, it will be clear that, as a rule, motions deduced in the E and D regions give the motion of the neutral air, while those in the F region indicate the electric field.

7.5.3 Spread-F, bubbles and F-region irregularities at low latitude

Scintillation may be particularly intense at low magnetic latitude, even in the UHF band (> 300 MHz). Equatorial scintillations have been reported on frequencies as high as 7 GHz, and at 4 GHz scintillations amounting to 9 dB have been observed. Their intensity and occurrence are greatest during the hours immediately after sunset, and around the equinoxes, though there are significant variations between measurements from different observing stations, no doubt due to geometrical factors and longitudinal variations of the equatorial F region. Equatorial scintillation is strongly correlated with the range-spread type of Spread-F, which is attributed to irregularities of electron density. The occurrence of this Spread-F is high at the equinoxes and at the December solstice but low from May to August. During the day it peaks about 2100 local time (LT).

Because of the severity of their radio propagation effects, the irregularities of the equatorial F region have been thoroughly investigated using the campaign approach, including rockets and satellites, coherent and incoherent radar, trans-ionospheric radio propagation, and photometers. This approach has shown that at times the equatorial ionosphere contains irregularities over a vast range of sizes from hundreds of kilometres to tens of centimetres, a range of 5 to 6 orders of magnitude.

In-situ measurements with rockets and satellites have revealed large reductions of plasma density in the equatorial F region by night. The depleted regions have sharp edges, and within them the ions have upward velocities of about 100 m/s as well as

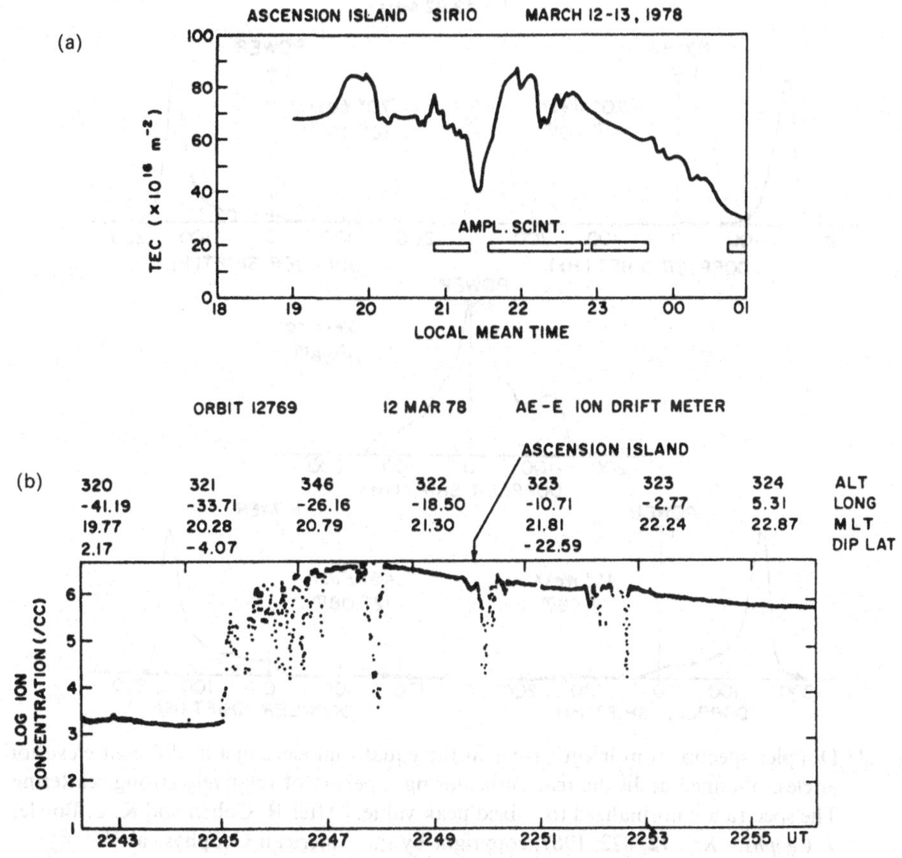

Fig. 7.42 (a) Fluctuations of electron content and associated scintillations observed at 137 MHz at Ascension Island on 12–13 March 1978. (b) Large biteouts of ion concentration observed on the satellite AE-E. (Reprinted with permission from S. Basu and S. Basu, *J. Atmos. Terr. Phys.* **43**, 473, copyright (1981) Pergamon Press PLC; after J. A. Klobuchar *et al.*, 1978, with data from J. P. McClure and W. B. Hanson)

some 20 m/s to the west. These *bubbles*, as they are known, are also detected by remote sensing techniques such as incoherent scatter radar and they are seen in electron-content data (Figure 7.42) where the depletion can be as much as 40%. The 630 nm airglow intensity is reduced in the same region. There is also an association with scintillations, which apparently develop in the sharp edges of the bubble. The depletions have typical dimensions of 100 km east–west and as much as 1200 km north–south, and they are aligned along the geomagnetic field. The cause of the rising bubbles has been explained by means of the Rayleigh–Taylor instability (Section 2.8.3).

7.5.4 Irregularities in the equatorial electrojet

Irregularities in the equatorial E region are closely associated with the equatorial electrojet. The electrojet is part of the global current system driven by winds in the neutral air, which is more intense over the magnetic equator because of the high value

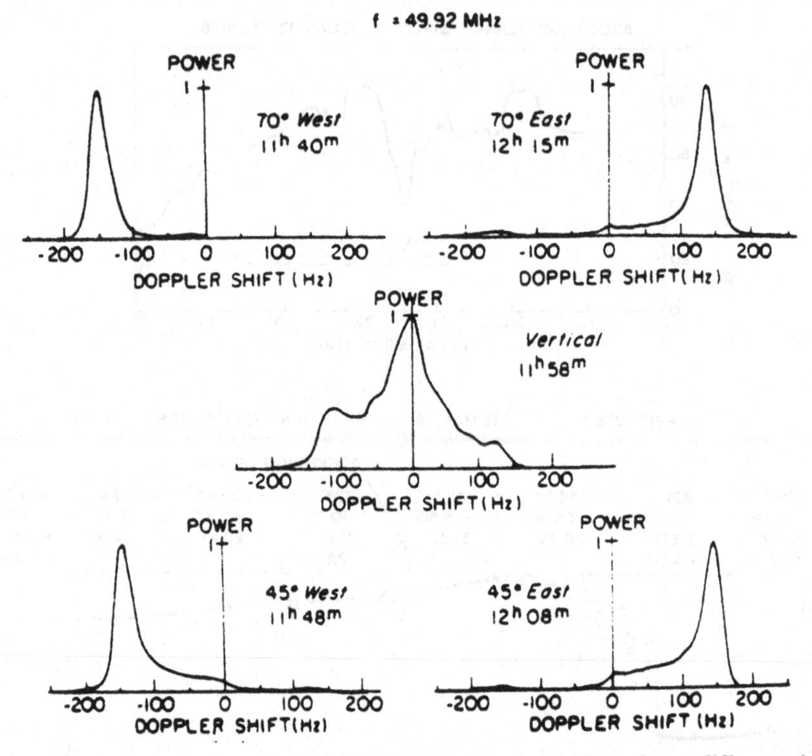

Fig. 7.43 Doppler spectra from irregularities in the equatorial electrojet at different elevation angles, obtained at Jicamarca, Peru, during a period of relatively strong scattering. The spectra are normalized to a fixed peak value. (After R. Cohen and K. L. Bowles, *J. Geophys. Res.* **72**, 822, 1967, copyright by the American Geophysical Union)

of the Cowling conductivity (Section 6.5.5). The east–west electric field over the equator is about 0.5 mV/m, but the vertical field (representing vertical polarization of the ionospheric layer) is much larger. The east–west drift velocity of the electrons is 400–600 m/s, and since this exceeds the ion-acoustic speed the two-stream instability may develop (Section 2.8.2), generating irregularities in the ionized medium.

Irregularities in the equatorial E region are seen as sporadic-E on ionograms, though in this case wind shear (Section 7.1.1) is not the cause. This type of sporadic-E does not blanket the returns from higher layers. The incidence of equatorial sporadic-E correlates with the intensity of the electrojet. Most investigations, though, have been by means of coherent scatter radar (Section 3.6.3). The radio frequency of the radar is not critical and can be from less than 10 MHz to more than 100 MHz; in effect the wavelength selects the size of the irregularities being observed. These irregularities are present day and night. They are stongly field-aligned, which means that the radar signal must meet the scattering region at right angles to the geomagnetic field direction. At the equator this condition can be satisfied by pointing the radar anywhere in the magnetic east–west plane.

Irregularities generated by the two-stream mechanism give echoes characterized by a narrow spectrum (Figure 7.43) and are known as 'Type 1 echoes'. The Doppler shift here is about 120 Hz (with a 50 MHz radar), which corresponds to the ion–acoustic speed of about 360 m/s. The Doppler shift changes sign according to whether the radar looks east or west, and it reverses between day and night, the electron flow being westward by day and eastward by night. It is a curious but significant fact that the Doppler shift is constant between zenith angles of about 45° and 70°. This shows that the speed of the irregularities 'saturates' at about the ion-acoustic speed and they travel in all directions (normal to the geomagnetic field) and are not merely swept along with the electron flow.

The second type of irregularity, giving 'Type 2 echoes', is explained by the gradient drift instability (Section 2.8.3). The relative motion of electrons and ions across the magnetic field, which is necessary for this instability to develop (Figure 2.17), is now provided by the electric current. The resulting vertical movements enhance irregularities in the electron density gradient or stabilize them, depending on the direction of the current. It is indeed observed that the echoes disappear if the daytime electrojet reverses from eastward to westward. Type 2 echoes show broader spectra than type 1, and there is no threshold current for their generation.

7.5.5 Travelling ionospheric disturbances

Acoustic–gravity waves are a phenomenon of the neutral air (Section 4.3), but the motions of the neutral particles may be communicated to the ionized component through collisions (i.e. by air drag). The nature of the ionosphere's response depends on the altitude. At the lower levels (where the collision frequency is larger than the gyrofrequency) the plasma will move with the neutral air, but higher up (collision frequency small in relation to gyrofrequency) ion motion is inhibited across the geomagnetic field and the only response allowed is to the velocity component parallel to the geomagnetic field. In the F region the ionospheric response is therefore biased, and by no means do observations of the ion motion at those altitudes represent those of the gravity wave.

Nevertheless, the most extensive body of information about gravity waves in the upper atmosphere has come from radio observations of the ionosphere. The principal methods are :

(a) ionosonde (Section 3.5.1) and, closely related to it, continuous sounding of virtual height at a constant frequency;
(b) HF Doppler (Section 3.5.1);
(c) electron content measurement (Section 3.5.2);
(d) incoherent scatter radar (Section 3.6.4–5).

The salient characteristic of these results is that the observed perturbations show a time difference between spaced observation sites, inviting the obvious interpretation of a propagating wave: hence the name *travelling ionospheric disturbance, TID* . When it comes to details, the information obtained depends on the technique. Some are more suited to spaced observations, giving velocity vectors in the horizontal plane. Vertical velocity components can be obtained from HF Doppler if several frequencies are used, since higher frequencies are reflected from higher levels. The reflection heights may be determined using an ionosonde. Though usually restricted to a single site, incoherent scatter radar has some considerable advantages through its ability to observe as a continuous function of height, including the topside, and to measure ion temperature and plasma velocity as well as electron density.

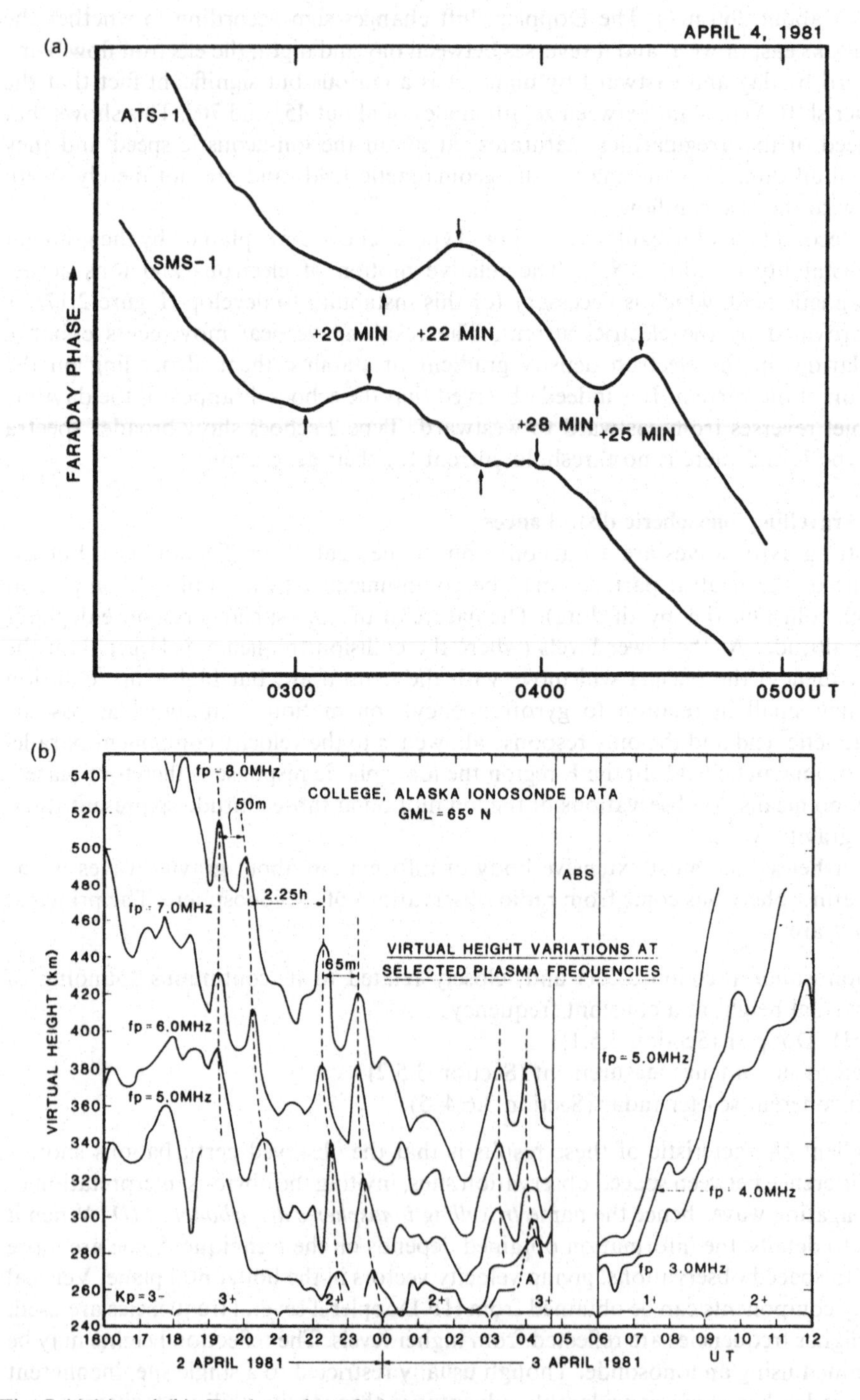

Fig. 7.44 (a) and (b). For legend see facing page.

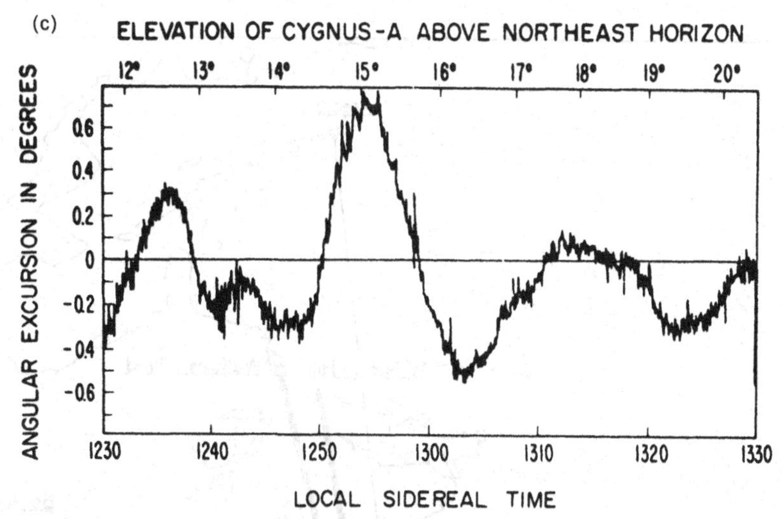

Fig. 7.44 Effects of travelling ionospheric disturbances. (a) Electron content from two geostationary satellite beacons received at a single ground station. The ionospheric crossing points were 350 km apart, and the time shift between the observed waves indicates an east–west velocity component. (b) Variations of virtual height at several frequencies from an ionosonde. The downward movement typical of gravity waves may be seen. (c) Effect of large irregularities on the apparent position of the radio star Cygnus-A, observed at 108 MHz, due to wedge refraction. ((a) and (b) reprinted with permission from R. D. Hunsucker and J. K. Hargreaves, *J. Atmos. Terr. Phys.* **50**, 167, copyright (1988) Pergamon Press PLC; (c) from R. S. Lawrence *et al.*, *Proc. IEEE* **52**, 4, 1964, © 1964 IEEE)

From a technique which measures as a function of height it is possible, using the observed period, to obtain the vertical wavelength. Thence, if model values are taken for the Brunt and acoustic cut-off frequencies (ω_b and ω_a) and for the speed of sound at the relevent altitude, Equation 4.36 allows one to estimate the horizontal wavelength and (knowing the period) the horizontal speed. Thus, quite a lot can be learned from a single technique, but for the most complete diagnosis it is best to use several techniques together, and some such campaigns have been mounted.

TIDs are not at all uncommon, and their occurrence had been noted for many years before 1961 when C. O. Hines explained them as gravity waves. In the examples of Figure 7.44, one is from electron content measurements, one from an ionosonde and the other shows the refraction of a radio star, also, no doubt, due to gravity waves in the terrestrial ionosphere. (See Section 3.5.2 for a discussion of *wedge refraction*.) An average TID produces about 1 % variation in the electron content, 5 % being rather a large wave, but these small changes are nevertheless readily detectable. Individual examples vary as to period, direction and speed, on which basis two main categories have been defined.

Waves with period longer than about 30 minutes show horizontal phase velocities of 400–1000 m/s and horizontal wavelengths exceeding a thousand kilometres, and are called *large-scale* waves. For a large scale wave $\omega \ll \omega_b$, and (see Equation 4.38) the wave propagates almost vertically, the particle motions being almost horizonal. Thus the horizontal phase velocity exceeds the speed in the direction of propagation.

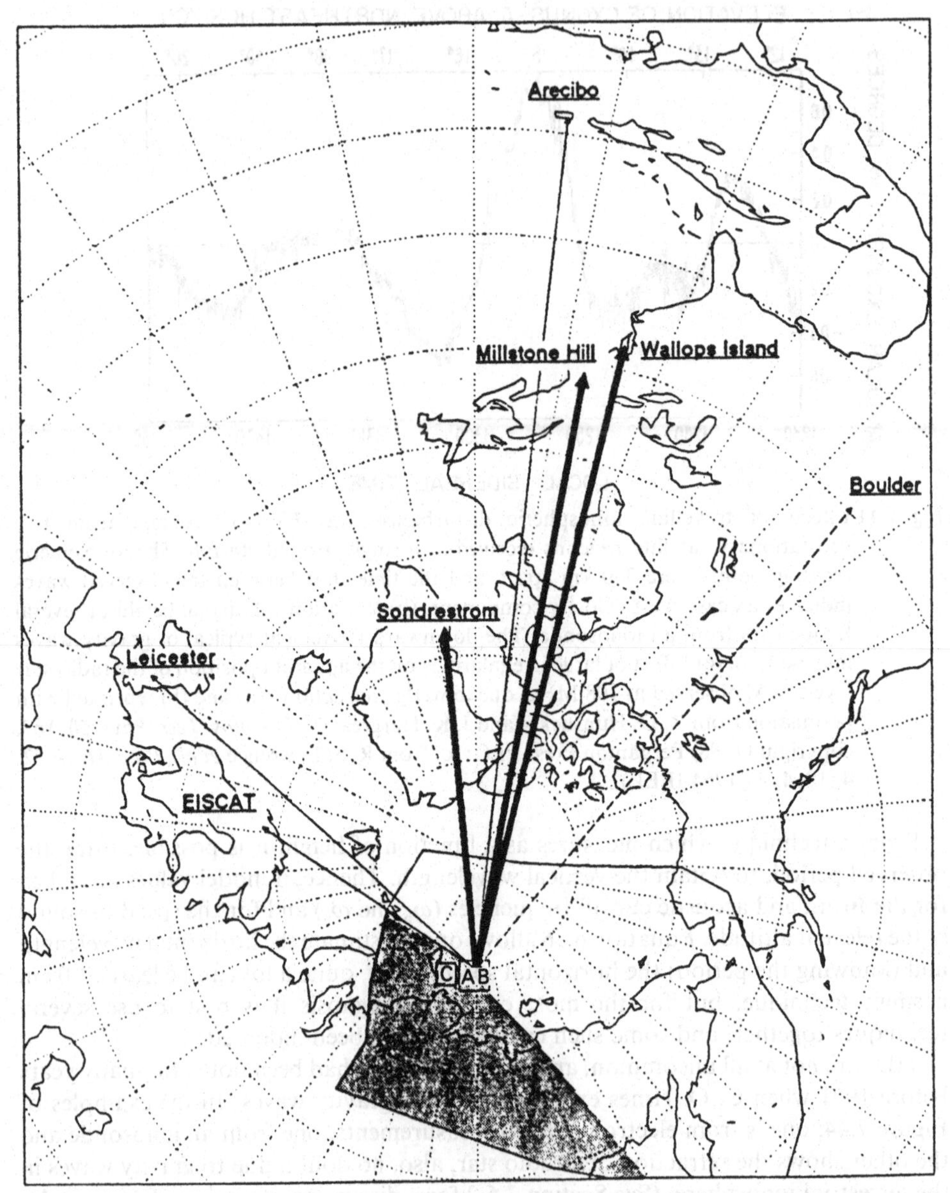

Fig. 7.45 Source region of a TID observed at many sites on 18 October 1985. (D. D. Rice *et al.*, *Radio Science* **23**, 919, 1988, copyright by the American Geophysical Union)

Large scale TIDs tend to travel from pole to equator, and the source is believed to be some aspect of auroral or geomagnetic activity at high latitude. There is statistical evidence for this association, though clear individual examples are not so easily found. Figure 7.45 illustrates one case when TIDs observed at a number of sites could be traced back to a common source. The ground-based observations (incoherent scatter radar, HF Doppler, ionosonde) placed the source within the shaded area, and satellite

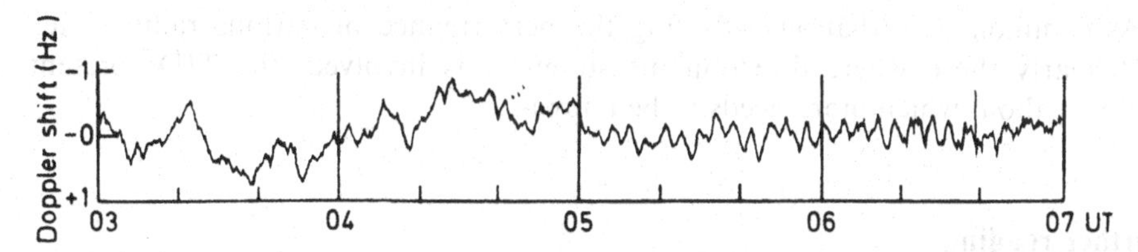

Fig. 7.46 HF Doppler record at 4 MHz, 14 June 1967, over Boulder, Colorado, during thunderstorms, hail and tornadoes in the neighbouring state of Nebraska. The rapid fluctuations are due to acoustic waves in the ionosphere. (After A. G. Jean and G. M. Lerfald, private communication)

observations of auroral emissions indicated a substorm in the region marked by A, B and C. A group of two or three waves, period 60–90 minutes, passed across the observing sites 2 to 7 hours later. At these various sites the horizontal speed was estimated as 425–720 m/s, the horizonal wavelength as 1600–3500 km, and the vertical wavelength as 360–500 km. Some of the sites gave an azimuth of arrival. This case study is interesting as a synthesis of individually noisy observations into a coherent overall picture supporting one idea about the nature of the source.

There is also a theoretical basis for asserting that the heating of the upper atmosphere during energetic particle precipitation and by the auroral electrojet (Section 8.4.2), and motions of the electrojet and of auroral arcs, may each act as sources. The electrojet appears to be the more favoured source, and there may be contributions both from its movement (the 'Lorentz contribution') and from its heating effect (the 'Joule contribution'). The relative magnitude of these contributions ('Lorentz/Joule') depends on the ratio of conductivity to current in the source region, and measurements of this ratio have been attempted by combining incoherent scatter radar measurements (which give the electron density and hence the conductivity) with ground-based magnetometer data (giving the current). It is on this evidence that 'Lorentz' seems to have the edge. The relative importance of heating by particles and of moving auroral structures is uncertain. In some cases, mid-latitude TIDs have been related to sharp-onset auroral precipitation events detected by the riometer technique (Section 8.3.6). The above mentioned case study would seem consistent with this.

Medium-scale waves have periods between about 15 minutes and an hour, horizontal speeds of 100 to 250 m/s and wavelengths of several hundred kilometres. Wavefronts are typically about 45° from the vertical, and the scale of coherence, at a few hundred kilometres, is smaller than for the large scale waves. The source of medium-scale waves is less certain than for the large-scale variety. Meteorological causes such as thunderstorms and jetstreams are favoured explanations.

Some HF Doppler observations of the ionosphere have shown periodicities of a few minutes, and these are thought to be due to acoustic waves. They propagate upward at 500–600 m/s, and occur in limited areas in the vicinity of severe weather in the troposphere. An example is shown in Figure 7.46. In this case the period of fluctuation was 3–5 min between 0500 and 0700 universal time (UT), and this is typical. These acoustic waves are filtered in the atmosphere, since the longest period is that determined by the acoustic resonance (Figure 4.24), and the shortest periods generated tend to be absorbed in the atmosphere.

As common perturbations affecting the performance of various radio systems, particularly those where direction measurement is involved, the TIDs present a problem about which more needs to be known.

Further reading

J. A. Ratcliffe and K. Weeks. The ionosphere. Chapter 9 in *Physics of the Upper Atmosphere* (ed. J. A. Ratcliffe). Academic Press, New York and London (1960).

H. Rishbeth and O. K. Garriott. *Introduction to Ionospheric Physics*. Academic Press, New York and London (1969). Chapter 5 on the morphology of the ionosphere.

K. Davies. *Ionospheric Radio*. Peregrinus, London (1990). Chapter 5 on morphology of the ionosphere, Section 9.3 on sudden ionospheric disturbances and Section 9.5 on storms.

A. P. Mitra. *Ionospheric Effects of Solar Flares*. Reidel, Dordrecht and Boston (1974).

S. Matsushita and W. H. Campbell. *Physics of Geomagnetic Phenomena*. Academic Press, New York and London (1977), particularly Chapter III-3 by T. E. Van Zandt on the quiet ionosphere and Chapter III-4 by R. Cohen on the equatorial ionosphere.

A. S. Jursa (ed.). *Handbook of Geophysics and the Space Environment*. Air Force Geophysics Laboratory, US Air Force, National Technical Information Service, Springfield, Virginia (1985). Chapter 9 on ionospheric physics, Section 10.3 on ionospheric modelling, and Chapter 10.7 on scintillations.

K. Rawer. Modelling of neutral and ionized atmospheres. *Encyclopaedia of Physics XLIX/7* (*Geophysics III part VII*, ed. K. Rawer), Springer-Verlag, Heidelberg, New York and Tokyo (1984).

R. C. Whitten and I. G. Poppoff. *Fundamentals of Aeronomy*. Wiley, New York, London, Sydney and Toronto (1971). Chapter 7 on ionospheric currents, and Chapters 8 and 9 on ionospheric behaviour.

S. Chapman and J. Bartels. *Geomagnetism*. Oxford University Press, Oxford, England (1940). Chapter 7, The solar daily variation on quiet days, S_q. Chapter 8, The daily magnetic variation, L.

P. Stubbe. Interaction of neutral and plasma motions in the ionosphere. *Encyclopaedia of Physics XLIX/6* (*Geophysics III part VI*, ed. K. Rawer), Springer-Verlag, Heidelberg, New York and Tokyo (1982).

R. W. Schunk and A. F. Nagy. Ionospheres of the terrestrial planets. *Reviews of Geophysics* **18**, 813 (1980).

J. D. Whitehead. Production and prediction of sporadic E. *Reviews of Geophysics and Space Physics* **8**, 65 (1970).

K. Davies. Recent progress in satellite radio beacon studies with particular emphasis on the ATS-6 radio beacon experiment. *Space Science Reviews* **25**, 357 (1980).

J. V. Evans. Ionospheric movements measured by incoherent scatter: a review. *J. Atmos. Terr. Phys.* **34**, 175 (1972).

H. G. Demars and R. W. Schunk. Temperature anisotropies in the terrestrial ionosphere and plasmasphere. *Reviews of Geophysics* **25**, 1659 (1987).

H. A. Taylor. Selective factors in Sun-weather research. *Reviews of Geophysics* **24**, 329 (1986).

J. J. Sojka. Global scale, physical models of the F region ionosphere. *Reviews of Geophysics* **27**, 371 (1989).

T. Beer. The equatorial ionosphere. *Contemp. Phys.* **14**, 319 (1973).

G. W. Prolss. Magnetic storm associated perturbations of the upper atmosphere: recent results obtained by satellite-borne gas analyzers. *Reviews of Geophysics* **18**, 183 (1980).

H. Volland. Dynamics of the disturbed ionosphere. *Space Science Reviews* **34**, 327 (1983).

H. Rishbeth. Thermospheric winds and the F-region: a review. *J. Atmos. Terr. Phys.* **34**, 1 (1974).

J. Taubenheim. Meteorological control of the D region. *Space Science Reviews* **34**, 397 (1983).

K. Labitzke. Stratospheric-mesospheric midwinter dynamics: a summary of observed characteristics. *J. Geophys. Res.* **86**, 9665 (1981).

B. H. Briggs. Ionospheric irregularities and radio scintillations. *Contemp. Phys.* **16**, 469 (1975).

J. A. Ratcliffe. Some aspects of diffraction theory and their application to the ionosphere. *Rep. Prog. Phys.* **19**, 188 (1956).

J. Aarons. Global morphology of ionospheric scintillations. *Proc. IEEE* **70**, 360 (1982).

K. C. Yeh and C. H. Liu. Radio wave scintillation in the ionosphere. *Proc. IEEE* **70**, 324 (1982).

S. Basu and S. Basu. Equatorial scintillations – a review. *J. Atmos. Terr. Phys.* **43**, 473 (1981).

G. S. Kent. Measurement of ionospheric movements. *Reviews of Geophysics and Space Physics* **8**, 229 (1970).

B. G. Fejer and M. C. Kelley. Ionospheric irregularities. *Reviews of Geophysics and Space Physics* **18**, 401 (1980).

R. D. Hunsucker. Atmospheric gravity waves generated in the high-latitude ionosphere. *Reviews of Geophysics and Space Physics* **20**, 293 (1982).

D. D. Rice, R. D. Hunsucker, L. J. Lanzerotti, G. Crowley, P. J. S. Williams, J. D. Craven and L. A. Frank. An observation of atmospheric gravity wave cause and effect during the October 1985 WAGS campaign. *Radio Science* **23**, 919 (1988).

S. H. Francis. Global propagation of atmospheric gravity waves – A review. *J. Atmos. Terr. Phys.* **37**, 1011 (1975).

H. Murata. Wave motions in the atmosphere and related ionospheric phenomena. *Space Science Reviews* **16**, 461 (1974).

C. R. Wilson. Infrasonic wave generation by aurora. *J. Atmos. Terr. Phys.* **37**, 973 (1975).

8

The ionosphere at high latitude

...he says in his sixth book that he observed...men of fire in heaven, who fought with lances, and who by this terrifying spectacle foretold the fury of the wars which followed. Yet I was with him in the same town, and I protest...that I saw nothing similar to his description, but only an appearance which is sufficiently common, in the form of pavilions in the sky flaming up and fading out again, as is usual with such meteors.

La Mothe le Vayer, 78th letter *De la Crédulité* (17th century).
Quoted by A. Angot in *The Aurora Borealis*, (1896)

8.1 Dynamics of the polar ionosphere

Chapter 5 described how the magnetosphere circulates as two regions, an inner one rotating daily with the Earth, and an outer one circulating under the influence of the solar wind. The polar ionosphere is connected by the geomagnetic field-lines to this outer region, and – since the field-lines are (almost) equipotentials – its circulation is essentially a projection of that of the outer magnetosphere.

8.1.1 F-region circulation

In the F region, where the ion–neutral collision frequency is small relative to the gyrofrequency, the plasma moves with the magnetic field-lines. Alternatively, we can say that the electric field which the solar wind generates across the magnetosphere (Section 5.5.3) is mapped into the F region along the equipotential field-lines. The polar-cap electric field so created (as measured by a stationary observer) then acts as the driving force for the F-region plasma (Sections 2.3.7 and 6.5.4). The integral of the electric field gives the total electric potential across the polar cap. This is nearly equal to the total potential across the magnetosphere between its dusk and dawn sides, and is an important parameter for the behaviour both of the magnetosphere and of the polar ionosphere. Its magnitude is around 60 kV.

We would expect the electric potential across the magnetosphere to depend on vB, where B is the flux density of the inter-planetary magnetic field embedded in the solar wind of velocity v. From satellite measurements it is found that the polar cap potential, ϕ, is given by an equation of the form

$$\phi = \mathrm{a} + \mathrm{bvB}\sin^2(\theta/2), \tag{8.1}$$

where θ is the angle defined in Figure 5.63. Various determinations have been made of

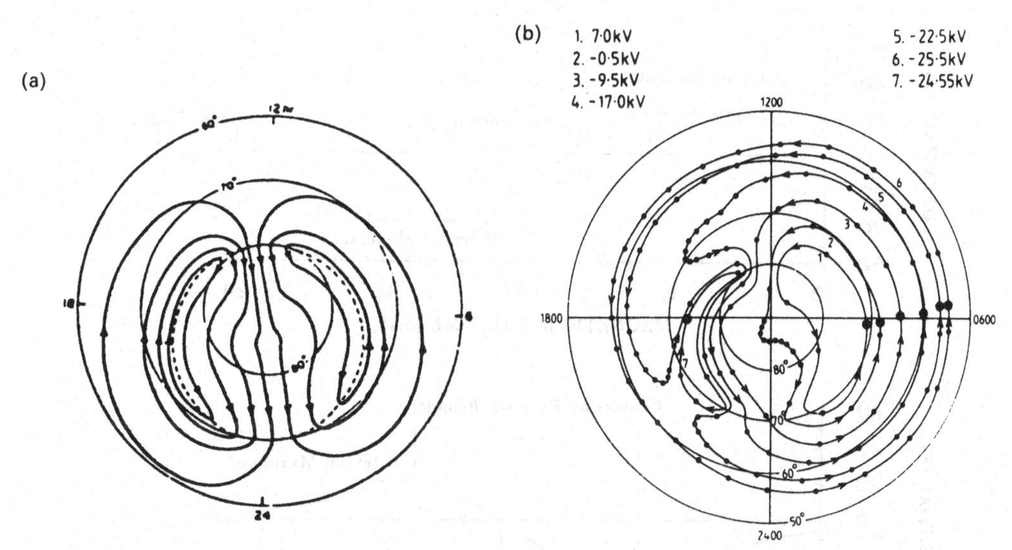

Fig. 8.1 Plasma convection at high latitude. (a) Polar convection pattern without co-rotation. (R. W. Spiro *et al.*, *J. Geophys. Res.* **83**, 4255, 1978) (b) Examples of convection paths of plasma at 300 km altitude in the northern hemisphere under the combined electric fields due to the magnetosphere and to co-rotation. The large dots indicate the starting points used in the calculation of the paths. The time between successive dots is one hour, except for the return to the starting point. Each path is an equipotential, whose value is indicated. The boundary of the polar cap is a circle (not marked) of radius 15°, centered 5° towards midnight from the geomagnetic pole. (After S. Quegan *et al.*, *J. Atmos. Terr. Phys.* **44**, 619, copyright (1982) Pergamon Press PLC)

a and b; a typical result is that if ϕ is in kV, v in km/s and B in nT, the data give b $\sim$ 0.04 and a $\sim$ 0. The formula has been verified up to about 120 kV, and it holds for both solar maximum and minimum conditions.

The basic flow pattern caused by the polar-cap electric field is simple enough: from the noon sector to the midnight sector directly over the pole, as sketched in Figure 8.1(a). The speed is typically several hundred metres per second. There is a return flow around the low-latitude edge of the polar cap, in the vicinity of the auroral oval, and that corresponds to the sunward flow of closed field-lines at the flanks of the magnetosphere. However the co-rotation effect, conveniently represented by the co-rotation electric field (Section 5.6.2), must also be included, and it distorts the flow pattern as shown in Figure 8.1(b). The two circulation cells are now different, and there is a marked distortion of the evening cell where the return flow and the co-rotation act in opposite directions. Some field-lines now follow long, complicated paths, while others may circulate endlessly in small vortices – except that the whole pattern must in any case be constantly changing in response to variations of the solar wind.

If the flow is regarded as $\mathbf{E} \times \mathbf{B}$ drift, it may be considered a consequence of a distribution of electric potential having a maximum on the morning side and a minimum on the evening side. Simplified but useful models can then be established, specified essentially by the magnitudes and locations of the maximum and the minimum. Figure 8.2 illustrates a typical example, showing the potential (a) as a

(a)

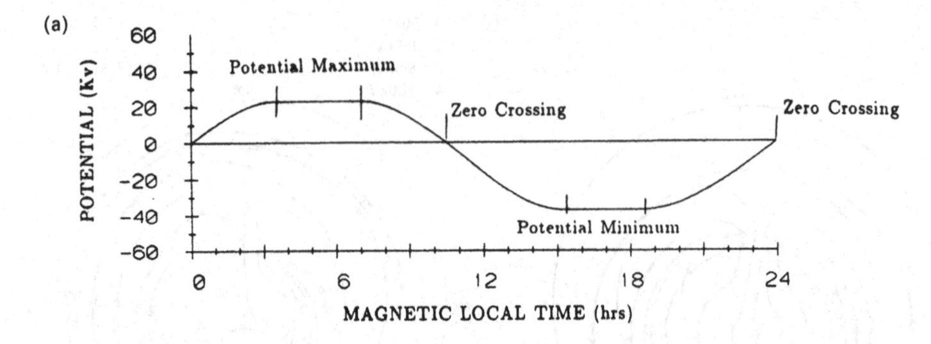

(b)

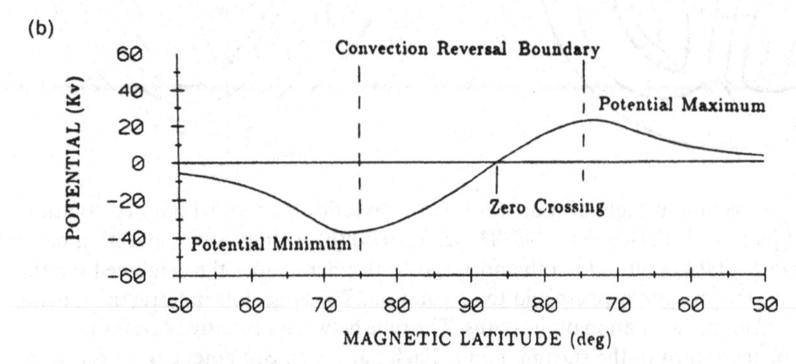

Fig. 8.2 Electric potential distribution between the magnetic pole and 50° latitude: (a) at constant magnetic latitude; (b) along a given magnetic meridian. (R. A. Heelis, private communication)

function of magnetic local time at constant magnetic latitude, and (b) along a meridian as might be observed by a satellite crossing the polar cap. The electric field is obtained from the gradient of the potential, and the velocity is given by $\mathbf{v} = \mathbf{E} \times \mathbf{B}/|\mathbf{B}|^2$ (Equation 2.40). Representations of this kind are useful as inputs to global models of thermosphere circulation (Sections 4.4, and 8.1.2).

The inter-planetary magnetic field (IMF) exerts a major influence. Because of stronger coupling at the magnetopause (Section 5.5.1), the magnetosphere circulates most strongly when the IMF has a southward component. The patterns in Figure 8.1 are for southward IMF. The east–west component also affects the circulation (Figure 8.3), presumably because of shifting connection regions at the magnetopause (Figure 8.4); in particular it alters the relative size of the circulation cells. The effect of the east–west IMF component (usually called B_y) should be opposite in northern and southern hemispheres and this prediction has been verified by HF coherent radar observations (Section 3.6.3).

When the IMF is northward the circulation pattern is more complicated. Distorted two-cell and three-cell patterns have been proposed. Since it takes several hours for the circulation to settle into a new pattern following a change in the IMF, and, since the IMF is always changing to some extent, there will obviously be some times when the polar circulation is in some ill defined state of transition.

Across some boundaries between circulation cells (particularly during northward IMF) the plasma speed will alter abruptly or may even reverse. As sketched in Figure

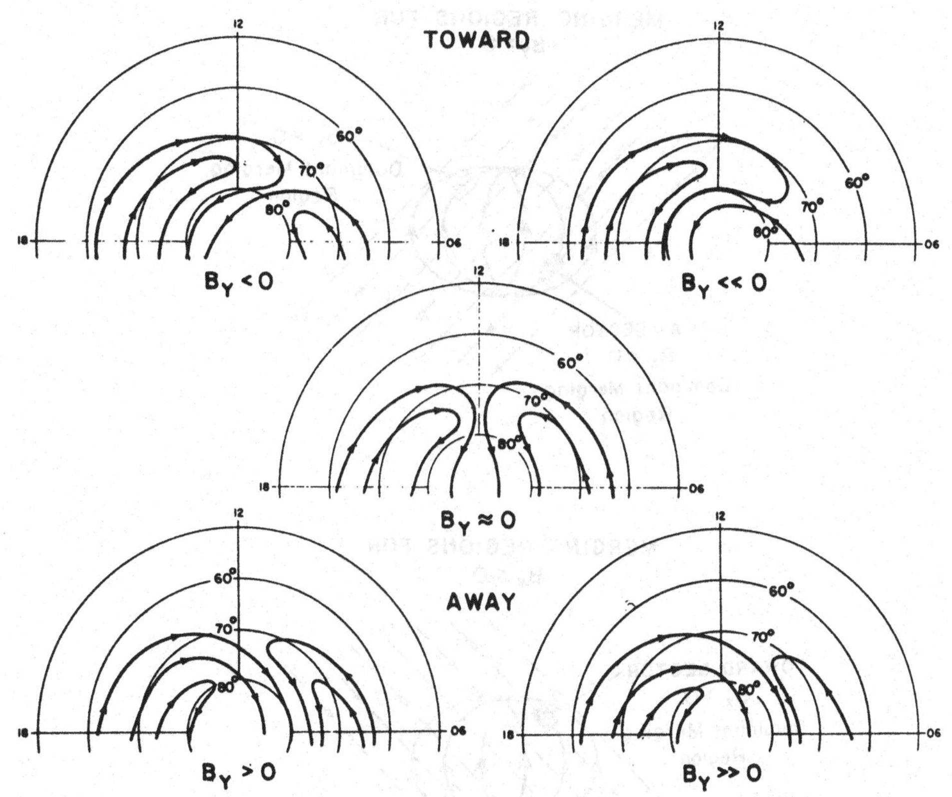

Fig. 8.3 Effect of the 'east–west' IMF component, B_y, on polar convection. The north–south component, B_z, is negative, and the radial component, B_x, is constant. (R. A. Heelis, *J. Geophys. Res.* **89**, 2873, 1984, copyright by the American Geophysical Union)

8.5, this implies a sharp change or reversal of electric field and of the Pedersen current (Section 6.5.5) in the E region. For continuity, current then flows up the field-lines (a Birkeland current), and the corresponding downward flow of electrons is probably the cause of the sun-aligned arcs observed in the polar cap when the IMF is northward (Section 8.2.1).

8.1.2 Interaction with the neutral air

As the plasma of the F region drifts with the magnetic field-lines over the polar cap, it interacts with the neutral air due to collisions between the ionized and neutral particles. The effect is similar to that of the thermospheric wind on the ionosphere (Section 7.1.3), except that the primary motion is now of the field-lines and the ions rather than of the neutral air. The theory of air drag and ion drag was considered in Section 7.2.4. Provided that the air is stationary or is moving more slowly than the magnetic field, the interaction tends to lift the F region as it approaches the magnetic pole from the day side, and to depress it as it moves to lower latitude on the night side. We have to remember, also, that the neutral air is already in motion due to the solar tide (Section 4.2.4) and the relative motion is what counts.

If a given pattern of plasma circulation is maintained for long enough the neutral air

MERGING REGIONS FOR
$B_y > 0$

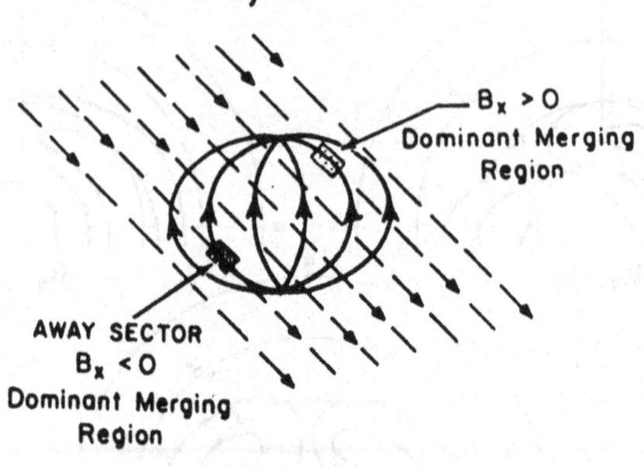

MERGING REGIONS FOR
$B_y < 0$

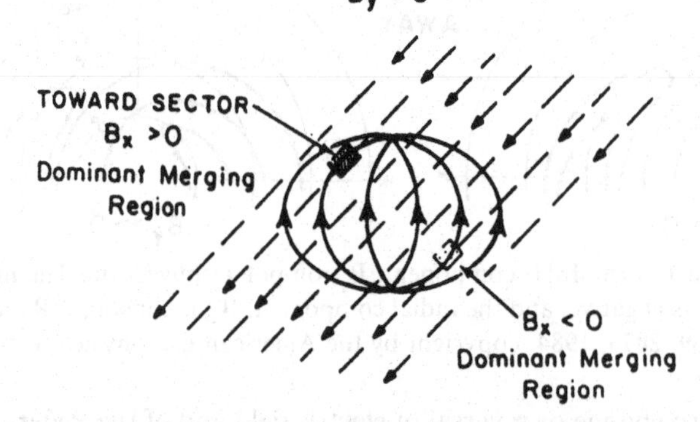

Fig. 8.4 Geometry of the IMF and geomagnetic field viewed from the Sun. Regions of preferred merging for different orientations of the IMF are indicated by the shaded boxes. The principal merging region changes its location according to the 'Sun–Earth' (B_x) and 'east–west' (B_y) components of the IMF. (R. A. Heelis, *J. Geophys. Res.* **89**, 2873, 1984, copyright by the American Geophysical Union)

will tend to catch up, and the thermospheric wind, otherwise due to the solar tide, will be modified. In general, therefore, the wind in the polar thermosphere is driven by two forces of quite different origin: the general circulation of the atmosphere due to solar heating, and the general circulation of the magnetosphere due to the solar wind. In an average situation these forces act together along the noon–midnight meridian over the poles, but they are opposed at dawn and dusk at somewhat lower latitudes where the magnetospheric flow drives plasma sunward. However, because of the time delay between a change in the plasma drift and the response of the neutral air (see Equation 7.16 and Table 7.2) the situation at any one time may well be unclear.

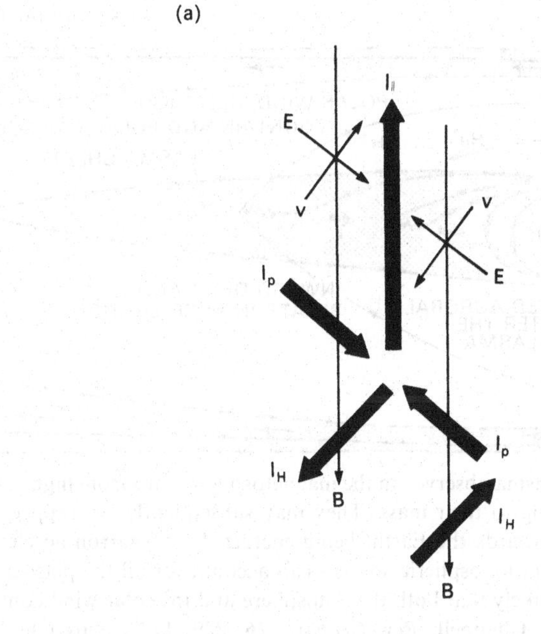

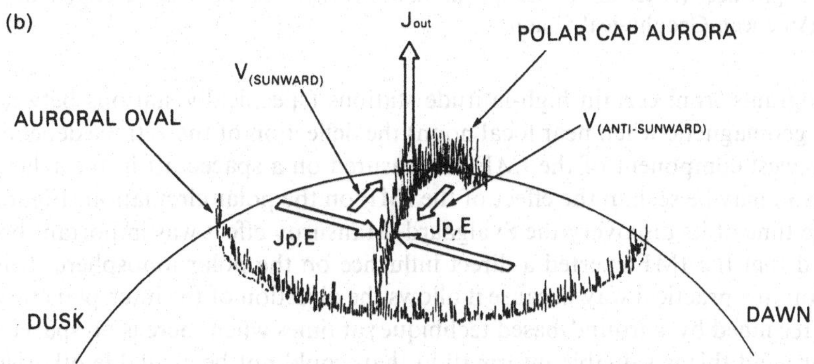

Fig. 8.5 (a) Field-aligned current due to plasma velocity shear. **B**: magnetic field. **v**: velocity. **E**: electric field. I_H, I_P, $I_\parallel$: Hall, Pedersen, field-aligned currents. (b) Field-aligned current associated with polar cap aurora at the boundary between circulation cells. (After H. C. Carlson *et al.*, *Adv. Space Res.* **8**, 49, copyright (1988) Pergamon Press PLC)

8.1.3 The S_q^p current system

In the dynamo region (Section 6.5.3), ion–neutral collisions are important but electron–neutral collisions are not: i.e. $\nu_i > \omega_i$ but $\nu_e < \omega_e$. Thus, the electrons move with the magnetic field-lines but the ions get left behind, and the result is an electric current oppositely directed to the field-line motion. This is the *S_q^p system* – that is, the polar part of S_q – and it is represented by the pattern of Figure 8.1(a) with the arrows reversed.

From the preceeding discussion we would expect S_q^p to be affected by the IMF, and such is the case. The effect was discovered some 20 years ago and is known as the *Svalgaard–Mansurov effect*. The key observation in the discovery was that in

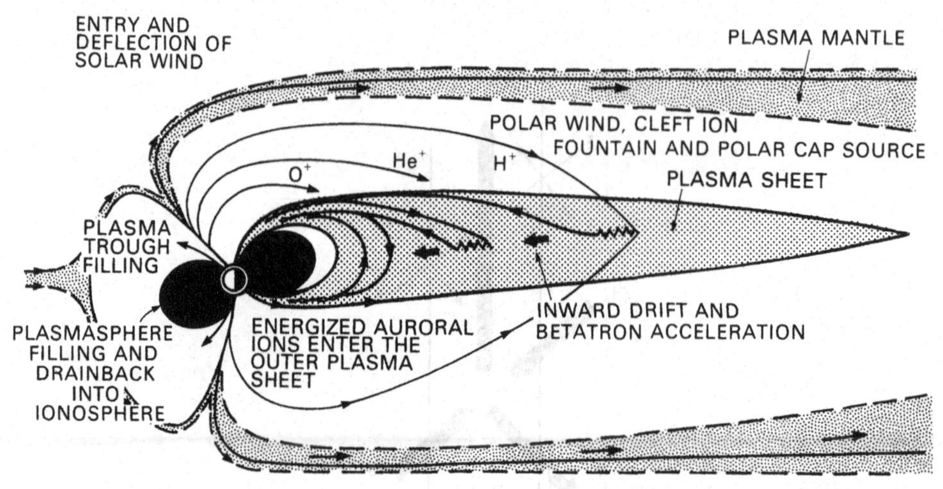

Fig. 8.6 Ionospheric sources of plasma observed in the magnetosphere. Ions from high latitudes tend to separate according to their mass. They may subsequently be trapped in the plasma sheet and drift towards the Earth, being energized by betatron acceleration. Computations suggest that ionospheric sources can account for all the plasma in the magnetosphere, but it is likely that both the ionosphere and the solar wind contribute in practice. (After C. R. Chappell, *Rev. Geophys.* **26**, 229, 1988, copyright by the American Geophysical Union)

magnetograms from certain high-latitude stations (specifically stations between 83° and 88° geomagnetic when near local noon) the deflection of the Z trace depended on the east–west component of the IMF as measured on a spacecraft in the solar wind. The reason may be seen in the effect of the IMF on the polar circulation (Figure 8.3).

At the time of its discovery the Svalgaard–Mansurov effect was important because it proved that the IMF exerted a direct influence on the polar ionosphere. It is also important in a practical way because it allows the direction of the inter-planetary field to be determined by a ground-based technique; at times when there is no spacecraft in the solar wind this is valuable information that would not be available otherwise.

8.1.4 Polar wind

The polar circulation carries field-lines through regions where they are open to the solar wind or go deep into the tail of the magnetosphere. For hot, light ions the scale height (Section 4.1.2) is very large, and if the field-lines are open there is no effective outer boundary. Thus the pressure at a great distance is effectively zero and the ions are too energetic to be held by gravity. While these conditions hold, plasma can flow away from the ionosphere continuously. This is the *polar wind* composed of ionospheric ions (H^+, He^+, O^+) and electrons. In theory the flow can even reach supersonic speeds, but the details depend on the flow speed at the assumed outer boundary. The term 'polar wind' is sometimes restricted to the supersonic regime, in which case subsonic flow would be called a 'polar breeze'.

There is still much to be learned about these polar outflows, and more observations are needed, but one important point, which has been established by satellite observations, is that the polar outflow is one of the sources of plasma in the

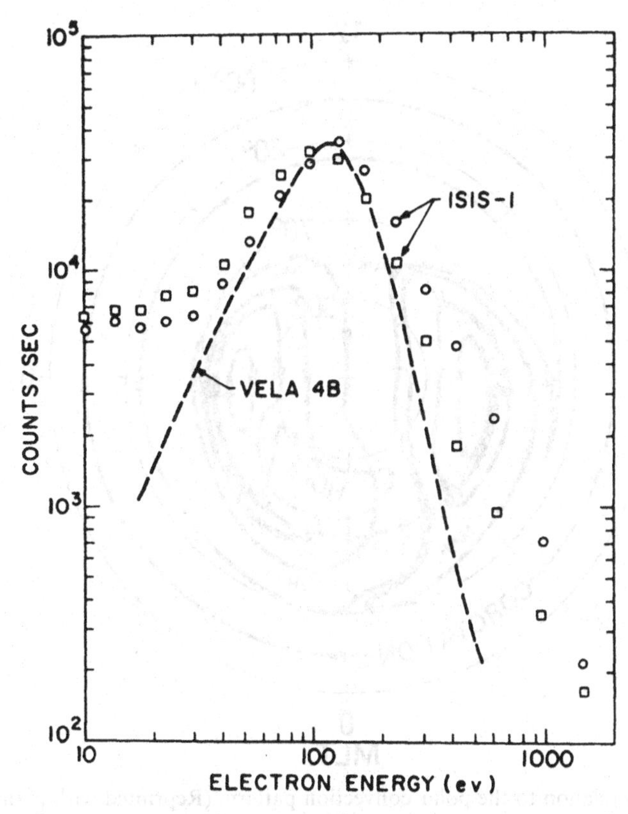

Fig. 8.7 Comparison between electron spectra measured on satellites in the ionospheric cusp region (ISIS-1) and in the magnetosheath (VELA-4B). (After W. J. Heikkila and J. D. Winningham, *J. Geophys. Res.* **76**, 883, 1971, copyright by the American Geophysical Union)

magnetosphere. The material is then subject to convection in the magnetospheric circulation and, it is estimated, reaches the tail plasma sheet within 50 R_E of the Earth.

A summary of terrestrial sources of magnetospheric plasma is shown as Figure 8.6.

8.1.5 The polar cusps

On the day side of the Earth are two regions, one in each hemisphere, where the geomagnetic field-lines provide a direct connection between the ionosphere and the magnetosheath (Section 5.4.4). In static models of the magnetosphere they correspond to the neutral points on the surface of the magnetosphere, marking the division between the closed field-lines at lower latitudes and those at higher latitudes that are swept back into the magnetotail. In the dynamic picture (Section 5.5) they are where the dayside field-lines open before being swept over the poles. The unique structure of these *cusps*, as sketched in Figure 5.37, makes them very interesting regions of the magnetosphere, but they also have significant effects within the ionosphere.

At ionospheric heights the cusp regions are recognized from two signatures:

(1) The appearance of charged particles with energies similar to those in the

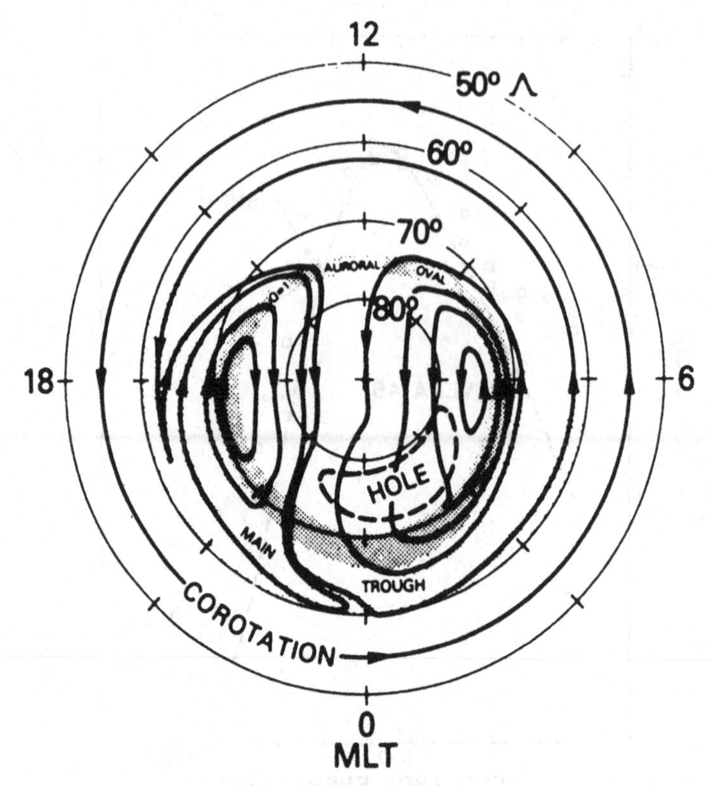

Fig. 8.8 Troughs in relation to the polar convection pattern. (Reprinted with permission from R. J. Moffett and S. Quegan, *J. Atmos. Terr. Phys.* **45**, 315, copyright (1983) Pergamon Press PLC)

magnetosheath. Figure 8.7 shows a remarkable agreement between spectra measured (though not at the same time) in the magnetosheath and in the ionosphere. The cusps are typically located near ±78° geomagnetic latitude, and are about 5° wide. According to the particle observations the cusps extend over all daylight hours and merge into the auroral oval (Section 8.3.1). There is also a second, smaller, region extending only a few hours from local noon. The particle flux from the magnetosheath is highly variable over short times (or over small distances, since these observations come from orbiting satellites).

(2) The enhancement of 630 nm luminous emissions, indicating low-energy excitation of the upper atmosphere (Section 6.4), and a reduction of those emissions typical of the aurora (Section 8.3.3). This latter feature is sometimes called the *noon gap*. The photometric observations have shown a considerable variation in the latitude of the cusp, from 84° under very quiet geomagnetic conditions to 61° when they are very disturbed.

The influx of particles from the magnetosheath enhances the density and temperature of the ionosphere in the cusps, and there is a greater degree of irregularity. Some of the ionospheric plasma also flows out to higher altitudes and into the magnetosphere, where its ionospheric origin has been recognized from its temperature and composition.

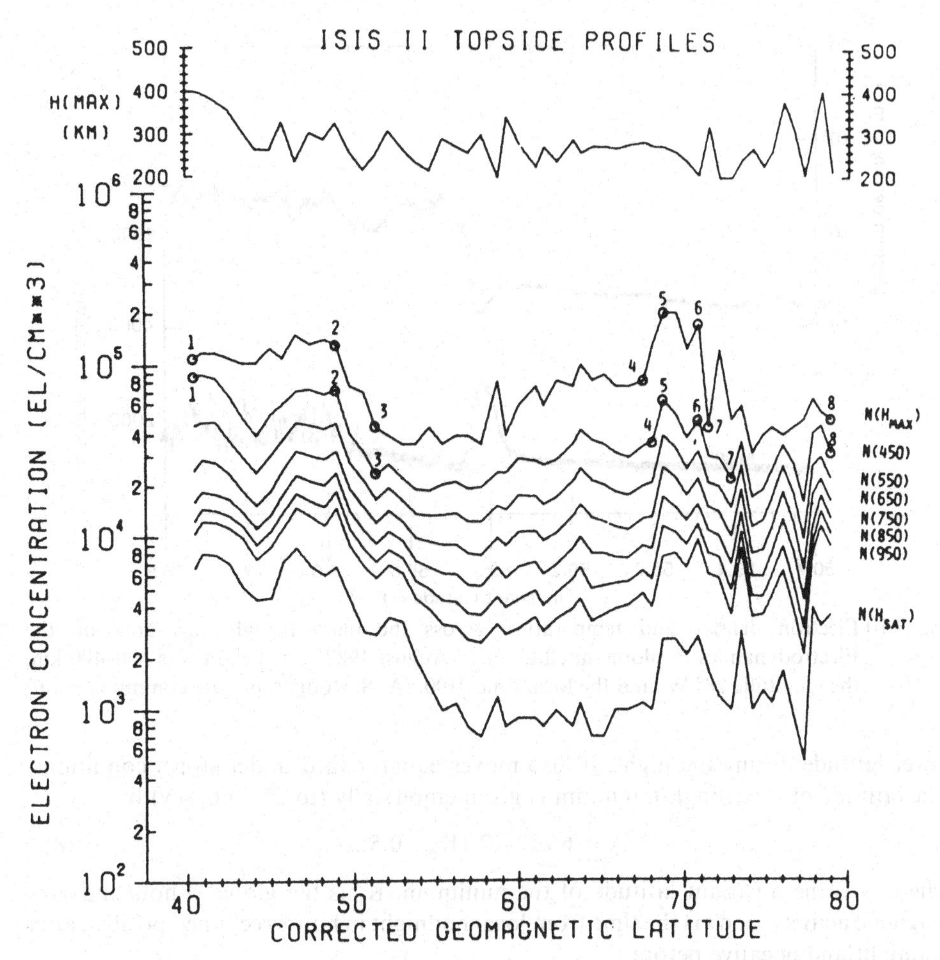

Fig. 8.9 Features of the main trough, from topside sounder observations. (M. Mendillo and C. C. Chacko, *J. Geophys. Res.* **82**, 5129, 1977, copyright by the American Geophysical Union)

8.1.6 Troughs

One of the consequences of the polar circulation is that the ionosphere tends to be depleted between the mid-latitude and high-latitude regions. Figure 8.8 shows the location of the depleted regions in relation to the polar circulation and to the auroral oval. The *main trough*, alternatively called the *mid-latitude trough*, was first detected with orbiting ionospheric satellites in the 1960s, when it was known as the 'Canadian border effect' since it appeared in many cases to coincide with the border between Canada and the USA! Subsequent work, both satellite- and ground-based, has amply confirmed the main trough as a major feature of the ionosphere, and one by no means confined to the western hemisphere. The main trough is interesting from several aspects. As observed in the ionosphere the trough is a nighttime occurrence at latitudes between 60° and 65° geomagnetic. It occurs in both hemispheres and in all seasons, though in summer it is only seen near midnight. The trough moves gradually towards

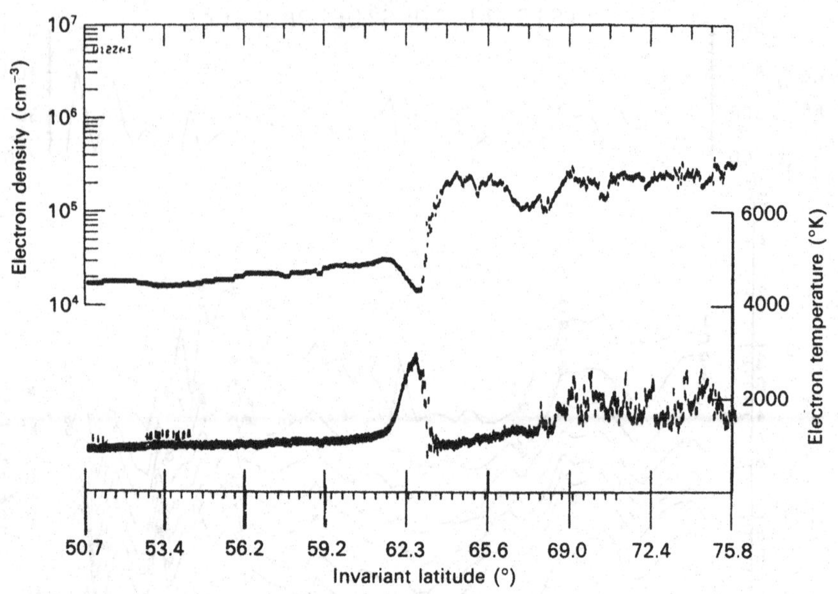

Fig. 8.10 Electron density and temperature across the main trough, measured on the Electrodynamics Explorer satellite on 17 August 1982. The height was 300–400 km, the longitude 28° W, and the local time 0100. (A. S. Rodger, private communication)

lower latitude during the night. It also moves equatorward under storm conditions. The latitude of the trough minimum is given empirically (to 2° accuracy) by

$$\Lambda = 65.2° - 2.1K_p - 0.5t, \tag{8.2}$$

where Λ is the invariant latitude of the minimum, K_p is the global 3-hour index of magnetic activity, and t is the time from local midnight in hours reckoned positive after midnight and negative before.

The detailed structure of the main trough varies considerably; Figure 8.9 illustrates typical features. The whole structure is quite broad, but the poleward edge, which is adjacent to the auroral oval, tends to be sharp and, depending on the technique, may be the most readily observed feature. The electron temperature is enhanced in the trough (Figure 8.10).

The trough is related in a general way to the plasmapause. Both are consequences of the change of circulation pattern between the inner and outer magnetospheres, but the correspondence is not exact because the immediate cause of the trough is the decay of ionization on flux tubes that have been away from the sunlit side of the Earth for too long a time. For example, the evening bulge in the plasmasphere (Section 5.6.2) does not show in the latitude of the trough. The 'hole' marked on Figure 8.8 is a depletion at higher latitude, not connected with the plasmapause but due to field-lines circulating only in darkness so that the ionization lost by recombination is not replenished.

At high altitudes, where H^+ is the principal ion, the trough is still observed. This *light-ion trough* usually occurs by day as well as by night. There are relationships, in some cases ill defined, with the heavy-ion trough and with the plasmapause. These

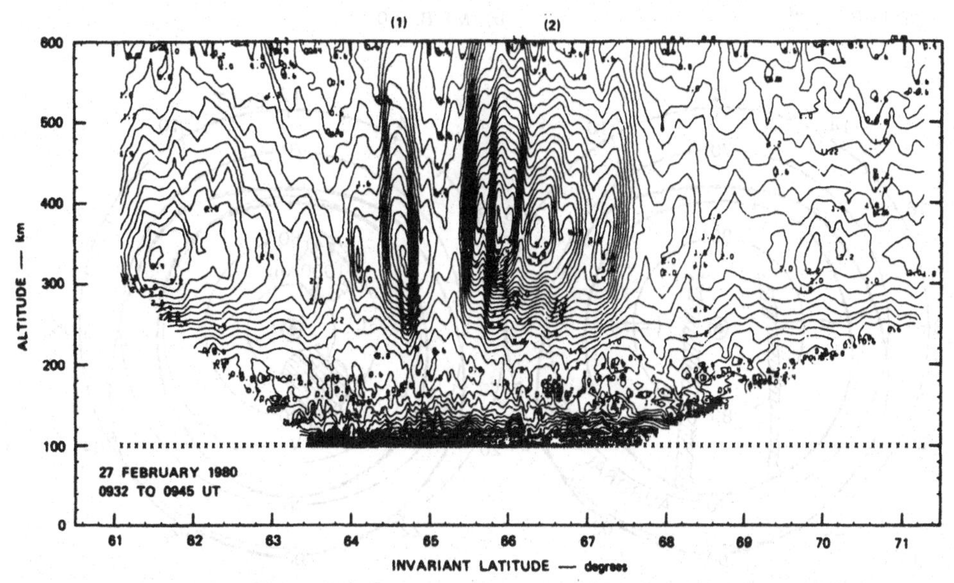

Fig. 8.11 Structure in the high-latitude ionosphere observed over Alaska by incoherent scatter radar during a latitude scan lasting 13 minutes. The contours of electron density are at 2×10^{10} m^{-3} interval. (M. C. Kelley *et al.*, *J. Geophys. Res.* **87**, 4469, 1982, copyright by the American Geophysical Union)

relationships are necessarily complicated, since they depend on the production and loss of ionization and on its flow between ionosphere and protonosphere, as well as on the polar circulation pattern, all of which depend to at least some extent on the state of geophysical disturbance. On the theoretical side, this is a topic which, because of its complexity, is ideally suited to computer modelling.

8.2 High-latitude irregularities

8.2.1 Blobs, enhancements or patches

Large regions of enhanced ionization are observed, by various techniques, in the polar cap and the auroral regions. An example from incoherent scatter radar is shown in Figure 8.11. The horizontal size of the patches is estimated as 50 to 1000 km, and they are remarkable for their high density, which is more typical of the daylight mid-latitude ionosphere and well above what we might expect to find in the polar ionosphere during the winter night. The patches can also be detected by the 630 nm airglow which they emit.

It seems clear that this type of enhancement is not produced locally, but was formed some distance away and has then drifted in the polar convection to the point of observation. Because the F region decays only slowly by recombination, the lifetime of the blobs should be quite long enough for them to cross the polar cap at a speed of several 100 m/s from a source on the day side. This possibility has been verified by computations which have also demonstrated how a change of polar circulation, for instance due to an increase in the solar wind flow, can detach plasma from the dayside cusp region and carry it over the pole into the midnight sector along a path such as in

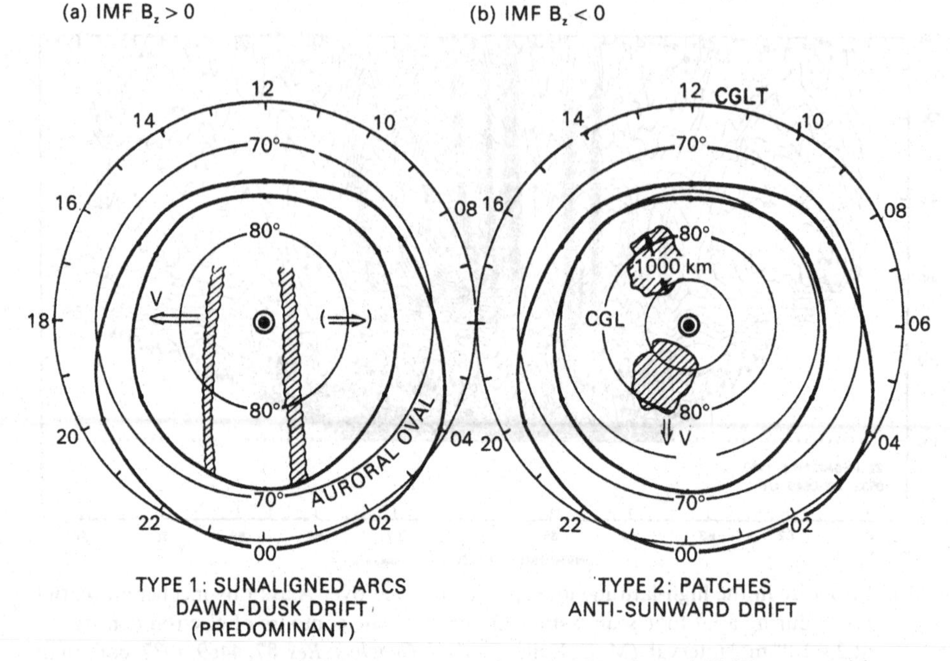

Fig. 8.12 Typical structures in the polar F region. (a) Northward IMF – arcs with noon–midnight alignment and dusk–dawn drift. (b) Southward IMF – patches drifting towards midnight. The coordinates are corrected geomagnetic latitude (CGL) and local time (CGLT), and the heavy lines indicate the position of the auroral oval. (After H. C. Carlson, private communication)

Figure 8.1b. What happens when the enhancement reaches the night sector is less clear, but it probably becomes stretched along the auroral zone in the return flow or merges into the mid-latitude ionosphere.

A different pattern is seen in the weaker circulation which occurs when the IMF has a northward component. At such times the airglow emissions form thin strips with noon–midnight alignment, and these drift slowly towards the dusk and dawn sides of the polar cap. Figure 8.12 compares the structures typically associated with northward and southward IMF.

In the auroral zone some of the enhancements peak low in the ionosphere, in the E region or the lower F region. These are probably a direct consequence of energetic particle precipitation at or near the point of observation.

8.2.2 Scintillation-producing irregularities

The irregularities of smaller scale produce scintillation phenomena in trans-ionospheric radio signals. At a radio frequency of 100 MHz the radius of the first Fresnel zone is about 1 km for a diffraction screen at a height of 300 km, and therefore amplitude scintillation is due to irregularities smaller than about 1 km. (See Section 2.6 for the theory of radio scintillation and Section 7.5.1 for a description of scintillation phenomena at middle and low latitudes.) Scintillation tends to be particularly severe at and around the auroral zone and it also occurs in the polar cap (Figure 7.40). The high-latitude scintillation zone is offset from the magnetic pole and shows a general correspondence with the auroral oval (Section 8.3.1), being nearer to the equator in the night sector. The occurrence and intensity maximize at night. In these scintillations the fading can be rapid, with periods as short as 1–3 s. The amplitude measured between the peaks and the dips may be as much as a factor of 6, or 15 dB in power.

The depth of scintillation also depends on the direction of propagation between sender (e.g. a satellite) and receiver (a ground station). The fading tends to become more severe with the obliquity of the ray because a longer path through the ionosphere is likely to include more irregularities. On the other hand, individual irregularities are strongly field-aligned and for this reason their effects are also enhanced along the local magnetic field direction.

In-situ measurements of electron density fluctuations can be made using satellite-borne probes, though the high velocity of an orbiting satellite limits the structure that can be resolved in this manner. In Figure 8.13, illustrating measurements of electron (or ion) density on an orbiting satellite, the fluctuations are as much as 20% of the mean. The figure also shows two patches (Section 8.2.1).

Cases have been identified where the small-scale irregularities which produce scintillations were located at edges of large-scale enhancements. Mechanisms such as the gradient drift and Kelvin–Helmholtz instabilities (Section 2.8) can cause a large patch to break up at the edges, thereby generating smaller ones. The situation appears to be that larger structures may progressively cascade to smaller ones with the passage of time.

8.2.3 E-region irregularities

The technique of coherent scatter radar, which was described in Section 3.6.3, is able to detect echoes from E-region irregularities and to determine their line-of-sight velocity from the Doppler shift of the returned signal. Using two stations (as in Figure 3.15) it is possible to obtain the velocity in magnitude and direction in the horizontal plane. Because the echoes are strongly aspect sensitive, the radars have to be carefully sited to the equatorward side of the auroral zone, so that where the beams intersect in the E region they are also normal to the geomagnetic field. As with scatter radar in general, the wavelength of the radar selects the size of the irregularities detected (e.g., the irregularity wavelength is half the radar wavelength for backscatter).

As in the equatorial region (Section 7.5.4) the high-latitude irregularities of the E region are associated with electric currents (the auroral electrojet – Section 8.4.2), and the radar echoes to which they give rise are again of two main types. Type 1 echoes have narrow spectra peaking near the ion acoustic speed, whereas Type 2 spectra are broader and peak at a smaller value of Doppler shift. They are respectively attributed to the two-stream and gradient drift mechanisms (Section 2.8). Both types may occur together. The Type 1 echoes are subject to a threshold of the electric field, consistent with the two-stream theory, but there are differences from those observed in the equatorial electrojet, most notably that the deduced velocity may far exceed the ion acoustic speed in the auroral zone but not in the equatorial zone. There is also a difference due to the observing geometry, since the radar would normally look almost along the direction of the current in the equatorial region but almost normal to the current at high latitude.

According to the simple theory – 'simple' being a relative term in this context! – the

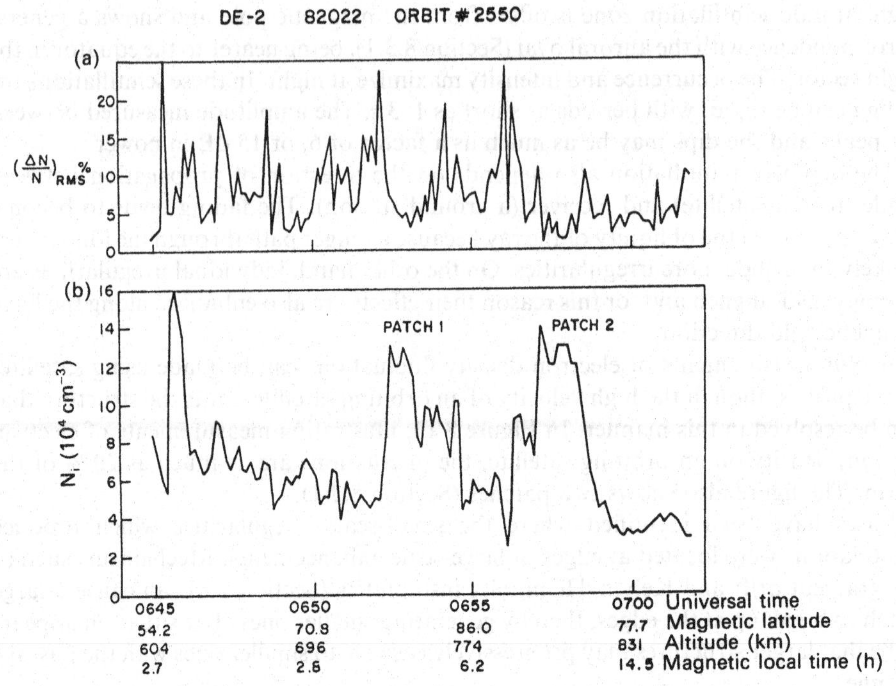

Fig. 8.13 (a) Relative irregularity and (b) ion density measured on a satellite crossing the polar cap. In (a) the variation ΔN was taken with respect to a linear least-squares fit to the electron density measured for 3 seconds; the plotted ΔN/N therefore refers to irregularities smaller than about 25 km. (After S. Basu *et al.*, *The Effect of the Ionosphere on Communication, Navigation, and Surveillance Systems* (ed. Goodman), p. 599. *Proc. Ionospheric Effects Symposium*, Springfield, Virginia, 1987)

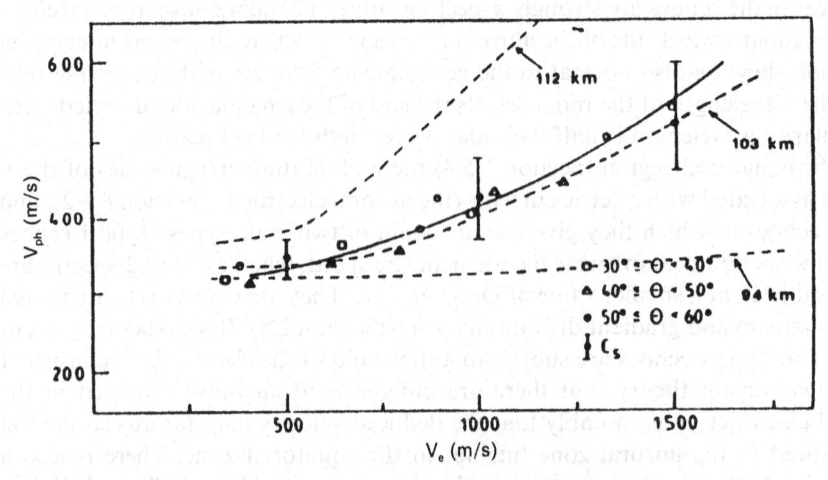

Fig. 8.14 The relation between the phase velocity of irregularities (v_{ph}) and the drift velocity of electrons (v_e) in the electrojet. Experimental results (——) are adapted from Nielsen and Schlegel (1985), and the theory (-----) from Robinson (1986). c_s is the ion-acoustic speed. (After C. Haldoupis, *Annales Geophysicae* 7, 239, 1989)

phase velocity of Type 1 echoes should vary in proportion to the electron velocity in the scattering region: $v = \mathbf{E} \times \mathbf{B}/|B|^2$ where $\mathbf{E}$ is the electric field. Hence the velocities of irregularities measured by coherent scatter radar have been interpreted as the electric field in the E region. However, direct comparison with electric fields measured by incoherent scatter radar showed that the relationship was not in fact linear (Figure 8.14). Modifications have consequently been made to the theory, essentially by including the increase of effective collision frequency (which comes into Equation 2.89) due to heating by the plasma waves. This non-linear theory appears to fit the observations.

8.3 The aurora

8.3.1 Introduction

The aurora is the most readily observed consequence of the dynamic magnetosphere and the most obvious characteristic of the high-latitude ionosphere. Mankind must have been looking at the northern and southern lights in the night sky, the *aurora borealis* and the *aurora australis*, for thousands of years, which surely puts them amongst the oldest known geophysical phenomena; though it is only in the last part of the 20th century that any proper explanation has become possible. The term 'aurora borealis' dates from 1621. There are detailed reports of auroral displays dating from 1716 and the first written work devoted entirely to the polar aurora was published in France in 1733, but accounts of lights in the night sky, which were frequently given a mystical or prophetic interpretation, go back to Greek and Roman times.

Early 'theories' range from the fanciful to some that were unwittingly near the truth: sunlight reflected from the polar snow and ice (wrong), 'elastic fluids' (vague), electric discharges (plausible). Only in the early 1950s was it proved that the immediate cause of the auroral emissions is excitation of atmospheric gas by energetic particles, and it was not until 1958, when rockets were fired into an aurora, that energetic electrons were identified as the primary source. An understanding of the source of those electrons has had to wait for a fairly detailed knowledge of the structure and physics of the magnetosphere. In fact that knowledge is not yet complete, though current ideas are well advanced and stand on firm ground.

The aurora comprises a group of upper-atmosphere phenomena, not only the emission of light. Each is a direct or an indirect consequence of the entry of energetic particles from the magnetosphere to the atmosphere:

(a) luminous aurora;
(b) radar aurora, the reflection of radio signals from ionization in the auroral region;
(c) auroral radio absorption, the absorption of radio waves in the auroral ionization;
(d) auroral X-rays, generated by the incoming particles and detected on high-altitude balloons;
(e) magnetic disturbances, due to enhanced electric currents flowing in the auroral ionization and detected on magnetometers;
(f) electromagnetic emissions in the very-low- and ultra-low-frequency bands,

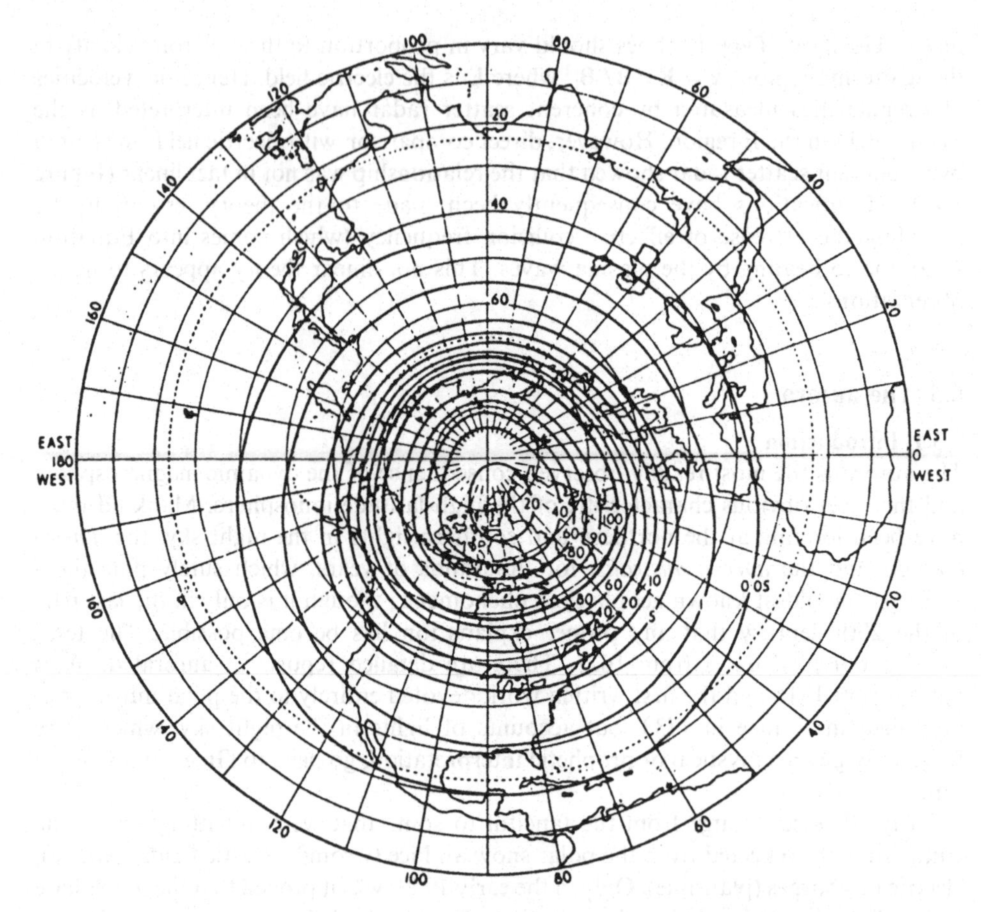

Fig. 8.15 Northern auroral zone, showing the percentage of good observing nights when aurora may be seen. (D. R. Bates, in *Physics of the Upper Atmosphere* (ed. Ratcliffe). Academic Press, 1960)

generated in the magnetosphere by wave–particle interactions (Section 9.1), and propagating to the ground for detection with a receiver or a sensitive magnetometer.

The auroral phenomena have several features in common. They are all related to solar activity in a general way, though usually without specific association with any particular solar event. The term *M region* was used from the 1930s to denote an unseen solar region causing aurora and magnetic storms, and for 40 years it served as a unifying hypothesis. It is now understood, of course, that the connection from the Sun is via the solar wind.

In general the auroral phenomena are highly structured in both space and time, with occurrence patterns that are essentially zonal. The classical picture of auroral occurrence (Figure 8.15) shows 100% occurrence at the peak of a zone centered between geomagnetic latitudes 65° and 70°, with the occurrence rate falling off on both the equatorward and the poleward sides. But in 1963, Y. I. Feldstein, analysing data

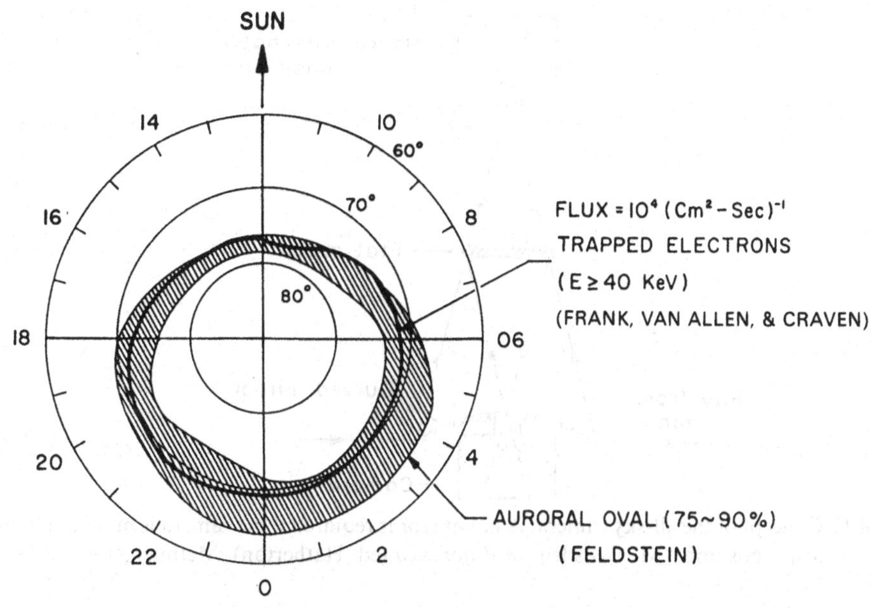

Fig. 8.16 Auroral oval in relation to the 40 keV trapping boundary. (S.-I. Akasofu, *Polar and Magnetospheric Substorms*. Reidel, 1968. Reprinted by permission of Kluwer Academic Publishers)

from the International Geophysical Year (1957–8), pointed out that the locus of the aurora at a fixed time is not circular but oval (Figure 8.16). The maximum is near 67° at midnight, increasing to about 77° at noon, and the classical zone is the locus of the midnight sector of the oval as the Earth rotates underneath it. To a first approximation the oval is fixed with respect to the Sun. The *auroral oval* is one of the important boundaries of geospace. In relation to the structure of the magnetosphere it is generally considered to mark the division between open and closed field-lines.

The third common feature of the auroral phenomena is that they all show substorm behaviour. They tend to occur in bursts, each lasting perhaps 30–60 min, which are separated by quiet intervals of several hours. In each substorm the auroral oval first becomes active in the midnight sector, and the activity then spreads in latitude and also to other local times. The auroral substorm is, of course, a consequence of the substorm in the magnetosphere, whose behaviour was discussed in Section 5.9, though the auroral substorm was in fact identified first. It will be considered further in Section 8.4.4.

8.3.2 Luminous aurora: distribution and intensity

The classical investigations of the aurora were limited to the luminosity, and before 1950 they proceeded along two largely independent lines. The purpose of morphological studies was to map the occurrence of the aurora in space and time, and to determine the details of the fine structure of individual forms. The second line of study was auroral spectroscopy, which was concerned with the emitted light, with its spectrum and its origin in photochemical processes – a topic which has strong affinities with airglow (Section 6.4).

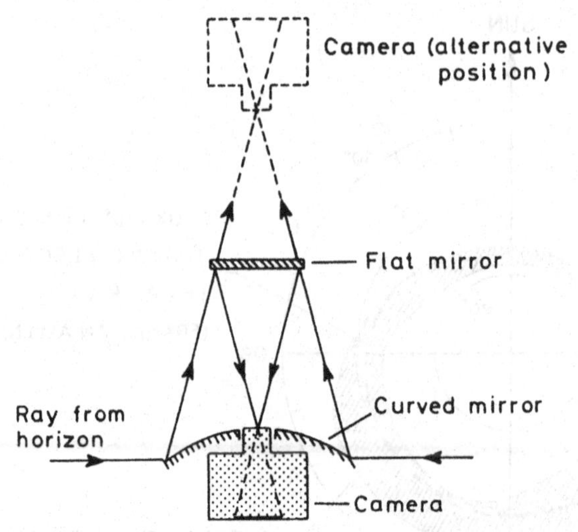

Fig. 8.17 Concept of the all-sky camera. A flat mirror is required if the camera is mounted in the lower position. (M. Gadsden, in *Antarctica* (ed. Hatherton). Methuen, 1965)

The luminous aurora is highly structured. Some features are as thin as 100 m, and time changes can be as rapid as 10 s^{-1}. The *all-sky camera,* first used during the 1950s and one of the basic instruments of auroral photography, is used for surveying auroral occurrence. This instrument (Figure 8.17) uses a convex mirror to obtain a picture of the night sky from horizon to horizon, and, typically, it would be operated automatically at regular intervals during every clear night.

Auroral structure is classified according to its general appearance, as in Table 8.1. When structure is present the height of the luminosity may be determined by triangulation. Between 1911 and 1943, C. Störmer made 12000 height determinations with spaced cameras and found that the lower borders of auroral forms were usually around 100–110 km. In some of the forms the luminosity is concentrated into a band only 10–20 km deep (Figure 8.18), and the lower edge in particular can be quite sharp. The brightness of a discrete arc typically falls to 10% within a few kilometres below the maximum, and to 1% a kilometre or two below that.

The intensity of an aurora may be rated on a scale of I to IV, as in Table 8.2, which also shows the standard that a visual observer might use for comparison, the equivalent in kilorayleighs (Section 6.4), and the approximate rate of energy deposition. For exact intensity measurement a photometer would be used, which may either be pointed in a fixed direction, for example to the zenith, or scanned so that the spatial distribution of intensity is determined as well. Neither scanning photometers nor cameras have sufficient sensitivity to record the most rapid fluctuations in the auroral light, but TV techniques, both monochrome and colour, are more sensitive and have been applied very successfully to dynamic auroral photography in recent years. The aurora is a phenomenon of great visual beauty, and, in addition to their considerable scientific value, these colour recordings enable a wider audience to experience some of the awe and fascination of a great auroral display.

Until the middle of the present century, auroral studies concentrated on the discrete forms, since they are spectacular and their fine and active structure makes them more

Table 8.1 *Classification of Auroral forms*

Forms without ray structure
Homogeneous ray structure: a luminous arch stretching across the sky in a magnetically east–west direction; the lower edge is sharper than the upper, and there is no perceptible ray structure
Homogeneous band: somewhat like an arc but less uniform, and generally showing motions along its length: the band may be twisted into horsehoe bends
Pulsating arc: part or all of the arc pulsates
Diffuse surface: an amorphous glow withoutdistinct boundary, or isolated patches resembling clouds
Pulsating surface: a diffuse surface that pulsates
Feeble glow: auroral light seen near the horizon, so that the actual form is not observed
Forms with ray structure
Rayed arc: a homogeneous arc broken up into vertical striations
Rayed band: a band made up of numerous vertical striations
Drapery: a band made up of long rays, giving the appearance of a curtain; the curtain may be folded
Rays: ray-like structures, appearing singly or in bundles separated from other forms
Corona: a rayed aurora seen near the magnetic zenith, giving the appearance of a fan or a dome with the rays converging on one point
Flaming aurora: waves of light moving rapidly upward over an auroral form

readily observed against the background light of the night sky. However, it was demonstrated in the early 1960s that a diffuse glow is usually present at the same time. This component, originally known as the *mantle aurora* but more recently as the *diffuse aurora*, contributes at least as much total light as the discrete forms, though it is more difficult to observe from the ground because of its low intensity per unit area. The nighttime discrete and diffuse aurorae map along the geomagnetic field into significant regions of the magnetotail. The diffuse aurora is generally associated with the central part of the plasma sheet, and the discrete forms, which appear poleward of the diffuse aurora, are thought to map to the edge of the plasma sheet or to an X-type neutral line (Section 5.5.2).

In recent years, satellite observations have provided much new information about the distribution of the luminous emissions. By observing from a distance, a 'camera' on a satellite can view a large part of the oval, or even the whole of it, at the same time – and of course the earth-bound observer's battle against poor seeing conditions does not arise. In these pictures the diffuse aurora tends to dominate, but, in addition, discrete forms are seen within the diffuse glow or poleward of it; they are not seen on the equatorward side. The general form of the oval is clearly observed (Figure 8.19), and moreover it is proved to be more than just a statistical distribution because the oval appears as a continuous band of light around the pole which is virtually always present, though its intensity varies greatly from time to time. This confirms the implication of the classical auroral zone (Figure 8.15) that the aurora is virtually a permanent feature of the polar atmosphere.

Table 8.2 *Intensity classification of the aurora*

Intensity	Equivalent to	Kilorayleighs	Energy deposition (erg cm^{-2} s^{-1})
I	Milky Way	1	3
II	Thin moonlit cirrus	10	30
III	Moonlit cumulus	100	300
IV	Full moonlight	1000	3000

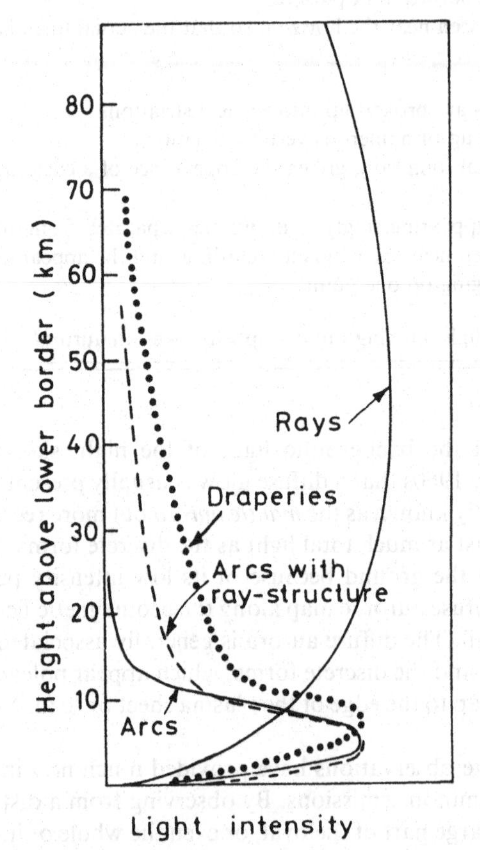

Fig. 8.18 Profiles of auroral luminosity along various forms. (B. Hultqvist, in *Physics of Geomagnetic Phenomena* (eds. Matsushita and Campbell). Chapman and Hall, 1967)

The latest auroral pictures from space are adding much detail and have shown that the oval form is not by any means the whole story. The details of the spatial distribution vary considerably from time to time. Sometimes an arc is seen to extend across the polar cap, connecting the day and night sides of the oval. This configuration is called a *θ-aurora*. Sometimes the morning side of the oval is quiet while the evening is active; sometimes the morning side is the more active. There are localized

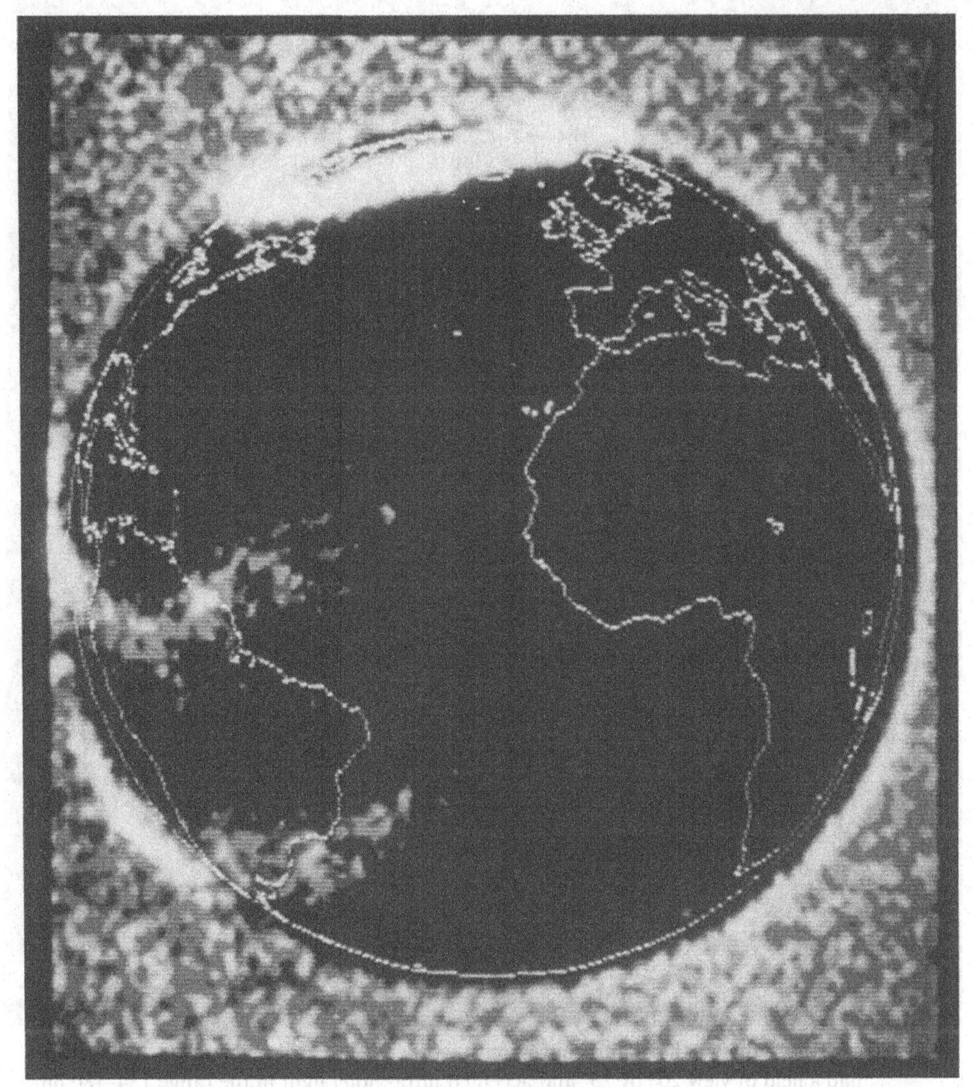

Fig. 8.19 The active auroral oval seen in the ultra-violet (118–165 nm) by the Dynamics Explorer I spacecraft on 16 February 1982. The picture was taken from an altitude of 3 Earth-radii above the dark side of the Earth. Also present are: (i) airglow bands north and south of the equator, extending from the evening (left) side; (ii) dayglow above the morning (right) limb of the Earth; (iii) resonant scattering of Lyman-α radiation in the atomic hydrogen of the exosphere. (L. A. Frank and J. D. Craven, University of Iowa, private communication)

brightenings that may expand, move eastward, or stay put. Some examples are shown in Figure 8.20. The variety of configurations and dynamics emphasizes the complexity of the auroral distribution and suggests that present classifications may be incomplete.

When the IMF is northward, luminous arcs extending for thousands of kilometres and aligned towards the Sun are observed in the polar caps. They are not bright

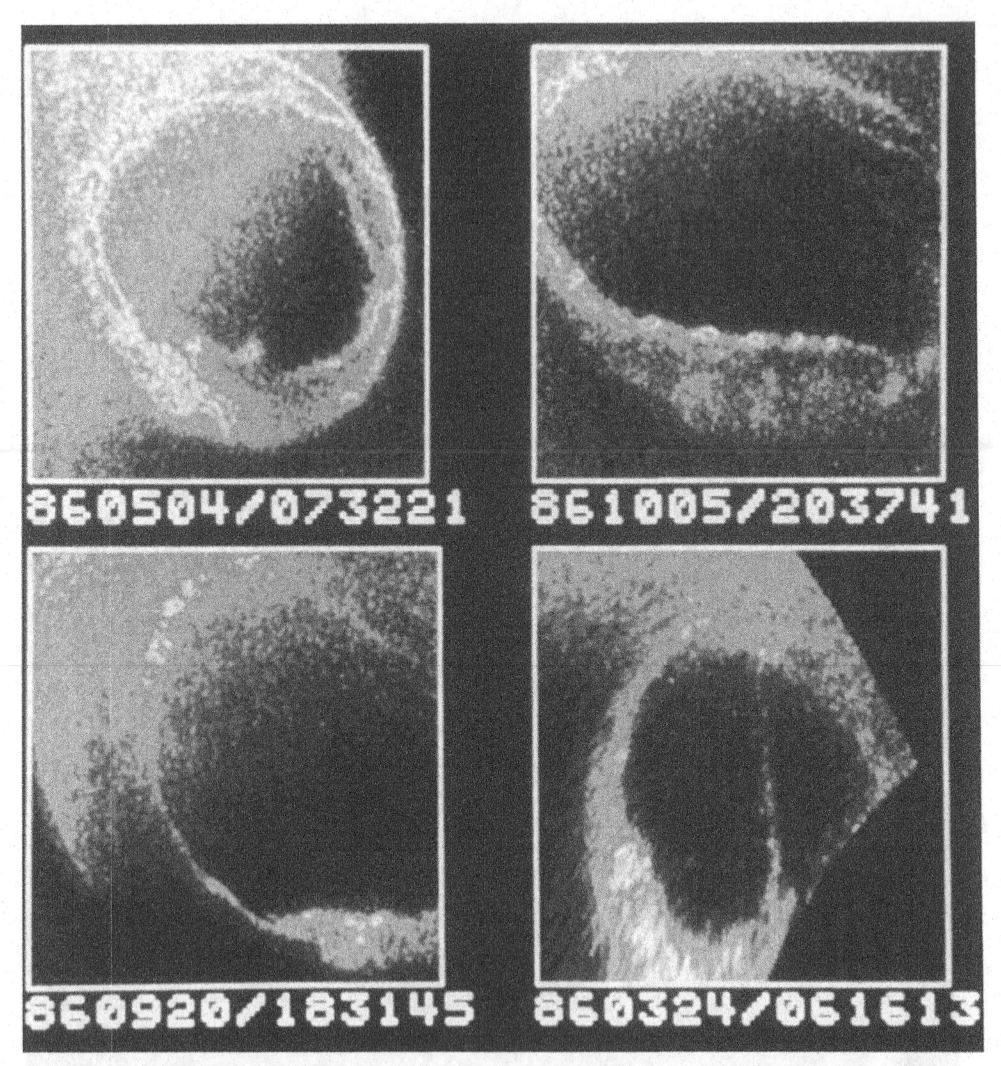

Fig. 8.20 Images of the northern auroral regions recorded by the Viking satellite. The camera had a field of view 20° by 25° and accepted ultra-violet light in the range 134–180 nm, emissions coming principally from molecular and atomic nitrogen. Each exposure lasted 1.2 s. The top left-hand image shows the whole auroral oval including the day side. The lower left-hand image is of a substorm in the midnight sector, with activity also around noon and faint arcs in the morning sector. This is a common type of image in the Viking data. During the last stages of a substorm, a series of regularly spaced bright spots, lasting 1–5 minutes, may appear along the poleward edge of the oval near the midnight sector, as illustrated in the top right-hand image. The fourth image shows a sun-aligned polar arc extending across from midnight (at the bottom) to noon. (Pictures and commentary from G. Enno, private communication)

(emitting only tens of rayleigh, against kilorayleighs for a normal aurora) but they can be detected with modern equipment and, at that low intensity, are observed about half the time. It appears, therefore, that they are almost always present when the IMF is northward. It is believed that these arcs are on closed field-lines and that they may be magnetically conjugate (i.e. occur simultaneously at opposite ends of field-lines in northern and southern hemispheres). The association between Sun-aligned arcs and velocity shears was mentioned in Section 8.1.1. The θ-aurora is also associated with a velocity shear, but it is much brighter and also much rarer than the common Sun-aligned arcs. It is not at present clear whether or not they are different phenomena.

8.3.3 Luminous aurora: spectroscopy

Auroral spectroscopy is based on the grating spectrograph, the image being recorded with a high speed camera, a TV system, a photomultiplier, photocell or infra-red converter. To observe a selected line a Fabry–Perot monochromator or an interference filter may be used. Rocket and satellite techniques have enabled the observations to be extended into the ultra-violet.

The auroral spectrum is complex. To the eye a bright aurora appears green or red, colours due to the atomic oxygen emissions at 557.7 and 630 nm respectively (See Figure 6.17). The 391.4 nm line of singly ionized molecular nitrogen (N_2^+) is present in the violet, and in an aurora of average intensity the overall effect is usually greyish yellow. Some rays are red at high altitude but green lower down. This is due to quenching of the ^1D state of atomic oxygen, as in the airglow (Section 6.4). These aurorae are called *Type A*. Some aurorae, *Type B*, have a red lower border. These occur at exceptionally low altitude when the incoming particles are unusually energetic, and the red light comes from enhanced molecular oxygen emissions, the green 557.7 nm line being quenched.

The interpretation of the auroral spectrum is anything but a simple exercise, since the lines are in general 'forbidden' according to the spectroscopists' parlance. This does not mean that the related transitions cannot occur at all, but that, because they defy the usual selection rules, they are relatively improbable. The green line was not identified until 1925, following the first accurate measurements of its wavelength two years previously. The next major advances came during the 1950s. In 1950, C. W. Gartlein identified the Hα line in the aurora, and the following year A. B. Meinel discovered a Doppler shift in the line. These observations showed beyond doubt that at least a part of the luminous auroral phenomenon is caused by the arrival of energetic protons from some higher level of the atmosphere. This accounts for the so-called *proton aurora*. Though it was not until several years later, when rockets were fired into aurorae, that the greater role of energetic electrons was discovered, the way was thereby cleared for the idea that the aurora is caused by energetic particles entering the atmosphere.

When the mechanisms of auroral luminosity are known, some of the lines may be employed for diagnostic purposes. The intensities of the N_2^+ emissions at 427.8 and 391.4 nm are proportional to the rate of ionization by the incoming electrons. The Doppler shift of the oxygen line at 630 nm can be applied to thermospheric wind measurements (Section 4.2.2).

8.3.4 Auroral electrons

The ionization of the upper atmosphere by energetic electrons entering from above was treated in Section 6.2.3, and if we know their energy and flux we can calculate many of the effects. A rocket flight, being of only a short duration, is most useful in

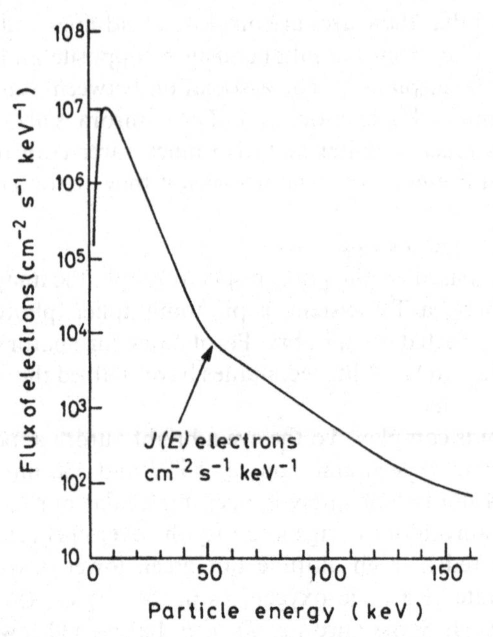

Fig. 8.21 A typical spectrum of auroral electrons (W. N. Hess, *Radiation Belt and Magnetosphere*. Blaisdell, 1968)

providing an essentially spot measurement of the flux, energy spectrum, and pitch angle as functions of the height. The first such flight, in 1960, showed a mono-energetic electron flux at 5 keV, but this is not typical. More usually the spectra cover a broad band of energies to 100 keV and beyond.

Some spectra approximate to a simple mathematical form such as the exponential,

$$N(E) = N_0 \exp(-E/E_0), \tag{8.3}$$

meaning that N(E) dE particles arrive per unit cross-sectional area per second in the narrow energy band between E and E+dE. The *characteristic energy*, E_0, might be 5–10 keV at the lower energies and 10–40 keV at higher energies. (It would be unusual for a single characteristic to hold over the entire measured spectrum.) Another useful form is

$$N(E) = N_0 E \exp(-E/E_0), \tag{8.4}$$

which corresponds to a Maxwell–Boltzmann distribution of energies and includes a maximum at $E = E_0$. Figure 8.21 shows a representative spectrum. Mono-energetic components may be superimposed on the background spectrum on some occasions.

Satellites in low orbit are able to map the spatial distribution of electron precipitation on the global scale (as in Figure 7.29). Such measurements can distinguish the division of the dayside auroral zone into two parts. The rapid latitude coverage provided by this type of orbit is convenient for integrating the total energy input from energetic electrons; the result of such estimates is that 0.1 to 0.2% of the energy of the solar wind intercepted by the magnetosphere appears as auroral particles. A geosynchronous satellite can monitor the electrons continuously at one location (relative to the Earth), and this orbit, at 6.6 R_E, is specially well placed for auroral

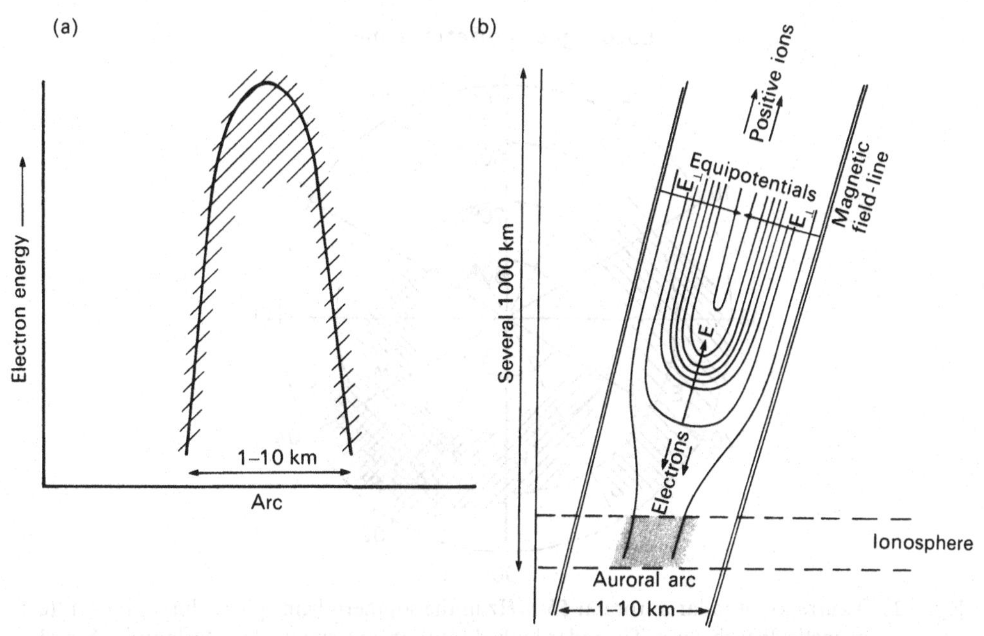

Fig. 8.22 (a) Variation of electron energy across an inverted-V event, as observed by a satellite passing above an auroral arc. (After D. A. Gurnett, in *Critical Problems of Magnetospheric Physics*. National Academy of Sciences, Washington DC, 1972) (b) Electric field structure above an auroral arc, creating an electric field parallel to the geomagnetic field. Electrons are accelerated downward, and positive ions upward. The acceleration is largest in the centre of the structure. (After G. Haerendel, in *The Solar Wind and the Earth* (eds. Akasofu and Kamide), p. 215. Reidel, 1987)

particle studies because it is on a field-line connected to the auroral zone. Provided there is no acceleration at lower altitude – which may not be strictly true – the geosynchronous flux can, via trapped particle theory, be related to the flux at the foot of the field-line.

One type of precipitation event observed from orbiting satellites is the *inverted-V*, so christened from the form of the spectrum as the satellite moves through the precipitation region. In an inverted-V the spectrum has a peak, and the energy at the peak is greatest in the middle of the region (Figure 8.22a). These events occur over auroral arcs and are connected with the mono-energetic components measured in some rocket flights. The pitch angle peaks along the magnetic field and the cause is believed to be acceleration by an electric field acting along the magnetic field, the potential distribution being as in Figure 8.22b. This is a case where acceleration does happen at a relatively low altitude.

While the magnitude of auroral effects obviously depends on the intensity of the electron flux, the nature of the effects depends more on the altitude and therefore on the energy of the electrons (see Figure 6.6). Particles of a few kiloelectron-volts stop in the E region, the level of greatest luminosity, where they also enhance the electron density; those with energies of tens of kiloelectron-volts penetrate to the D region and enhance the otherwise weak ionization at the base of the ionosphere. We now consider some consequences.

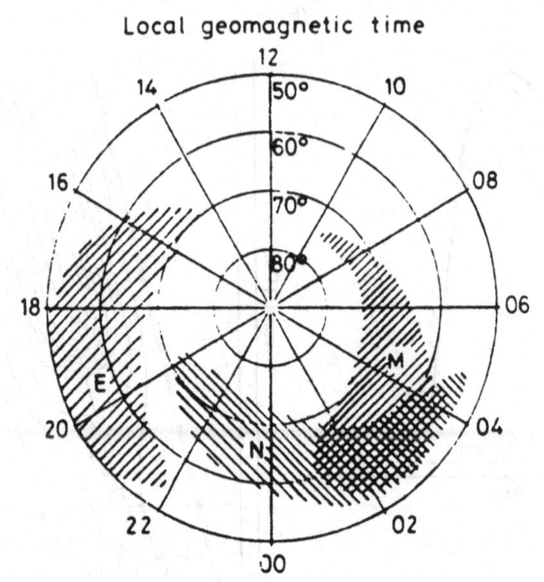

Fig. 8.23 Occurrence of radar aurora at 55 MHz in the southern hemisphere during moderate magnetic disturbances. The radar looked south from a site in New Zealand. E, N and M, stand for evening, night and morning. (After R. S. Unwin, *J. Atmos. Terr. Phys.* **28**, 1167, copyright (1966) Pergamon Press PLC)

8.3.5 Radar aurora

Given that one ion–electron pair is produced for every 35 eV of energy dissipated, and that the effective recombination rate in the E region is about 10^{-13} m^3/s (10^{-7} cm^3/s), it is easily shown that the E region electron density may be increased to several times 10^{12} m^{-3} during auroral activity. According to Equation 3.9, such electron densities reflect radio waves of frequency below 20 MHz. The auroral ionization may be detected also at VHF (> 30 MHz) and at UHF (> 300 MHz), when the echoes are due to scattering from irregularities in the ionization, as discussed in Section 8.2.3.

The echoes come from the E region of the ionosphere, generally between 100 and 130 km and at heights similar to those of the luminous aurora. As was pointed out, these irregularities are field-aligned and the echoes are aspect sensitive, being received most strongly when the radar is in a plane normal to the field-line. The radar aurora may be observed over a wide range of latitude from subauroral to polar, and there is a strong dependence on the level of magnetic disturbance.

The echoes received by an auroral radar may be classified into 'diffuse' and 'discrete'.The first type is the more common. They last for at least 10 minutes and have a wide extent in latitude (50–500 km) and longitude (several thousand kilometres). They occur primarily in the afternoon–evening sector and the early morning hours. The discrete echoes are short lived, show spatial structure, and cover a smaller area. They prefer the midnight sector. Figure 8.23 illustrates the spatial occurrence of radar aurora. Echoes from the polar cap also show diffuse characteristics. The echoes discussed in Section 8.2.3 belong to the diffuse type in the auroral classification.

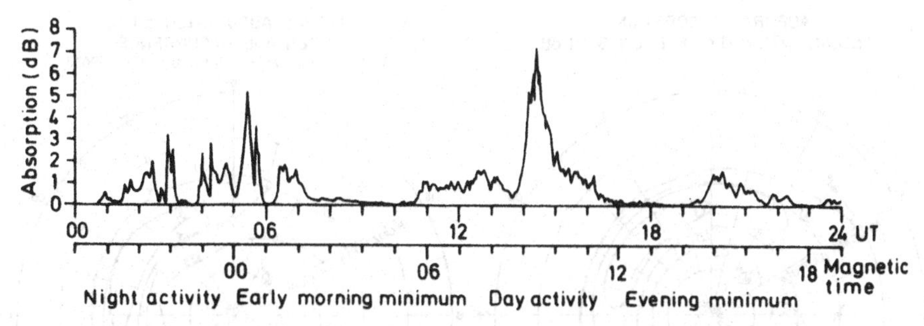

Fig. 8.24 Auroral radio absorption observed on one day, 15 October 1963, at Byrd Station, Antarctica, with a 30 MHz riometer. The descriptions below the axis refer to the typical behaviour; the evening minimum was not respected on the day shown! Note the difference of structure between the night and day activity. (After J. K. Hargreaves, *Proc. Inst. Elect. Electronics Engr.* **57**, 1348, 1969, © 1969 IEEE)

8.3.6. Auroral radio absorption

The enhanced ionization of the D region absorbs radio waves by the mechanism described in Section 3.5.1. Ionosondes in the auroral zone can suffer from the *black out* condition (when no echoes come back) due to the auroral absorption. However, a riometer (Section 3.5.2) operating at a frequency near 30 MHz is less sensitive and can measure the attenuation of the cosmic radio noise at such times; thus this has become the principal technique for absorption measurement at high latitude.

The amount of the absorption is up to a few decibels, values over 10 dB being rare. The absorption is structured in time, with periodicities of minutes to hours, and the preferred occurrence times are pre-midnight and pre-noon (Figure 8.24) though the pattern can be quite different on individual days. The statistics indicate an occurrence zone more circular than oval (Figure 8.25). The absorption is lower in the atmosphere than is the auroral luminosity, and it peaks around 90 km by night and 75 km by day, though again with much variation. In the horizontal, the absorption tends to occur in dynamic patches extending several hundred kilometres, some but not all of which show extension in the east–west direction. Association with the luminosity is with the diffuse rather than the discrete forms.

8.3.7 X-rays

There are some striking similarities between the phenomena of auroral radio absorption and of the X-rays which may be detected on high altitude balloons during auroral activity. Both maximize in the morning sector at a latitude some 10° equatorward of the auroral oval, and there are strong similarities of time structure as well.

Auroral X-rays were first detected in the mid-1950s by means of *rockoons* – small rockets launched from balloons – but the modern technique is to use a large balloon to carry the detectors up to about 30 km altitude. The balloon remains aloft for up to several days, moving in the prevailing wind. Space–time resolution has been achieved by launching several balloons at intervals. The X-rays detected at this height have energies exceeding 30 keV and they are generated by the *bremsstrahlung* process

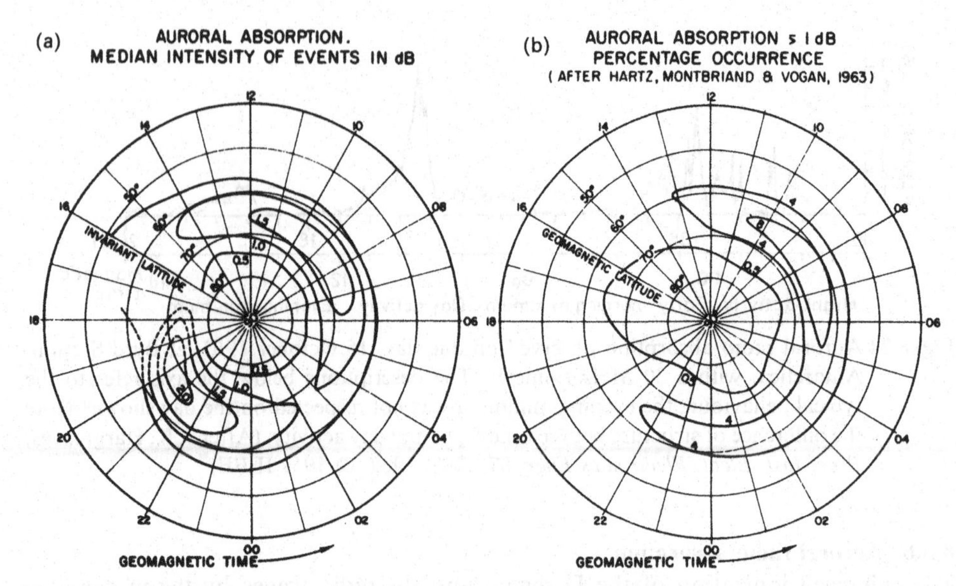

Fig. 8.25 (a) Median intensity of auroral absorption events in dB. (J. K. Hargreaves and F. C. Cowley, *Planet. Space Sci.* **15**, 1571, 1967) (b) Percentage occurrence of auroral absorption exceeding 1 dB. (After T. R. Hartz *et al.*, *Can. J. Phys.* **41**, 581, copyright (1963) Pergamon Press PLC) The diagrams differ because the night events are shorter in duration than the day events.

(Section 6.2.3) when the primary electrons are stopped by collisions with neutral molecules. Figure 8.26 shows an example of the X-radiation observed from high altitude balloons; note the similarity of the structure to that of the radio absorption measured simultaneously with riometers.

To a large extent the X-ray balloons and the riometers are complementary techniques for studying the same basic phenomenon. The riometer can provide complete data continuity and is cheaper. The balloons enable the flux and spectrum to be measured, and have greater sensitivity to the rapid fluctuations and to spatial fine structure.

8.3.8 VLF and ULF emissions

We have referred to VLF whistlers which propagate between hemispheres along the geomagnetic field, when the energy originates in lightning flashes and the signals are dispersed with respect to radio frequency by the plasma along the propagation path. Whistlers provide an important diagnostic technique for measuring electron densities in the outer plasmasphere (Section 3.5.4). Not all radio noises have whistler characteristics, however. They may occur repeatedly in the same form instead of showing the increasing dispersion typical of the whistler, and this, coupled with their alternate appearance at geomagnetically conjugate points, indicates that these signals are actually being regenerated at each passage through the magnetosphere. In this case the energy comes from the charged particle population of the magnetosphere through mechanisms (Section 9.1) that permit an exchange of energy between particles and waves.

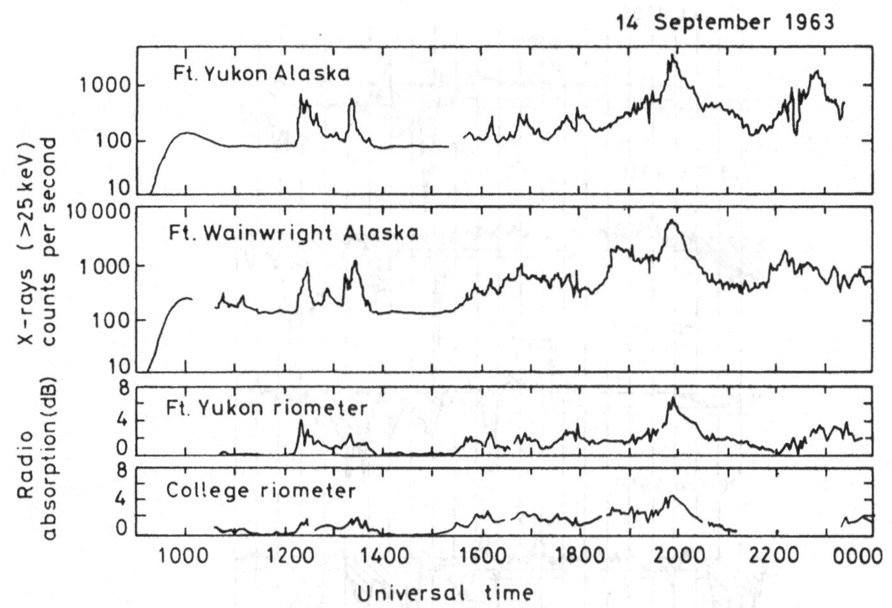

Fig. 8.26 Auroral X-rays observed on balloons over Alaska, with simultaneous riometer measurements. (J. R. Barcus and R. R. Brown, *J. Geophys. Res.* **71**, 825, 1966, copyright by the American Geophysical Union)

Electromagnetic noises from the magnetosphere cover a wide band of frequency, from below 1 kHz to above 100 kHz, and are of various types. Some have well defined frequency–time characteristics, while others are relatively unstructured and sound like random noise when played back through a loudspeaker. These two types are usually described respectively as *chorus* and *hiss*, the first because of a resemblence to birdsong at dawn. The occurrence of both chorus and hiss increases with the level of magnetic and auroral activity.

Emissions occurring at lower frequencies are recorded on magnetograms as *micropulsations*. Their periods are measured in seconds and minutes, and the waves are hydromagnetic rather than electromagnetic. Some of these, also, are related to auroral particle precipitation. They can provide information on wave–particle interactions in the magnetosphere, and have some diagnostic applications.

VLF and ULF emissions are considered further in Section 9.2.

8.4 Magnetic storms and substorms at high latitude

8.4.1 Magnetic bays

In Section 7.4.2 we described the classical magnetic storm as detected by magnetometers at low latitudes, a consequence of the ring current. At high latitudes the storm appears quite different. Magnetograms recorded in and near the auroral zone show a series of *bay* events like those in Figure 8.27. Their magnitude can amount to 1000 nT, and the perturbation tends to be positive before midnight and negative afterwards. Where the perturbations change sign is called the *Harang discontinuity*.

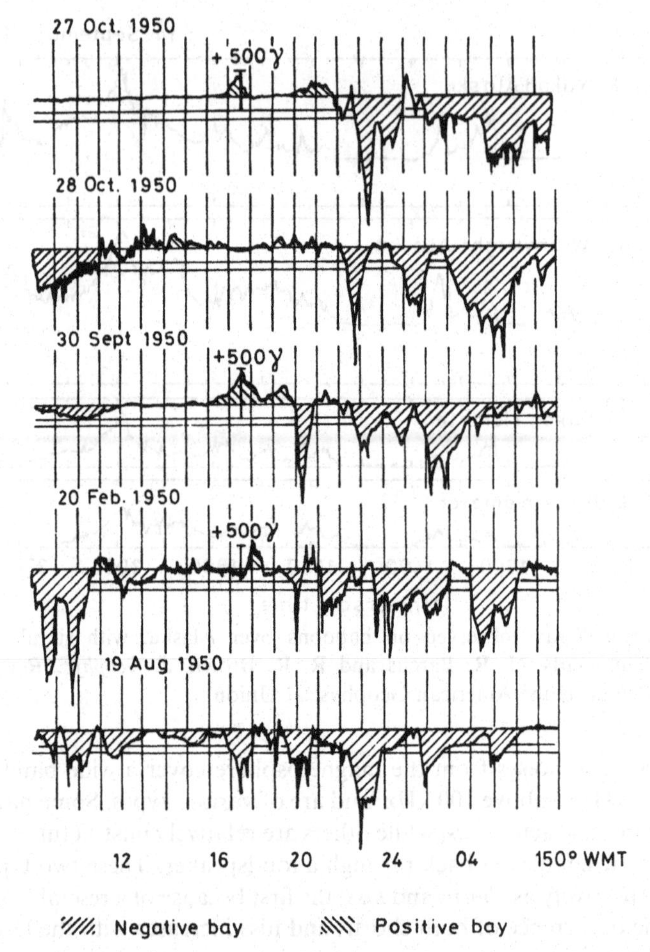

Fig. 8.27 Negative and positive bays seen on magnetograms at College, Alaska. The time zone is that of the 150°W meridian. (After M. Sugiura and S. Chapman, *Abh. Akad. Wiss. Gottingen Math. Phys. K*l *Spec. Issue* **4**, 53, 1960)

8.4.2 Auroral electrojets

The magnetic bay results from an electric current, the *auroral electrojet*, flowing along the auroral zone in the E-region. To explain the sign of the bay the current has to be eastward before midnight and westward afterwards: i.e. converging on the midnight meridian. Obviously there must also be return currents for continuity, and the exact form of the overall current system has been the subject of much investigation over the years. S. Chapman's original interpretation, which assumed, of course, that the currents only flowed horizontally, showed two electrojets with return currents at higher and lower latitudes as in Figure 8.28. This pattern was obtained by averaging the daily magnetic variations during the first two days of a number of storms, a procedure that concentrates the inferred current into the auroral zone. Figure 8.29 shows a current system for a magnetic substorm; this version accounts for the

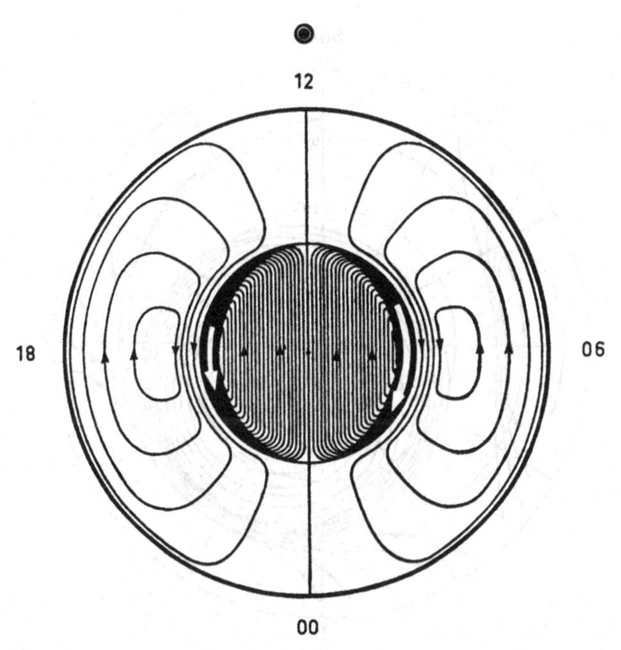

Fig. 8.28 Chapman's original SD current system. The SD analysis takes magnetic disturbance vectors observed simultaneously at a number of stations and infers a current system which could give rise to them. (S.-I. Akasofu, *Polar and Magnetospheric Substorms*. Reidel, 1968. Reprinted by permission of Kluwer Academic Publishers)

relatively greater intensity on the morning side of midnight. Figures 8.29 and 8.28 are related much as the auroral oval is to the auroral zone.

Appreciation of the Birkeland current (Section 5.8.4) has changed the whole approach to current modelling, and in a fundamental way because it is now permissible to involve the magnetosphere. With this possibility opened up, a two-dimensional system is seen to be only an *equivalent current system*. It is not possible, in fact, to derive a unique three-dimensional current system from ground-based observations alone, which still leaves the modeller with some scope for imagination. The equivalence between vertical and horizonal currents in terms of their magnetic effects is demonstrated in Figure 8.30. In (a), two Birkeland currents are connected by an electrojet; also, (a) is equivalent to (b) + (c), where (c) is the classical electrojet model with all current horizontal. Now, (b) may be divided into (b1) and (b2), spacially separated, each comprising a vertical current plus a horizontal spreading current in the ionosphere. It can be shown that neither (b1) nor (b2) produces any magnetic effect at the ground. Therefore a ground-based magnetometer cannot distinguish between (a) and (c). On the other hand, (b1) and (b2) do have magnetic effects above the ionosphere and therefore Birkeland currents can be detected from satellites, as we have seen (see Figure 5.55).

Having regard to the observational evidence on field-aligned currents, present models of the auroral electrojets resemble that shown in Figure 5.66 in which the electrojets are Hall currents between two Birkeland sheets, the outer of which connects to the magnetotail during a substorm.

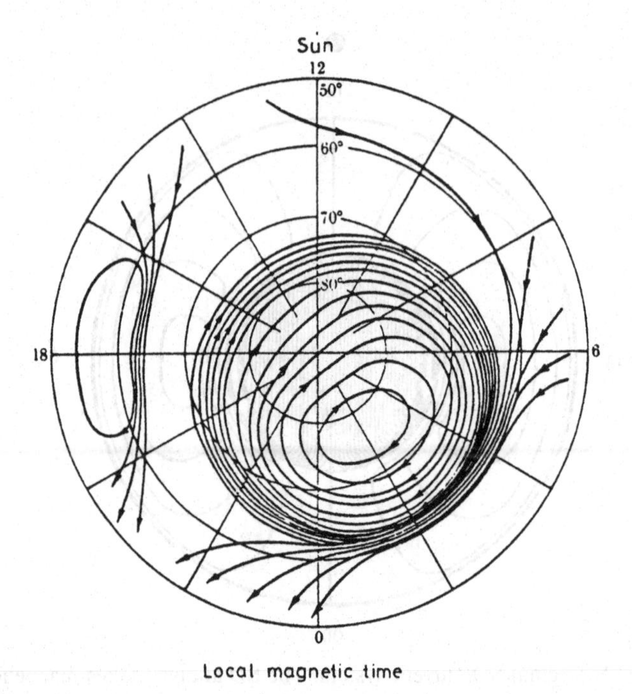

Fig. 8.29 Equivalent current system of a magnetic substorm. The concentrations of current lines in the early morning and near 1800 LT would appear as electrojets. (S.-I. Akasofu and S. Chapman, *Solar–Terrestrial Physics*. By permission of Oxford University Press, 1972)

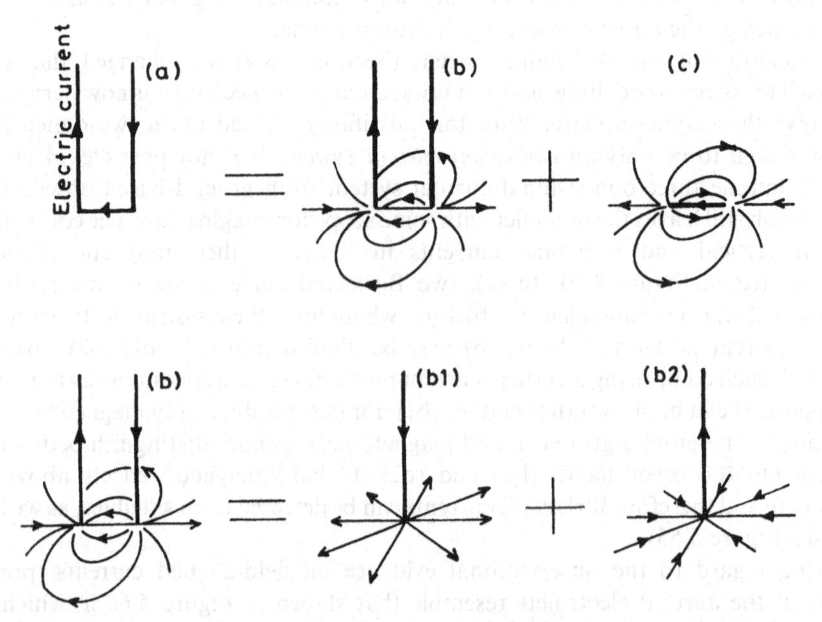

Fig. 8.30 The equivalence of different current systems as detected by a magnetometer on the ground. (N. Fukushima, *Rep. Ionosphere Space Res. Japan* **23**, 209, 1969)

8.4.3 Magnetic indices

The magnetic bays illustrated in Figure 8.27, and the electrojets causing them, are part of the substorm phenomenon. As such they also serve a practical function as the basis of important indices of geophysical activity. In a field of study comprising many phenomena, all depending in part on an external and somewhat intangible influence like the solar wind, indices provide a common reference point and a basis for comparison between different observations. The disturbance level is part of the circumstances of an observation along with season, time of day, latitude and so on. A number of indices are regularly compiled from magnetic records and published. The most useful are probably those known as K_p, A_p, and AE.

K_p and A_p

K_p is based on the range of variation within 3-hour periods of the day (00–03 UT, 03–06 UT, and so on) observed in the records from about a dozen selected magnetic observatories. After local weighting, and averaging, the K_p value for each 3 hours of the day is obtained on a scale from 0 (for 'very quiet') to 9 (for 'very disturbed'). The scale is quasi-logarithmic, and the integer values are sub-divided into thirds by use of the symbols + and −: thus, 3, 3+, 4−, 4, etc.

A_p is a daily index, obtained from the same basic data, but converted to a linear scale (the 3-hour a_p) and then averaged over the day (00–24 UT). The value of the intermediate a_p is approximately half the range of variation of the most disturbed magnetic component measured in nanoteslas. The relationship between K_p and a_p is as in Table 8.3. K_p is conveniently represented as a *Bartels musical diagram*, as in Figure 8.31, in which form it shows the recurrence tendency of the 'M-regions' and has some predictive value.

The first solar wind measurements led to an empirical relation between K_p and solar wind speed:

$$v(\text{km/s}) = (8.44 \pm 0.74)\Sigma K_p + (330 \pm 17), \tag{8.5}$$

where ΣK_p is the sum of the 8 K_p values over a UT day. This was an important result in the task of relating terrestrial geophysical disturbances to the solar wind, though later work has produced more sophisticated relations, such as Equation 5.49, that go beyond it and take the IMF also into account.

AE

The magnetic observatories contributing to K_p and A_p are situated at various latitudes and longitudes, but with a preponderance in the higher middle latitudes, i.e. the equatorward side of the auroral zone. To achieve an index more tightly related to the auroral regions and to provide better time resolution, AE was invented. The magnetograms from observatories at several different longitudes around the auroral zone are superimposed and the upper and lower envelopes are read. The difference between these envelopes is AE. Values are published at an hourly interval in printed reports, and they are available at a 2-minute interval by special request. AE measures the activity level of the auroral zone irrespective of local time, and it is particularly valuable as an indicator of magnetic substorms. This index is obviously more

Table 8.3

K_p	a_p
0	0
1	3
2	7
3	15
4	27
5	48
6	80
7	140
8	240
9	400

sophisticated than K_p, and its preparation requires a correspondingly greater effort so that the values do not generally become available for a year or more. K_p, on the other hand, can be obtained through the World Data Centers within a few days.

8.4.4 Substorm in the luminous aurora

The tendency for auroral phenomena to occur in bursts has been realized since the time of K. Birkeland in the opening years of the 20th century. The terms 'polar magnetic substorm' and 'auroral substorm' were first used by S. Chapman, and the concept that there is some fundamental event causing auroral and magnetic disturbances within a more extended 'storm' has thus been around for many years. However it is since about 1964, when S.-I. Akasofu published extensive studies of all-sky camera pictures from the International Geophysical Year, that the substorm has been generally accepted as the central unifying concept of the auroral phenomena.

The aurora tends to be active for about an hour at a time, with quiescent periods of 2–3 hours between. There is also a dynamic aspect relating to the spatial distribution. Akasofu's representation is illustrated in Figure 8.32. The sequence begins with a quiet arc that brightens and moves poleward, forming a bulge. Active auroral forms appear in the bulge, equatorward of the original arc. This is called *break-up* or the *expansion phase*. The oval is now broader than before near local midnight, but the polar cap contained within the oval is smaller. At the same time active auroral patches move eastward towards the morning sector and other forms travel westward towards the evening. The westward movement is called the *westward travelling surge*. After 30 min to an hour the night sector recovers and the substorm as a whole dies away.

Satellite observations using downward pointing photometers have confirmed the general picture derived from the ground. Figure 8.33 shows a substorm break-up. It is found that the extent of the bright area, A_B, is related to the area contained within the visible oval, A, by

$$A_B = 0.05(A - A_0)^2, \qquad (8.6)$$

where A_0 is the area of the polar cap during quiet times preceding the substorm. Considering the oval as the boundary between open and closed field-lines, this equation relates the rate of energy dissipation during a substorm to the mag-

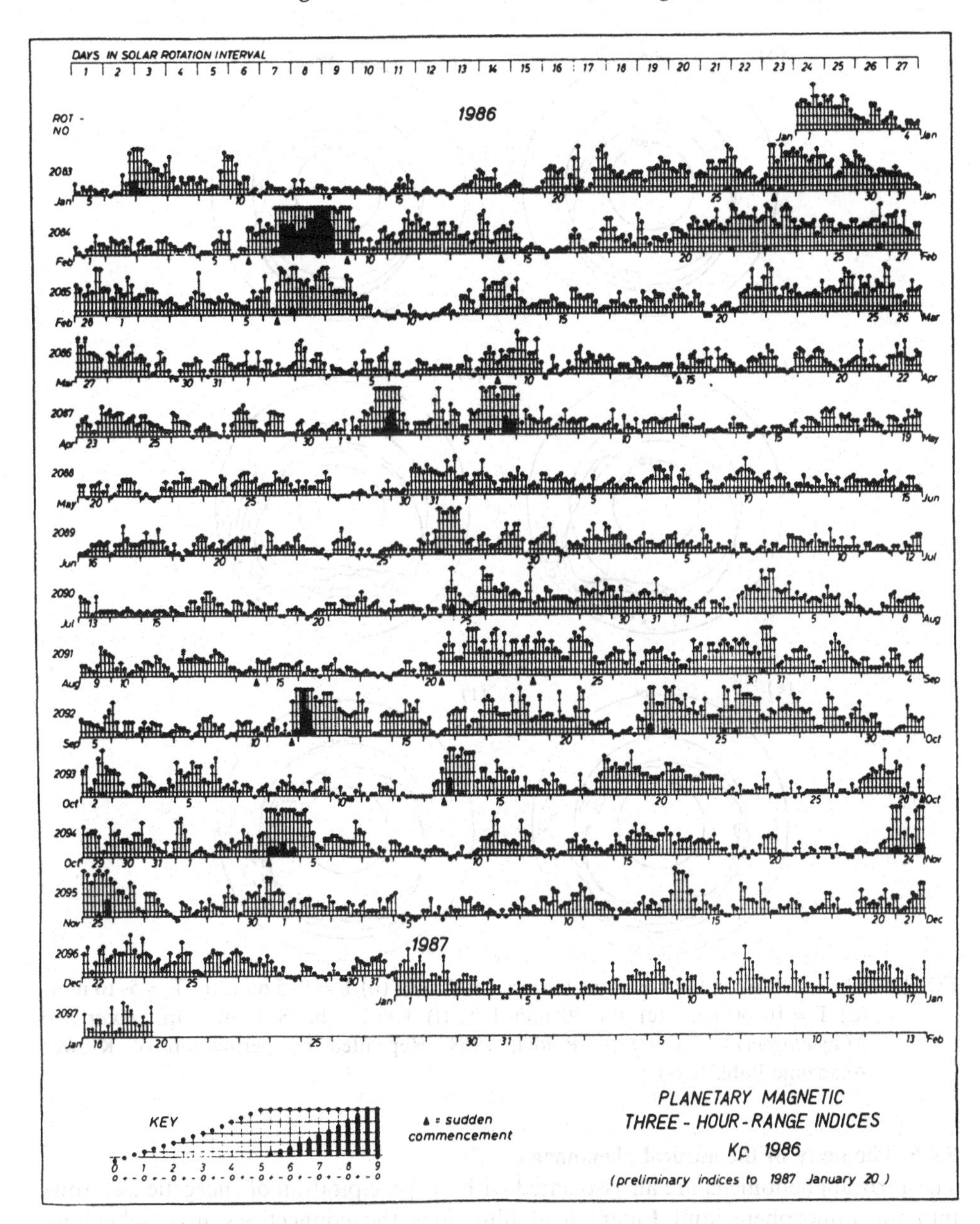

Fig. 8.31 A Bartels diagram of K_p. (*Solar–Geophysical Data.* World Data Center A, February 1987)

netosphere's energy excess over some threshold. This is consistent with the concept of the substorm driven by the reconnection of open field-lines in the magnetotail (Section 5.9.1).

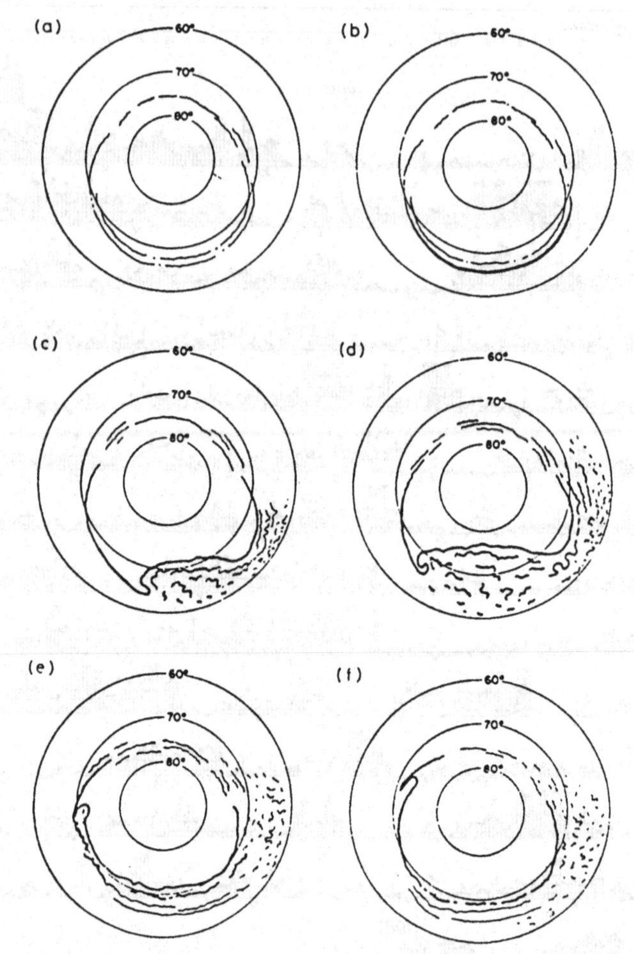

Fig. 8.32 The substorm in the luminous aurora. (a) T = 0; (b) T = 0–5 min; (c) T = 5–10 min; (d) T = 10–30 min; (e) T = 30 min–1 h; (f) T = 1–2 h. (S.-I. Akasofu, *Polar and Magnetospheric Substorms*. Reidel, 1968. Reprinted by permission of Kluwer Academic Publishers)

8.4.5 The unity of the auroral phenomena

The auroral phenomena are all associated with the precipitation of energetic electrons into the atmosphere, and Figure 8.34 illustrates the connections in a schematic manner. They all occur in one of two high latitude zones (Figure 8.35). In the inner one, corresponding to the luminous oval, occur

luminosity
sporadic-E (ionograms)
spread-F (ionograms)
soft X-rays
impulsive micropulsations
negative bays (magnetometers)

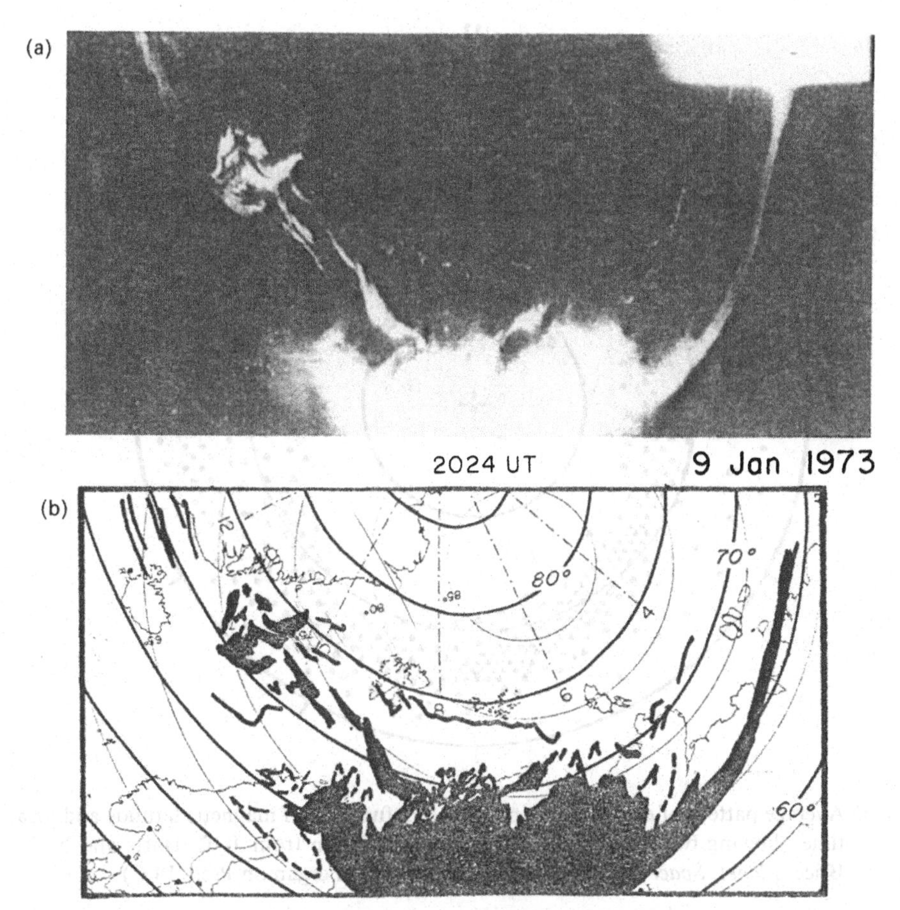

Fig. 8.33 Aurora seen from space (DMSP satellite) at the maximum of a substorm: (a) photograph; (b) interpretation. (S.-I. Akasofu, *Space Sci. Rev.* **16**, 617, 1974. Reprinted by permission of Kluwer Academic Publishers)

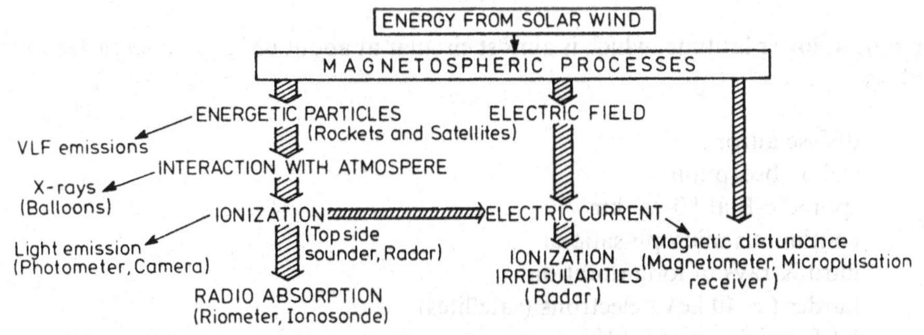

Fig. 8.34 Some links between auroral phenomena. Techniques of observation are shown in brackets. (J. K. Hargreaves, *Proc. Inst. Elect. Electronics Engr.* **57**, 1348, 1969, © 1969 IEEE)

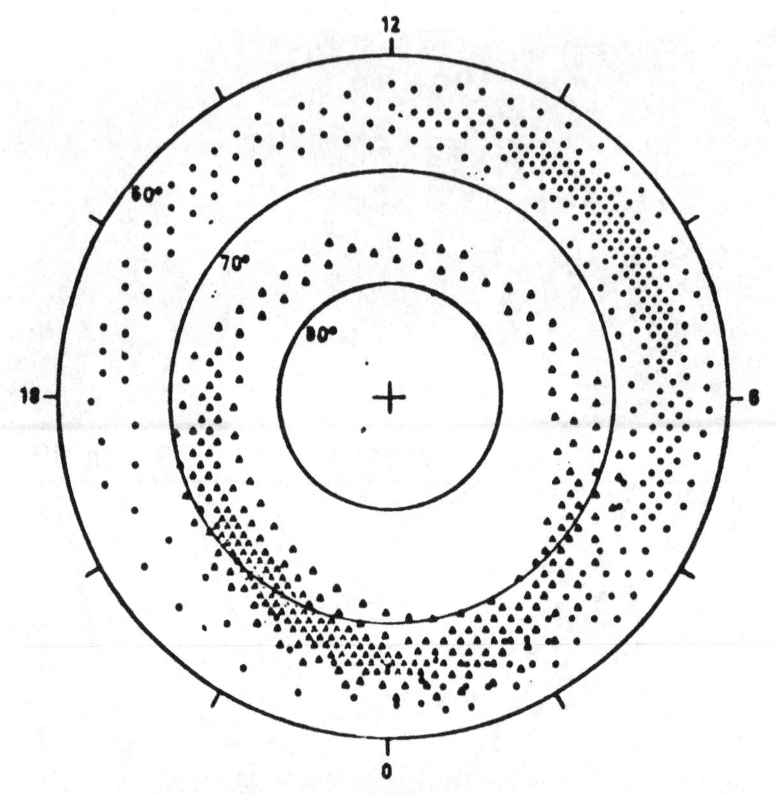

Fig. 8.35 Average pattern of electron precipitation as a function of magnetic latitude and local time, showing two zones. (Reprinted with permission from T. R. Hartz and N. M. Brice, *Planet. Space Sci.* **15**, 301, copyright (1967) Pergamon Press PLC)

soft but intense electron fluxes (satellites)
high frequency (> 4kHz) VLF hiss
rapid fading of VHF scatter signals

The zone at lower latitude, which is almost circular at about 65° geomagnetic latitude, displays

diffuse aurora
radio absorption
sporadic-E at 80–90 km
continuous micropulsations
hard X-rays of long duration
harder (> 40 keV) electrons (satellites)
VLF emissions < 2 kHz
slow fading of VHF scatter signals

The phenomena of the oval are of shorter duration than those at lower latitude. Occurrence is greatest by night and by day in the two zones respectively. In both zones

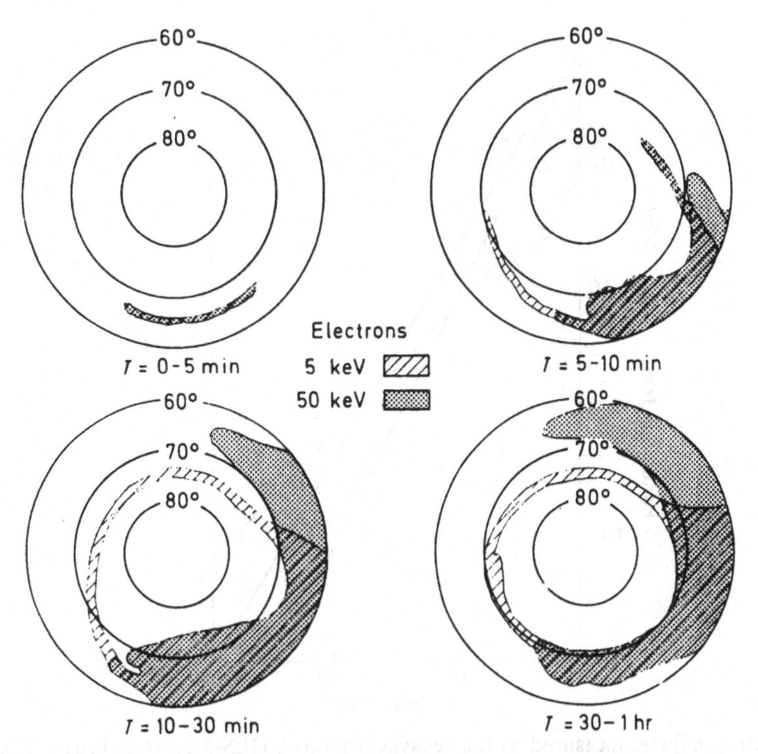

Fig. 8.36 Typical development of electron precipitation in a substorm. Note that the two zones are distinct on the day side. (S.-I. Akasofu, *Polar and Magnetospheric Substorms*. Reidel, 1968. Reprinted by permission of Kluwer Academic Publishers)

the phenomena are sporadic and dynamic. Figure 8.36 shows a unified picture of the substorm in the two zones, as represented by the fluxes of 5 and 50 keV electrons.

8.5 Polar cap events

8.5.1 Introduction

Unlike the aurora, with its long history of observation and speculation, the polar cap event was discovered in modern times as a consequence of one outstanding example. A large solar flare (3+) on 23 February 1956 was followed by polar radio blackouts that lasted for several days; at the same time, ground-based cosmic ray monitors detected a large increase in the cosmic ray intensity. Studying the effects on VHF forward-scatter radio links, D. K. Bailey showed that the radio effects were due to enhanced ionization in the D region of the polar ionosphere, and that the cause of this added ionization was a flux of energetic protons presumably released from the Sun at the time of the flare. Subsequent studies using riometers to measure the absorption (Section 3.5.2) showed that the effects were confined to high magnetic latitudes, not zonal like the aurora but covering the whole polar cap. Thus they became known as *polar cap absorption* (*PCA*) *events*.

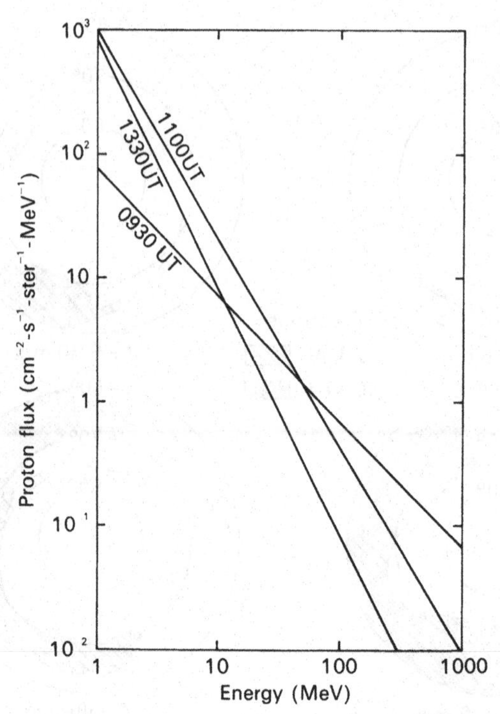

Fig. 8.37 Proton fluxes measured by the geosynchronous GOES-5 on 16 February 1984, fitted by spectra of form $E^{-\alpha}$. (Data from F. C. Cowley, NOAA, Boulder, Colorado, private communication)

PCA events occur at the rate of several each year on average. The most energetic are also detected at the ground by cosmic ray counters at the rate of one every year or two. (There were 15 between 1942 and 1967. The earlier events were identified by going back through old records, and therefore it is possible that some of the older ones have gone undiscovered.) The first recognized *ground-level event* (*GLE*) occurred on 28 February 1942. (The associated flare has another claim to fame as the source of the first solar radio noise to be recorded at Earth.)

Since the early 1960s it has been possible to observe solar protons in space, and the monitoring of energetic protons from satellites is now a routine matter. As might be expected, the satellite monitors find some events that are not seen by ground-based methods. Most solar flares emit protons at the lower energies, up to 10 MeV in this context. At energies of several tens of megaelectron-volts the flux reaching the Earth's vicinity far exceeds that from galactic cosmic rays, though at the highest energies, greater than 1 GeV, the galactic particles predominate. An example of proton spectra observed at geosynchronous orbit during a proton event in 1984 is shown in Figure 8.37. The production and release of energetic protons appears to be a normal part of the solar flare phenomenon, and flares causing PCA and GLE at the Earth probably differ from others more in degree than in nature. A proton must have initial energy of 30 MeV to produce radio absorption at 50 km, and of over 1 GeV to reach the ground (Figure 6.7).

In addition to their radio absorption effects in the polar ionosphere, affecting riometers, ionosondes and VHF forward-scatter circuits, the proton influx alters the propagation conditions for VLF waves because the Earth-ionosphere waveguide (Section 3.5.3) is changed; this provides rather a sensitive monitor of the event. Despite the name *solar proton event*, it should be appreciated that other particles, α-particles and heavier nuclei, also arrive, the proportions being typical of those in the solar atmosphere.

In comparison with some of the phenomena of geospace the proton event is relatively staightforward – protons leave the Sun, travel to Earth, find their way into the polar atmosphere, and are lost there with the consequent production of ionization. The details are reasonably well understood, and they introduce a number of important ideas. Scientifically, studies of proton events fall into three areas:

(a) the occurrence of proton events in relation to other solar phenomena, because this bears on the source and on the production mechanism;
(b) the propagation between Sun and Earth – from the Sun to the boundary of the magnetosphere, and then motions within the magnetosphere;
(c) their interaction with the atmosphere, which provides a means of studying D-region processes.

Proton events also have some practical importance. Radio links can be severely disrupted in the polar regions, including effects on aircraft communications. They are also a potential radiation hazard to space travellers.

Item (a) of this list was covered in Section 5.2.5. The other items are considered further in the following sections.

8.5.2 Propagation from Sun to Earth

A solar flare typically lasts for tens of minutes, but a proton event at the Earth may go on for several days (Figure 8.38). Also, most events do not begin at the Earth until several hours after the source flare has commenced. A proton of 10 MeV travels at 4.4×10^4 km/s and should reach the Earth in an hour if it comes in a straight line. Clearly, some delay is introduced into the propagation of the protons from Sun to Earth.

Three kinds of explanation may be considered.

(a) Particle emission from the Sun might continue after the visible flare has ended. However this cannot be the explanation because some flare regions have been seen to move round behind the Sun while the proton event continued. If the protons were coming directly from that region they should vanish when the region goes out of sight.

(b) Particles might be stored by trapping in the geomagnetic field, to be deposited gradually into the atmosphere. But this is not the case because satellites have observed that the duration of the proton event in space is the same as that of the PCA in the atmosphere.

(c) The storage might take place in the space between the Sun and the Earth. Evidence that the inter-planetary magnetic field has an influence is that (1) as a rule, flares near the eastern limb of the Sun (the left-hand side as seen from the Earth's northern hemisphere) do not produce PCAs at the Earth, and (2) the time delay between the flare and the PCA increases with the eastern longitude of the flare. These facts suggest

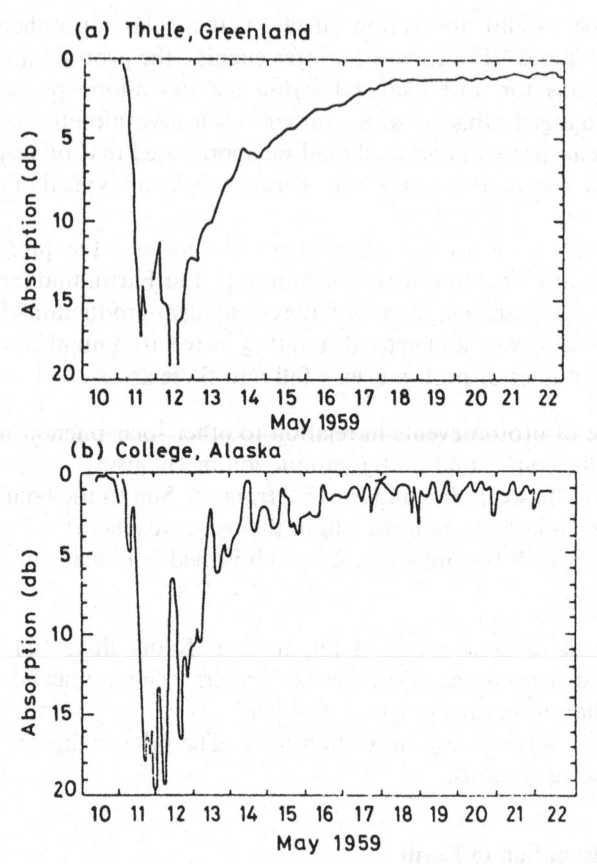

Fig. 8.38 A PCA recorded by riometers at (a) Thule, Greenland and (b) College, Alaska. (G. C. Reid, in *Physics of Geomagnetic Phenomena* (eds. Matsushita and Campbell). Academic Press, 1967)

that protons from a western heliographic longitude have better access to the Earth, and the obvious explanation is that the cloud of protons is guided by the IMF whose spiral form would produce just such effects (Figure 8.39). Another significant observation is that, although their source is a small region of the solar disc, the protons are found to be travelling in all directions by the time they approach the Earth.

The gyroradius (Equation 2.26) of a 1 GeV proton in an IMF of 5 nT is less than one hundredth of the distance from Sun to Earth. Therefore the IMF can guide particles even of this high energy over that distance. Protons with less energy gyrate in tighter loops and are sensitive to local irregularities as well as to the general form of the IMF. Proton propagation through the IMF therefore becomes like a process of diffusion in which the bulk velocity of the cloud is much less than that of individual particles. By this approach it is possible to account for the initial delay and the duration of proton events, and for the isotropy of the particles near the Earth. Since the IMF is more irregular when the Sun is more active it might be expected that the proton flux will be more strongly affected at such times, and it is indeed observed that the delay between a solar flare and the related PCA is greatest at times of high solar activity.

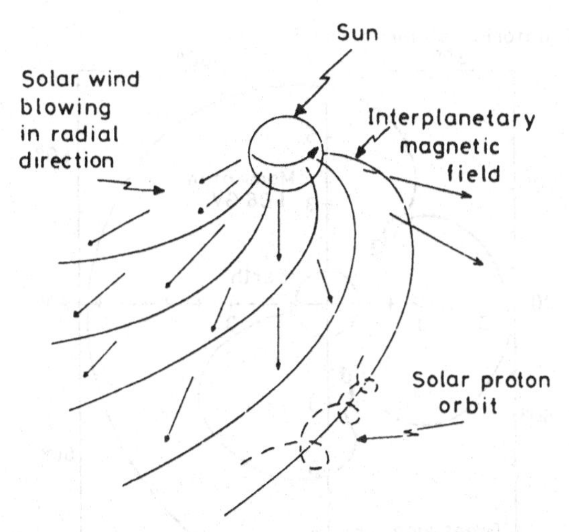

Fig. 8.39 Influence of the interplanetary magnetic field on solar protons. (K. G. McCracken, in *Solar Flares and Space Research* (eds. De Jager and Svestka). North-Holland, 1969. Elsevier Science Publishers)

The association between solar proton emission and Type IV radio bursts was noted in Section 5.2.5. If it is assumed that the radio burst and the protons leave the Sun at the same time, the observed time delay gives the proton propagation time. This time appears to be shorter (about 1 h) for strong events and longer (about 6 h) for weak ones.

8.5.3 Proton propagation in the magnetosphere – Störmer theory

After reaching the magnetopause the protons must pass through the geomagnetic field to reach the atmosphere. The theory of charged particle trajectories in the geomagnetic field was worked out by C. Störmer for his studies of the aurora. The Störmer theory is not relevant to auroral particles, because their energies are too low. However the theory is valid for cosmic rays and for solar protons.

A charged particle in a magnetic field tends to follow a spiral path whose radius of curvature ($r_B = mv/Be$) is proportional directly to the velocity and inversely to the magnetic flux density. Because solar protons are of relatively high energy the magnetic field changes significantly over one gyration and therefore the simplification of assuming an almost uniform field, as in trapping theory (Section 5.7.2), cannot be made. It is still true, though, that particles travelling almost along the magnetic field are deviated least, and consequently the polar regions are the most accessible. To reach the equator, the least accessible region, a proton has to cross field-lines all the way down to the atmosphere; charged particles may do this if they are sufficiently energetic, but the equatorial region is effectively forbidden to typical protons of solar origin.

Since the radius of gyration in a given magnetic field depends on the momentum per unit charge (mv/e), it is convenient to discuss particle orbits in general in terms of a parameter, *rigidity*:

$$R = Pc/ze, \tag{8.7}$$

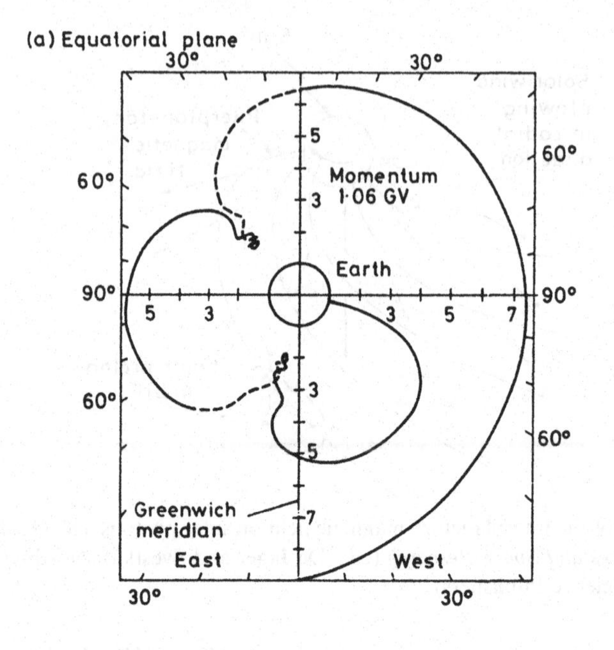

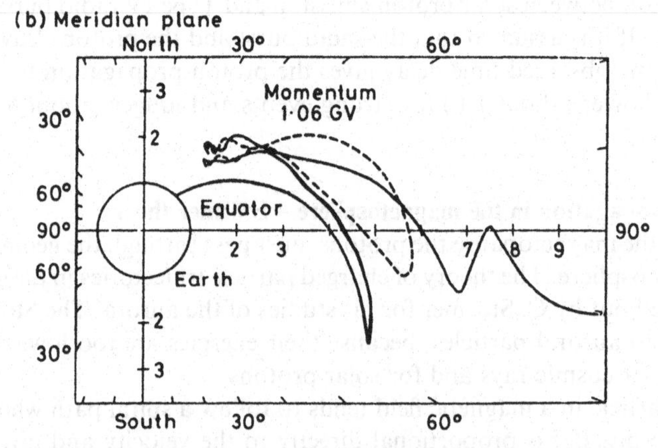

Fig. 8.40 An example of a proton trajectory in the geomagnetic field. (After K. G. McCracken *et al., J. Geophys. Res.* **67**, 447, 1962, copyright by American Geophysical Union)

where P is the momentum, c the speed of light, z the atomic number, and e the electronic charge taken positive. The advantage of this parameter is that all particles with the same value of R will follow the same path in a given magnetic field.

The trajectory of a proton in the geomagnetic field can be very complicated, even if a simple dipole field is assumed (Figure 8.40), but Störmer's analysis simplified matters by defining 'allowed' and 'forbidden' regions which could or could not be reached by a charged particle approaching the Earth from infinity. To reach magnetic latitude λ_c in a dipole field, the rigidity of the particle must exceed a *cut-off rigidity*, R_c, where

$$R_c = 14.9 \cos^4 \lambda_c, \tag{8.8}$$

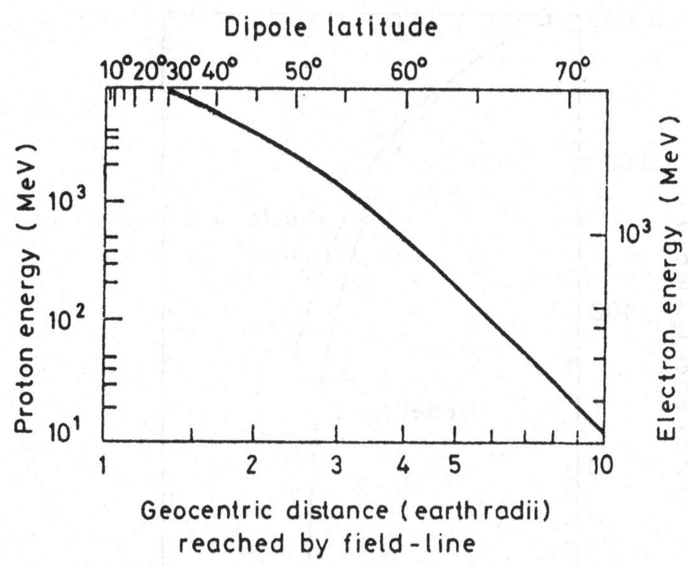

Fig. 8.41 The Störmer cut-off latitude for protons and electrons. (After T. Obayashi, *Rep. Ionosphere Space Res. Japan* **13**, 201, 1959)

and is measured in GV (10^9 V). That is, particles of rigidity R_c reach latitudes λ_c and above. At latitude λ_c one would receive particles of rigidity equal to and greater than R_c. Figure 8.41 plots the Störmer cut-off latitude against energy for both protons and electrons.

The extent of the forbidden regions from the Earth is approximately equal to the *Störmer unit*, given by

$$C_{st} = (M/Br_B)^{\frac{1}{2}} = (Me/P)^{\frac{1}{2}}, \qquad (8.9)$$

where M is the Earth's dipole moment, B the magnetic flux density, r_B the gyroradius, and P the momentum of the particle. For example, for a solar-wind proton of 1 keV, C_{st} exceeds 100 R_E; therefore a solar-wind proton cannot be expected to reach the surface of the Earth. For electrons the Störmer unit is larger and is always at least several hundred Earth-radii for electrons with energy up to 1 MeV.

To calculate the trajectory of a proton through the geomagnetic field the procedure is to imagine a proton with negative charge projected upward from the point of impact, since its trajectory is exactly the reverse of that of the incoming positively charged particle with the same energy. From a set of such computations it is possible to work out the directions in space from which the particles reaching a given place at a given time must have come. We have said that most protons are isotropic near the Earth, but the more energetic ones, those exceeding 1 GeV and causing ground-level events, are found to originate more than 50° west of the Earth–Sun direction. This is further evidence for the influence of the IMF on trajectories.

According to riometer observations the absorption region during the main part of a typical PCA event is essentially uniform and symmetrical over the polar caps down to about 60° geomagnetic latitude. Störmer theory says that the protons should have energies exceeding 400 MeV, but direct observations of the particles have shown that

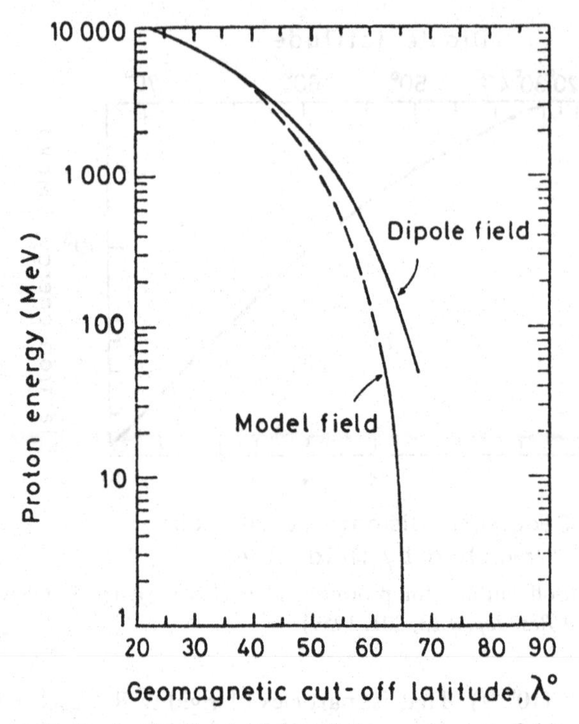

Fig. 8.42 Cut-off latitudes for dipole and realistic geomagnetic fields. (G. C. Reid and H. H. Sauer, *J. Geophys. Res.* **72**, 197, 1967, copyright by the American Geophysical Union)

the cut-off rigidity at the edge of the polar cap is significantly less than the Störmer value. A ring current (Section 5.8.3) and the distortion of the geomagnetic field from dipolar form by the solar wind can both reduce the cut-off. Figure 8.42 shows the difference between the dipole and a more realistic field. The cut-off is reduced still further if a magnetic storm, which enhances the ring current and moves the magnetopause inward, occurs while a PCA event is in progress.

The spatial distribution of radio absorption is not always uniform during the early and late phases of a PCA. The absorption usually develops first near the geomagnetic poles and extends to the polar caps some hours later. Towards the end of the event it is likely that the PCA will be contaminated by auroral electrons related to a magnetic storm, and a concentration into the auroral zone is to be expected.

Although the main features of the PCA are understood, there are some effects whose cause is not clear. Some events show a reduction in the absorption over several hours near local noon. This *midday recovery* is localized near to the boundary of the polar cap and probably indicates a local change of cut-off. There also appears to be a seasonal variation in the occurrence rate which is unlikely to be due to the Sun. Fewer events seem to occur in northern hemisphere winter than at other times of year (Figure 8.43), and those events that do occur then are weaker and have longer delay times with respect to the associated flare. This cannot be explained as an asymmetry between northern and southern hemispheres because those events that have been observed in both hemispheres show good magnetic conjugacy. The statistics may have some bias

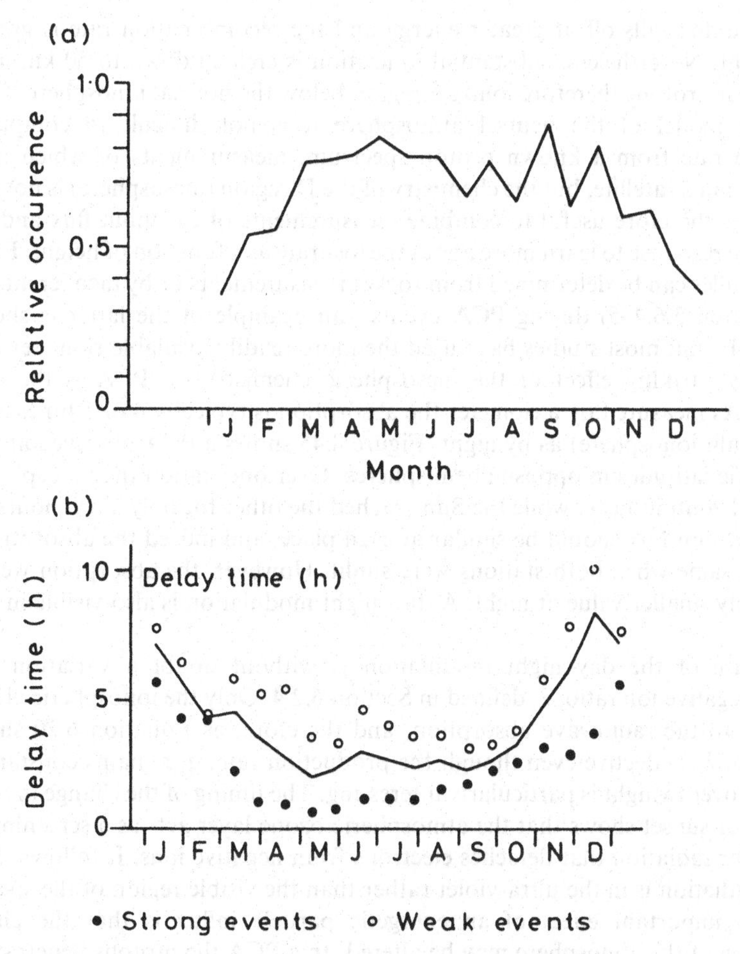

Fig. 8.43 Seasonal effects in PCA. (a) Fraction of flares with Type IV radio bursts that also produce PCA. (b) Seasonal variation of delay time for PCA, for weak and strong events. (After B. Hultqvist, in *Solar Flares and Space Research* (eds. De Jager and Svestka). North-Holland, 1969. Elsevier Science Publishers)

because most PCA reports come from riometers in the northern hemisphere and a riometer is less sensitive to PCA ionization when the upper atmosphere is not illuminated by the Sun (Section 8.5.4). But this hardly explains Figure 8.43(b), and there is still room for an effect depending on the orientation of the magnetosphere with respect to the Earth–Sun direction.

8.5.4 Atmospheric effects of solar protons

When an energetic proton enters the atmosphere it loses energy in collisions with the neutral molecules and leaves an ionized trail. The rate of ionization was discussed in Section 6.2.3. If an event contains particles of 1–100 MeV we expect to find effects over heights of 35–90 km (Figure 6.8). Effects tend to be smaller at the higher energies because the flux falls off at greater energy and the recombination rate is greater at lower height. Nevertheless, substantial ionization is created down to 50 km in some events. Solar protons therefore ionize a region below the normal ionosphere. Given a reasonable model of the neutral atmosphere it is not difficult to compute the production rate from a known proton spectrum, measurements of which may be available from a satellite, but the chemistry of the D region (mesosphere) is not so well known. It is therefore useful to combine measurements of the input flux and of the ionospheric response to learn more about the loss rate as a function of height. Electron density profiles can be determined from rocket measurements or by incoherent scatter radar (Section 3.6.4–5) during PCA events – an example of the latter is shown in Figure 8.44 – but most studies have used the more readily available riometer data.

The most striking effect of the mesospheric chemistry on PCA is the diurnal variation. As measured by a riometer the absorption is typically 4 or 5 times as large by day (sunlit ionosphere) as by night. Figure 8.45 shows a PCA observed at similar geomagnetic latitudes in opposite hemispheres. Over one station the ionosphere was illuminated continuously, while the Sun reached the other for only a few hours of the day. The proton flux should be similar at each place, and indeed the absorption was almost the same when both stations were sunlit. However, the absorption went to a considerably smaller value at night. A day–night modulation is also visible in Figure 8.38b.

The cause of the day–night modulation is without doubt a variation of the electron/negative ion ratio, λ, defined in Section 6.2.4. Only the ionospheric electrons contribute to the radiowave absorption, and therefore, as Equation 6.20 shows, a variation of λ is effective even though the production rate, q, remains constant. The behaviour over twilight is particularly interesting. The timing of the change in relation to sunrise or sunset shows that the atmospheric ozone layer acts as a screening layer for the solar radiation that detaches electrons from negative ions. It follows that the effective radiation is in the ultra-violet rather than the visible region of the spectrum.

Another important effect of an energetic particle influx is that the chemical composition of the atmosphere may be altered. In a PCA the protons penetrate deep into the atmosphere and the effects can be at a relatively low altitude. The ionization results in secondary electrons with energies of 10s and 100s of electron-volts that can dissociate and ionize molecular nitrogen to produce atoms and ions of atomic nitrogen. The N and N^+ then react with O_2 to give nitric oxide, NO, which in turn acts to destroy ozone as discussed in Section 4.1.5 (Reactions 4.21). The NO here acts as a catalyst, being destroyed in the first reaction of the pair but regenerated in the second. Ozone loss may also result from 'odd-hydrogen' species such as OH, which can be formed in the region of the mesosphere below about 85 km where water vapour is sufficiently abundant (Figure 4.8).

These processes go on continuously with the arrival of galactic cosmic rays, and almost continuously with electron precipitation in the auroral zone. However, it has been estimated that the total production of NO during one large PCA event can exceed the annual production by cosmic rays. The great proton event that occurred in August 1972 had a measurable effect on the ozone concentration in the stratosphere, and changes in the upper mesosphere have been detected in weaker events more recently. Depletion of the ozone by as much as 50% has been observed between heights of 70 and 80 km.

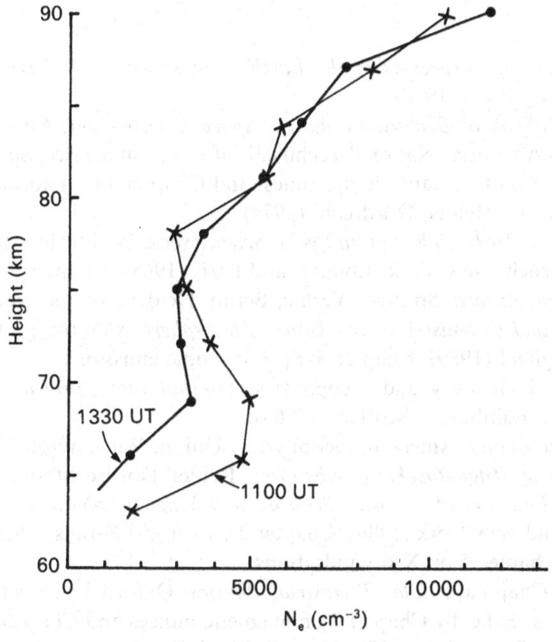

Fig. 8.44 Electron density profiles measured by incoherent scatter radar during a polar cap event on 16 February 1984. (Reprinted with permission from J. K. Hargreaves *et al.*, *Planet. Space Sci.* **35**, 947, copyright (1987) Pergamon Press PLC)

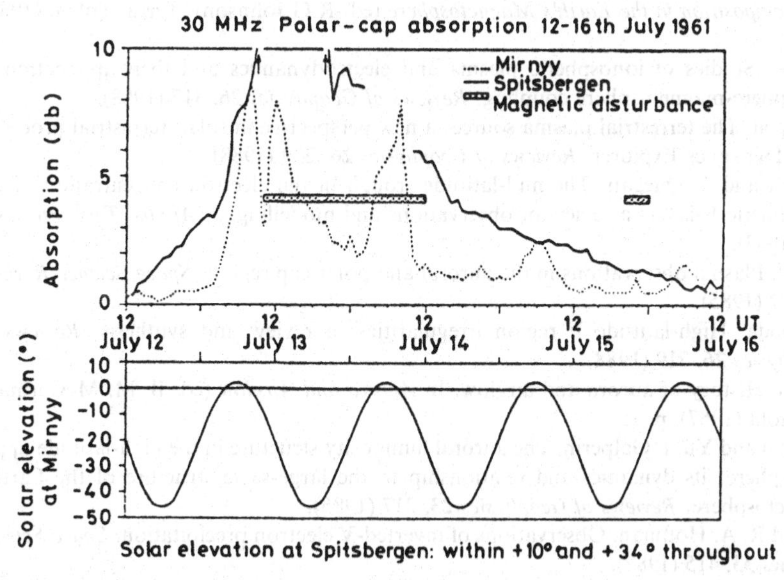

Fig. 8.45 Polar cap absorption at two high-latitude stations. (Reprinted with permission from C. S. Gillmor, *J. Atmos. Terr. Phys.* **25**, 263, copyright (1963) Pergamon Press PLC)

The radiation hazard to space travellers from proton events will be considered in Section 10.4.2.

Further reading

Y. Kamide. *Electrodynamic Processes in the Earth's Ionosphere and Magnetosphere.* Kyoto Sangyo University Press (1988).

A. S. Jursa (ed.). *Handbook of Geophysics and the Space Environment.* Air Force Geophysics Laboratory, US Air Force. National Technical Information Service, Springfield, Virginia (1985). Chapter 8 on high-latitude dynamics, and Chapter 14 on aurora.

A. Vallance Jones. *Aurora.* Reidel, Dordrecht (1974).

T. N. Davis. The aurora. In *Introduction to Space Science* (eds. W. N. Hess and G. D. Mead). Gordon and Breach, New York, London and Paris (1968). Chapter 5.

A. Omholt. *The Optical Aurora.* Springer-Verlag, Berlin, Heidelberg and New York (1971).

M. H. Rees. *Physics and Chemistry of the Upper Atmosphere.* Cambridge University Press, Cambridge, England (1989). Chapter 8, Spectroscopic emissions.

International Union of Geodesy and Geophysics. *International Auroral Atlas.* Edinburgh University Press, Edinburgh, Scotland (1963).

R. H. Eather. *Majestic Lights.* American Geophysical Union, Washington, DC (1980).

S.-I. Akasofu. *Physics of Magnetospheric Substorms.* Reidel, Dordrecht and Boston (1977).

L. L. Lazutin. *X-ray Emission of Auroral Electrons and Magnetospheric Dynamics.* Springer-Verlag, Berlin and New York (1986). Chapter 2 on auroral X-rays, Chapter 4 on auroral electrons, and Chapter 5 on X-ray pulsations.

S.-I. Akasofu and S. Chapman. *Solar-Terrestrial Physics.* Oxford University Press, Oxford, England (1972), specifically Chapter 7 on magnetic indices and Chapter 8 on substorms.

National Geophysical Data Center, Boulder, Colorado. *Solar-Geophysical Data.* Report published monthly, with an Explanation annually.

W. J. Raitt and R. W. Schunk. Composition and characteristics of the polar wind. In *Energetic Ion Composition in the Earth's Magnetosphere* (ed. R.G.Johnson). Terra, Tokyo (1983), p 99.

R. A. Heelis. Studies of ionospheric plasma and electrodynamics and their application to ionosphere-magnetosphere coupling. *Reviews of Geophysics* **26**, 317 (1988).

C. R. Chappell. The terrestrial plasma source: a new perspective in solar–terrestrial processes from Dynamics Explorer. *Reviews of Geophysics* **26**, 229 (1988).

R. J. Moffett and S. Quegan. The mid-latitude trough in the electron concentration of the ionospheric F-layer: a review of observations and modelling. *J. Atmos. Terr. Phys.* **45**, 316 (1983).

J. F. Fennell. Plasma observations in the auroral and polar cap region. *Space Science Reviews* **42**, 337 (1985).

R. T. Tsunoda. High-latitude F region irregularities; a review and synthesis. *Reviews of Geophysics* **26**, 719 (1988).

S. Chapman. History of aurora and airglow. In *Aurora and Airglow* (ed. B. M. McCormac). Reinhold (1967), p. 15.

Y. I. Feldstein and Yu. I. Galperin. The auroral luminosity structure in the high-latitude upper atmosphere: its dynamics and relationship to the large-scale structure of the Earth's magnetosphere. *Reviews of Geophysics* **23**, 217 (1985).

C. S. Lin and R. A. Hoffman. Observations of inverted-V electron precipitation. *Space Science Reviews* **33**, 415 (1982).

J. R. Kan. Towards a unified theory of discrete auroras. *Space Science Reviews* **31**, 71 (1982).

S.-I. Akasofu. A study of auroral displays photographed from the DMSP-2 satellite and from the Alaska meridian chain of stations. *Space Science Reviews* **16**, 617 (1974).

L. A. Frank and J. D. Craven. Imaging results from Dynamics Explorer 1. *Reviews of Geophysics* **26**, 249 (1988).

D. W. Swift. Mechanisms for auroral precipitation: a review. *Reviews of Geophysics* **19**, 185 (1981).

J. S. Boyd. Rocket-borne measurements of auroral electrons. *Reviews of Geophysics and Space Physics* **13**, 735 (1975).

J. K. Hargreaves. Auroral absorption of HF radio waves in the ionosphere: a review of results from the first decade of riometry. *Proc.IEEE* **57**, 1348 (1969).

W. L. Imhof. Review of energetic (>20 keV) bremsstrahlung X-ray measurements from satellites. *Space Science Reviews* **29**, 201 (1981).

B. Hultqvist. The aurora. *Space Science Reviews* **17**, 787 (1975).

G. Rostoker. Geomagnetic indices. *Reviews of Geophysics and Space Physics* **10**, 935 (1972).

Y. Kamide and S.-I. Akasofu. Notes on the auroral electrojet indices. *Reviews of Geophysics* **21**, 1647 (1983).

N. Fukushima and Y. Kamide. Partial ring current models for worldwide geomagnetic disturbances. *Reviews of Geophysics and Space Physics* **11**, 795 (1973).

J. M. Wilcox. Inferring the interplanetary magnetic field by observing the polar geomagnetic field. *Reviews of Geophysics and Space Physics* **10**, 1003 (1972).

A. Nishida and S. Kokubun. New polar magnetic disturbances: S_q^p, SP, DPC and DP2. *Reviews of Geophysics and Space Physics* **9**, 417 (1971).

N. A. Saflekos, R. E. Sheehan and R. L. Carovillano. Global nature of field-aligned currents and their relation to auroral phenomena. *Reviews of Geophysics* **20**, 709 (1982).

S.-I. Akasofu, B.-H. Ahn and G. J. Romick. A study of the polar current systems using the IMS meridian chains of magnetometers. *Space Science Reviews* **35**, 337 (1983).

Y. Kamide. The relationship between field-aligned currents and the auroral electrojets: a review. *Space Science Reviews* **31**, 127 (1982).

O. A. Troshichev. Polar magnetic disturbances and field-aligned currents. *Space Science Reviews* **32**, 275 (1982).

G. Rostoker. Polar magnetic substorms. *Reviews of Geophysics and Space Physics* **10**, 157 (1972).

G. C. Reid. Polar-cap absorption – observations and theory. *Fundamentals of Cosmic Physics* **1**, 167 (1974).

B. Hultqvist. Polar cap absorption and ground level effects. In *Solar Flares and Space Research* (eds. C. De Jager and Z. Svestka), North-Holland (1969), p 215.

D. J. Hoffmann and H. H. Sauer. Magnetospheric cosmic-ray cutoffs and their variations. *Space Science Reviews* **8**, 750 (1968).

L. J. Lanzerotti. Solar energetic particles and the configuration of the magnetosphere. *Reviews of Geophysics and Space Physics* **10**, 379 (1972).

A. P. Mitra. D-region in disturbed conditions, including flares and energetic particles. *J. Atmos. Terr. Phys.* **37**, 895 (1975).

9

Magnetospheric waves

For while the tired waves, vainly breaking,
Seem here no painful inch to gain,
Far back, through creeks and inlets making,
Comes silent, flooding in the main.
A. H. Clough (1819–61), 'Say not the Struggle naught availeth'

9.1 Wave generation by magnetospheric particles

In this chapter we are concerned with waves in the magnetosphere, and particularly with those waves that are generated within the medium. An oscillating charge radiates an electromagnetic wave at its frequency of oscillation – the action of a radio antenna is one familiar example. Similarly, a charged particle that gyrates in the geomagnetic field radiates a circularly-polarized wave at the gyration frequency. By this means, energy is transferred from the particle to the wave. It is also possible for energy to be coupled from a wave to a particle, and this aspect will be covered in Section 9.4.

If the energy transfer between a specific particle and a specific wave is to be efficient, there must be a condition of resonance, meaning that the particle and the wave must maintain a constant phase relation for a sufficient period of time. The force (**F**) on a charged particle due to the electric (**E**) and magnetic (**B**) fields of an electromagnetic wave is

$$\mathbf{F} = q(\mathbf{E}+\mathbf{v}\times\mathbf{B}),$$

where q is the charge on the particle and **v** its velocity. The change of energy during the interaction is $\Delta W = \mathbf{F}\cdot\Delta\mathbf{s}$ where $\Delta\mathbf{s}$ is the distance travelled. Hence,

$$\begin{aligned}\Delta W &= \mathbf{F}\cdot\Delta\mathbf{s} = \mathbf{F}\cdot\mathbf{v}\Delta t \\ &= q(\mathbf{E}+\mathbf{v}\times\mathbf{B})\cdot\mathbf{v}\Delta t \\ &\approx q\mathbf{E}\cdot\mathbf{v}\Delta t \qquad (9.1)\end{aligned}$$

since $\mathbf{v}\times\mathbf{B} \ll \mathbf{E}$ in an electromagnetic wave. The message of Equation 9.1 (and rather an obvious point anyway) is that the maximum energy transfer occurs when the electric field is in the same direction as the particle velocity ($\mathbf{E}\cdot\mathbf{v}$) and when that is maintained for as long as possible (Δt). If this matching is not achieved, energy transferred to the wave during one half-cycle ($\mathbf{E}\cdot\mathbf{v}$ positive) will be returned to the particle during the next half-cycle ($\mathbf{E}\cdot\mathbf{v}$ negative) and there will be no net gain.

Several wave–particle resonances are possible in the magnetosphere. We shall illustrate the point by means of a transverse resonance that involves the component of

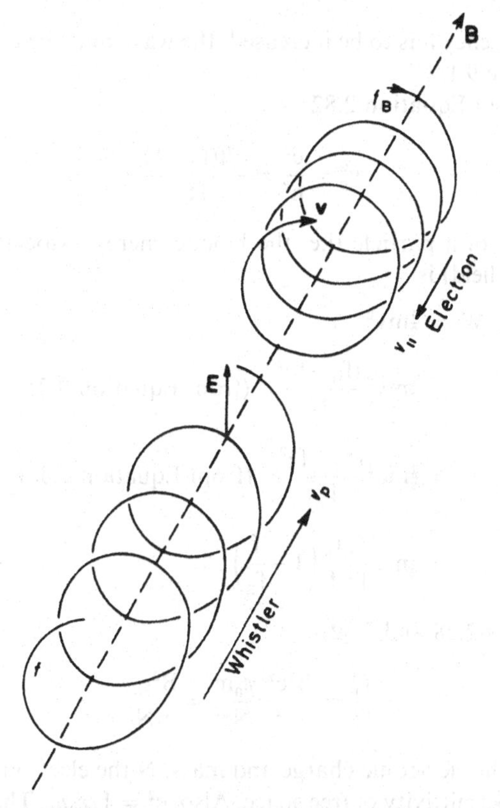

Fig. 9.1 Spiral motions of electron and whistler about the geomagnetic field. For effective interaction the electric vector of the whistler must be maintained parallel to the velocity of the electron.

particle velocity transverse to the geomagnetic field ($\mathbf{v}_\perp$). The particle may resonate with a whistler-mode wave (Section 2.7.2), and a Doppler shift is necessary to achieve a matching of velocities. This particular interaction is therefore called the *Doppler-shifted cyclotron resonance.*

First note that if $\mathbf{E}\cdot\mathbf{v}$ is to be constant the wave and the particle must rotate around the geomagnetic field in the same direction. This condition is met for electrons and whistler waves, which both rotate clockwise as seen by a stationary observer looking along the field (Figure 9.1).

The next condition is that wave and particle must preserve a constant relationship as they travel along the geomagnetic field; that is, an observer moving with the particle must see the wave frequency to be the same as that of the gyration frequency of the particle. The electron gyrates at f_B by definition, but according to theory (Equation 2.82) the whistler frequency may be any $f < f_B$ (both as seen by a stationary observer). We have to invoke Doppler shift to bring these two frequencies together. If the wave and the particle velocities along the field are v_p and $v_\parallel$, the Doppler shift $(f_B - f)$ required to bring the frequency of the wave to that of the particle is given by

$$\frac{f_B - f}{f} = \frac{|v_\parallel|}{|v_p|}. \tag{9.2}$$

Since the wave frequency has to be increased, the wave must be travelling towards the particle, as in Figure 9.1.

We can get v_p from Equation 2.82:

$$v_p^2 = \frac{c^2}{n^2} = \frac{c^2 f(f_B - f)}{f_N^2}. \tag{9.3}$$

The parallel energy of a particle (i.e. the kinetic energy associated with its velocity along the magnetic field) is

$$\begin{aligned} W_\parallel &= \tfrac{1}{2} m v_\parallel^2 \\ &= \tfrac{1}{2} m v_p^2 \frac{(f_B - f)^2}{f^2} \quad \text{(from Equation 9.2)} \\ &= \tfrac{1}{2} m c^2 \frac{(f_B - f)^3}{f_N^2 f} \quad \text{(from Equation 2.82)} \\ &= \tfrac{1}{2} m c^2 \frac{f_B^2 f_B}{f_N^2 f}\left(1 - \frac{f}{f_B}\right)^3. \end{aligned} \tag{9.4}$$

But, from Equations 2.28 and 2.32,

$$\frac{f_B^2}{f_N^2} = \frac{B^2 e^2}{m^2} \cdot \frac{\varepsilon_0 m}{N e^2} = \frac{B^2 \varepsilon_0}{m N}, \tag{9.5}$$

where e and m are the electronic charge and mass, N the electron density (of the cold plasma) and ε_0 the permittivity of free space. Also, $c^2 = 1/\varepsilon_0 \mu_0$. Therefore, substituting in Equation 9.5,

$$W_\parallel = \frac{B^2}{2\mu_0 N} \cdot \frac{f_B}{f}\left(1 - \frac{f}{f_B}\right)^3. \tag{9.6}$$

This expression gives the parallel energy of the electrons that resonate with a whistler wave of frequency f, in terms of the ratio of wave frequency to gyrofrequency (f/f_B) and the magnetic energy density per electron of the cold plasma ($B^2/2\mu_0 N$).

The Doppler-shifted cyclotron resonance between energetic electrons and electromagnetic waves propagating in the whistler mode is of some importance in the physics of the magnetosphere. (See also Section 9.4.) Equation 9.6 states a criterion which involves the parallel energy of the particles, not their total energy. To decide which particles will be the most effective in generating VLF emissions we recall that the flux of particles is generally greatest at the lowest energies (Section 5.7.3), and therefore we can say that the greatest contribution is expected from the electrons of the lowest energy that can satisfy Equation 9.6. Along a given field-line the term $B^2/2\mu_0 N$ is smallest in the region of the equatorial plane, and it is usually assumed, therefore, that this is the the main source region. That being so, the value of f_B appropriate to a given magnetic latitude is that for the nose of that field line. Hence, f/f_B can be calculated as a function of magnetic latitude or L-value for given values of the radiofrequency of the emission. Figure 9.2a illustrates the variation of $W_\parallel N$ (Equation 9.6) with L and f. (The resonance energy varies inversely with the electron density if other quantities are constant.) Then, taking models of the electron density as a

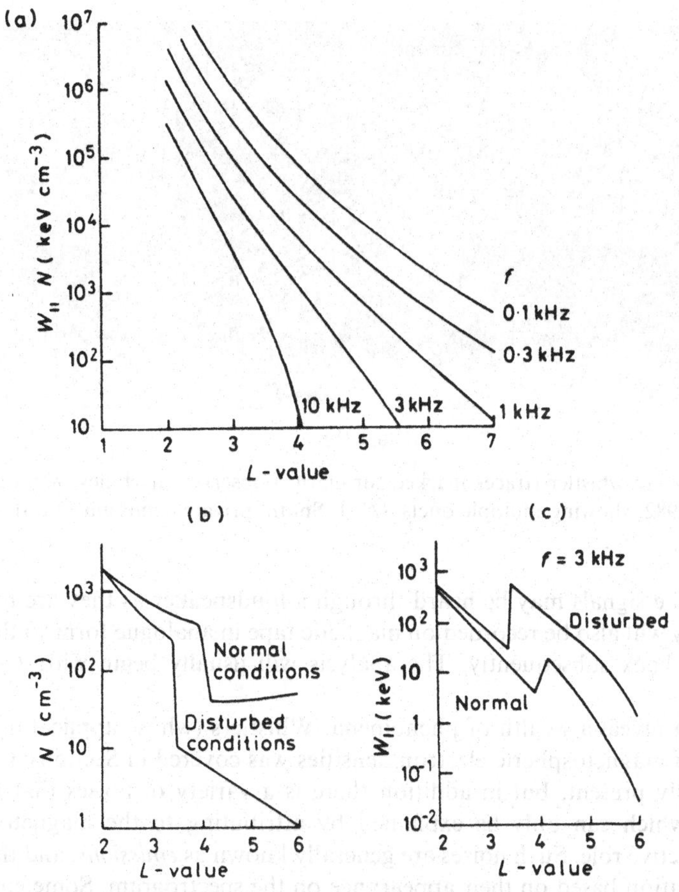

Fig. 9.2 (a) Variation of $W_{\parallel}N$ with f and L for the condition of transverse resonance at the equator. (b) Realistic models for the distribution of N lead (c) to estimates of $W_{\parallel}$ for a selected frequency. (After M. J. Rycroft, *Planet. Space Sci.* **21**, 239, copyright (1973) Pergamon Press PLC; and M. J. Rycroft, *NATO Advanced Study Institute on ELF/VLF Radio Wave Propagation.* Reidel, 1974. By permission of Kluwer Academic Publishers)

function of L (Figure 9.2b) it is possible to deduce the energy of electrons which produce most of the emissions at a given frequency (Figure 9.2c).

9.2 Observations from the ground

One might think that to investigate magnetospheric waves it is necessary to observe from a satellite. Satellite observations are certainly important, but some of the propagating waves can be studied effectively from the ground and such observations have been carried out for many years.

9.2.1 VLF signals

It is not difficult in principle to detect VLF radio signals from the magnetosphere. The equipment comprises just a large loop antenna feeding signals to an amplifier sensitive over the audio frequency range; it helps if the receiver can be operated in a quiet radio

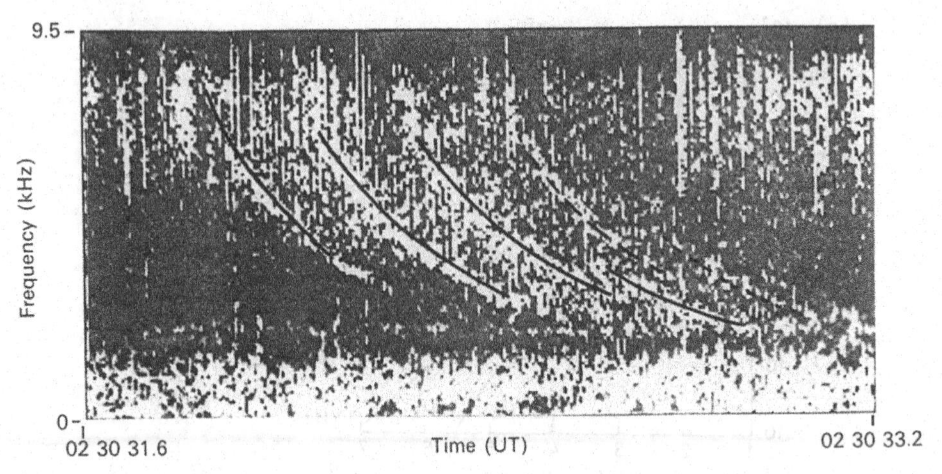

Fig. 9.3 Plasmasphere whistler (traces marked for clarity) observed at Halley, Antarctica, 11 August 1982, showing multiple ducts. (A. J. Smith, private communication)

environment. The signals may be heard through a loudspeaker as they are received, and usually they will also be recorded on magnetic tape in analogue form so that they may be played back subsequently. The analysis will usually begin with a spectral analysis.

These spectra reveal a wealth of phenomena. Whistlers (whose application to the measurement of magnetospheric electron densities was covered in Sections 3.5.4 and 5.6.2) are usually present, but in addition there is a variety of noises that are not whistlers and which can only be explained by attributing to the magnetospheric medium some active role. Such noises are generally known as *emissions*, and there is a further classification based on their appearance on the spectrogram. Some emissions are *discrete* because they are of short duration and show a well defined spectrum. Less structured emissions, which often continue for some time, are called *hiss* or *chorus*.

Whistlers and the magnetosphere

Detailed study of whistlers reveals information about ducts in the magnetosphere and about the motion of the plasmapause.

For whistlers to be observed at the ground there are three requirements:

(a) a lightning flash;
(b) one or more ducts through the magnetosphere;
(c) suitable propagation conditions beneath the ionosphere from the ends of the ducts to the source of the flash and to the receiving station.

The typical spectrum of Figure 9.3 shows both whistlers and hiss over the band 0–10 kHz. The spheric that produced the whistler is not seen, and indeed was probably in the other hemisphere. The structured nature of the whistler indicates that some structure in the plasmasphere was able to act as a duct, guiding the wave energy along

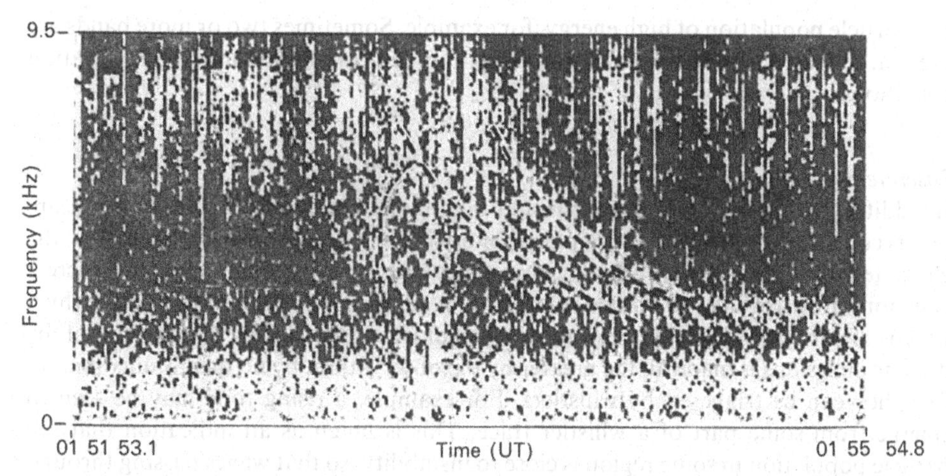

Fig. 9.4 Whistler observed at Halley, Antarctica, 11 August 1982, showing the effect of the plasmapause. The first element to arrive has travelled outside the plasmapause, while the other two have been ducted within the plasmasphere. (A. J. Smith, private communication)

several adjacent paths. Ducts may persist for some hours, evidence for which is the repetition of a particular form over a number of successive whistlers. The whole may drift in L value during this time, indicating an inward or outward movement of the plasmasphere. In some cases it may appear that a structured duct is just inside the plasmapause, and in such a case the outermost element, having the lowest nose frequency, can be taken to mark the location of the plasmapause.

A better event for plasmapause determination is illustrated in Figure 9.4. It sometimes happens that, of a group of whistlers, some travel within the plasmasphere but others are just outside. The second type fall to the left of the plasmasphere whistlers on the spectrogram because they have travelled through a region with sharply lower electron density, but their noses appear at a lower frequency because that region is further out in the magnetosphere. Whistlers of this type can be particularly useful for determining the position of the plasmapause at a given time, though in fact they may be rather faint and not easily measured. VLF signals received from ground-based senders provide an alternative method.

Hiss and chorus

The noise-like signal seen at the lower frequencies, as in Figure 9.3, is magnetospheric in origin, has propagated to the ionosphere in the whistler mode and then to the receiving station by subionospheric propagation. This noise is generally called *hiss* if unstructured and *chorus* if showing a degree of structure (see Section 8.3.8.) – the noise in Figure 9.3 is an example of the latter. The hiss or chorus commonly covers a band up to some limiting frequency, typically 2 to 5 kHz, which frequency tends to increase gradually into the morning hours. Using the theory of Doppler-shifted cyclotron resonance, it is possible to draw inferences about the energetic particles causing the emissions observed. A band restricted to a low frequency can be attributed to a particle population of high energy, for example. Sometimes two or more bands are present, in which case the presence of a gap may indicate unfavourable propagation conditions.

Discrete emissions

In addition to the ubiquitous chorus and hiss, many emissions of discrete form are observed in the spectra. These are plainly not whistlers because they do not show the characteristic falling tone; indeed, rising tones are the most common. The discrete emissions are attributed to limited interaction regions in the magnetosphere, in which groups of particles transfer energy into particular wave frequencies. The interest of the discrete emissions is made all the greater by the observation, by no means uncommon, that they can be triggered by whistlers. For example, a rising tone may be seen to emerge from some part of a whistler trace. This is taken as an indication that the particle population in some region is close to instability, so that waves passing through it are readily amplified and a small injection of wave energy is all that is needed to stimulate the emission.

9.2.2 Micropulsations

Regular or quasi-periodic fluctuations in a geophysical phenomenon are generally called *pulsations*. The periodicity can be as short as a second or as long as an hour; the train of waves might include only a few cycles or many. The best known pulsations are those appearing on magnetograms – often called *micropulsations*. They are small, less than 10^{-4} of the total geomagnetic field. Their immediate cause is in the electrojet currents flowing in the ionospheric E region overhead, but the true origin is generally thought to be Alfvén and other waves in the magnetosphere.

The instruments used in ground-based micropulsation research are of three types, providing:

(a) a direct recording of the magnetic elements at high sensitivity and high time resolution: i.e. from a sensitive magnetometer;
(b) a recording of the rate of change of the magnetic field using a large induction loop or a loop with a magnetic core; or
(c) a recording of the electric currents induced in the ground by variations of the geomagnetic field (*earth currents*).

Micropulsations are classified according to their period and duration, as in Table 9.1. Examples are shown in Figures 9.5–9.7. The impulsive type (Pi) occurs mainly in the evening sector, whereas those occurring in the morning and in daylight hours tend to be of the more persistent Pc type.

Many of these pulsations are similar at magnetically conjugate points, which proves that the magnetosphere is a major influence. Pc oscillations are generated at the surface of the magnetosphere or within the magnetosphere, and they propagate in a hydromagnetic mode. Pc 1 is attributed to bunches of particles, probably protons, oscillating back and forth between mirror points (Section 5.7.2). At a given latitude there is a characteristic period which is related to the ion gyrofrequency at the equatorial crossing of the field-line, and a resonance between protons and ion–

Table 9.1 *Micropulsation classification*

Continuous and regular		Irregular	
Type	Period(s)	Type	Period(s)
Pc 1	0.2–5	Pi 1	1–40
Pc 2	5–10	Pi 2	40–150
Pc 3	10–45		
Pc 4	45–150		
Pc 5	150–600		

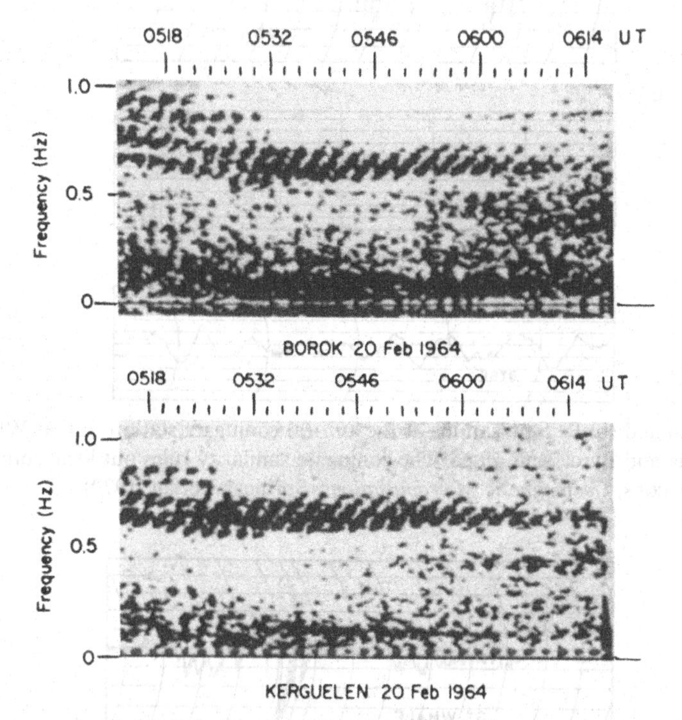

Fig. 9.5 Frequency–time display of Pc 1 micropulsations observed at the conjugate stations Borok (USSR), and Kerguelen (Indian Ocean). The fine structure of rising tones is characteristic of Pc 1. (After V. A. Troitskaya, in *Solar–Terrestrial Physics* (eds. King and Newman). Academic Press, 1967)

cyclotron waves (Section 2.7.2) – which rotate in the same sense – is probably involved. This interaction is analogous to that between electrons and whistler waves. The modulation seen in Figure 9.5, which attracted the name *pearls*, is caused by a bunching of particles in the troughs of the wave, and the modulation period is related to the inter-hemispheric bounce period.

Pc 2–5 are explained as various oscillation modes in the magnetosphere. *Toroidal* waves propagate along the lines of force, and *poloidal* waves propagate across them.

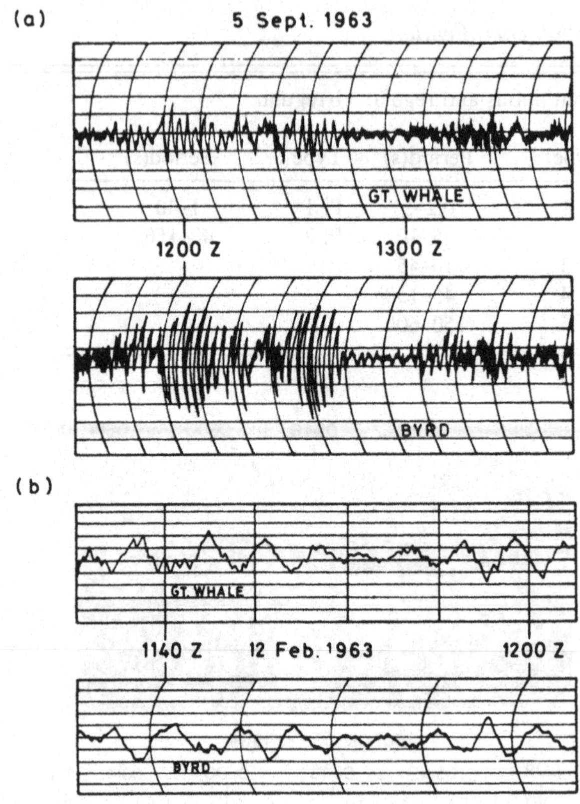

Fig. 9.6 (a) Pc 4 and (b) Pc 5 pulsations at the auroral conjugate stations Great Whale River, Canada and Byrd, Antarctica. The conjugate similarity rules out local sources. (After J. A. Jacobs, *Geomagnetic Micropulsations*. Springer-Verlag, 1970)

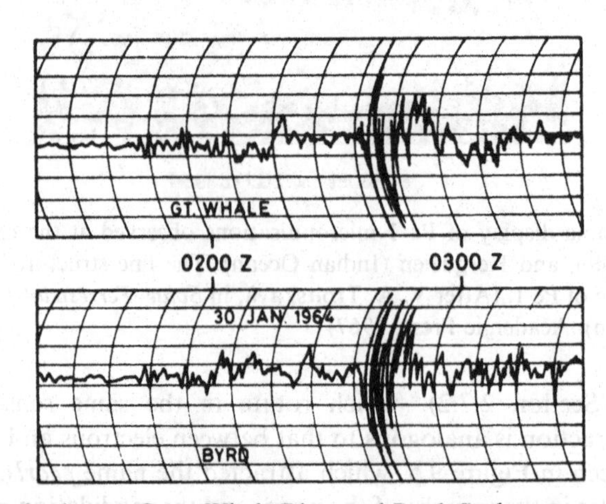

Fig. 9.7 Irregular pulsations at Great Whale River and Byrd. Such events are usually substorm associated, and there is a degree of similarity between conjugate points though not to the extent found in Pc events. (After J. A. Jacobs, *Geomagnetic Micropulsations*. Springer-Verlag, 1970)

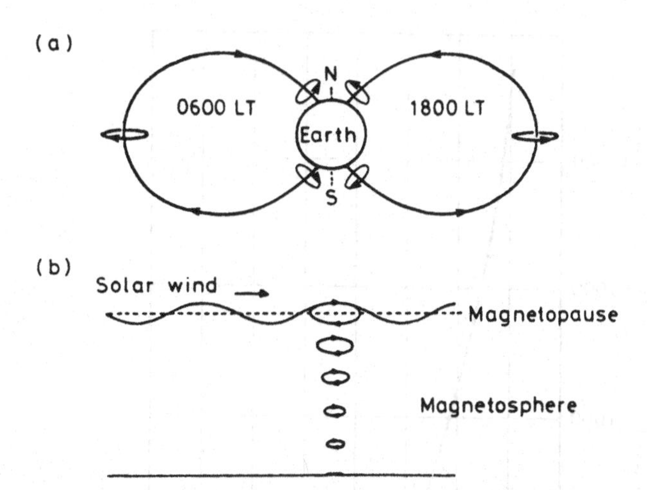

Fig. 9.8 (a) The sense of Pc 5 polarization, as viewed from the Sun. (After T. Nagata *et al.*, *J. Geophys. Res.* **68**, 4621, 1963) (b) Possible generation of Pc 5 at the magnetopause. (J. A. Jacobs, *Geomagnetic Micropulsations*. Springer-Verlag, 1970)

Pc 5 waves are thought to be a toroidal oscillation of the whole field-line between conjugate points. The oscillations show opposite polarizations in the morning and the evening, as if generated by waves travelling over the surface of the magnetosphere (Figure 9.8). The polarization also reverses between conjugate points. Particle precipitation into the auroral zones can also show modulation, and Pc 4 and 5 pulsations are commonly seen in riometer data (Section 8.3.6) from high-latitude sites. There is evidence that some waves in this range are generated in the ring current (Section 5.8.3).

Micropulsations in the Pc 3–4 range can be used as diagnostic tools since the period depends on the electron density; the effect of the plasmapause can be seen. To take a simple example, suppose that a group of field-lines at L = 3 is caused to vibrate. Then a wave travels along the field-line in the Alfvén mode such that the length of the field-line is half a wavelength. At L = 3 this distance is 56000 km. The Alfvén wave speed is given by Equation 2.78, and an average proton density of 100 cm^{-3} taken with an average magnetic flux density of 3×10^{-6} Wb/m^2 would give a speed of 6.5×10^3 km/s. The resonance period comes to 17 s.

In fact the particle density and the magnetic flux density vary along the field-line. A more exact treatment shows that the fundamental period, T, of a hydromagnetic wave standing on a field-line between co-latitudes θ_0 is given by

$$T = \frac{8\pi^{\frac{1}{2}} R_E^4}{M \sin^8 \theta_0} \int_{\theta_0}^{\frac{\pi}{2}} \rho^{\frac{1}{2}} \sin^7 \theta \, d\theta, \tag{9.7}$$

where M is the Earth's dipole moment, θ the co-latitude of a point on the field-line, θ_0 the co-latitude of its foot, R_E the radius of the Earth, and ρ the plasma density. Figure 9.9 illustrates the variation of period with co-latitude for a model magnetosphere with an assumed $\rho = 6.5 \times 10^{-22}$ g/cm^3, equivalent to 40 $protons/cm^3$.

The effect of the plasmapause is a sharp change of frequency corresponding to the step in particle density; Pc 4 tend to arise inside the plasmasphere and Pc 3 outside.

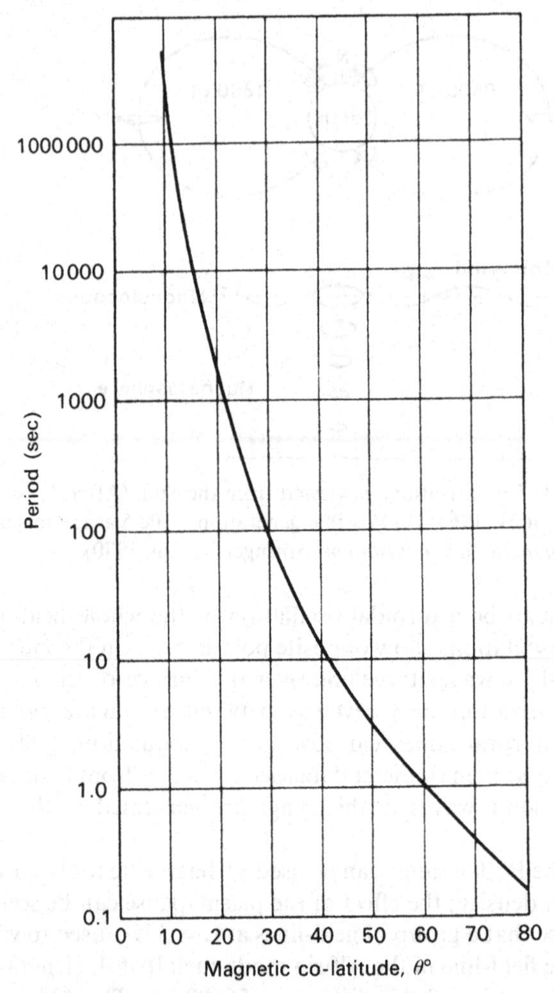

Fig. 9.9 Fundamental period of toroidal oscillations assuming a dipole field and uniform plasma density. (After K. O. Westphal and J. A. Jacobs, *Geophys. J.* **6**, 360, 1962, copyright by the American Geophysical Union)

The location of the resonant field-lines can be identified from the polarization, which is linear at the resonant latitude and elliptical in opposite senses at higher and lower latitudes.

Figure 9.10 shows an analysis of a pulsation event in which a 55 s periodicity showed maxima at two latitudes, one within the plasmasphere and the other in the trough just outside the plasmapause. Detailed study of this event, which was also observed on a geosynchronous satellite, suggests that the $L = 3$ peak was due to a fundamental resonant frequency within the plasmasphere with somewhat less than 1000 particles/cm^3 over the equator. The peak at $L = 5$ is attributed to a second harmonic, the plasma density over the equator being 55 particles/cm^3.

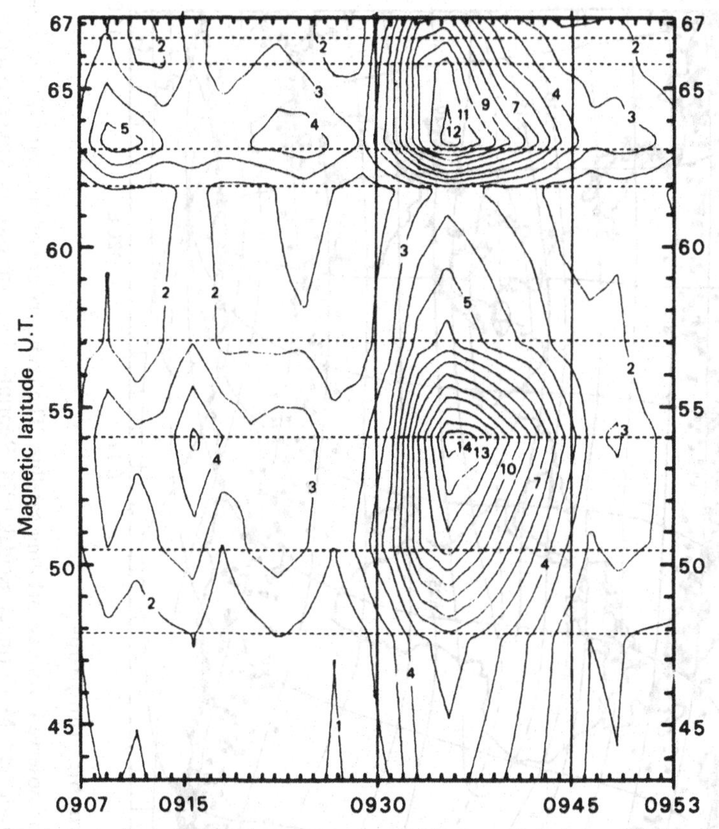

Fig. 9.10 Variation of magnetic H-component amplitude as functions of latitude and time for a 55 s pulsation observed at European stations on 29 July 1975. This pulsation had maxima at magnetic latitudes 54° and 63°, respectively inside and outside the plasmasphere. The contours are marked in nanotesla. (Reprinted with permission from H. Gough and D. Orr, *Planet. Space Sci.* **34**, 863, copyright (1986) Pergamon Press PLC)

9.3 Space observations

Satellite and ground-based observations are in many ways complementary. From a space vehicle it is possible to observe waves that do not reach the ground because they do not propagate or because they have been absorbed. Even at VLF and ULF, however, satellites have much to add to what can be measured from the ground.

Global maps

For example, they are particularly useful for mapping global distributions. Figure 9.11 is a survey of VLF occurrence mapped from a satellite. It shows bands of emissions peaking at 50–65° magnetic latitude, generally known as the *mid-latitude emissions*. These are thought to be generated by the Doppler-shifted cyclotron mechanism, as outlined above. (Compare the 'mid-latitude zone' of particle precipitation on Figure 7.29.) Other emissions are received at high magnetic latitudes, and these are

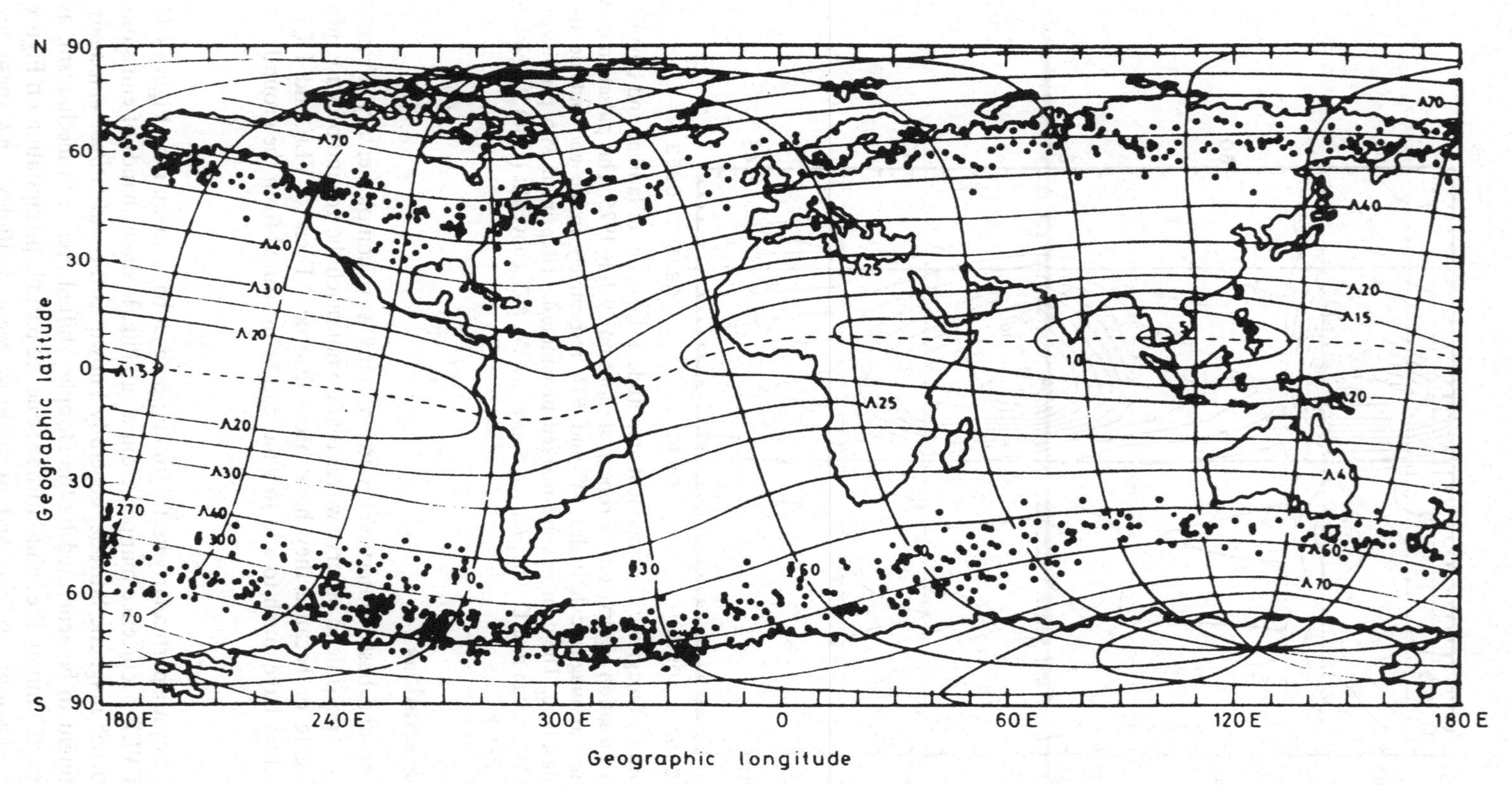

Fig. 9.11 Global distribution of maxima of 3.2 kHz emissions observed from an orbiting satellite. The contours are of invariant latitude and longitude. (K. Bullough *et al.*, in *Magnetospheric Physics* (ed. McCormac). Reidel, 1974. Reprinted by permission of Kluwer Academic Publishers)

distinguished from the mid-latitude variety by different spectral characteristics. Since they maximize at 73–79° magnetic latitude they must arise on open magnetic field-lines and therefore the transverse cyclotron resonance cannot be the cause. A longitudinal resonance with soft (< 100 eV) electrons, producing Cerenkov radiation, may be the cause.

Co-rotating zones

Another interesting result from the satellite work is the existence of co-rotating zones of enhanced emission. One would normally expect to find the occurrence of trapped radiation phenomena related to the position of the Sun, but these enhancements rotate with the Earth and have liftimes of 12, or even 24, hours. They are probably due to a local increase of electron density in the plasmasphere, reducing the local value of the quantity $B^2/2\mu_0N$. As noted in Section 9.1, this would lower the threshold energy for the generation of emissions by trapped electrons and thereby enhance the intensity.

Plasmaspheric hiss

Plasmaspheric hiss, a band of noise in the extremely-low-frequency (ELF) band (3–3000 Hz) having maximum intensity at a few hundred hertz, is thought to have a similar cause. Plasmaspheric hiss is confined to a limited range of L values just inside the plasmapause, where the zone ends abruptly. It occurs at all local times but is most intense in the afternoon sector. This is where energetic electrons following nearly circular drift paths around the Earth will encounter the plasmasphere bulge (Section 5.6.2). In that region, $B^2/2\mu_0N$ will be reduced and so the emission intensity will be increased.

Auroral kilometric radiation

Auroral kilometric radiation (*AKR*) occurs on auroral field-lines and is related to auroral arcs. It is most intense between 100 and 400 kHz (wavelengths around 1 or 2 km). At the altitudes concerned these frequencies are well above both the plasma frequency and the gyrofrequency (f_N and f_B), and the signals can leave the Earth but cannot propagate down to the ground. Satellite measurements have shown that the polarization is right circular and the signal is therefore an extraordinary wave (Section 2.5.1). It is produced by a cyclotron resonance with electrons. Direction finding from a satellite has shown that the source is on a field-line which has a bright aurora at its foot; the AKR is generated at the altitude where the gyrofrequency is equal to the frequency of the emission.

Auroral hiss and Z-mode radiation

As we have seen, hiss is an emission, produced by auroral electrons, which propagates in the whistler mode. When observed *in situ*, its frequency goes as high as the plasma frequency, this being the limit according to magneto-ionic theory when $f_N < f_B$, as is the case at high altitude over the polar regions. The cut-off is sharp enough to make this a useful method of measuring the electron density. Figure 9.12 gives the first determination of electron density as a function of height over the polar cap, which was obtained by this method. Hiss observed above the ionosphere appears to be related to upward moving electrons with energy about 50 eV.

At higher frequencies, between the cut-off frequency for the O-mode

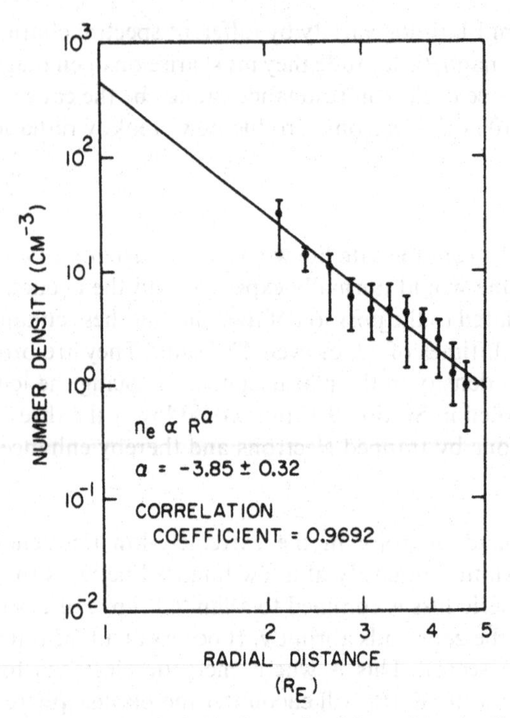

Fig. 9.12 Electron density at high latitude determined from the cut-off frequency of auroral hiss. The variation of electron density with distance has been expressed by a power law. (A. M. Persoon *et al.*, *J. Geophys. Res.* **88**, 10123, 1983, copyright by the American Geophysical Union)

$(-f_B/2+[(f_B/2)^2+f_N^2]^{\frac{1}{2}})$ and the so-called 'upper hybrid frequency' $(f_N^2+f_B^2)^{\frac{1}{2}}$, another band of noise called *Z-mode radiation* is observed. This is an ordinary wave, named by analogy with the Z-trace that sometimes occurs on ionograms (Section 3.5.1). It can exist with a wide bandwidth only if $f_B > f_N$. Thus it is a phenomenon confined to the low-density region over the polar ionosphere, and it cannot propagate into regions where its frequency exceeds the local upper hybrid frequency. The source of the Z-mode radiation is not clear.

Myriametric radiation

Terrestrial myriametric radiation (*TMR*), also known as *continuum radiation*, is an emission with a wavelength of tens of kilometres generated near the plasmapause by electrostatic oscillations (Section 2.7.3) at the upper hybrid frequency $(f_N^2+f_B^2)^{\frac{1}{2}}$. The radiation is not actually a continuum but is composed of many closely spaced lines. As generated, this emission could not escape from the magnetosphere and the energy has to be converted to a propagating ordinary wave in order to escape. The theory predicts that the TMR will be emitted into space as two beams, and spacecraft observations appear to confirm this prediction.

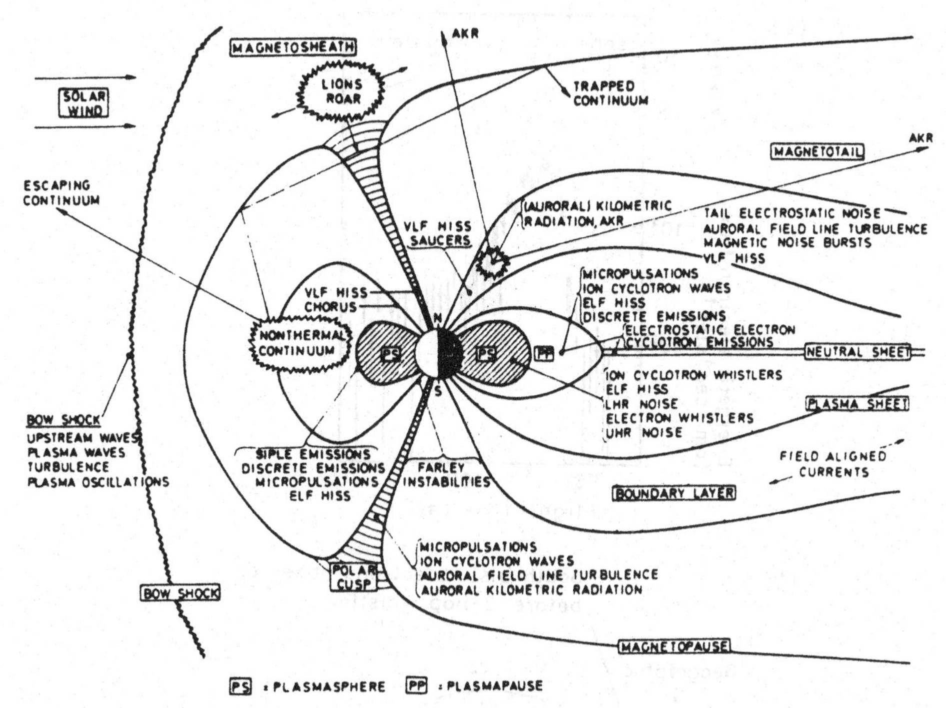

Fig. 9.13 Illustrating the multitude of waves observed in various regions of the magnetosphere. (S. D. Shawhan, *Rev. Geophys. Space Phys.* **17**, 705, 1979, copyright by the American Geophysical Union)

Summary

Figure 9.13 is a summary of magnetospheric waves and the regions with which they are associated. Not all these waves are properly understood and we make no attempt here to review the topic comprehensively. The diagram will serve to indicate something of its complexity.

9.4 Effects of waves on particles

9.4.1 Particle precipitation by waves

The interaction between waves and particles works both ways. When a gyrating electron feeds energy into a VLF wave by transverse resonance its velocity transverse to the magnetic field ($v_\perp$) is reduced while the parallel velocity ($v_\parallel$) is unchanged. Therefore the pitch angle ($\alpha = \tan^{-1}(v_\perp/v_\parallel)$) is reduced and the particle moves nearer to the loss cone (Section 5.7.2). If a particle receives transverse energy its pitch angle increases. A spectrum of VLF signals therefore alters the pitch-angle distribution of an assembly of trapped particles, some of which will enter the loss cone to be lost to the atmosphere at the next bounce. For example, plasmaspheric hiss (Section 9.3) is believed to be a major cause of the precipitation of trapped energetic particles from the outer Van Allen zone.

One possibility raised by this mechanism is that electron precipitation events might be initiated by lightning flashes or other VLF sources on the ground, and there is

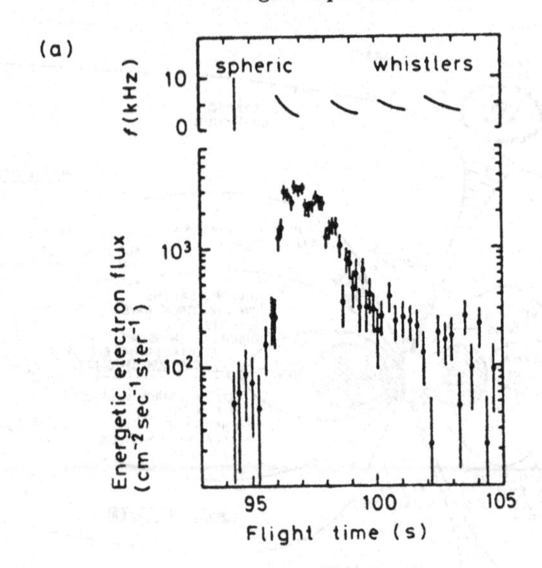

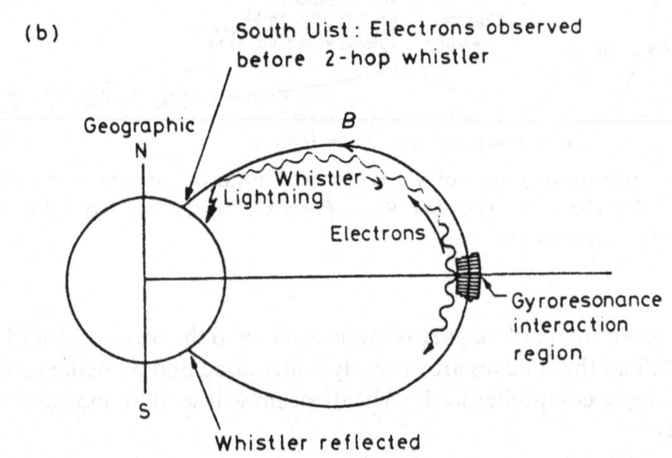

Fig. 9.14 (a) An electron (>45 keV) precipitation event observed on a rocket, just after a spheric due to a lightning stroke but before the arrival of the whistlers. (b) A possible explanation, in which the electron event is produced as the first-order whistler reaches the equatorial plane. (After M. J. Rycroft, *Planet. Space Sci.* **21**, 239, copyright (1973) Pergamon Press PLC)

growing evidence that this happens. In the Doppler-shifted cyclotron interaction the whistler and the electron have to be travelling in opposite directions along the magnetic field-line, and therefore the precipitation should appear in the same hemisphere as the lightning flash which caused the whistler, though a higher order whistler could presumably produce an effect in the other hemisphere. Figure 9.14 shows a precipitation event observed from a rocket which could well be an example of such a case.

Some observations have suggested that lightning flashes may, in some instances, be triggered by electron precipitation due to VLF noise in the magnetosphere. The suggested mechanism invokes an increase of atmospheric conductivity above the

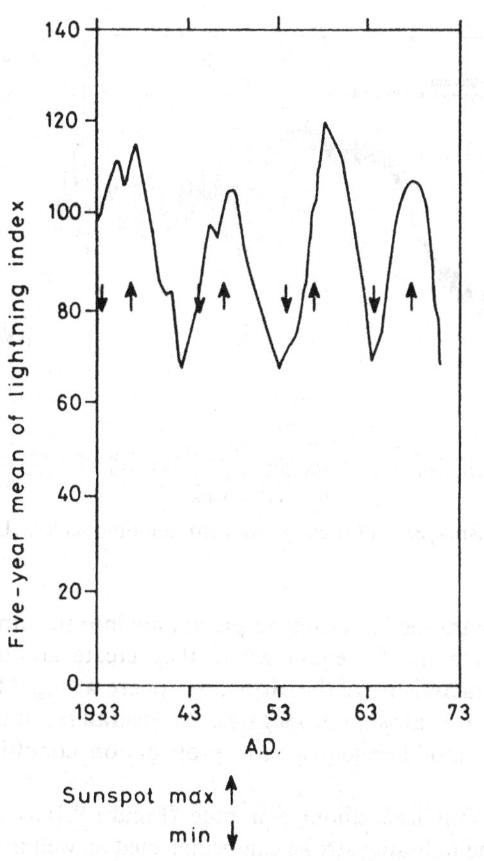

Fig. 9.15 Variations in the incidence of lightning in relation to maxima and minima of the sunspot cycle. (After J. W. King, personal communication; data from Stringfellow, 1974)

cloud, due to bremsstrahlung X-rays (Section 6.2.3) from the precipitating electrons. Processes of this kind, if verified, will be of interest in relation to the question of 'Sun–weather relationships': i.e. whether solar phenonema (acting, presumably, through magnetospheric processes) exert an influence on the weather of the terrestrial troposphere. Empirical relationships in that area of science have often proved insubstantial and mechanisms have been hard to find, but a connection between the incidence of lightning and the 11-year sunspot cycle is one of the more convincing results (Figure 9.15). Further evidence bearing on that topic could be quite important.

9.4.2 The Trimpi event

One well established consequence of a wave–particle interaction is the *Trimpi* event (so named after its discoverer). It is a curious phenomenon at first sight: a flash of lightning produces a short duration change in the amplitude of the signal received from a VLF or LF transmitter. The mechanism is that the lightning stroke emits electromagnetic energy in the VLF band, some of which propagates into the magnetosphere in the whistler mode. Through the cyclotron resonance mechanism

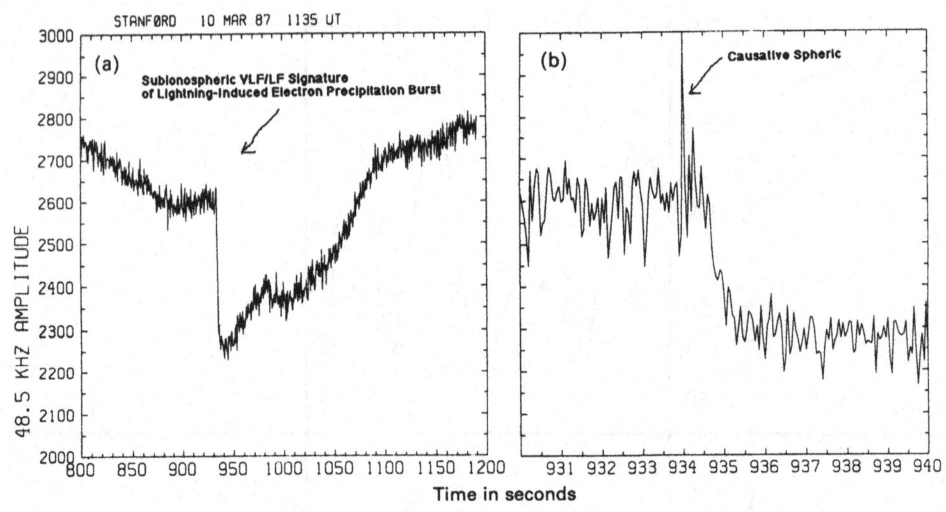

Fig. 9.16 A Trimpi event displayed on (a) coarse and (b) fine time scales. (U. S. Inan, private communication)

this whistler causes some trapped electrons to precipitate into the atmosphere. Those with enough energy reach the D region where they create additional ionization sufficient to alter the characteristics of the earth–ionosphere waveguide (Section 3.5.3) in which the VLF or LF transmission propagates. The overall result is that the energy from a lightning stroke modifies ionospheric propagation conditions via a magnetospheric mechanism.

The typical Trimpi event lasts about a minute (Figure 9.16a), and the spheric received directly from the lightning stroke can be detected as well if the observations have fine enough time resolution (Figure 9.16b). These events are only observed at night, and the electrons concerned have energies of 100–200 keV. Often only the amplitude of the subionospheric VLF or LF wave is altered, but phase changes have been noted in some cases.

9.4.3 Controlled injection of whistler waves

The mechanisms generating chorus and hiss are not clear beyond a general agreement that they involve an exchange of energy between VLF waves and trapped energetic electrons through the Doppler shifted cyclotron resonance (Section 9.1) which, as we have noted, works both ways. Some insight into the processes has come from experiments which inject VLF signals into the magnetosphere in a controlled manner. One of the facilities used in this work was briefly described in Section 3.7.3.

These experiments have shown that emissions with frequency/time characteristics similar to those of natural emissions can be produced by radiating a continuous wave of constant frequency (near 3 kHz, for example) into the magnetosphere in the whistler mode. An example is shown in Figure 9.17. To learn more about the conditions which determine the amplification and stimulation of waves in the magnetosphere a variety of modulations have been applied to the transmitted signals, for instance amplitude or frequency modulation, or the addition of sidebands or noise. If the transmitted signal is weak it is found that the intensity received in the other hemisphere is proportional

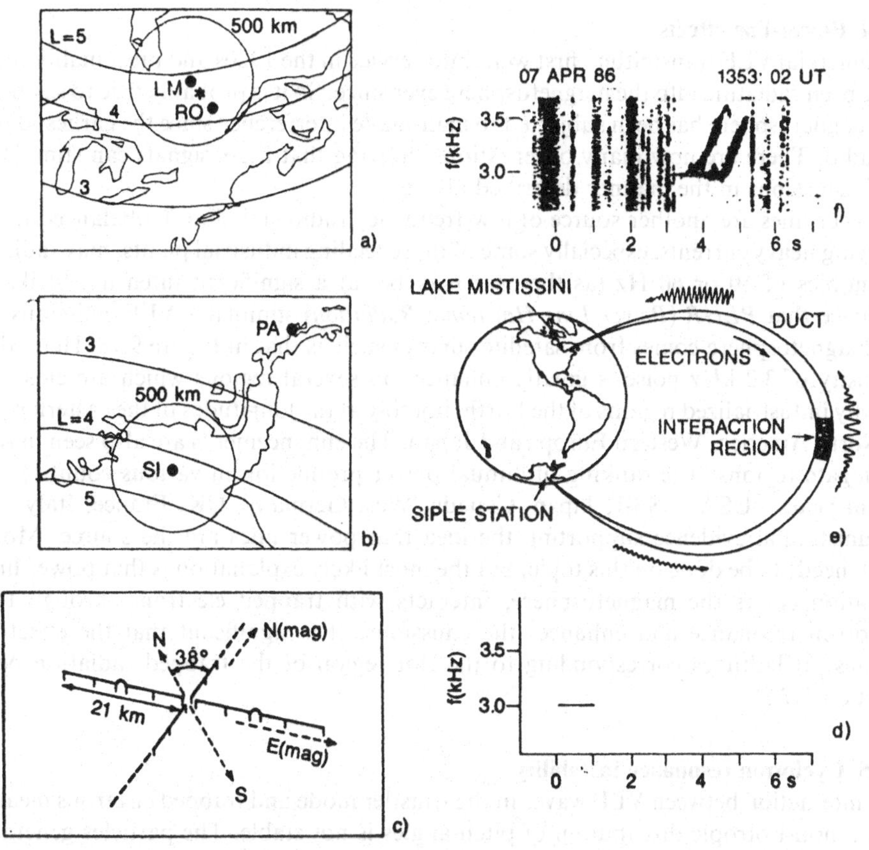

Fig. 9.17 Wave injection experiment between (b) Siple Station, Antarctica (SI) and (a) Lake Mistissini, Canada (LM), using a 42 km antenna (c). A 1-second pulse (d) travelled along two ducts (e), and interaction with trapped electrons amplified the wave and triggered rising emissions which were received in the northern hemisphere (f). (After R. A. Helliwell, *Rev. Geophys.* **26**, 551, 1988, copyright by the American Geophysical Union)

to that transmitted. But with increasing amplitude a threshold is reached at which the amplitude of the received signal increases exponentially. Within about a second the signal reaches a saturation level that seems to be independent of the intensity of the original transmission. Shortly afterward, an emission with rising frequency arrives – the frequency typically increasing from about 3 to 5 kHz in less than 1 second.

Added noise raises the threshold for exponential signal growth, and, if two waves are transmitted on adjacent frequencies separated by less than 20 Hz, growth is suppressed. These experiments demonstrate the importance of phase coherence if waves in the magnetosphere are to undergo amplification in the medium or if an emission is to be generated. Noise tends to suppress these processes because it lessens the coherence. There are indications that chorus and hiss are closely related phenomena, and that VLF energy is readily interchanged between these more and less structured forms.

9.4.4 Power-line effects

Commercial VLF transmitters first went into service in the 1920s and presumably they have been radiating into the magnetosphere ever since. So it is probably true to say that the magnetosphere has been subjected to man-made interference since the earliest days of radio. There are now many observations showing that these signals can stimulate VLF emissions in the manner described above.

Power lines are another source of low-frequency radio emission. Unbalanced lines carrying heavy currents, especially some of those feeding industrial plants, may radiate harmonics of 50 or 60 Hz (as the case may be) at a significant intensity. Striking evidence that *PLHR* (*Power Line Harmonic Radiation*) stimulates VLF emissions in the magnetosphere comes from satellite surveys such as that in Figure 9.18. Here, the intensity of 3.2 kHz noise is greatly enhanced in several regions which are close to heavily industrialized regions of the Earth, notably at the longitudes of the eastern part of North America, Western Europe and Japan. The enhancements are also seen in the conjugate regions. The ranking of annual power production in various countries in recent years – USA, USSR, Japan, Canada, West Germany, UK, France, Italy – is circumstantial evidence supporting the idea that power lines are the source. More work needs to be done on this topic, but the most likely explanation is that power line radiation enters the magnetosphere, interacts with trapped electrons through the cyclotron resonance and enhances the emissions. It is significant that the effect is greatest at latitudes corresponding to the slot region of the trapped radiation belt (Section 5.7.3).

9.4.5 Cyclotron resonance instability

The interaction between VLF waves in the whistler mode and trapped electrons means that a non-isotropic distribution of pitch angles is not stable. The particles generate waves and then the waves alter the particle pitch angles, with the result that the pitch angles spontaneously redistribute themselves and try to become more isotropic. The rate at which the redistribution occurs depends on the initial anisotropy, which is quantified by a parameter

$$A = (T_\perp/T_\parallel) - 1, \tag{9.8}$$

where $T_\perp$ and $T_\parallel$ are respectively the particle temperatures perpendicular and parallel to the geomagnetic field. The value of A is zero for an isotropic distribution, and positive if the kinetic energy of the particles is greater across the field than along it.

The practical significance of the redistribution of pitch angles arises from the existence of a loss cone (Section 5.7.2) from which particles are lost at the next bounce. For this reason the pitch angle distribution is always anisotropic to some extent. Pitch angle diffusion operates to move more particles into the loss cone and they are deposited to the atmosphere in their turn. We therefore have a mechanism by which trapped particles can be spontaneously dumped into the atmosphere (Figure 9.19).

As we have seen, the cyclotron resonance mechanism depends on the generation of waves by gyrating particles. This wave energy can be lost in various ways, and the instability can grow only if the rate of generation exceeds the rate of loss. The theory shows that the condition for the instability to grow is that the anisotropy, A, must exceed a value

$$A_c = \omega/(\omega_B - \omega), \tag{9.9}$$

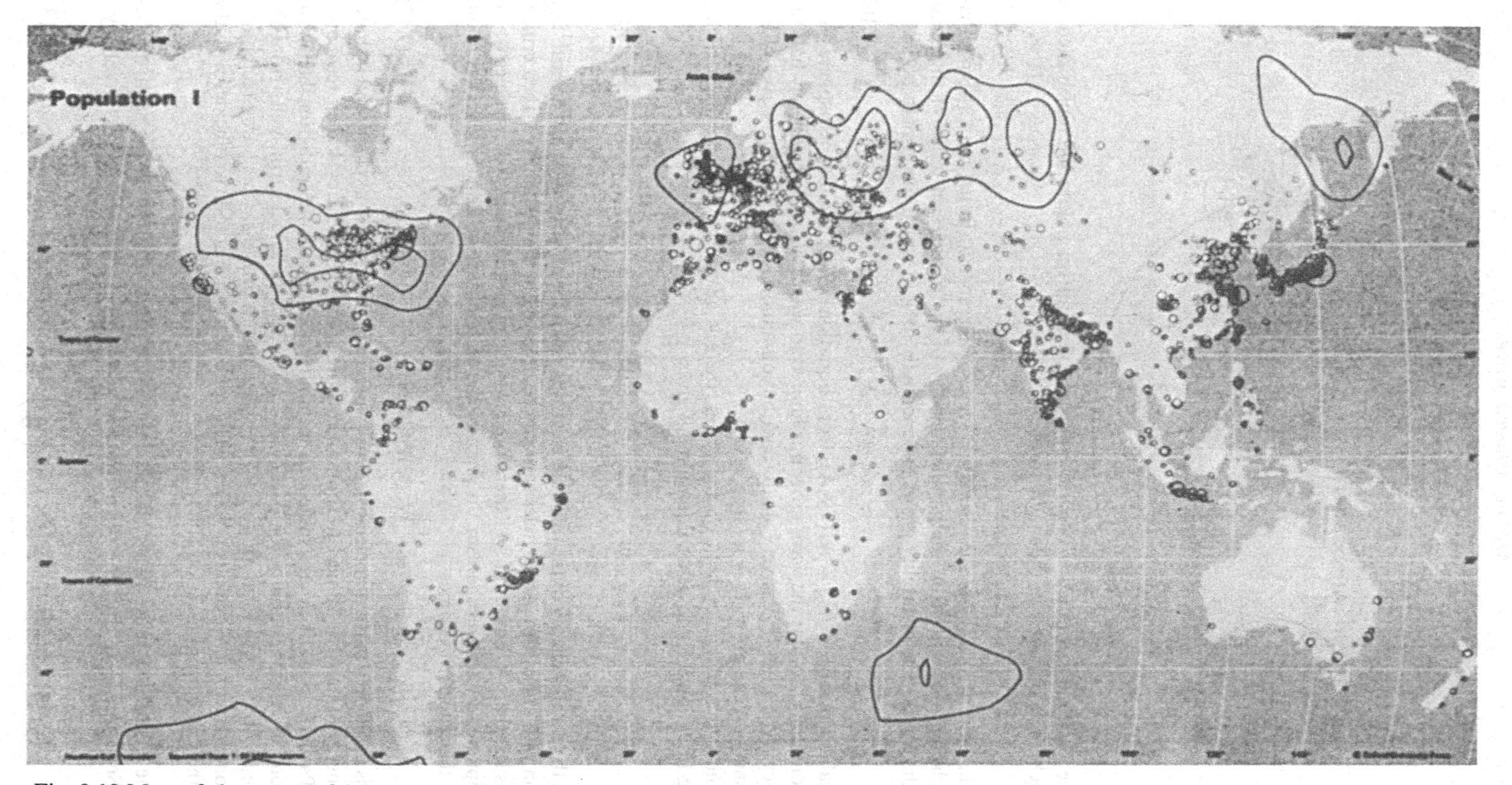

Fig. 9.18 Map of the zones of 3.2 kHz emission observed by the satellite Ariel-4. The circles show towns of at least 100 000 population. The enhancements correspond with the slot region in latitude and with the most industrial countries in longitude. (Reprinted with permission from K. Bullough *et al.*, *J. Atmos. Terr. Phys.* **47**, 1211, copyright (1985) Pergamon Press PLC)

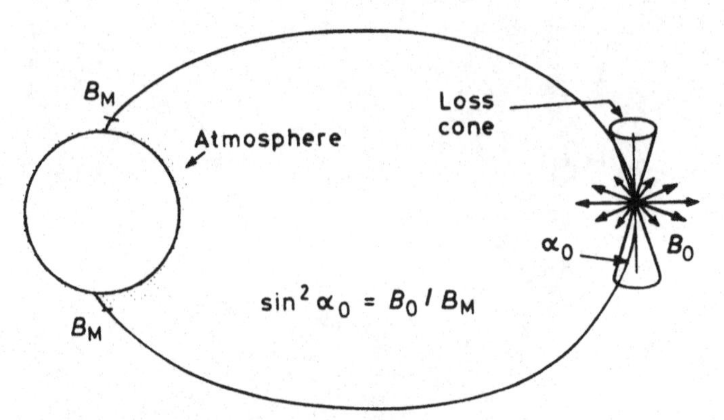

Fig. 9.19 The loss cone is emptied as particles approach their mirror points and replenished by pitch-angle diffusion in the magnetosphere.

where ω is the radio frequency of the resonating waves. Reference to Equation 9.2 shows that this is like placing a condition on the parallel velocity of the particles in terms of the phase velocity of the waves. The instability grows exponentially ($e^{\gamma t}$), in which the growth rate is

$$\gamma \approx \pi \omega_B \eta (A - A_c). \tag{9.10}$$

At the critical anisotropy, $\omega/\omega_B = A/(1+A)$, from Equation 9.9. In the magnetosphere, A is typically about unity, so that the wave growth is limited to waves of frequency below half the gyrofrequency (typically). If A is very small (distribution almost isotropic) only waves of low frequency (small ω) can grow. If A is large, $\omega \to \omega_B$.

In addition to this frequency limit on the waves, there is an energy limit on the particles. Rearranging Equation 9.6 and using Equation 9.9 gives

$$W_\parallel = \frac{B^2}{2\mu_0 N} \cdot \frac{1}{A_c(1+A_c)}, \tag{9.11}$$

which states the parallel energy of the resonant electrons when the anisotropy is just critical. For a distribution with anisotropy A, particles can only be involved in the instability if their critical anisotropy is less than A. Equation 9.11 therefore gives the minimum energy of particles which can be precipitated by this mechanism. In practice the second part of Equation 9.11 is about unity, when the criterion becomes

$$W_\parallel \gtrsim B^2/2\mu_0 N. \tag{9.12}$$

Figure 9.20 is based on a particular electron density profile in the equatorial plane, and it shows the lowest energy of particles subject to spontaneous precipitation by the cyclotron resonance instability. Particles > 100 keV are subject to instability throughout the radiation belts, but there is a zone of relative stability for particles of a few tens of kiloelectron-volts from the plasmapause to 7 or 8 Earth-radii. (Figure 9.20 applies also to protons, which are precipitated through resonance with ion–cyclotron waves.) The zone of relative stability coincides with the location of the ring current (Section 5.8.3).

When the instability is at work the variations of pitch angle may be regarded as diffusion (Section 2.2.5), and the diffusion coefficient (it may be shown) is given by

$$D_\alpha \sim \omega_B^2 \left(\frac{\delta B}{B}\right)^2 \Delta t, \tag{9.13}$$

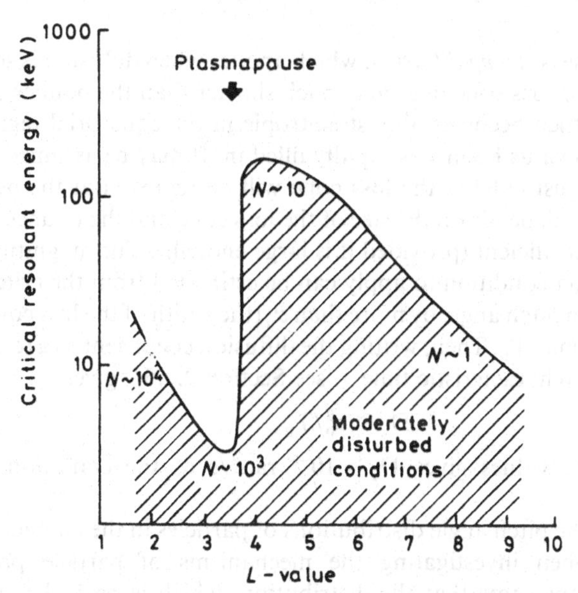

Fig. 9.20 Minimum energy of particles subject to precipitation by the cyclotron resonance instability, assuming an equatorial electron density (N) profile. (R. M. Thorne, in *Critical Problems of Magnetospheric Physics*. National Academy of Sciences, Washington DC, 1972)

where δB is the amplitude of the magnetic vector of the VLF noise, B is the flux density of the geomagnetic field, and Δt is the time for which the electrons and the waves remain in resonance.

Weak diffusion

Irrespective of its physical cause, pitch angle diffusion has two limiting cases according to the speed of the diffusion in relation to the bounce time. The particles undergoing diffusion are mirroring back and forth in the geomagnetic field. They are lost near the mirror points, while the redistribution of pitch angles occurs near the equatorial plane. If the diffusion is slow, so that a particle goes through several bounces while moving into the loss cone, we have *weak diffusion*, and this can be likened to the conduction of heat along a rod. Suppose that heat (particles) is supplied to one end (pitch angle 90°) and removed (particles precipitated) at the other (pitch angle 0°), which is held at a constant temperature (loss cone). The rate of heat transfer (particle precipitation) depends on the heating rate (arrival of fresh particles from a source) and on the thermal conductivity of the rod (diffusion coefficient). In the case of the particles, the diffusion coefficient is controlled by the intensity of the interacting waves (e.g. Equation 9.13). From this analogy it can be seen that the form of the pitch angle distribution does not depend on the intensity of the particle source, nor on the value of the diffusion coefficient. There will be relatively few particles in the loss cone at any one time.

Strong diffusion

The other limiting case is *strong diffusion*, which applies when diffusion is so rapid that particles move into the loss cone in a time much shorter than the bounce period. The particle distribution then becomes almost isotropic in the equatorial region because the hole left at the previous bounce is rapidly filled in. It may be assumed that all the particles which have just entered the loss cone will be removed at the next bounce. Therefore the loss rate depends on the size of the loss cone and the bounce period, but not on the diffusion coefficient (provided it is large enough). The magnitude required for the strong diffusion condition to apply can be estimated from the criterion that a particle must diffuse in pitch angle by more than α_0 (the width of the loss cone) in a time less than the bounce time, T_B. Then, writing the diffusion coefficient as a (characteristic length)^2 divided by a (characteristic time) – see Section 2.2.5 – gives

$$D_\alpha \gtrsim \alpha_0^2/T_B. \tag{9.14}$$

Putting in reasonable values gives $D_\alpha \gg 10^{-3}$ rad^2/s as the condition for strong diffusion.

Measurements of the pitch angle distributions of particles in the magnetosphere are clearly important when investigating the mechanisms of particle precipitation. Spacecraft observations show that the distribution, which is peaked normal to the field-line for much of the time, tends to the isotropic during intense precipitation.

The cyclotron resonance instability also provides a mechanism for limiting the flux of trapped particles in the magnetosphere. If the trapped flux increases, so does the intensity of wave turbulence and this in turn increases the diffusion coefficient. The precipitation rate therefore increases disproportionately and eventually the number of particles in the trapping region is limited, though actual numbers are subject to the resonance condition already discussed (Figure 9.20). The limitation of trapped particle flux is supported by satellite data. During magnetic disturbances more particles are injected, probably from the magnetotail, but during the most intense storms the flux comes up to a high but constant level that is not exceeded.

Further reading

S.-I. Akasofu and S. Chapman. *Solar–Terrestrial Physics*. Oxford University Press, Oxford, England (1972). Chapter 6 on waves in the magnetosphere.

Ya. L. Al'pert. *The near-Earth and Interplanetary Plasma, Vol 2, Plasma Flow, Plasma Waves and Oscillations*. Cambridge University Press, Cambridge, England (1983). Part 2 on waves and oscillations.

L. R. Lyons and D. J. Williams. *Quantitative Aspects of Magnetospheric Physics*. Reidel, Dordrecht (1984). Chapter 5 on wave-particle interactions.

R. A. Helliwell. *Whistlers and Related Ionospheric Phenomena*. Stanford University Press, Stanford, California (1976).

J. A. Jacobs. *Geomagnetic Micropulsations*. Springer-Verlag, New York, Heidelberg and Berlin (1970).

S. Matsushita and W. H. Campbell. *Physics of Geomagnetic Phenomena*. Academic Press, New York and London (1967), particularly Chapter IV-4 by W. H. Campbell on 'Geomagnetic pulsations', and Chapter V-1 by J. W. Dungey on 'Hydromagnetic waves'.

W. J. Hughes. Hydromagnetic waves in the magnetosphere. In *Solar-Terrestrial Physics* (eds. R. L. Carovillano and J. M. Forbes). Reidel, Dordrecht, Boston and Lancaster (1983).

A. Nishida. *Geomagnetic Diagnosis of the Magnetosphere*. Springer-Verlag, New York (1978). Chapter V, 'Magnetosphere as a resonator'.

C. F. Kennel. Consequences of a magnetospheric plasma. *Reviews of Geophysics*., **379** (1969).

R. M. Thorne. Wave–particle interactions in the magnetosphere and ionosphere. *Reviews of Geophysics and Space Physics* 1., 291 (1975).

A. C. Fraser-Smith. ULF/lower-ELF electromagnetic field measurements in the polar caps. *Reviews of Geophysics* **2**., 497 (1982).

D. Orr. Magnetic pulsations within the magnetosphere: a review. *J. Atmos. Terr. Phys*. **35**, 1 (1973).

Special Issue. IAGA symposium on micropulsations theory and new experimental results. *Space Science Reviews* **1**., 331 (1974).

W. J. Hughes. Pulsation research during the IMS. *Reviews of Geophysics* **20**, 641 (1982).

R. R. Heacock and R. D. Hunsucker. Type Pi 1–2 magnetic field pulsations. *Space Science Reviews* **28**, 191 (1981).

R. L. Arnoldy, L. J. Cahill, M. J. Engebretson, L. J. Lanzerotti and A. Wolfe. Review of hydromagnetic wave studies in the Antarctic. *Reviews of Geophysics* **26**, 181 (1988).

D. J. Southwood and W. J. Hughes. Theory of hydromagnetic waves in the magnetosphere. *Space Science Reviews* **35**, 301 (1983).

D. A. Gurnett and U. S. Inan. Plasma wave observations with Dynamics Explorer 1 spacecraft. *Reviews of Geophysics* **26**, 285 (1988).

S. D. Shawham. The menagerie of geospace plasma waves. *Space Science Reviews* **42**, 257 (1985).

Special Issue. Workshop on controlled magnetospheric experiments. *Space Science Reviews* **15**, 751 (1974).

R. A. Helliwell. VLF wave stimulation experiments in the magnetosphere from Siple Station, Antarctica. *Reviews of Geophysics* **26**, 551 (1988).

Special issues on power line radiation and its coupling to the ionosphere and magnetosphere. *Space Science Reviews* **35** (1–2), (1983).

10

Technological applications of geospace science

'I see you're admiring my little box', the Knight said in a friendly tone. 'It's my own invention – to keep clothes and sandwiches in. You see I carry it upside-down, so that the rain ca'n't get in.'
Lewis Carroll, *Through the Looking Glass* (1871)

10.1 Introduction

Science and engineering are related activities with different objectives. The purpose of science is to gain knowledge, and the essence of scientific achievement is intellectual rather than practical. To a scientist the knowledge and the ideas are what matter most. Engineering, on the other hand, is all about practical things. The result of successful engineering endeavour is a machine, a device or a scheme for performing some specific task. What matters to the engineer is that the machine, device, etc., should work well, and he/she will draw on any area of knowledge or experience to achieve this. Some of that knowledge might be science based; some might not.

Having drawn the distinction, we should at the same time recognize that there are strong links between these activities. Science relies on instruments and computers, the products of the engineer, and it should be abundantly clear from Chapter 3 how much the progress of geospace has depended on the development of techniques. And although there is no law that engineering must be science based, it draws heavily on scientific knowledge in practice. It would be an unusual engineer who relied entirely on historical practice or intuition. It is the purpose of this chapter to consider the impact of geospace science on practical activities within the province of the engineer.

Table 10.1 indicates the kinds of user whose activities are affected by phenomena of the geospace environment.

All solar–terrestrial relations are driven by solar behaviour, though modified by events in inter-planetary space and by terrestrial factors. The resulting variations occur over a wide range of time scales. Some geospace applications are for the very long term in the sense that the nature and the properties of the upper atmosphere determine once and for all the systems that can be supported. For example, the very existence and general properties of the ionosphere present the possibility of ionospheric radio propagation and set the band of radio frequencies that can be used. On another planet the situation would be quite different.

The solar flare and its immediate consequences take place in minutes and hours. The same applies to the isolated substorm and the auroral phenomena. Proton events tend

Table 10.1 *Activities affected by geospace phenomena*

	Geomagnetic activity	Solar radio interference	Solar particle emissions	Solar phenomena
Satellite operations				
Orbital variations	✓			
Command and control anomalies	✓	✓		
Communications with ground stations	✓	✓		
Aviation				
Communication at mid-latitudes (at VHF)		✓		
Communication in the polar cap (at HF)	✓		✓	
Navigation (at VLF)	✓		✓	
Polar flights at high altitude	✓		✓	
Electric power distribution	✓			
Telephone communication over long lines	✓			
High-frequency radio communications	✓		✓	
Pipeline operations	✓			
Geomagnetic prospecting	✓			
Scientific studies by satellite	✓		✓	✓
Scientific studies by rocket	✓		✓	✓
Scientific studies from the ground	✓		✓	✓

to last for a few hours though their terrestrial effects can go on for some days. Most magnetic and ionospheric storms also continue for several hours to a day or two. The medium term variability of the Sun due to the solar rotation and the evolution of active regions, coronal holes and the solar wind, modulates the frequency and intensity of disturbances over periods of weeks and months. Then the 11-year and 22-year cycles and their long-term modulation (Section 5.2.4) impose a slow variation, not only in the particle effects but also in the output of electromagnetic emission. The relevant time scale extends from years to thousands of years.

Our particular concern in the present chapter is with the effects of these variations, and with the problem of forecasting them. We begin with communications, which includes radio propagation, satellite communications, navigation and systems for remote sensing. The effects of solar activity on power lines and on long metal pipelines at high latitudes will be considered. Space operations, with and without humans, are also subject to interference and risk from solar activity. Finally we turn to the services that have been set up to monitor and predict geospace perturbations in order to minimize the amount of disruption to operational systems and to allow them to work as efficiently as possible.

10.2 Communications

10.2.1 Radio propagation predictions

Taken as a whole, communications is probably the world's largest industry. It includes telephony and telegraphy (both civil and military), broadcasting, navigational aids, and aircraft and ship communications. Once, virtually all distant communications were by means of ionospheric radio. Today there are better alternatives for some services – submarine cables and satellite relays – but ionospheric propagation is still very important for other services such as broadcasting, ship and aircraft communications and navigation, for military uses and in the less populated regions of the world. Point-to-point systems (cable and satellite) may be unsuitable where one of the terminals is mobile or the service has to cover an area. To some extent the vulnerability of hardware in space has deflected interest back to HF radio, which is now seen as a robust alternative in case of hostilities or other disaster. Although ionospheric radio now carries a smaller proportion of the total traffic than was once the case, it is carrying more than ever before in absolute terms. In consequence the radio spectrum is more or less saturated and seems likely to remain so.

The effectiveness of the ionosphere as a radio reflector depends on the radio frequency of the transmitted signal, and therefore it is necessary to know the behaviour of the ionosphere in order to choose the best frequency. We know that the ionosphere experiences considerable variation, for example from day to night and from summer to winter. To the extent that the variations are regular they may be included in predictions of future conditions, and on these predictions rests the planning of radio services that use ionospheric propagation. The first quantity predicted is the *maximum usable frequency*, or *MUF*. At vertical incidence MUF = f_EF2 (by definition of the highest frequency reflected from the ionosphere), but MUF > f_EF2 at oblique incidence when transmitter and receiver are separated – as they must be in a communication circuit. Figure 10.1 shows an example of a world map of MUF(Zero) for propagation involving the F2 layer. Maps of this kind are now prepared on a regular basis using a computer program (see Section 7.1.8). Propagation is usually best at a frequency just less than the MUF because the absorption increases at lower frequencies.

MUF data are essentially long term, and they are used to plan new circuits and to select working frequencies for existing ones. In overseas broadcasting, for example, the organization will have been allotted several frequencies in the HF band. Since it will publish in advance its programmes and the frequencies to be used, accurate predictions are very important if the service is to be received clearly in the intended service area.

The short-term variability of the ionosphere and changes due to solar events pose a more difficult problem. This is handled differently in different countries; Figure 10.2 shows a weekly forecast produced in the USA. This reports conditions from the previous week, forecasts conditions for the coming week, and up-dates the long term predictions of MUF and sunspot numbers. The forecast also includes a *quality figure*, defined from 'useless' to 'excellent' on a scale of 0 to 9; 5 means 'fair' and 7 is 'good'. The likelihood that ionospheric storms (Section 7.4.3), solar flare effects (Section 7.1.5) and PCA (Section 8.5.4) will occur are also predicted.

Disturbances such as the SID, the PCA, and the ionospheric storm have all been

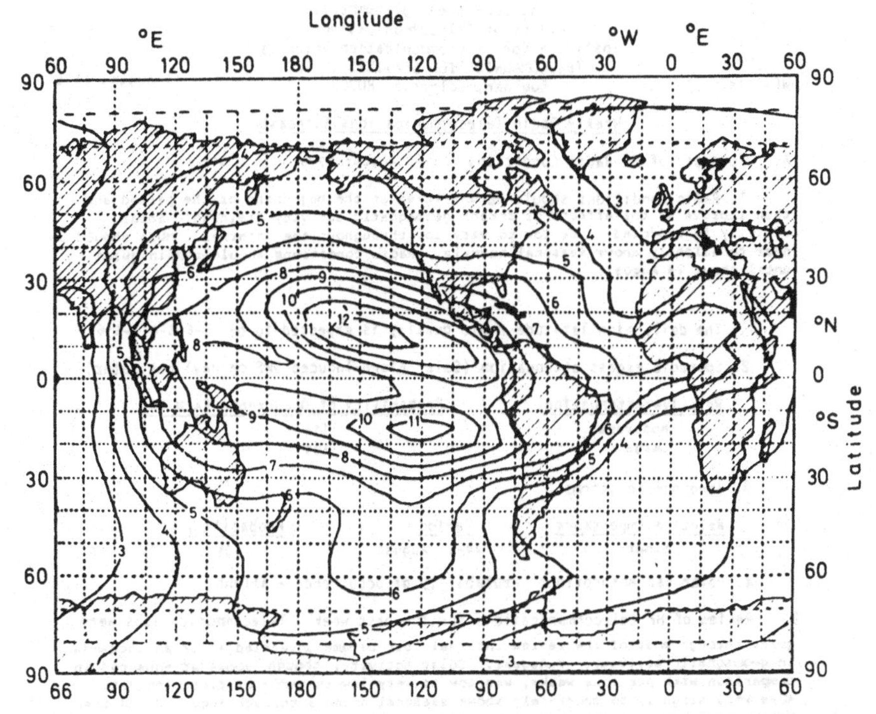

Fig. 10.1 World map of monthly median MUF(Zero)F2 – identical to f_EF2 – at 0000 UT in April when the sunspot number is 10. (W. Roberts and R. W. Rosich, *Report OT/TRER* 13. Institute of Telecommunication Sciences, Boulder, Colorado, 1971)

discussed as phenomena of the ionosphere. Their effects on radio propagation are summarized in Table 10.2. When the occurrence of a particular type of event is important to a user, a rapid warning can be sent by telex or telephone, or it may be incorporated in a service broadcast such as the standard time and frequency transmission (for example WWV in the USA or MSF in the UK). However, a warning sent over a circuit that is already affected serves little purpose, and this can be a difficulty – for example, an SID warning sent by HF radio would be of little value to the aircraft for which it was intended!

The performance of HF communication circuits is also affected by irregularities such as sporadic-E (Section 7.1.1), scintillations (Section 7.5.1, 8.2.2), spread-F (Section 7.5.1) and travelling disturbances (Section 7.5.5), which focus or de-focus the signal, cause fading, or allow propagation over two or more paths simultaneously with consequent interference. An extreme example is 'flutter fading' at 10–100 Hz which may occur on trans-equatorial paths.

One effect of travelling ionospheric disturbances is to degrade the accuracy of HF direction finders, a major application of which is to search-and-rescue operations following ship or aircraft disasters at sea. Whereas a modern direction finder may in principle be accurate to 0.1 degree, in practice the errors may amount to several degrees due to ionospheric irregularities. It appears difficult to predict these errors in detail, but the simultaneous monitoring of TIDs, for example by the HF Doppler technique, allows for some correction.

U.S. Department of Commerce
Office of Telecommunications
Institute for Telecommunication Sciences
TELECOMMUNICATION SERVICES CENTER
Boulder, Colorado 80302

WF- 614 11 August 1976

Weekly Radio Telecommunications Forecast

A. Forecast of HF Radio Conditions 12-18 August 1976

1. Radio conditions should be unsettled at the beginning of the period and then inprove as the effects of recent geomagnetic activity diminish. Solar activity is expected to be low to very low throughout the interval. MUFs should be near to slightly above seasonal normals. Radio conditions should remain quiet until about 24 August.

The daily high latitude radio quality is expected to be: 5-6-6-6-6-6-6

2. Sporadic excess attenuation (Solar Flare-Induced SWF on daylight paths)

Relative Attenuation	Probability of Occurrence Each Day
Moderate	15%
Large	10%

3. Ionospheric storms:

Relative Importance	Period	Probability
Minor	24-27 August	50%

4. Polar Cap Absorption: Probability of occurrence-- slight

B. Review of HF telecommunications for the past week. Telecommunications were quiet through most of the review interval, but became unsettled after an increase in geomagnetic activity. 9 August. Solar Activity, though somewhat enhanced in comparison with previous weeks, was low to very low in terms of flare activity. MUFs were slightly to moderately above seasonal normals through August 8 and then became normal to slightly below normal .

High Latitude Observed Radio Quality

August	4	5	6	7	8	9	10
Whole day index	6	6	6	6	6	6	5
6-hour indices	6666	6666	6776	6666	6666	6566	5466
Geomagnetic Activity A_{FR}	5	8	5	5	5	14	7
1700Z 2800 MHz Flux (provisional)	80	82	82	81	80	81	81

C. The following effective solar activity indices (12 month moving average Zurich sunspot numbers) are for use with the new "Ionospheric Predictions" OT/TRER 13.

1976-1977	AUG	SEP	OCT	NOV	DEC	JAN	FEB	MAR	APR	MAY
	9	8	8	7	7	6	6	5	5	5

D. Semimonthly revised ionospheric predictions. The following factors for 16-31 Aug. apply to monthly median predictions for August based on the solar activity index, SS No. 8 .

	UT 00	06	12	18
U.S. East Coast	.90	1.15	.95	1.00
U.S. West Coast	1.00	1.00	1.00	1.05
Europe	1.00	.90	1.00	.90
Alaska	.90	.90	.90	.90
Central Pacific	.95	1.00	.90	.90
Central Asia	.90	.95	.95	.90
S.E. Asia	1.00	1.00	.95	.85
Japan	1.05	.95	.85	.95
M.East	.90	1.00	.95	.90

Fig. 10.2 Example of a weekly forecast of radio propagation conditions produced in the USA.

Table 10.2 *Disturbances to ionospheric radio propagation*

Disturbance	Ionospheric effect	Propagation effect	Typical duration
SID	Increased D-region electron density by day	Increased absorption; loss of signal	$\frac{3}{4}$h
Storm	Increased D-region electron density at high latitude	Loss of signal on polar routes	Several hours at a time, over several days
	Decreased F-region electron density, world-wide*	Loss of signals near MUF; narrowing of useful radio spectrum	1 day
PCA (Polar cap absorption)	High D-region electron density over polar caps	Loss of polar communications	Several days

*The effect increases with latitude and may exceed 30% at latitudes greater than 45° (see Fig. 7.23).

10.2.2 Satellite communications

In communications involving links between satellites and ground stations the most serious ionospheric effect is scintillation (Sections 7.5.1, 8.2.2). For accurate data transmission it is necessary to maintain an adequate signal-to-noise ratio for the longest possible time. In the example shown in Figure 7.38 the signal faded by 6 dB or more below the mean leavel for 8.3% of the total time. Therefore a circuit with a signal margin of 6 dB would operate at little more than 90% efficiency (which would be considered very poor) under these scintillation conditions. The net effect of scintillation, thus, is to require an increase in the transmitted power, which increases the cost of the system. The designer of satellite communications must therefore know about the scintillations in order to achieve the required reliability most economically.

Scintillation effects due to the ionosphere invariably fall off as the radio frequency increases, and satellite communications typically operate at about 1.6 GHz (L band). However, scintillations may still present a problem, particularly in the equatorial regions. L-band scintillation in 1979, a time of high solar activity, produced fading of at least 20 dB for 30% of the time at the peak of the equatorial anomaly (Section 7.3). In 1982, when solar activity was lower, the fading was still at least 5 dB for 30% of the time. The collected information on the occurrence and intensity of scintillation has been amalgamated into empirical models to aid the design of satellite communication systems.

The rotation of the plane of polarization by the Faraday Effect (Section 3.5.2) is usually assumed negligible at GHz frequencies, but even here it may become significant when the electron content reaches the high values that can occur occasionally, for instance during a high sunspot maximum. For example, an electron content of 10^{18} m^{-2} rotates the polarization by about 30° at 1.6 GHz. The practical effect is that if orthogonal polarizations are being used independently, some cross coupling between the signals will be caused. If only one polarization is being used there will be a loss of received signal. The solution is to transmit circularly-polarized waves, but this may be undesirable because of the weight penalty.

10.2.3 Navigation, positioning and timekeeping

In principle it is possible to determine the position of a point on the Earth's surface by measuring its absolute distance from three satellites at known positions in space, or its relative distance from four satellites. Suppose that sychronized radio pulses are transmitted from each satellite and are received at the point whose position is to be determined. Comparing the arrival times of two of the signals defines a surface relative to the positions of the sources. (In the special case that the signals arrive together this surface is the plane bisecting and normal to the line joining the sources.) Comparing three signals locates the receiver on a line, and the fourth signal fixes the required location at a point. If the altitude is already known, or only latitude and longitude are required, then three signals will suffice. For engineering reasons pulsed systems may be inconvenient and the same result can be achieved by transmitting a continous wave with one or more sinsuoidal modulations. The receiver then has to measure phase differences rather than time differences, but the principles are the same and both approaches can be treated in terms of time measurement.

Ground-based navaids

The principle of position determination by radio was first applied during the Second World War, and since then various purely ground-based navigational aids (navaids) have been brought into service. Over short distances these systems are very accurate, but to reach beyond a few hundred kilometres ionospheric propagation is required and the variability of the ionosphere then imposes a limit to the accuracy. For long range navigation the VLF band (3–30 kHz) is used. Because reflection is from the D region, propagation is more stable at VLF than would be the case at higher frequencies reflected from the E region. Nevertheless there is a large day–night variation (Section 7.1.4) which alters the time delay over a given path, and there can be perturbations during solar flares (Section 7.1.5) and aurorae (Section 8.3).

The *Omega* navigational aid uses eight transmitters in the VLF band between 10 and 14 kHz and provides a world-wide position fixing system for ships and aircraft. Some ionospheric variations, such as the day–night change, are regular enough to be partly compensated for, but disturbances such as solar flares and PCA (Section 8.5.4) can cause positional errors of several kilometres. Solar flares affect mainly low- and middle-latitude propagation paths across the daylit hemisphere, while polar cap events affect polar paths, particularly at night. The after-effect of the D-region storm (Section 7.4.4) will also affect middle and high latitudes, again most strongly at night. In addition there is some random variation due to ionospheric variations of smaller scale – notably travelling disturbances (Section 7.5.5) – and these errors can amount to several hundred metres.

Satellite-based navaids

The errors introduced by the ionosphere are greatly reduced in satellite navigation systems which use radio frequencies well above the ionospheric penetration frequency.

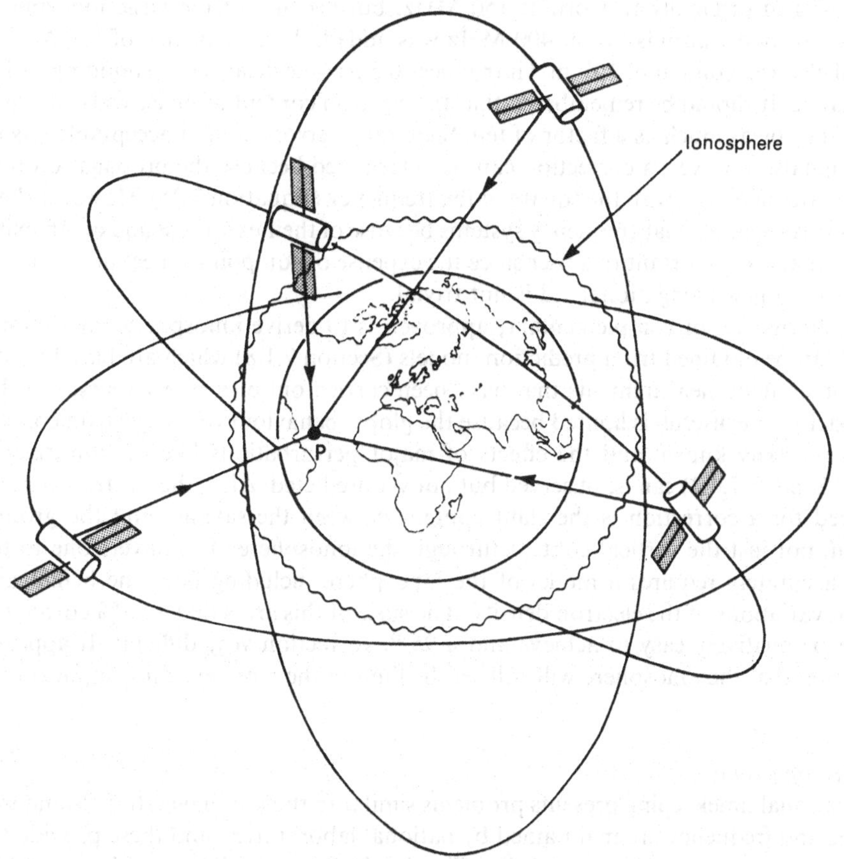

Fig. 10.3 A navigation system using four satellites, orbiting with 12-h period at 20 000 km. The transmissions would be near 1600 MHz to minimize propagation effects. (After J. A. Klobuchar, private communication.) Reception of signals from four satellites can fix the position of a point, P, in latitude, longitude and altitude.

The Navstar system, also known as the Global Positioning System (GPS), is now (1990) being implemented for world-wide navigation and is to be available to many categories of user. A partial system has been in use and undergoing tests for several years, and the first satellite of the operational system was launched in early 1989. When complete the system will consist of 21 satellites in circular orbits at altitude 20 200 km, inclination 55°, and with a 12-hour period. The principle is illustrated in Figure 10.3. The accuracy is intended to be about 10 m for military users and 90 m for anyone else. The signal will be transmitted at a frequency near 1.6 GHz, at which an ionospheric electron content (Section 3.5.2) of 4×10^{17} m^{-2} will introduce a transmission time error of 15 ns and a range error of 5 m. This will be accurate enough for most navigational purposes, though corrections may be needed for applications to geodesy and surveying needing accuracies to 1 m or better.

Older systems that are still in use transmit on lower frequencies and the ionospheric effects are correspondingly larger. The Navy Navigational Satellite System (NNSS) was designed originally to work at 150 MHz, but because of the large ionospheric effects a second transmission at 400 MHz was added. At a frequency of 150 MHz a typical electron content of 10^{17} m^{-2} introduces 0.5 μs time delay, corresponding to 150 m in range. It should be remembered that the electron content often exceeds 10^{17} m^{-2}, sometimes by as much as a factor of ten. Such range errors are not acceptable. Using both signals, however, a correction can be determined because the propagation time delay varies inversely with the square of the frequency (Equation 3.25). However there is some bias against dual-frequency systems because of their extra cost and complexity, factors that are important to a user since the expense of equipping a fleet of aircraft or ships with a new navigational aid is not trivial.

An alternative, or complementary, approach is to derive ionospheric corrections. These can be obtained from prediction models (Section 7.1.8) which are based on the electron content measurements that have been carried out over many years. But this approach, while useful, is limited because the global behaviour of the electron content is not precisely known and the effects of major perturbations like the ionospheric storm (Section 7.4.3) can be observed but not yet predicted. Also, the electron content required for a correction is the slant content between the satellite and the ground station, not just the vertical content through the ionosphere. To convert one to the other accurately requires a model of the ionosphere including both the height and spacial variations of the electron density. The state of this art is that a 50% correction would be relatively easy to achieve and a 95% correction very difficult. It appears, therefore, that the ionosphere will still set the limit in the more exacting applications.

Time comparisons

International timekeeping presents problems similar to those of navigation. Standards of time and frequency are maintained by national laboratories, and these provide the basis of timekeeping in those countries. Clearly, these standards have to be compared so that a common international standard may apply. One method is to carry crystal clocks between countries so that they may be compared side by side, and another method is to exchange time signals by radio propagation. The VLF band, which offers the most stable mode of ionospheric propagation, has been used for this purpose. However, as time standards become more precise, with the primary standards now achieving errors as low as 10 ns per day (equivalent to 1 s in 250 000 years), more accurate means are needed, indicating satellite transmissions. As with navigation, ionospheric uncertainties will provide the ultimate limit.

A scientific application of precise time comparison arises in *very long baseline interferometry* (*VLBI*), a technique which permits observing baselines of continental or inter-continental dimensions to be achieved in radio astronomy. In VLBI, the signals received at these widely separated sites are recorded with timing signals derived from local clocks, which have to be synchronized to less than 1 ns.

10.2.4 Effects on remote sensing systems

Some aspects of remote sensing, in the broader meaning, go back to the original purpose of radar: the detection and 'observation' of distant objects. In recent years the ever increasing sophistication of electronics and satellite technology have opened up some new applications of the radar principle, and the performance of these systems can be affected by the ionosphere. That is, the limit to the performance achieved in practice may be set by the propagation medium rather than by the hardware.

Altimetry

The idea of the *satellite altimeter* is to determine the altitude of the orbiting vehicle from the time delay of pulses reflected from the surface of the Earth. Since the orbit can be accurately determined, it then becomes possible to measure the height variations of the terrain, the shape of the Earth (the 'geoid'), ice and glacier levels, and ocean waves. A typical satellite altitude for this purpose will be 1000 km, and so the signal will twice traverse the whole of the bottomside ionosphere plus part of the topside. Operating at about 14 GHz (wavelength about 2 cm) the theoretical height resolution is a few centimetres. Ionospheric delay (Section 3.5.2) for an electron content of 10^{18} m^{-2} is about 40 cm, and therefore correction or compensaton is needed for best accuracy. It is possible to determine the ionospheric effect with a dual frequency system and thereby make corrections.

Synthetic aperture radar

A 'synthetic aperture' is constructed by moving a small antenna to several adjacent points, and then building up the signal (in both amplitude and phase) bit by bit to achieve the effect of a larger antenna. It is an important technique in radio astronomy, since the radio sky does not change and a large aperture can be synthesized over many days.

A *synthetic aperture radar* (*SAR*) may be carried on an aircraft or a satellite to observe the Earth's terrain for cartography, geological mapping, glaciology and other purposes. The general configuration is that the radar antenna points obliquely downward, illuminating the ground to one side of the vehicle as well as forward and backward. Resolution to the side is obtained from the time delay, and fore and aft resolution (along the line of flight) comes from the Doppler shift of the echoes. However, since an object is observed (with changing Doppler shift) throughout the time the radar is passing over, it is possible with a coherent system to sum these signals and so improve the resolution. SAR has been used from aircraft for many years, and on them the synthetic aperture can be much longer than the aircraft.

Satellite-borne SARs have been flown experimentally since the late 1970s, and the technique is being developed for operational use in the 1990s. The frequencies will be between 1 and 10 GHz, but the signals will pass through the ionosphere which will therefore affect the performance, particularly due to irregularities whose effect is to limit the distance over which the aperture can be synthesized. Resolutions will be 10s or 100s of metres.

OTH radar

Over-the-horizon (*OTH*) *radars*, using one or more ionospheric reflections, have been developed in the HF band (e.g. 20 MHz) to extend the range of radar to several thousand kilometres, so that the movement of ships or aircraft can be detected over wide areas including oceans. They can also detect ascending missiles, echoes being returned both from the solid body and from the exhaust gases. The ionized trails from meteors can also be observed. The range accuracy may be as good as 10–20 km, but of course this depends on knowing the path taken by the ray. Angular resolution may be better than 1° (i.e. 50 km at 3000 km range), but this will deteriorate if the ionosphere is 'tilted' due to day–night changes or travelling ionospheric disturbances (Section 7.5.5). The line-of-sight velocity of the target may be determined from the Doppler shift of the frequency of the echo, but this requires stability in the ionosphere. Also, the radar will be of little use if its transmissions are absorbed in an active D region (Section 8.3.6). An OTH radar has much in common with a coherent radar for ionospheric studies (Section 3.6.3). The field-aligned structures that the ionospheric radar seeks to observe could, however, be a source of clutter to an OTH radar, producing a large unwanted signal that may obscure that being sought.

10.3 Power lines, pipelines and magnetic prospecting

During magnetic storms (Section 8.4), strong electric currents flow in the high-latitude ionosphere, and these induce currents in the ground beneath and in any long conductor such as a power line. These currents can amount to 50–100 A, which may trip the overload protectors, or, if flowing through the windings of a transformer, can produce core saturation and adverse consequential effects. The electrical distribution system may be disrupted. In recent years electrical distribution networks have been interconnected over greater distances, and this interdependence makes the problem of disruption during magnetic storms more serious. High-latitude regions such as Scandinavia, Canada, and the northern USA are the most affected. During the great magnetic storm of 13 March 1989, admittedly one of the two or three largest storms on record, the whole of the Quebec electric power system, a capacity of 9000 MW, was out of operation for 9 hours. Outages of this kind are plainly very serious, which is why the electric power authorities require warnings and forecasts of geomagnetic storms in the management of the distribution system.

Another kind of long line that may be affected by induced currents is a pipeline used to carry oil or gas. The effect of the current in this case is to cause corrosion. The Alaska oil pipeline has been studied from this point of view. Although 1280 km long, its electrical resistance from end to end is less than 10 Ω. The pipeline runs north to south, crossing the auroral zone. The induced electric field is generally east–west, but inhomogeneities in the ground modify the field and produce a north–south component. The pipe is earthed at frequent intervals and currents flow in the pipe, effectively to neutralize the induced field in the ground. Corrosion occurs at the points where the current flows to ground.

The disturbance of the geomagnetic field is a direct interference with magnetic prospecting. The technique of magnetic prospecting for minerals from aircraft requires geomagnetically quiet conditions, and warnings of the imminence of magnetic disturbance are needed if wasted flights are to be avoided.

10.4 Space operations

10.4.1 Effects on satellites

Field perturbations can affect the operation of satellites for communications or other purposes because some satellites are oriented in space by reference to the geomagnetic field. During quiet periods the orientation of the field is well known, but large errors can occur if orientation manoeuvres are attempted during a stormy period. In a large storm it is possible that the magnetopause moves inside the geosynchrous orbit (at 6.6 R_E), and then a satellite may find itself in the solar wind whose magnetic field (Section 5.3.4) is entirely different from that of the Earth in both magnitude and direction.

The presence of fluxes of electrons in the magnetosphere causes spacecraft to become electrically charged. This does no harm in itself, but there can be harmful consequences when electrical discharges occur between different parts of the spacecraft. Components may suffer physical damage, but more serious in practice are the effects of the electromagnetic radiation from the discharge, pulses that are picked up by the spacecraft's command receiver and may, in effect, present the vehicle with false orders. The result may be that the satellite is put out of operation until the operators on the ground can detect and correct the malfunction. Such failures are known as *spacecraft anomalies*.

It is observed that some spacecraft suffer anomalies at certain times of day, while others show no such preference. Solar radiation therefore plays a part in some cases only, and the detail of the spacecraft design is also a factor. One significant element is the flux of trapped energetic particles in the magnetosphere, based on the evidence that the number of anomalies per day depends strongly on the level of geophysical activity. (The correlation is best if one uses an index that averages the activity over several days.) The most affected part of the construction seems to be dielectric material mounted on the surface of the vehicle, since this can stop, and gather the charge from, electrons in the energy range 20–200 keV, particles whose flux varies strongly with geomagnetic activity (Section 5.7.3).

Finally, the life of a satellite depends on the air density, which varies with solar activity. For satellites that orbit at heights of a few hundred kilometres, a longer life can be expected over solar minimum than over solar maximum. Large, isolated, magnetic storms also increase the rate of orbital decay and make it more difficult to predict the exact place and time of re-entry.

10.4.2 Effects on space travellers

Without doubt the most hazardous geophysical event for people in space is the solar proton event (Section 8.5). Several events of this kind, each lasting 1–3 days, occur each year on the average, and the largest events could inflict serious radiation damage on any astronaut caught in space without adequate shielding. While no serious effects are known to have occurred so far the risk is well appreciated. During the Apollo series of flights to the Moon in the early 1970s, for example, a careful watch was kept on solar activity and on the emergence of any solar regions likely to produce proton flares. The shielding on the Apollo spacecraft was designed to minimize (though not entirely remove) the dose of radiation received by those within. The astronauts were most vulnerable when on the surface of the Moon, and it would have been necessary to recall them to the orbiting command module had a proton event commenced. Even so, some radiation sickness may have resulted. Longer flights, such as to Mars, which are likely to meet several proton events *en route*, will require heavier shielding.

During Space Shuttle flights in low orbit the exposure will be greatest while the vehicle is over the polar caps. To take an example, the flight made on 12–14 April 1981 was nearly affected by a proton event. Protons of 50–500 MeV started to arrive on 10 April and continued until 14 April. The flux had peaked before the launch, but protons of energy > 10 MeV were present throughout the flight. Had an EVA (extra-vehicular activity) been planned it would have had to be cancelled. The problems will be greater for operations at greater altitude and of longer duration.

Somewhat nearer home for most ordinary mortals is the risk of being irradiated while in an aircraft over the polar regions. At 30 000 ft, a typical altitude for subsonic aircraft, the proton fluxes are small, but supersonic aircraft fly at twice that altitude and the radiation could be significant during a large event. During the great event of August 1972 the dose at 65 000 ft reached 400 mrem/h, which might be acceptable for astronauts or others in high-risk occupations but is outside the conservative safety limits imposed for the benefit of fare-paying civilians. Fortunately, the effects can be avoided by bringing the aircraft down to a lower altitude. Even though the hazard may be avoided by altering the flight plan, this does stress the need for a reliable monitoring, prediction and warning service.

10.5 Activity monitoring and forecasting

10.5.1 The users

There are several groups of people who need to know about geophysical occurrences as soon as they occur, and preferably sooner. The scientific community is itself an important user of warnings and forecasts. Since geophysical events cannot be served up to order, experiments designed to study some kind of event work best if the experiment can be switched on as soon as a suitable event begins if not before. A rocket experiment is a good example, and the launch decision may well be influenced by ground-based monitoring; it is also helpful to have a prediction of when a suitable event is likely to occur. In other cases an experiment may be modified, for instance switched to higher resolution, during a desirable event. Even where experiments are not started or modified, it is helpful to have prompt information about significant events so that those periods may be selected for detailed study.

The other half, approximately, of the market for warnings and forecasts comes from practical activities such as those discussed in Sections 10.1 to 10.4 above. Table 10.3 summarizes the effects of, and users affected by, four kinds of common geophysical disturbance.

10.5.2 Monitoring systems

Monitoring means keeping an eye on things, and practically any instrument that observes some aspect of the geophysical environment continuously could provide an input to geospace monitoring. Some scientists effectively do their own monitoring using the instruments to which they have local access. At the 'official' level there is a world-wide organization of 'Warning Centers'. These are part of the International

Table 10.3 *Practical consequences of disturbances*

Event	Effect	Users
Geomagnetic storm	Induced earth currents	Electric power companies
	Field perturbations	Magnetic surveyors
	Enhanced particle fluxes in space	Satellite operators
Ionospheric storm	Ionospheric changes	Radio communications
		Ship and airlines
Solar flare (X-rays)	Sudden ionospheric disturbance	Radio communicators
		Ship and airlines
Polar cap event (protons)	Damaging radiation	Space agencies
		Airlines

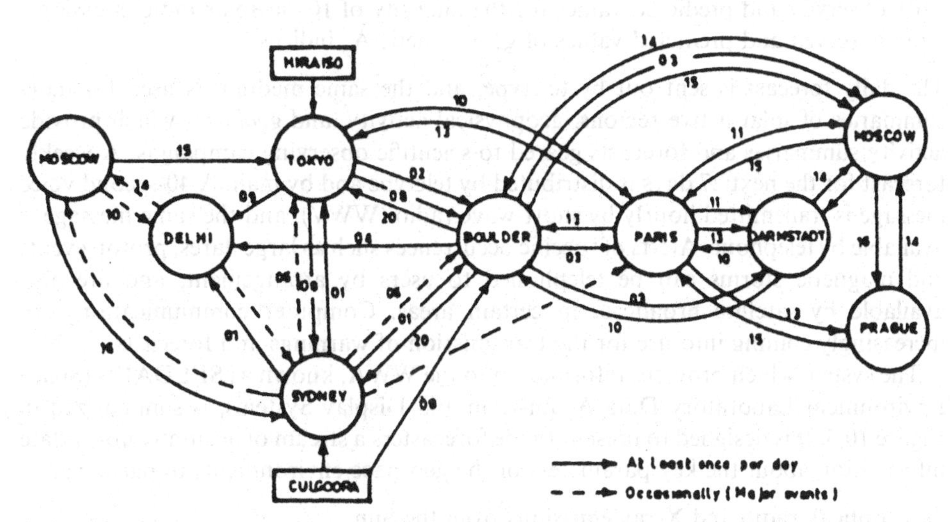

Fig. 10.4 Daily interchange of messages between Regional Warning Centers. (*COSPAR Information Bulletin*, August 1989)

Ursigram and World Days Service (IUWDS), which is a permanent service of the International Scientific Radio Union (URSI). Via the International Council of Scientific Unions (ICSU), URSI comes within the ambit of the United Nations Organization (UNO). The IUWDS was formed in 1962, but was based on earlier arrangements for the rapid exchange of solar and geophysical data built up since about 1930. The present arrangement comprises one World Warning Agency (WWA) in Boulder, Colorado, USA, Regional Warning Centers in Moscow, Paris, Darmstadt, Tokyo and Sydney, and Associate Regional Warning Centers in some other countries. Each regional centre serves customers in its own region with forecasts and alerts as required, but the data are shared daily between the centres so that full information is available to all (Figure 10.4).

In forecasting, the procedure is that each regional centre makes its own forecast which is passed to the WWA, and the personnel at the WWA then make up a 'consensus forecast' taking account of all the information available as well as all the regional forecasts. The consensus is then sent to each regional centre to be used as may be appropriate in the advice sent out from there.

The WWA in Boulder, Colorado, also serves as regional centre for the Western Hemisphere. It issues forecasts and reports several times a day, and the principal forecast, issued at 2200 UT, covers the following topics :

(a) report of solar features and activity, and of immediate geophysical effects during the preceding 24 hours, and a forecast of solar activity for the next 72 hours;
(b) a report of geomagnetic activity over the past 24 hours, including proton observations at satellite altitudes, and a forecast in general terms of geomagnetic activity for the next 72 hours;
(c) probabilities for the occurrence of X-ray flares and proton flares during the coming 72 hours, and a forecast of the probability of polar cap absorption occurring during the next 24 hours;
(d) observed and predicted values for the intensity of 10 cm solar radio emission;
(e) observed and predicted values of geomagnetic A_p indices.

The daily forecast is sent out by teletype, and the same medium is used for daily summaries of solar active regions, geophysical activity, and *geoalerts* which provide activity summaries and forecasts geared to scientific observing campaigns. A weekly forecast for the next 27 days is distributed by teletype and by mail. A 40-second voice message is transmitted hourly by short-wave radio (WWV), and the same message is available by telephone. Alerts of specific occurrences such as large flares, proton events and magnetic storms can be telephoned to users by arrangement, and are also available by satellite broadcast in certain areas. Computer communications are increasingly coming into use for the transmission of warnings and forecasts.

The system which provides information to the WWA, known as SELDADS (Space Environment Laboratory Data Acquisition and Display System), is summarized in Figure 10.5. It is designed to present to the forecasters a stream of accurate, up-to-date information about the key parameters of the geospace environment: in particular

optical, radio and X-ray emissions from the Sun,
the state of the solar wind,
the arrival of energetic solar protons near the Earth,
observable terrestrial effects, such as changes in the ionosphere, the aurora detected by optical, radio, or radar methods, and geomagnetic field perturbations.

SELDADS receives information in several ways. Reports of solar observations in Hα light (636.3 nm) are received by teletype from sites of the Global Solar Flare Patrol, and the images from several telescopes may be displayed in real time so that the forecaster may see for him/herself how the sunspots develop. The readings of X-rays, protons and the geomagnetic field from a geosynchronous satellite are received by telemetry. In detecting terrestrial effects, high-latitude monitoring is particularly important. SELDADS receives the readings from a chain of magnetometers and riometers in Alaska which are transmitted to the Agency every 12 minutes.

The data collected from this and other systems are required in the first instance for the warning and forecasting operations, the essence of which is the rapid dissemination

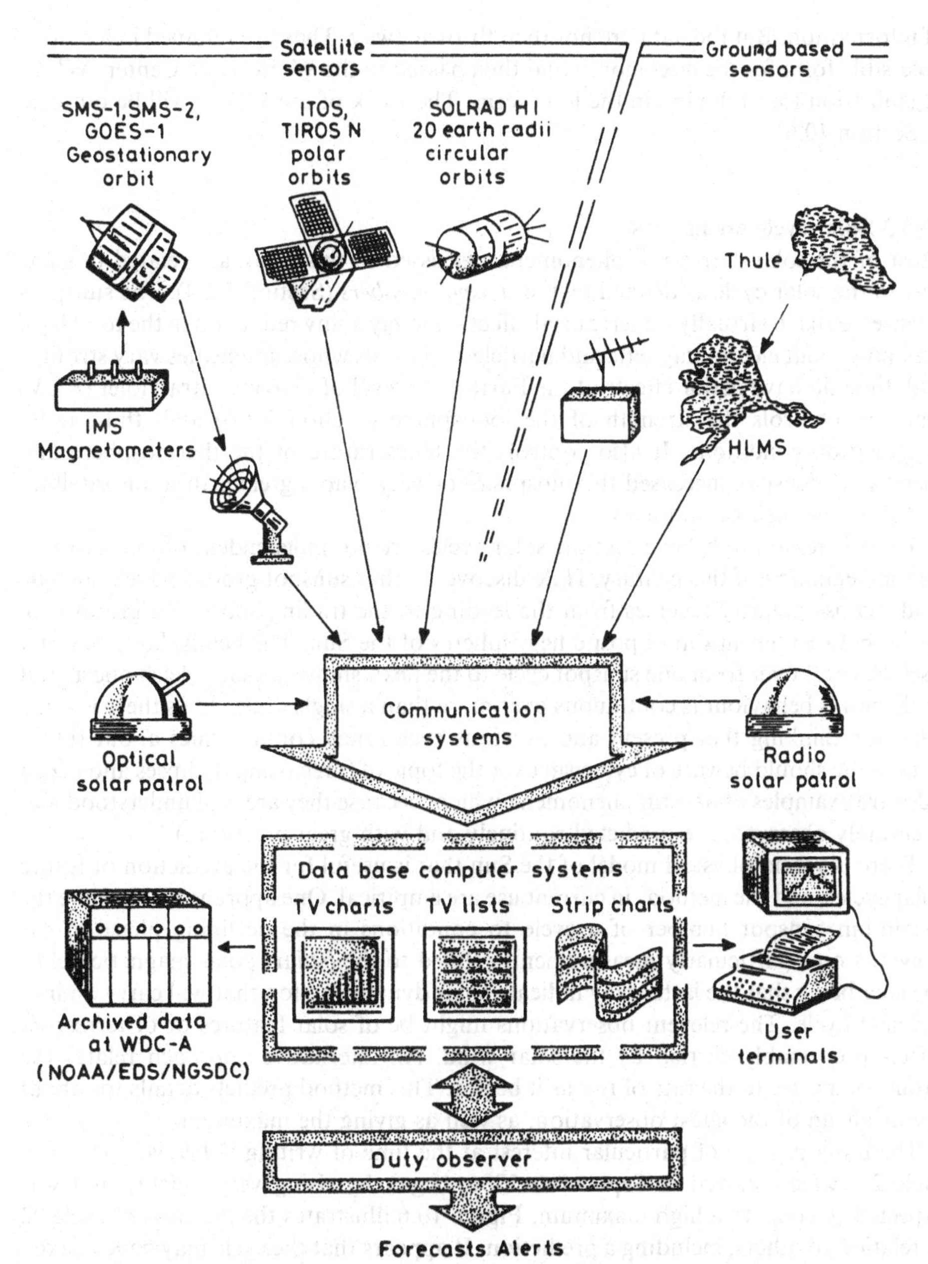

Fig. 10.5 The SELDADS approach to monitoring the space environment. Satellite and ground-based sensors are included (HLMS stands for High Latitude Monitoring Station), as well as inputs from optical and radio solar patrols. The output data are available to local users, the forecasters, and for archiving. (D. J. Williams, *Report ERL* 357-*SEL*. Environmental Research Labs., Boulder, Colorado, 1976)

of information. But the data are not then thrown away. They are retained in a readily accessible form for the next month and then passed to the World Data Center (WDC) organization for archiving in the long term. The work of the WDCs will be outlined in Section 10.6.

10.5.3 Solar cycle predictions

Most of the solar–terrestrial phenomena discussed in this book are related in some way to the solar cycle as defined by the *sunspot numbers* (Section 5.2.4). The sunspots themselves have virtually no terrestrial effect – merely a tiny reduction in the total light emission – but electromagnetic and particle radiations whose intensities vary strongly with time do have major effects at the Earth. The level of extreme ultra-violet (EUV) emission controls the strength of the ionosphere (Section 7.1.6) and, thus, radio propagation conditions. It also controls the temperature of the thermosphere and thereby its density; increased thermosphere density places greater drag on satellites and shortens their orbital lives.

There is reason to believe that the solar cycles are not independent of one another. At the beginning of this century, Hale discovered that sunspot groups have magnetic fields whose polarity reverses from the leading to the trailing spots of a group and, again, between groups in opposite hemispheres of the Sun. The hemispheric polarity itself changes over from one sunspot cycle to the next, showing that at least one aspect of the Sun's behaviour is continuous over more than a single cycle. Thus there is some basis for thinking that present and previous cycles may contain clues about future ones. (One should beware of cynicism over the topic of forecasting. Eclipses and ocean tides are examples of natural phenomena which, because they are well understood and accurately observed, are predicted routinely and with great precision.)

There is yet no physical model of the Sun that is useful for the prediction of future solar cycles, and the methods in current use are empirical. One approach is to relate the maximum sunspot number of a cycle to conditions in the declining phase of the previous one, particularly measurements related to the (solar) polar magnetic fields, the idea being that the latter may indicate solar dynamo action that will carry on into the next cycle. The relevent observations might be of solar features or of terrestrial effects presumably carried by the solar wind. An alternative approach relates the profile of a cycle to the rate of rise as it begins. This method predicts details for about a year ahead of the latest observation, as well as giving the maximum.

These matters are of particular interest at the time of writing (1989/90). The new cycle 22, which started in September 1986, began by rising very quickly and was expected to come to a high maximum. Figure 10.6 illustrates the progress of cycle 22 in relation to others, including a prediction. It appears that the cycle may have peaked in 1989, but a full report will have to be postponed until the next edition of this book!

10.5.4 Flare forecasting

The prediction of solar flares is at the heart of short-term solar–terrestrial predictions. Flares may be observed in white light (i.e. using a filter that darkens the solar disc but does not alter its colour), but they are most prominent in Hα light (656.3 nm) which comes from the chromosphere; current methods of flare prediction depend heavily on

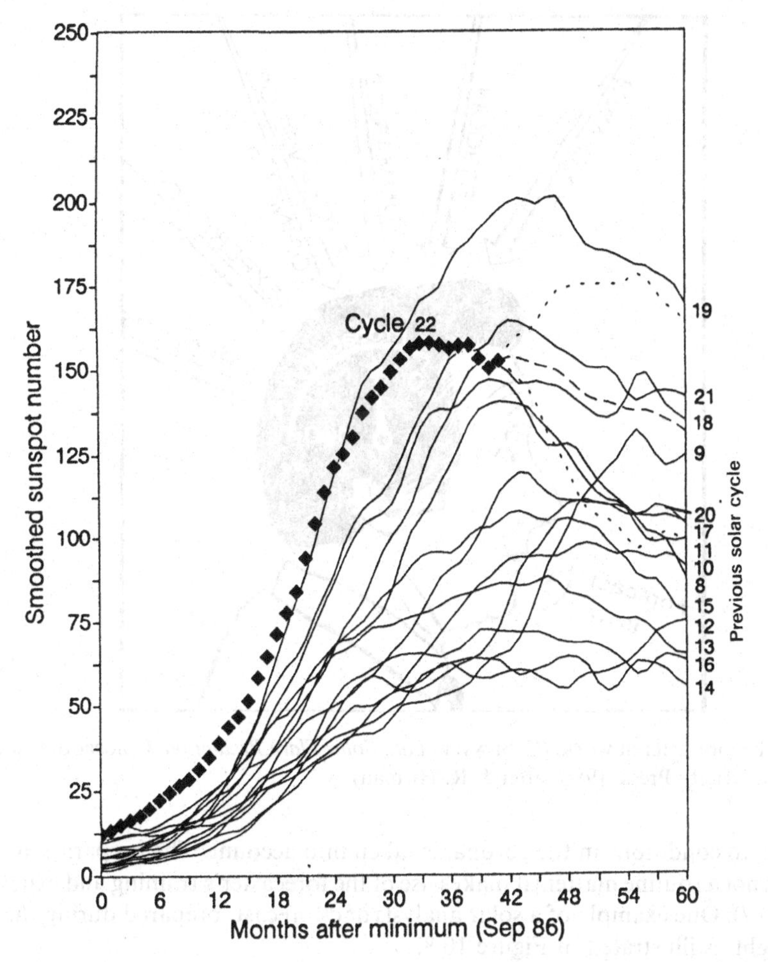

Fig. 10.6 The progress of solar cycle 22, compared with previous cycles. The dashed and dotted lines show the prediction and the 90% prediction interval. (G. R. Heckman, private communication)

observations of features such as filaments and plages in Hα light. The structure of the magnetic field in active regions of the Sun may be inferred from the appearance of these features. The forecaster inspects the solar disc (normally displayed on a monitor screen) for neutral lines where the magnetic field reverses, particular attention being paid to active regions near the lines of polarity reversal. Each region is tested against a number of criteria, such as:

the development of new magnetic fields;
the merging of active regions;
reversal of polarities from the normal;
twisting of the neutral line.

Reference to past activity assists the interpretation of the tests. Ancilliary information, such as solar radio flares, X-ray emissions, and optical activity at the limb of the Sun,

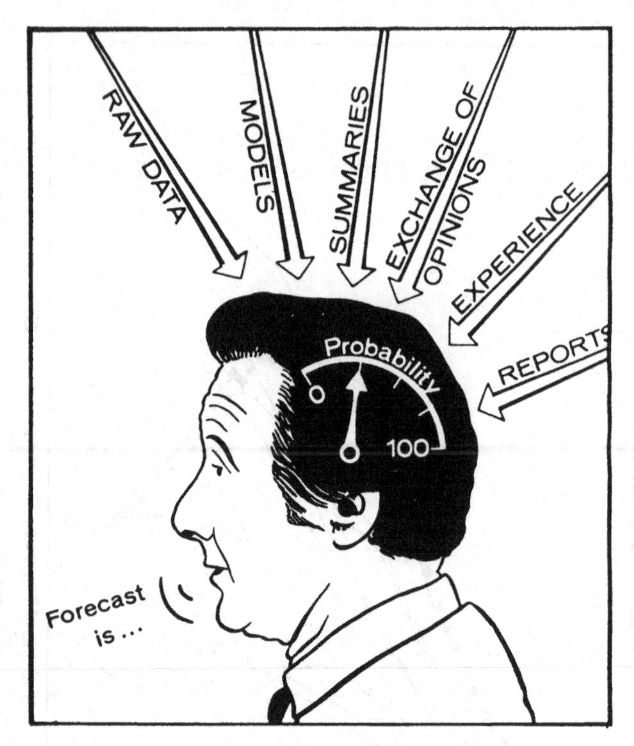

Fig. 10.7 The forecaster at work. (C. Sawyer *et al.*, *Solar Flare Prediction*. Colorado Associated University Press, 1986; after J. R. Hirman)

all relating to conditions in the corona, is taken into account. The preparation of the forecast is not a routine matter; it makes use of the forecaster's training and experience (Figure 10.7). One example of a solar analysis and forecast, prepared during the 1973 Skylab flight, is illustrated in Figure 10.8.

An important part of the flare forecast is the probability that a flare with a specified X-ray flux will occur within the next 24 hours. The classification shown in Table 10.4 has been in use since 1969. The daily forecast gives the probability (on a scale of 0–1) of flare occurrence at the M and X levels.

The accuracy of the flare forecasts is assessed by comparing them with subsequent events. Of course, a forecaster who was merely guessing would score some successes, but the forecasts are actually much better than that. Predictions of whether or not a given solar region will flare eventually are very good, but predicting the actual time of the flare is more difficult. About 65% of days have at least one flare of class M or X, and the probability figure in the daily forecasts has an r.m.s. error of 0.22. Class C and smaller flares are too common (nearly every day) for forecasting at that level to be useful. Once a flare has been observed, useful predictions may be made of some of the consequences. For example, proton events, including their sizes, can be predicted with an accuracy of about 80%.

So-called *expert systems* are proving to be useful in flare forecasting. In an expert system the knowledge of an expert is encoded in a computer program. Tests show that the system does as well as a trained forecaster (which is considerably better than

Table 10.4 *X-ray flare classification*

Class	Energy flux in 1–8 Å measured near the Earth (W/m²)
X	$\geqslant 10^{-4}$
M	10^{-5}–10^{-4}
C	10^{-6}–10^{-5}
B	10^{-7}–10^{-6}
A	10^{-8}–10^{-7}

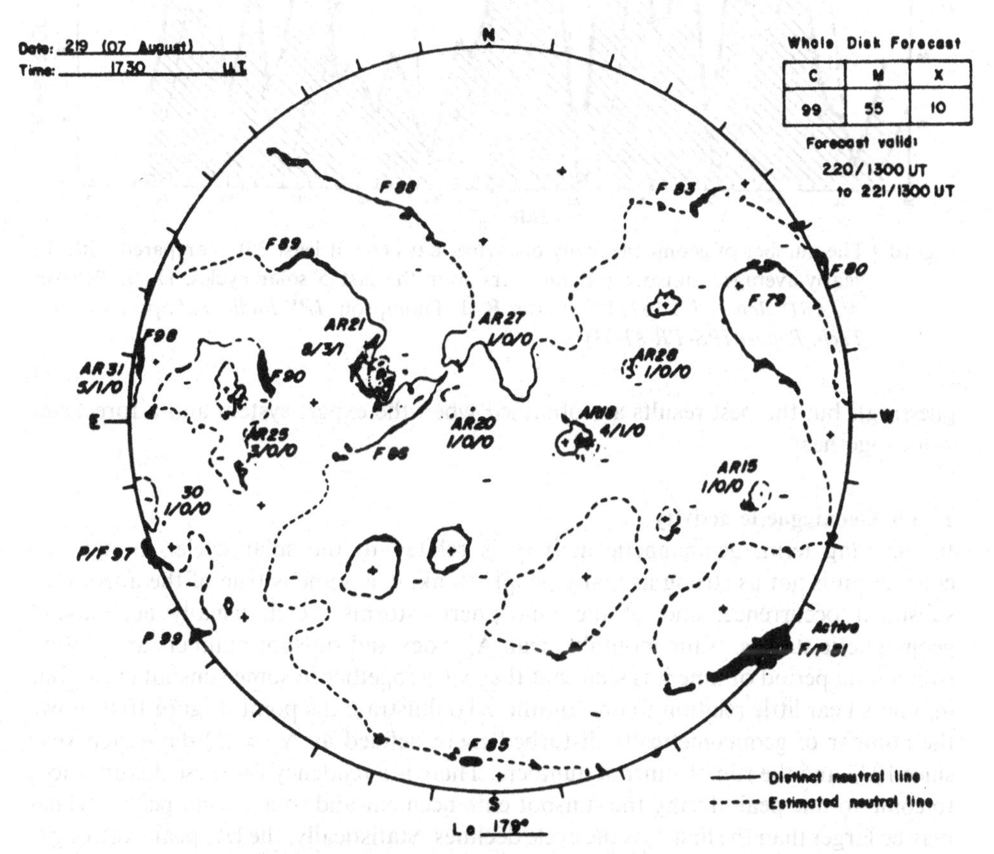

Fig. 10.8 Solar analysis and flare forecast made on 7 August 1973, during a Skylab flight. The analysis shows magnetic polarities, neutral lines and active regions. The flare forecast gives the percentage chance that a flare of Class C, M or X will occur during a 24-hour period. (J. R. Hirman, personal communication)

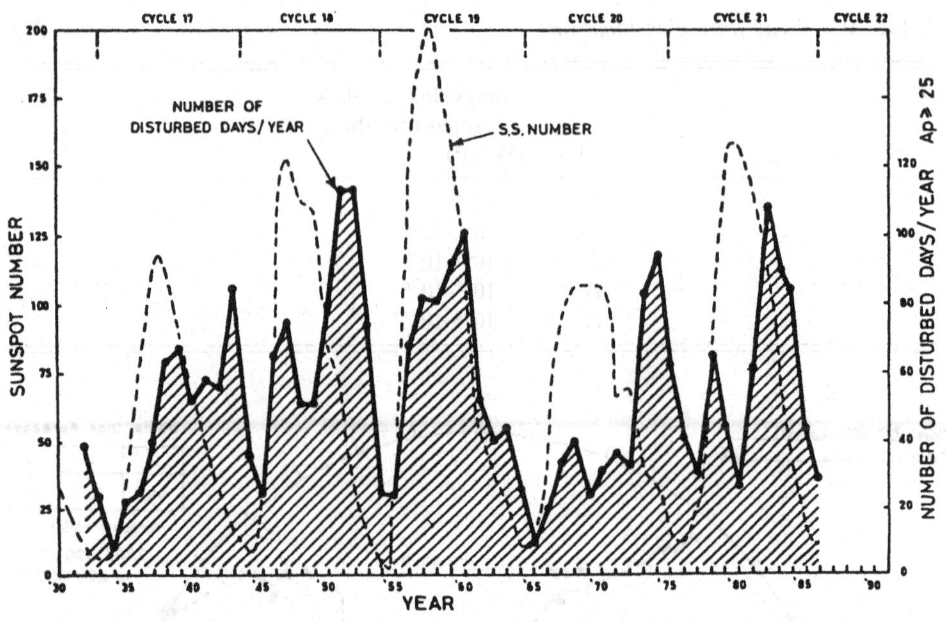

Fig. 10.9 The number of geomagnetically disturbed days (A_p at least 25), compared with the yearly-averaged sunspot (ss) numbers over the last 5 solar cycles. (J. S. Belrose, *AGARD Report LS*-162, 1988; after R. J. Thompson, *IPS Radio and Space Services Tech. Report IPS-TR*-87-03)

guessing), but the best results are obtained when the expert system and a forecaster work together!

10.5.5 Geomagnetic activity

In the long term, geomagnetic activity is related to the solar cycle, though the connection is not as strong as many people think. The same is true of the aurora, of substorm occurrence, and of the ionospheric storms which usually accompany geomagnetic storms. If the monthly mean A_p index and sunspot numbers are plotted over a long period of time it is seen that they vary together in some sunspot cycles but in others bear little relation to one another. To illustrate the point, Figure 10.9 shows the number of geomagnetically disturbed (here defined as $A_p \geqslant 25$) days each year since 1932 and the run of sunspot numbers. There is a tendency for these disturbances to come to one peak during the sunspot enhancement and to a second peak – which may be larger than the first – as the cycle declines. Statistically, the late peak is stronger in even numbered cycles, and Figure 10.9 is consistent with this. So while one may anticipate that a larger cycle will bring more storms, there is no certainty when the peak will occur or how large it will be. Geomagnetic activity involves the solar wind as well as the Sun, and we should not be surprised that its prediction is more difficult.

In the shorter term, the best method of predicting geomagnetic activity uses the recurrence tendency due to solar rotation. This depends on the solar active regions persisting for more than one solar rotation, and the method works best in the declining phase of the cycle. The recurrence tendency was illustrated in Figure 8.31. In the rising

Table 10.5 *Classification of geomagnetic disturbances*

Description	K index	A index
Quiet	0–2	< 8
Unsettled	3	8–15
Active	4	16–29
Minor storm	5	30–49
Major storm	6	50–99
Severe storm	7–9	100–400

part of the cycle more of the magnetic storms are related to flares (though one school of thought believes the source to be a coronal hole near the flare rather than the flare itself), so storm prediction comes back to flare prediction during those years.

For prediction purposes, magnetic disturbance is classified into 6 categories, as in Table 10.5.In these terms the predictions are accurate on 40% of occasions and come within one category 89% of the time. However, this test disguises the fact that the major and severe storms, which occur infrequently but have the most severe effects, are not so accurately predicted.

10.6 Data services

The science of geospace is necessarily based on observations, for although experiments (both in the laboratory and in the geospace medium) are possible to some extent, and both modelling and theory are vital elements, the central fact must be what actually happens in the solar–terrestrial environment. This natural behaviour is what we must observe and explain. The geospace observer setting out to investigate some problem, having no doubt read all the relevant books and studied the pertinent papers, will specify equipment, location, mode of operation, data rate, and so on, in order to optimize the observations for the purpose in hand. Yet this prodigious effort might come to nought if nature fails to produce the event needed – or not quite, because something else of interest might occur instead! If so, should these observations be discarded because they did not fill the original request?

It is of the essence of solar–terrestrial science that some observations are made for long periods to monitor what goes on, that some observations accrue that are not as intended and cannot be immediately explained, and that some phenomena can never be explained without a sufficient battery of techniques being brought to bear simultaneously. Many of the observations are relevant to the needs of engineering users (as we have seen) and – one has to say this – they are all expensive. Data therefore have value in several ways, and, as the geophysics community has long recognized, they should not readily be thrown away if their quality is good.

To facilitate the exchange of data between scientists and to ensure that the most important data are preserved for the future, an organization of World Data Centers (WDCs) was set up under ICSU (the International Council of Scientific Unions). The scheme began in 1957 for the International Geophysical Year (IGY) (1957–8) because it was foreseen that much new data was about to be generated and that for maxixmum

Table 10.6. *Disciplines covered by the various World Data Centers (WDCs)*

	Data Center				
Discipline	A	B	C1	C2	D
Astronomy					✓
Glaciology	✓	✓	✓		✓
Meteorology	✓	✓			✓
Nuclear radiation				✓	
Renewable resources and environment					✓
Oceanography	✓	✓			✓
Marine geology and geophysics	✓	✓			
Tsunami, mean sea level and tides		✓			
Solar-terrestrial physics (or activity)	✓	✓	✓	✓	
Rockets and satellites	✓	✓			
Airglow				✓	
Aurora				✓	
Cosmic rays				✓	
Geomagnetism			✓	✓	
Ionosphere				✓	
Solar activity			✓		
Solar radio emissions				✓	
Space sciences					✓
Sunspot index			✓		
Solid-Earth geophysics	✓	✓			✓
Earth tides			✓		
Geology					✓
Recent crustal movements			✓		
Rotation of the Earth	✓	✓			
Seismology	✓				✓

The disciplines most relevant to geospace have been grouped together.
Information from *Guide to the World Data Center System*, issued by the Secretariat of the ICSU Panel on World Data Centers, NOAA, Boulder, Colorado, 1987, 1989.

scientific benefit these data should be open to the entire community of geophysical scientists. The IGY arrangements proved effective, and studies based on IGY data continued for years after the period of IGY special observations had ended. The WDCs have continued since then and have expanded (though not enough) in the effort to keep pace with the growth of science. They are funded by the various countries in which they are situated, and their policy is to make available their data to all bona fide scientists at minimum cost.

The disciplines covered by the WDCs are listed in Table 10.6. There are four Centers, each covering a different geographical region. WDC-A is in the USA, WDC-

B is in the USSR, WDC-C is divided between C1 in various Western European countries and C2 in India and Japan, and WDC-D is in China. Generally, the holdings of a particular Center are distributed amongst several national institutions, and they vary as to size and degree of specialization. The geographical distribution is also intended for the convenience of the users, and each Center in fact assumes an international role since the data are exchanged between them (though the former practice of exchanging all data has been abandoned for practical reasons) and the archives of each are open to any scientific investigator regardless of national origin.

Data are held in various forms. The earlier data tend to be on paper or film, while more recent sets are, to an increasing extent, in digital form on magnetic tape or optical media (e.g. 'compact disk' as used for domestic sound reproduction).

The Data Centers perform various additional services, one particularly valuable one being the publication of data summaries such as *Solar–Geophysical Data*, a monthly report from WDC-A. This is a prompt publication, much of the data being only one or two months old and therefore still 'warm' – i.e. still of current interest and timely for the analysis of an experimenter's own recent work. From time to time special reports are compiled, to bring together in one volume a data collection, perhaps of data of one type taken over a number of years, or of different kinds of measurements made during some particularly interesting geophysical event. Such reports assist the progress of science through the facts they disseminate and by helping to keep working scientists aware of what is going on.

It would be wrong to see data services as merely a subsidiary operation, a sort of tidying up after the interesting work has been done. Geospace is a global science in which communications are vital. The scientists must be able to cooperate and they need a common base of information to do so. This field of work has a tradition of data sharing and of collaboration across national frontiers, and that has to be encouraged if the subject is to move forward. Data services are an integral part of this process. Like monitoring, they deserve better support than they usually get.

Further reading

G. S. Ivanov-Kholodny and A. V. Mikhailov. *The Prediction of Ionospheric Conditions*. Reidel, Dordrecht, Boston, Lancaster and Tokyo (1986).

K. Davies. *Ionospheric Radio*. Peregrinus, London (1990). Section 8.7 on applications to space systems, Section 9.8 on forecasting, and Chapter 12 on the prediction of HF radio propagation.

A. A. Kolosov *et al. Over the Horizon Radar*. Artech House, Norwood, Massachusetts (1987). (Translated from the Russian, originally published 1984.)

S. A. Hovanessian. *Synthetic Array and Imaging Radars*. Artech House, Dedham, Massachusetts (1980).

C. De Jager and Z. Svestka (eds.). *Solar Flares and Space Research*. North-Holland (1969). Part 5 on forecasting solar activity.

P. S. McIntosh and M. Dryer (eds). *Solar Activity Observations and Predictions*. MIT Press, Cambridge, Massachusetts (1972).

C. Sawyer, J. W. Warwick and J. T. Dennet. *Solar Flare Predictions*. Colorado Associated University Press, Boulder, Colorado (1986).

R. P. Demaro. Navstar – the all-purpose satellite. *IEEE Spectrum* **18** (5), 35 (May 1981).

G. F. Earl and B. D. Ward. The frequency management system of the Jindalee over-the-horizon backscatter HF radar. *Radio Science* **22**, 275 (1987).

G. C. M. Reijnem. Major legal developments in regard to space activities 1957–1975. *Space Science Reviews* **19**, 245 (1976).

H. B. Garrett. The charging of spacecraft surfaces. *Reviews of Geophysics* **19**, 577 (1981).

J. Feynman and X. Y. Gu. Prediction of geomagnetic activity on time scales of one to ten years. *Reviews of Geophysics* **24**, 650 (1986).

Index